Economic Commission for Europe
Geneva

ECONOMIC SURVEY OF EUROPE

2002 No. 1

Prepared by the
SECRETARIAT OF THE
ECONOMIC COMMISSION FOR EUROPE
GENEVA

UNITED NATIONS
New York and Geneva, 2002

NOTE

The designations employed and the presentation of the material in this publication do not imply the expression of any opinion whatsoever on the part of the Secretariat of the United Nations concerning the legal status of any country, territory, city or area, or of its authorities, or concerning the delimitation of its frontiers or boundaries.

UNITED NATIONS PUBLICATION

Sales No. E.02.II.E.7

ISBN 92-1-116803-1
ISSN 0070-8712

Copyright © United Nations, 2002
All rights reserved
Printed at United Nations, Geneva (Switzerland)

CONTENTS

Page

Explanatory notes x
Abbreviations xi
Preface xiii

Part One

MACROECONOMIC TRENDS AND PROSPECTS IN THE ECE REGION

Chapter 1   THE ECE ECONOMIES IN SPRING 2002 3

     1.1   Introduction 3

     1.2   Western Europe and North America 4
           (i)   The current outlook 4
           (ii)   EMU: the currency changeover and the macroeconomic policy framework 7

     1.3   The transition economies 10
           (i)   Recent developments 10
           (ii)   The short-term outlook 12

Chapter 2   THE GLOBAL CONTEXT AND WESTERN EUROPE 15

     2.1   The global context 15
           (i)   Overview 15
           (ii)   North America 19
           (iii)   Japan 24

     2.2   Western Europe 25
           (i)   Euro area 25
           (ii)   Other western Europe 34

     2.3   The two pillars of the ECB monetary strategy 35

     2.4   Some reflections on the "weakness" of the euro 39
           (i)   Possible factors behind the euro's weakness 39
           (ii)   Conclusions 42

     2.5   Overview of growth patterns in industrialized countries in the 1990s: the role of demand factors 42
           (i)   Growth performance in the 1990s 42
           (ii)   Demand factor contributions 43

Chapter 3   THE TRANSITION ECONOMIES 51

     3.1   Macroeconomic policy 51
           (i)   Monetary policy 51
           (ii)   Fiscal policy 56
           (iii)   Economic distress in Poland 63
           (iv)   Can Russia be a regional growth engine? 69
           (v)   Is there a risk of "currency board contagion" for the transition economies? 74

			Page
3.2	Output and demand		76
	(i)	Patterns of output and demand in 2001	76
	(ii)	Eastern Europe and the Baltic states	80
	(iii)	Commonwealth of Independent States	86
	(iv)	Openness and the cyclical behaviour of selected east European and Baltic economies	88
3.3	Costs and prices		95
	(i)	Consumer prices in 2001	96
	(ii)	Producer prices and labour costs in industry in 2001	99
	(iii)	Sources of inflation in the transition economies of eastern Europe and the Baltics, 1991-2001	102
3.4	Labour markets		112
	(i)	Employment and unemployment in 2001	112
	(ii)	Changes in the structure of unemployment, 1998-2001	115
	(iii)	The changing patterns of manufacturing employment, 1993-2000	119
3.5	Foreign trade and payments		127
	(i)	Current account developments	127
	(ii)	International trade	129
	(iii)	Trade specialization by stage of production in eastern Europe and the Baltic states, 1996-2000	137
	(iv)	External financing, FDI and debt issues	149

Part Two

POLICIES FOR ADJUSTMENT AND GROWTH

Chapter 4	TECHNOLOGICAL ACTIVITY IN THE ECE REGION DURING THE 1990s		161
4.1	Introduction		161
4.2	Convergence and divergence in per capita income levels in the 1990s		162
4.3	Education and technological learning		164
4.4	R&D activity in the ECE region during the 1990s		167
	(i)	The structure of R&D	167
	(ii)	Shifting spending priorities or improving efficiency?	169
4.5	Inventive activity in the ECE region		170
4.6	Innovative activity in Europe		172
4.7	International technology transfer and domestic spillovers in eastern Europe		175
4.8	Is there a role for science and technology policy?		176
Chapter 5	ALTERNATIVE POLICIES FOR APPROACHING EMU ACCESSION BY CENTRAL AND EAST EUROPEAN COUNTRIES		181
5.1	Introduction		181
5.2	Catching up and EMU accession		182
	(i)	Income gap – structural gap	182
	(ii)	Equilibrium real appreciation and inflation in catching-up economies	183
	(iii)	Macroeconomic constraints and risks during the catching-up period	184
	(iv)	Policy challenges	185

			Page
5.3	The role of the exchange rate regime in a catch-up process		186
	(i) Flexible exchange rates as a policy tool		187
	(ii) The role of nominal anchors		187
	(iii) The exchange rate mechanism (ERM-2) as a flexible tool for managing convergence		188
5.4	EMU accession scenarios		189
	(i) The policy framework of EMU accession		189
	(ii) The timing of accession		191
	(iii) The choice of accession strategy		192
5.5	Conclusions		193

Part Three

SOCIAL DIMENSIONS OF ECONOMIC DEVELOPMENT

Chapter 6 NEW FORMS OF HOUSEHOLD FORMATION IN CENTRAL AND EASTERN EUROPE: ARE THEY RELATED TO NEWLY EMERGING VALUE ORIENTATIONS? 197

6.1	Introduction	197
6.2	The European Values Surveys of 1999	199
6.3	Which values matter?	201
6.4	The footprints of selection and adaptation: what to expect?	203
6.5	Measurement and profiles: do we find the footprints of selection and adaptation?	205
6.6	Finer distinctions	208
6.7	Changes in value orientations during the 1990s	213
6.8	Conclusions	215

Part Four

STATISTICAL APPENDIX

STATISTICAL APPENDIX 217

LIST OF TABLES

Table		Page
1.1.1	Annual changes in real GDP in the ECE region, 1999-2002	3
1.2.1	Real GDP in the ECE market economies, 2000-2002	5
1.3.1	Annual changes in real GDP in eastern Europe, the Baltic states and the CIS, 1999-2002	11
2.1.1	Quarterly changes in real GDP in the major seven economies, 2000QIV-2001	17
2.1.2	Annual changes in real GDP and major expenditure items in the United States, 2000-2001	21
2.1.3	Annual changes in real GDP and major expenditure items in Japan, 2000-2001	25
2.2.1	Annual changes in the major expenditure items on GDP in western Europe, North America and Japan, 2000-2001	28
2.2.2	Contributions of major expenditure items to annual changes in real GDP in western Europe, 2000-2001	29
2.2.3	Inflation in western Europe and North America, 1999-2001	30
2.2.4	Unemployment in western Europe and North America, 1999-2001	31
2.2.5	General government budgetary positions in the European Union, 2000-2002	34
2.5.1	Macroeconomic indicators for industrialized countries, 1970-2000	43
2.5.2	Changes in expenditure items and their contribution to real GDP growth in industrialized countries, 1990-2000	44
2.5.3	Changes in expenditure items and their contribution to real GDP growth in major economies, 1990-2000	45
3.1.1	Short-term interest rates in selected east European, Baltic and CIS economies, 1999-2001	54
3.1.2	Monetization in selected east European, Baltic and CIS economies: share of monetary aggregates in GDP, 1997-2001	56
3.1.3	Consolidated general government deficits and their sources of financing in eastern Europe, the Baltic states and the CIS, 1998-2002	58
3.1.4	Consolidated general government current revenue in eastern Europe, the Baltic states and the CIS, 1999-2001	61
3.1.5	Consolidated general government expenditure in eastern Europe, the Baltic states and the CIS, 1999-2001	63
3.2.1	GDP and industrial output in eastern Europe, the Baltic states and the CIS, 1999-2001	77
3.2.2	Contribution of final demand components to real GDP growth in selected east European, Baltic and CIS economies, 1999-2001	81
3.2.3	Real domestic demand components in selected east European, Baltic and CIS economies, 1999-2001	83
3.2.4	Investment outlays in selected east European, Baltic and CIS economies, 1999-2001	84
3.2.5	Retail trade in the east European, Baltic and CIS economies, 1999-2001	84
3.2.6	Contributions of exports and net exports to cumulative real GDP growth during two recent recoveries in eastern Europe and the Baltic states	90
3.2.7	Simple correlations between trade ratios and GDP in eastern Europe and the Baltic states, 1995QI-2001QII	93
3.2.8	Correlations between quarterly rates of change of exports and imports in eastern Europe and the Baltic states, 1995QI-2001QII	94
3.3.1	Consumer prices in eastern Europe, the Baltic states and the CIS, 2000-2001	96
3.3.2	Producer prices, wages and unit labour costs in industry in eastern Europe, the Baltic states and the CIS, 2000-2001	99
3.3.3	Contributions to changes in the GDP and domestic demand deflators in selected east European and Baltic economies, 1998-2001	106
3.4.1	Total employment and registered unemployment in eastern Europe, the Baltic states and the CIS, 1999-2001	113
3.4.2	Male and female activity rates in selected central and east European economies, 1985, 1997-2000	119
3.4.3	Cumulative changes in manufacturing employment by branches in selected central European economies, 1993 and 2000	121
3.4.4	Changes in manufacturing employment by branches, 1993 and 2000	124
3.4.5	Breakdown of employment in manufacturing in selected central European economies and the EU, 1993 and 2000	125
3.4.6	Cross-country similarity indices of manufacturing employment in selected central European economies and the EU, 1993 and 2000	125
3.4.7	Relative country employment specialization by branch, 1993 and 2000	126
3.4.8	Main features of employment specialization in selected central European economies, 2000	126
3.5.1	Current account balances of the ECE transition economies, 2000-2001	128
3.5.2	Trade performance and external balances of the ECE transition economies, 2000-2001	130
3.5.3	Foreign trade of the ECE transition economies by direction, 1999-2001	131
3.5.4	Changes in the volume of foreign trade in selected transition economies, 1998-2001	132
3.5.5	CIS countries' trade with CIS and non-CIS countries, 1999-2001	137
3.5.6	Composition of east European and Baltic trade by stages of production, 1996-2000	143
3.5.7	Revealed comparative advantage by stage of production, 1996 and 2000	145
3.5.8	Types of trade among the east European and Baltic countries and with the EU, 1996-2000	147
3.5.9	Average ratio and coefficient of variation of export to import unit values in trade with EU, 1996, 1998 and 2000	149
3.5.10	Net capital flows into the ECE transition economies, 2000-2001	150
3.5.11	Net capital flows by type of capital into eastern Europe, the Baltic states and selected members of the CIS, 1999-2001	151
3.5.12	Selected external financial indicators for eastern Europe, the Baltic states and the CIS, 1999-2001	152
3.5.13	Foreign direct investment in the ECE transition economies, 2000-2001	154
3.5.14	Fitch credit ratings for the transition economies and changes in 2000-2002	156

Table		Page
4.3.1	Educational attainment of the workforce in the ECE region, 2000	164
4.3.2	Gross enrolment ratios in the ECE region, 1989 and 1996	165
4.3.3	Expenditure per student in the ECE region by level of education, 1998	166
4.4.1	R&D intensity in the ECE region, 1981-2000	168
4.4.2	Gross expenditure on R&D (GERD) by source of financing and performing sector in the ECE region, 1999	170
4.5.1	Patenting activity in the ECE region, 1990-2000	172
4.6.1	Innovative activity in European firms, 1994-1996	173
4.6.2	Structure of innovative expenditure in manufacturing and services in the ECE region, 1996	174
4.7.1	International royalties and licence fees in the ECE region, 2000	175
4.8.1	Government budget appropriations or outlays for R&D (GBAORD) in the ECE region, 2000	178
5.2.1	The speed of catching up by the CEEC-10, 1996-2004	183
5.2.2	Structural indicators in the CEEC-10, 1995-2000	184
5.4.1	Indicators of nominal convergence of CEEC-10 with the Maastricht criteria, 1996-2001	190
6.2.1	European Values Surveys, 1999: sample size and relative proportion of household positions in three regional groups of countries	200
6.2.2	Unmarried cohabitation of women in the transition economies in the 1990s	201
6.5.1	European Values Surveys, 1999: overview of 80 values used in the current analysis	206
6.7.1	Trends in selected comparable items among respondents aged 18-49, three groups of countries with transition economies, 1990 and 1999	214
A.1	Real GDP in western Europe, North America and Japan, 1987-2001	219
A.2	Real private consumption expenditure in western Europe, North America and Japan, 1987-2001	220
A.3	Real general government consumption expenditure in western Europe, North America and Japan, 1987-2001	221
A.4	Real gross domestic fixed capital formation in western Europe, North America and Japan, 1987-2001	222
A.5	Real total domestic expenditures in western Europe, North America and Japan, 1987-2001	223
A.6	Real exports of goods and services in western Europe, North America and Japan, 1987-2001	224
A.7	Real imports of goods and services in western Europe, North America and Japan, 1987-2001	225
A.8	Industrial output in western Europe, North America and Japan, 1987-2001	226
A.9	Total employment in western Europe, North America and Japan, 1987-2001	227
A.10	Standardized unemployment rates in western Europe, North America and Japan, 1987-2001	228
A.11	Consumer prices in western Europe, North America and Japan, 1987-2001	229
B.1	Real GDP/NMP in eastern Europe, the Baltic states and the CIS, 1980, 1988-2001	230
B.2	Real total consumption expenditure in eastern Europe, the Baltic states and the CIS, 1980, 1988-2001	231
B.3	Real gross fixed capital formation in eastern Europe, the Baltic states and the CIS, 1980, 1988-2001	231
B.4	Real gross industrial output in eastern Europe, the Baltic states and the CIS, 1980, 1988-2001	232
B.5	Total employment in eastern Europe, the Baltic states and the CIS, 1980, 1988-2001	233
B.6	Employment in industry in eastern Europe, the Baltic states and the CIS, 1989-2001	234
B.7	Registered unemployment in eastern Europe, the Baltic states and the CIS, 1990-2001	235
B.8	Consumer prices in eastern Europe, the Baltic states and the CIS, 1990-2001	236
B.9	Producer price indices in eastern Europe, the Baltic states and the CIS, 1990-2001	237
B.10	Nominal gross wages in industry in eastern Europe, the Baltic states and the CIS, 1990-2001	238
B.11	Merchandise exports of eastern Europe, the Baltic states and the CIS, 1980, 1989-2001	239
B.12	Merchandise imports of eastern Europe, the Baltic states and the CIS, 1980, 1989-2001	240
B.13	Balance of merchandise trade of eastern Europe, the Baltic states and the CIS, 1980, 1989-2001	241
B.14	Merchandise trade of eastern Europe and the Russian Federation, by direction, 1980, 1989-2001	242
B.15	Exchange rates of eastern Europe, the Baltic states and the CIS, 1980, 1990-2001	243
B.16	Current account balances of eastern Europe, the Baltic states and the CIS, 1990-2001	244
B.17	Inflows of foreign direct investment in eastern Europe, the Baltic states and the CIS, 1990-2001	245

LIST OF CHARTS

Chart		Page
2.1.1	Annual changes in real GDP in the world economy, 1980-2001	16
2.1.2	Changes in key official interest rates in the major industrialized countries, January 1999-February 2002	17
2.1.3	International share prices, January 1999-January 2002	18
2.1.4	World commodity prices, January 2000-February 2002	19
2.1.5	Crude petroleum prices, January 2000-February 2002	19
2.1.6	Quarterly changes in real GDP and major expenditure items in the United States, 1998-2001	20

Chart		Page
2.1.7	Consumer confidence and retail sales in the United States, January 1993-January 2002	21
2.1.8	Output and capacity utilization in manufacturing industry in the United States, January 1997-January 2002	22
2.1.9	Consumer prices in the United States, January 1999-January 2002	22
2.1.10	Employment and unemployment in the United States, January 1998-February 2002	23
2.1.11	Unit labour costs in the United States, 1998-2001	23
2.1.12	Average monthly nominal short-term and long-term interest rates in the United States, January 1998-January 2002	24
2.1.13	Nominal and real effective exchange rates of the dollar, January 1995-February 2002	24
2.1.14	Exchange rate of the yen, January 1999-January 2002	25
2.2.1	Quarterly changes in real GDP and major expenditure items in the euro area, 1998-2001	26
2.2.2	Manufacturing output and capacity utilization in the euro area, 1997-2001	27
2.2.3	Business and consumer confidence in the European Union, January 1995-February 2002	27
2.2.4	Consumer prices in the euro area, January 2000-January 2002	30
2.2.5	Employment and unemployment in the euro area, 1993-2001	31
2.2.6	Unit labour costs in the euro area, 1995-2001	32
2.2.7	Average monthly nominal short-term and long-term interest rates in the euro area, January 1998-February 2002	32
2.2.8	The exchange rate of the euro, January 1999-January 2002	33
2.2.9	Monetary conditions index (MCI) for the euro area, January 1999-December 2001	33
2.3.1	M3 growth and money market rates in the euro area, 1999-2001	37
3.1.1	Real effective exchange rates in selected east European and Baltic economies, 1995-2001	55
3.1.2	Real exchange rates against the Russian rouble in selected CIS economies, 1999-2001	57
3.1.3	Fiscal balance and its change from the previous year in eastern Europe, the Baltic states and the CIS, 2000-2001	59
3.1.4	Changes in the fiscal balance and domestic absorption in eastern Europe, the Baltic states and the CIS, 2000-2001	60
3.1.5	Contributions to Poland's GDP growth, 1991-2001QIII	64
3.1.6	Consumption and investment growth in Poland, 1991-2001QIII	65
3.1.7	The current account and the volume of exports in Poland, 1993-2001QIII	65
3.1.8	The volume of retail sales in Poland, 1997-2001	66
3.1.9	The average profitability of Polish enterprises, 1995-2001	66
3.1.10	Monthly exports of crude oil from Russia, 1996-2001	71
3.1.11	Brent crude price, real exchange rate of the rouble and Russia's real revenue from oil exports, 1996-2001	72
3.1.12	Annual change in Russia's real revenue from oil exports and its components, 1997-2001	72
3.1.13	Russia's real revenue from oil exports and the rate of GDP growth, 1996-2001	73
3.1.14	Russia's real revenue from oil exports and the general government fiscal balance, 1996-2001	73
3.1.15	Changes in the real exchange rate of the Russian rouble, 1995-2001	74
3.2.1	Trend and dispersion of monthly changes in industrial production in transition economies by subregions, January 1997-December 2001	78
3.2.2	Dynamics of monthly industrial production in transition economies and the European Union, centred three-month moving averages, January 1998-December 2001	79
3.2.3	Annual rates of change of manufacturing output by NACE industries in selected east European and Baltic economies, 1999-2001	85
3.2.4	The openness of 11 east European and Baltic economies, 1991-2000	89
3.2.5	The contribution of exports to cumulative real GDP growth and openness in eastern Europe and the Baltic states	91
3.2.6	The contribution of net exports to cumulative real GDP growth and openness in eastern Europe and the Baltic states	92
3.2.7	Partial correlation coefficients between export ratios and real effective exchange rates (REERs) versus the partial correlations between export ratios and foreign (EU) activity, 1996QI-2000QII	94
3.3.1	Components of consumer prices in eastern Europe, the Baltic states and the CIS, 1998-2001	97
3.3.2	Consumer and industrial producer prices in eastern Europe, the Baltic states and the CIS, 1998-2001	101
3.3.3	Change in the GDP and domestic demand deflators in selected east European and Baltic economies, 1991-2001	108
3.3.4	Change in the export and import deflators in selected east European and Baltic economies, 1991-2001	109
3.3.5	Share of imports in domestic demand in selected east European and Baltic economies, 1991-2001	110
3.3.6	The contributions of domestic costs and import prices to the total domestic inflation rate in selected east European and Baltic economies, 1991-2001	111
3.4.1	Unemployment rates in selected central and east European economies, 1998-2001	115
3.4.2	Dynamics of male and female unemployment rates in selected central and east European economies, 1998-2001	117
3.4.3	Ratio of female to male unemployment rates, in selected central and east European economies, 1998QII and 2001QII	118
3.4.4	Total and youth unemployment rates in selected central and east European economies, 2001QII	120
3.4.5	Total unemployment rate and share of women, youth and long-term unemployed in total unemployment in selected central and east European economies, 2001QII	120
3.4.6	Output and employment in manufacturing in selected central European economies, 1993-2000	122
3.4.7	Share of manufacturing in total employment in selected central European economies, 1993 and 2000	123
3.4.8	Job creation and job destruction in manufacturing industries in selected central European economies, 1993-2000	124
3.4.9	Relative specialization in manufacturing employment, the four countries combined vis-à-vis the EU average, 1993 and 2000	127

Chart		Page
3.5.1	Specific western demand for selected transition economies' exports, 1998-2002	133
3.5.2	Exports and imports by commodity groups, January-September 2000 and January-September 2001	134
3.5.3	Monthly dollar exports and imports in selected transition economies, 2000-2001	135
3.5.4	Geographical concentration of east European and Baltic countries' exports, 1993-2000	138
3.5.5	Commodity structure of east European and Baltic countries' exports to the world, 1996-2000	139
3.5.6	Commodity concentration of east European and Baltic countries' exports, 1993-2000	140
3.5.7	Changes in the structures of exports from eastern Europe and the Baltic states, 1993-2000	141
3.5.8	Share of intra-industry in total trade in manufactures between EU and east European and Baltic countries and in trade with CEFTA-7, 1993-2000	148
3.5.9	Yield spreads on the international bonds of selected transition and other emerging market economies, 2000-February 2002	157
4.2.1	Real GDP per capita in current PPPs in the region, 1993 and 2000	163
4.2.2	High-technology exports and per capita GDP in the ECE region, 1999	163
4.3.1	Public expenditure per secondary student in relation to GDP per capita in the ECE region, 1995	166
4.3.2	Public expenditure per tertiary student in relation to GDP per capita in the ECE region, 1995	166
4.4.1	Real GDP and GERD growth rates in the European Union and the United States, 1986-2000	169
4.4.2	Real GDP and GERD growth rates in Hungary, 1991-2000	169
4.4.3	Research intensity and GDP per capita in the UNECE region, 1991 and 1999	171
4.4.4	Researchers per thousand of the labour force and GDP per capita in the ECE region, 1991 and 1999	171
4.5.1	Resident patents per million of the population and per capita GDP in the ECE region, 1996-1998 average	172
4.6.1	Manufacturing innovation and per capita GDP in Europe, 1996	173
4.6.2	Manufacturing innovation and R&D intensity in Europe, 1996	174
4.7.1	High-technology imports and per capita GDP in the ECE region, 1999	176
5.2.1	Productivity growth and inflation differentials in CEEC-8 and the EU, 1987-1998	185
5.3.1	Nominal exchange rates of CEEC-5 currencies against the deutsche mark, 1996-2001	188
5.3.2	Interest rate premia on the Czech koruna, Hungarian forint and Polish zloty, 1995-2001	189
6.4.1	Flow chart of life-course development and hypothesized changes in value orientations stemming from selection-adaptation mechanism	204
6.5.1	Number of positive (non-conformist) net deviations for 80 items, European Values Surveys, 1999: pooled results for eight western, seven central and five east European countries	207
6.5.2	Number of positive (non-conformist) net deviations for groups of items and countries, European Values Surveys, 1999	208
6.6.1	Correspondence between household positions and 80 non-conformist values, European Values Surveys, 1999: results for eight west European countries	210
6.6.2	Correspondence between household positions and 80 non-conformist values, European Values Surveys, 1999: results for seven central European countries	211
6.6.3	Correspondence between household positions and 80 non-conformist values, European Values Surveys, 1999: results for five east European countries	213

LIST OF BOXES

Box		Page
3.1.1	Why are fiscal deficits not measured accurately?	62
3.3.1	Contributions to changes in the GDP and domestic demand deflators	104
3.5.1	Indicator of revealed comparative advantage	142

ISIC	International Standard Industrial Classification
IT	information technology
MCA	Multiple Classification Analysis
MCI	monetary conditions index
MNC	multinational corporation
MPC	Monetary Policy Council
NACE	Nomenclature générale des activités économiques dans les Communautés européennes (General Industrial Classification of Economic Activities within the European Communities)
NASA	National Aeronautics and Space Administration
NATO	North Atlantic Treaty Organization
NBER	National Bureau of Economic Research, Inc.
n.e.c.	not elsewhere classified
n.e.s.	not elsewhere specified
NMP	net material product
OECD	Organisation for Economic Co-operation and Development
OPEC	Organization of the Petroleum Exporting Countries
OPT	outward processing trade
PPI	producer price index
PPP	purchasing power parity
PSA	production sharing agreement
RCA	revealed comparative advantage
RCD	revealed comparative disadvantage
R&D	research and development
RFE/RL	Radio Free Europe/Radio Liberty
RPIX	retail price index excluding mortgage interest payments
SETE	south-east European transition economies
SGP	Stability and Growth Pact
SITC	Standard International Trade Classification
SNA	System of National Accounts
S&T	science and technology
TACIS	Technical Assistance for the Commonwealth of Independent States (of the EU)
TBP	technology balance of payments
TFR	total fertility rate
TNC	transnational corporation
UIP	uncovered interest parity
UNCTAD	United Nations Conference on Trade and Development
UNECE	United Nations Economic Commission for Europe
USSR	(former) Union of Soviet Socialist Republics
VAT	value added tax
WIIW	The Vienna Institute for International Economic Studies
WTO	World Trade Organization

PREFACE

The present *Survey* is the fifty-fifth in a series of annual reports prepared by the secretariat of the United Nations Economic Commission for Europe to serve the needs of the Commission and of the United Nations in reporting on and analysing world economic conditions.

Until 1997 the *Economic Survey of Europe* was issued once a year as was the *Economic Bulletin for Europe*, the secretariat's second publication which focused on trade and payments issues. At its 52nd Session, in April 1997, the Commission decided to replace these two publications with an annual *Survey* of several issues. In 1998 and 1999 there were three issues each year. There are now two issues a year published in April and November.

The Survey is published on the sole responsibility of the Executive Secretary of ECE and the views expressed in it should not be attributed to the Commission or to its participating governments.

The analysis in this issue is based on data and information available to the secretariat in late March 2002.

Economic Analysis Division
United Nations Economic Commission for Europe
Geneva

PART ONE

MACROECONOMIC TRENDS AND PROSPECTS IN THE ECE REGION

CHAPTER 1

THE ECE ECONOMIES IN SPRING 2002

1.1 Introduction

There was a progressive slowdown in the rate of expansion of the global economy in the course of 2001, and a parallel deterioration of the short-term outlook. World output is estimated to have increased by some 2.5 per cent in 2001 compared with a rise of 4.7 per cent in 2000, and the volume of world merchandise trade stagnated. The dominant feature was the synchronous cyclical downturn – the first since 1974-1975 – in the three major economies, the United States, Japan and Germany, which ended in recession. For the industrialized countries as a whole, real GDP increased by only 1 per cent in 2001, down from 3.7 per cent in 2000, the most rapid deceleration in real GDP since 1973-1974.[1]

In the ECE region, real GDP rose by only 1.7 per cent in 2001, against 4.2 per cent in 2000 (table 1.1.1). This considerable slowdown masks, however, a striking resilience of the transition economies to the deterioration in the external economic environment. In eastern Europe, real GDP rose on average by 3.2 per cent in 2001. In Russia, the economic boom lost some momentum, but the annual increase in real GDP still amounted to 5 per cent, down from 9 per cent in 2000. The strength of the Russian economy was a major factor behind the overall buoyancy of economic activity in the CIS.

Global economic developments were overshadowed by the terrorist attacks in New York and Washington D.C. on 11 September 2001. These occurred at a time when the United States economy and the other major economic regions were in a fragile state and thought to be close to a cyclical turning point. The general effect of the attacks has been to worsen the economic outlook, at least in the short term. However, their generally depressing impact on consumer and business confidence throughout the world economy appears to have waned in early 2002.

There is a broad consensus among economic forecasters that economic growth in the United States will recover in the course of 2002, and that in its wake the rate of economic growth in the rest of the world will strengthen as well. The short-term economic outlook, however, is still highly uncertain, not least because of the persistence of the very large domestic and external imbalances in the United States economy, which pose a major risk to a sustained cyclical recovery.

UNECE argued a year ago that a domestic demand-led recovery in the United States could turn out to be a mixed blessing for the world economy because it would only postpone the inevitable readjustment needed to redress these large imbalances and potentially increase the risk of an abrupt and disruptive adjustment.[2] The ideal environment for a smoother adjustment to take place would, of course, be sustained and strong growth in the rest of the world economy in combination with restrained domestic demand growth in the United States. Given the enfeebled state of the Japanese economy this implies especially a much stronger economic performance in western Europe. This may be difficult to bring about, however, as it would require a more positive attitude of monetary policy towards economic growth as well as a more flexible framework for fiscal policy in the euro area. Nevertheless, it is becoming clear that the macroeconomic imbalance in the world economy is a

TABLE 1.1.1

Annual changes in real GDP in the ECE region, 1999-2002
(Percentage change over previous year)

	1999	2000	2001 [a]	2002 [b]
ECE region	3.2	4.2	1.7	1.8
Western Europe	2.2	3.5	1.3	1.4
European Union	2.6	3.4	1.7	1.3
Euro area	2.7	3.4	1.6	1.2
North America	4.2	4.2	1.2	1.6
United States	4.1	4.1	1.2	1.6
Eastern Europe [c]	1.5	3.8	3.2	2.8
CIS	4.5	8.3	6.2	4.8
Russian Federation	5.4	9.0	5.0	4.3
Memorandum items:				
Europe (east and west)	2.1	3.5	1.5	1.5
Europe (east and west) and CIS	2.4	4.2	2.1	2.0

Source: Tables 1.2.1 and 1.3.1 of this *Survey*.

Note: Weights for the calculation of regional aggregates were derived from 1996 GDP data converted from national currency units into dollars using purchasing power parities.

[a] Preliminary estimate.
[b] Forecast.
[c] Including the Baltic states.

---

[1] Real GDP rose by 0.5 per cent in 1974, down from 6.2 per cent in 1973.

[2] UNECE, *Economic Survey of Europe, 2001 No. 1*, p. 5.

serious policy concern for the United States.[3] Rather than provide a stimulus to United States exports, the euro area still appears to be looking to domestic demand in the United States as the main source of its own growth.

1.2 Western Europe and North America

(i) The current outlook

In the spring of 2002, there are increasing signs that the pronounced cyclical downturn of 2001 has started to bottom out. The short-run outlook, however, remains very uncertain and the prospects are for only a gradually strengthening recovery in 2002.

In the United States, economic conditions showed signs of improving in early 2002. The sustained fall in industrial activity since the beginning of 2001 petered out into a small increase in output in the first two months of 2002. This improvement is also reflected in a marked rise of the Institute for Supply Management[4] index in February. Consumer confidence has been volatile, but surged in March after a small decline in February. Nevertheless, retail sales growth was sluggish in early 2002, hiring conditions in the labour markets remained weak, and the number of persons claiming unemployment insurance in March remained high. The Conference Board's index of leading indicators stagnated in February 2002, following consecutive increases in the four preceding months.

The consensus of forecasters is now for an annual increase in real GDP in the United States of about 1.5 per cent in 2002. This annual average masks expectations of a somewhat more pronounced strengthening of growth in the second half of 2002. Moreover, such a growth rate is unlikely to lead to any significant reduction in excess capacities in the business sector in 2002. These, in combination with a meagre growth of profits, will continue to depress business fixed investment which, for the year as a whole, is expected to be less than in 2001.

It is not clear what progress has been made in the high-tech sector in adjusting to the sharp decline in business spending on high-technology equipment. The sharp cutback in expenditures on ICT equipment reflects to some extent the downward revisions of expected rates of returns on these assets. This more realistic assessment of profit prospects could also restrain demand for these products in 2002.

In contrast, economic activity will be supported by the completion and partial reversal of the large cuts in business inventories that occurred in the course of 2001 in response to the sharp deterioration in sales prospects. The growth of private consumption is likely to be relatively weak in 2002, partly reflecting the balance sheet adjustments required by the fall in the personal savings rate to a very low level, the loss in net wealth triggered by the fall in equity prices, and the steep rise in the burden of debt-servicing since the mid-1990s (and which is approaching its previous peak of end 1986). In addition, the growth of disposable incomes will be restrained by weak labour market conditions, although this will be partly offset by fiscal policy measures. No significant support is expected from exports given the overall weakness of overseas demand. However, there could be some feedback effects if a gradual strengthening of domestic demand spilled over to other major economies, boosting their growth and, in turn, stimulating demand for United States products.

Domestic demand will be supported by the considerable monetary stimulus which is already in the pipeline, although this has not fed through to interest rates at the longer end of the maturity spectrum and banks have tightened borrowing conditions. On 19 March 2002, the Federal Reserve decided to keep its target for the federal funds rate unchanged at 1.75 per cent judging that the risks are now balanced between the long-run goals of price stability and sustainable economic growth. The background to this decision was information pointing to a stronger rate of economic growth based on a marked swing in inventory investment. In addition, and despite the failure of Congress to agree to the fiscal stimulus package proposed in the wake of 11 September, increased government spending will partly offset the overall weakness of private sector demand.

In the euro area, the fall in real GDP in the final quarter of 2001 is generally expected to be followed by a small increase in economic activity in the first quarter of 2002. The confidence of consumers, industrial managers and producers of services has improved somewhat in the first two months of 2002. This contrasts, however, with increasing pessimism in the retail trade sector. As for the United States, a reversal of the inventory cycle is expected to support domestic demand. The growth of private household consumption is likely to remain weak with the impact of adverse developments in the labour markets on disposable incomes being partly offset by the expected fall in the rate of inflation. Fixed investment is expected to remain sluggish in view of weak sales prospects and relatively large margins of spare capacity in industry. Surveys of business investment plans made

---

[3] Europe's reliance on the United States as an engine of growth was alluded to also in the controversy about the decision of the United States administration to impose tariffs on steel imports. As was pointed out, the "United States economy appears poised for a recovery that will once again help other nations regain growth, including for steel industries". R. Zoellick, "The reigning champions of free trade", *Financial Times*, 13 March 2002. The author is the United States trade representative. An earlier warning that strains in international trade relations could spread from steel to other commodities if the EU and Japan failed to reflate their economies was made by Grant Aldonas, the United States Under-Secretary of Commerce for International Trade. "We have told people over time that if you don't see stronger growth abroad you end up seeing friction on the trade account. There is only so much patience you have when you are talking about very serious macroeconomic issues that have been out there for a long time", *Financial Times*, 11 March 2002, p. 1.

[4] This index was formerly known as the Purchasing Managers' Index.

in the autumn of 2001 point to a fall in the volume of industrial investment by 5 per cent in 2002.[5] In line with the expected profile of the United States recovery, economic activity is expected to strengthen in the second half of the year. The main assumption behind this scenario for the euro area is a gradual but sustained strengthening of domestic demand in the United States, which will spill over via higher exports to domestic consumption, and subsequently business investment, in the euro area. Euro area exports will also be supported by the relatively strong rate of expansion forecast for the transition economies.[6] For the year as a whole, real GDP in the euro area is forecast to increase by only about 1.25 per cent, down from 1.6 per cent in 2001. As a result of this low rate of growth, the level of employment can be expected to more or less stagnate and the average annual unemployment rate to edge up by about half a percentage point to 8.9 per cent.

Economic growth in Germany, the largest economy of the euro area, was only 0.6 per cent in 2001, the smallest increase since 1993. Little improvement is expected in 2002, with the annual growth rate forecast at about 0.75 per cent (table 1.2.1). The general government deficit rose to 2.7 per cent of GDP in 2001, close to the 3 per cent ceiling established in the Stability and Growth Pact (see below). This narrowly circumscribes any scope for discretionary fiscal measures designed to support economic growth.

Among the other member countries of the euro area, economic growth in France is expected to hold up somewhat better than in Germany and Italy, partly because of a more expansionary fiscal policy. Other national growth rates in 2002 are forecast to range from 1.1 per cent in Austria, Belgium and the Netherlands to some 3.75 per cent in Ireland (table 1.2.1).

In the euro area, the stance of monetary policy was tightened in October 2000 when the first signs of a cyclical slowdown were emerging. This was followed by a long wait-and-see period despite increasing indications of a serious global economic slowdown. The stance of monetary policy was eased only hesitantly, and rather late, in May 2001 followed by further reductions in the main refinancing rate in the second half of the year. The cumulative lowering of the main refinancing rate in 2001 amounted to only 1.5 percentage points, from 4.75 per cent in October 2000 to 3.25 per cent in early November 2001 against the background of the worst global economic downturn since the first oil price crisis of the early 1970s.[7] A more rapid response to the cyclical weakness in early 2001 would have improved growth prospects for 2002. In fact, the ECB's main refinancing rate, which has remained unchanged since November 2001, is still 0.25 percentage points above its level in November 1999. (This change is also reflected in money market rates.) In view of moderate inflationary expectations and a sizeable increase in the output gap,[8] there is still room for a further lowering of official interest rates.

The average general government budget deficit in the euro area rose to 1.1 per cent of GDP in 2001 and a further increase to 1.4 per cent is forecast for 2002. This largely reflects the operation of the automatic stabilizers. The cyclically adjusted deficit is forecast to fall slightly to 1.1 per cent of GDP in 2002, down from 1.3 per cent in 2001. This average masks expansionary measures (mainly tax cuts) in a number of countries (Belgium, Finland,

TABLE 1.2.1

Real GDP in the ECE market economies, 2000-2002
(Percentage change over previous year)

	2000	2001 [a]	2002 [b]
France	3.6	2.0	1.4
Germany	3.0	0.6	0.7
Italy	2.9	1.8	1.1
Austria	3.0	1.1	1.1
Belgium	4.0	1.3	1.1
Finland	5.6	0.7	1.3
Greece	3.8	4.1	3.2
Ireland	11.5	6.5	3.7
Luxembourg	7.5	4.0	3.0
Netherlands	3.5	1.5	1.1
Portugal	3.4	1.7	1.3
Spain	4.1	2.8	1.8
Euro area	3.4	1.6	1.2
United Kingdom	3.0	2.3	2.0
Denmark	3.0	1.3	1.4
Sweden	3.6	1.4	1.6
European Union	3.4	1.7	1.3
Cyprus	5.1	3.7	2.8
Iceland	3.6	1.1	2.4
Israel	6.4	-0.5	–
Malta	5.4	-0.3	-0.3
Norway	2.3	1.4	2.3
Switzerland	3.0	1.3	1.1
Turkey	7.2	-7.3	2.6
Western Europe	3.5	1.3	1.4
Canada	4.4	1.5	1.4
United States	4.1	1.2	1.6
North America	4.2	1.2	1.6
Japan	2.4	-0.5	-1.1
Total above	3.6	1.0	1.1
Memorandum items:			
4 major west European economies	3.1	1.6	1.2
Western Europe and North America	3.9	1.2	1.5

Source: National statistics; OECD, *Economic Outlook*, No. 70 (Paris), December 2001; Consensus Economics, Inc., *Consensus Forecasts* (London), various issues.

[a] Preliminary estimate.

[b] Forecast.

---

[5] European Commission, *Business and Consumer Survey Results*, February 2002 [http://europa.eu.int].

[6] See sect. 1.3 below.

[7] For comparison, the federal funds rate was lowered by a cumulative 4.75 percentage points to 1.75 per cent in the course of 2001.

[8] The OECD estimates that the average annual output gap for the euro area will increase by 1 percentage point to 1.5 per cent in 2002.

France, Ireland, Luxembourg and the Netherlands), which are offset by tax increases in others (Austria, Germany and Italy). For the euro area as a whole, the fiscal policy stance in 2002 will be broadly neutral. The expected change in the cyclically adjusted primary balance (which excludes interest payments) is in the same direction. In Germany, the budget deficit rose to 2.7 per cent of GDP in 2001, close to the ceiling of 3 per cent prescribed by the Stability and Growth Pact, and is forecast to remain more or less unchanged in 2002. In France, the budget deficit is forecast to rise above 2 per cent of GDP in 2002. German fiscal policy is seen to be broadly neutral in 2002 although the recessionary environment suggests the need for a discretionary fiscal stimulus. Indeed, the same could also be said for the euro area as a whole.

Outside the euro area, in the United Kingdom, a relatively moderate slowdown in the rate of economic growth to 2 per cent is forecast for 2002 (down from 2.3 per cent in 2001). This mainly reflects continued vigorous growth in private household consumption and a large increase in public sector spending. Moreover, there are increasing concerns about the sustainability of the recent surge in private household debt, which is at record levels relative to income. Partly in reaction to this, the Bank of England's Monetary Policy Committee has left its base rate unchanged at 4 per cent since November 2001 in order to avoid any further stimulus to borrowing, although inflation is forecast to continue undershooting its 2.5 per cent target.

In western Europe as a whole, real GDP is forecast to increase by 1.4 per cent in 2002, largely a reflection of weak domestic demand and the external environment.

Risks to the outlook

The outlook for the global economy, including Europe, is crucially dependent on the assumption that there will be a sustained and gradually strengthening recovery in the United States, led by domestic demand. This is expected to stimulate domestic activity in the rest of the world, including Europe, via exports and the spillover effects from increasing business and consumer confidence in the United States.

However, the signs of a cyclical upturn in the United States economy in early 2002 could well turn out to be a false dawn. The spending behaviour of private households and the balance sheet adjustments that they consider desirable or necessary in the face of increased job insecurity, high debt service burdens, a very low savings rate and a substantial loss in net financial assets, are crucial for the outcome.

The shallow recession, moreover, has not led to a correction of the sizeable external imbalance of the United States economy. The current account deficit fell only slightly in 2001 and is still more than 4 per cent of GDP. This is, of course, mainly the mirror image of the considerable excess of private sector investment over private savings (which has led to the accumulation of high levels of private sector debt). As a domestic demand-led recovery in the United States can be expected to lead to a further deterioration of the United States external imbalance[9] there is a risk that financial markets will feel increasingly uncomfortable with such a tendency. This could trigger a sudden reversal of capital flows and a sharp fall in the exchange rate of the dollar which, on a trade-weighted basis, is close to a 16-year high against other major currencies. The other side of the coin would be a strong appreciation of the euro, which would act as a brake on export growth and be likely to bring a cyclical upswing in western Europe to a premature end.

Another major uncertainty is how spending on high-technology goods will respond to an improved outlook for growth. There is increasing scepticism about the contribution of ICT goods to the increase in United States productivity in the second half of the 1990s. This performance appears to have been not broadly based but rather concentrated in a few sectors.[10] It is also not clear to what extent the massive spending on these products has generated the expected high rates of return or, in some cases, any return at all. The frustrated expectations of companies could lead to a much lower growth of IT spending in the year ahead with subsequent repercussions on profitability and equity valuations in the IT sector. This, in turn, is likely to have negative feedback effects on private consumption and business investment.

More generally, current levels of private sector indebtedness in the major industrial countries are quite high given the stage of the business cycle. A weak or aborted recovery in the second half of 2002 would test the profit expectations built into current equity prices. Any disappointment could trigger a sharp fall in these prices and a further deterioration in the balance sheets of households, the corporate sector and financial institutions in the major industrial countries.[11]

---

[9] This reflects the empirical finding that United States imports respond more strongly to a strengthening of domestic activity than United States exports to changes in foreign economic activity. This "income asymmetry" implies that even if the United States and its major trading partners have the same rate of economic expansion, there will still be a widening of the trade deficit. C. Mann, *Is the U.S. Trade Deficit Sustainable?* (Institute for International Economics, Washington, D.C., 1999), p. 124.

[10] According to a McKinsey study, most of the productivity gains between 1995 and 1999 originated in only 6 out of 59 economic sectors and the role of information technology was relatively small with the most important factors being innovation (including, but not limited to, IT and its applications), competition and, to a lesser extent, cyclical demand factors. McKinsey Global Institute, *US Productivity Growth 1995-2000*, [http://www.mckinsey.com], and R. Gordon, "Does the "new economy" measure up to the great inventions of the past?", *Journal of Economic Perspectives*, Vol. 14, No. 4, Fall 2000, pp. 49-74.

[11] IMF, *Global Financial Stability Report* (Washington, D.C.), March 2002 [www.imf.org].

Other sources of downside risks come from the lingering financial sector problems in Japan and uncertainty over the evolution of the price of crude oil. The price of Brent crude rose above $24 a barrel in the first half of March 2002 for the first time since the events of 11 September. This reflects the discipline of OPEC member countries in adhering to their agreed cuts in production as well as expectations of a strengthening of world output growth in the second half of 2002. In addition, there has been upward pressure on oil prices due to fears of interruptions to oil supplies in the Middle East as a result of a possible conflict between Iraq and the United States. The reaction to such developments of Russia, the second largest exporter of crude oil behind Saudi Arabia (Russia is not member of OPEC), could also have an impact on global oil supply and prices.

(ii) EMU: the currency changeover and the macroeconomic policy framework

The changeover to euro-denominated coins and notes in the 12 member states of the euro area at the beginning of 2002 was very successful from a logistic point of view.[12] Until the end of February 2002, the traditional national currencies circulated alongside the euro.[13] However, by mid-January 2002, virtually all cash transactions were already being conducted in euro notes and coins.[14] This strong demand for euro notes and coins reflected the high transaction costs involved when using the old national currencies alongside the new common one. The approximately 300 million inhabitants of the euro area can now use the same notes and coins for all payments across the member states.[15]

The smooth introduction of euro banknotes and coins was the final step in the long and difficult process of creating the EMU. Although this changeover has received much attention from the media and the public at large, its economic significance has been more limited. Since 1 January 1999, with the irrevocable fixing of national exchange rates, the national currency units were already nothing more but non-decimal subunits of the (then virtual) euro.

Overall, the macroeconomic effects of the changeover are likely to have been small and, in any case, difficult to gauge. An immediate economic effect is the increased transparency of prices across countries of the euro area. This, it is believed, is likely to increase competitive pressures faced by companies and could potentially reduce the observed variation in prices for internationally traded products across the euro area. Prices will continue to vary, however, given the differences in national indirect tax rates, transport costs and other costs of wholesale and retail distribution. In addition, prices reflect competitive conditions in local markets. And prices for non-tradeable goods and services will continue to reflect the relative levels of productivity in the tradeable sectors and the related real wage levels in the various countries. Similarly, inflation in the euro area will continue to diverge across countries, partly reflecting differential rates of growth of demand in the short run and, in the longer run, the differential impact of real income convergence across countries on the prices of non-tradeables – the so-called Balassa-Samuelson effect.[16]

The common currency could, moreover, by abolishing exchange rate volatility in bilateral trade among member states of the euro area and through other channels such as reduced transaction costs and closer integration of product markets, have a significant effect on bilateral trade among the member countries of the euro area. Although recent empirical research has found a strong expansionary impact of currency unions on trade among their members, there is considerable uncertainty about the order of magnitude involved.[17] It may also be surmised that this more intensive trade reflects not only the influence of the common currency but also of other factors. In any case, these findings suggest that membership in a monetary union is likely to significantly affect the level and composition of intraregional trade in the longer run.

Apart from its economic dimension, the currency changeover has also been seen as a tangible symbol of European integration that could help to foster a sense of common identity among Europe's citizens and be a catalyst for further economic and political reforms in the European Union. In this sense, the euro is much more than a means of payment – it is seen as propelling further progress in European integration. This notion, of course, is consistent with the tradition that the integration of Europe has progressed mainly through economic initiatives.

---

[12] Some 10 billion euro banknotes were printed to replace the national banknotes of the 12 participating states. In addition some 5 billion notes were printed as logistical stocks to ensure a smooth banknote changeover in 2002. The total value of these nearly 15 billion banknotes amounted to some €633 billion. In addition, to replace the national coins in the 12 countries, about 52 billion coins were minted with a total value of €15.75 billion.

[13] In the Netherlands, this changeover period lasted only until 28 January 2002. In Belgium and France, the corresponding deadlines were 9 February and 17 February 2002, respectively.

[14] ECB, "Update on the euro cash changeover", *Press Release*, 18 January 2002 [www.ecb.int/press/02/].

[15] The euro has also become the legal tender of Andorra, Monaco, San Marino and the Vatican.

[16] UNECE, "Inflation and interest rate differentials in the euro area", chap. 2.5, pp. 59-63, and "Economic transformation and real exchange rates in the 2000s: the Balassa-Samuelson connection", chap. 6, pp. 227-239, *Economic Survey of Europe, 2001 No. 1*.

[17] It has been estimated that a common currency expands trade within a range of 50 to 300 per cent. For the higher estimate see A. Rose, "One money, one market: the effect of common currencies on trade", *Economic Policy*, Vol. 30, April 2000, pp. 7-45. For the lower estimate see T. Persson, "Currency unions and trade: how large is the treatment effect?", *Economic Policy*, Vol. 33, October 2001, pp. 435-448.

The first three years of EMU have not been easy from a macroeconomic policy perspective. At the start of EMU, monetary policy had to cope with the fall-out from the 1998 financial crises and the marked depreciation of the euro. In addition, there was a surge in oil prices which gathered momentum in the course of 1999 and, when this shock abated, the global economy was hit first by the abrupt ending of the United States economic boom and thereafter by the terrorist attacks of 11 September 2001. But this series of shocks also serves to highlight one of the advantages of EMU compared with the hard EMS. It may be surmised that these events could have created considerable tensions within the ERM, the former exchange rate mechanism that linked the national currencies of the EU member states, with the risk of severe exchange rate crises as in 1992-1993.

This does not mean, of course, that everything was fine with the operation of macroeconomic policy during the first three years of EMU. The macroeconomic policy framework is, in fact, quite complex, combining a single monetary policy with (currently) 12 national fiscal policies. Many implications of this framework still need to be better examined and understood. This pertains especially to the interaction of fiscal and monetary policy and the implications of tight fiscal rules for the process of economic adjustment to asymmetric shocks at the individual country level.

The reliance of the ECB on a two-pillar monetary policy strategy, moreover, has not contributed to transparency and public understanding of the ECB's interest rate decisions (see chapter 2.3 below). In fact, the first pillar – the reference value for M3 – has, not surprisingly, been a poor guide for monetary policy in the short run. This reflects, *inter alia*, the unstable demand for money in the short run and special factors such as the recent flight to liquidity. Indeed, the first pillar has been more of a barrier to effective communication with the public. It has also created confusion abut the effective role of changes in money supply in the conduct of monetary policy; repeated efforts were made, for example, to explain that the overshooting of the reference value was due to special factors and therefore did not require a policy reaction.

The second pillar of the ECB's monetary strategy – the "assessment of the inflation outlook" – suffers from the lack of an explicit inflation forecast although, in view of the weakness of the first pillar, the ECB must be pursuing de facto a policy of inflation targeting. This also affects the transparency of policy and weakens the accountability of the Bank. Moreover, the lack of clarity of monetary policy is enhanced by the ECB's asymmetric definition of price stability as a year-on-year increase in consumer prices for the euro area "of below 2 per cent". The ECB has, however, excluded *deflation,* i.e. a fall in the price level. This would imply that the target range is between 0 and 2 per cent, but there is no evidence that policy is focused on the middle of the range. An inflation target of 1 per cent would in any case be rather low for at least three reasons: the downward rigidity of nominal prices and wages; upward biases in the price index due to a lack of adequate adjustments for quality improvements of products; and finally the need to allow for relative price changes between the European countries on account of the Balassa-Samuelson effect. Also, with a fixed base index, low rates of inflation could reflect a large proportion of relative price changes, apart from the Balassa-Samuelson effect. Taking all these factors into account suggests that the annual inflation target should be raised to 2.5 per cent.[18]

The ECB could not be expected to inherit the Bundesbank's credibility and it has to build its reputation by a clear justification of its policy decisions and by demonstrating that it is not subject to political pressures. The lack of a transparent framework and strategy for monetary policy, however, has created considerable uncertainty about the ECB's policy reaction function, i.e. whether and when it will react to deviations of inflation from its target and to changes in output gaps. This was especially striking in the first half of 2001, when there was a broad consensus that the ECB reacted "too little, too late" to counter the cyclical slowdown in the euro area. One way to improve the transparency and communication of monetary policy would be to set a symmetric inflation target. Another would be to integrate the monitoring of changes in money supply into the second pillar. A good understanding of the ECB's policy reaction function is also important for EU governments. Efforts to further reduce structural budget deficits within the framework of the Stability and Growth Pact (SGP) need to find an appropriate offset in a correspondingly more accommodative interest rate policy in order to ensure an appropriate policy mix for the euro area as a whole.

In the EMU, individual countries no longer possess the exchange rate as an instrument of adjustment to asymmetric shocks. This means that macroeconomic stabilization at the national level has to rely entirely on fiscal policy. This is a cause for concern since the rules of the SGP may constrain the flexible use of this instrument when most needed. At the same time, the SGP is also seen as a coordination device to help ensure an appropriate policy mix for the euro area as a whole, but the extent to which this can be effective is an open question.[19]

The SGP, which was adopted by the European Council in Amsterdam in June 1997, is unique in that sovereign countries have committed themselves to adhere to a set of common fiscal rules and a multilateral

---

[18] H. Sinn and M. Reuter, *The Minimum Inflation Rate for Euroland,* NBER Working Paper, No. 8085 (Cambridge, MA), January 2001.

[19] For a detailed discussion, see A. Brunila, M. Buti and D. Franco (eds.), *The Stability and Growth Pact* (New York, Palgrave, 2001).

surveillance mechanism.[20] The main rationale for the SGP is to provide additional protection for the ECB from political pressure for an inflationary debt bailout as a consequence of profligate fiscal policies and an unsustainable rise in public debt. This would suggest the need for limits to be placed on levels of public debt, but the SGP focuses exclusively on budget deficits.[21] Another reason for the SGP is that excessive borrowing by a member country could lead to higher interest rates for the euro area as a whole and affect the exchange rate of the euro.[22]

The SGP stipulates that the national general government budgetary positions should in the medium term be "close to balance or in surplus". Only under "exceptional and temporary" conditions is the government budget deficit allowed to exceed a ceiling of 3 per cent of GDP. This would be the case if the deficit results from a severe economic downturn, for which an annual decline in real GDP by 0.75 per cent will, as a rule, be taken as a reference point. In "normal times" a deficit larger than the reference value is regarded as excessive and will trigger the "excessive deficit procedure", which can lead to sanctions and financial penalties. The implicit assumption is that a budgetary position of close to balance or in surplus will provide sufficient room for the operation of the automatic stabilizers without breaching the 3 per cent reference value during normal cyclical fluctuations. The Pact therefore makes an implicit distinction between the cyclical and structural components of the deficits. The conclusion is that the medium-term target for budgetary positions should de facto be a target for the cyclically adjusted budget balance.[23] This is widely regarded as a better measure of the short-term fiscal policy stance than the actual deficit.

Calculations based on postwar business cycles suggest that a budgetary position close to balance or in surplus should, in principle, provide sufficient room for the operation of the automatic stabilizers during normal cyclical downturns.[24] However, alternative estimates based on stochastic simulations suggest that the medium-term fiscal target stipulated by the Pact is unnecessarily restrictive in countries where the budget deficit is less sensitive to the cycle. The conclusion is that a structural deficit target of around 1 per cent of GDP would be adequate for almost all countries in the EMU.[25] Such a less restrictive target would at the same time alleviate a constraint on borrowing to finance public investment, which has been one of the main victims of fiscal consolidation in the 1990s.

In any case, all such calculations are surrounded by large margins of uncertainty, if only because the ongoing structural changes in an economy can significantly affect the sensitivity of changes in government net revenues to cyclical conditions. Past experience may also be a poor guide to present policy because in the monetary union the exchange rate is no longer available as an adjustment instrument and this puts a correspondingly larger burden on fiscal stabilization. All this points to the need for a flexible interpretation of the Pact. (It should also be recalled that the 3 per cent ceiling was based on the ratio of public investment to GDP in the 1980s.) It would also be appropriate to shift the focus of the Pact away from targets for actual budget balances to explicit targets for cyclically adjusted balances.[26] This would require, however, agreement on how to calculate them.

Recent developments in Germany have illustrated that progress in fiscal consolidation is strongly dependent on where the country stands in the business cycle. Any attempt in the past year by the German government to adhere to the fiscal targets established in the Stability Programme would have had procyclical effects with the risk of making the budgetary position even worse.

Given its focus on the medium term of balanced budget targets in the individual member countries, the SGP is more restrictive than the fiscal convergence criteria of the Maastricht Treaty.[27] In fact, the rules of the SGP are tantamount to a medium-term "one-size-fits-all" fiscal policy with the same fiscal rule applying to each country.[28] This is combined with a one-size-fits-all monetary policy.

This framework has important implications for the adjustment that is necessary in the event that a balanced budget is incompatible with domestic macroeconomic balance in a given country.[29] In the case of Germany,

---

[20] The Pact includes, besides the European Council resolution adopted in Amsterdam, two Council regulations: one on the surveillance of budgetary positions and the coordination of economic policies; the other dealing with the procedure for responding to excessive deficits.

[21] M. Canzoneri and B. Diba, "The SGP: delicate balance or albatross?", in A. Brunila et al., op. cit., pp. 53-74.

[22] S. Eijffinger and J. de Haan, *European Monetary and Fiscal Policy* (Oxford, Oxford University Press, 2000), chap. 4.

[23] It has been pointed out, however, that a structural budgetary position close to balance or in surplus does not necessarily imply that the fiscal position is sustainable because it may still not exclude the risk of a rising debt-to-GDP ratio. P. Brandner, L. Diebalek and H. Schuberth, *Structural Budget Deficits and Sustainability of Fiscal Positions in the European Union,* Öesterreichische Nationalbank Working Paper, No. 26, February 1998.

[24] M. Buti and A. Sapir (eds.), *Economic Policy in EMU* (Oxford, Oxford University Press, 1998), pp. 102-137.

[25] R. Barrel and K. Dury, "Will the SGP ever be breached?", in A. Brunila et al., op. cit., pp. 235-255.

[26] A first step in this direction was taken at the Göteborg European Council when it was agreed that: "Cyclically adjusted budgetary positions should move towards, or remain, in balance or surplus in the coming years …", *Presidency Conclusions*, Göteborg European Council, 15 and 16 June 2001, item 34 [http://europa.eu.int/council/].

[27] This should also be considered by the accession countries given the urgent need to improve their public infrastructure.

[28] C. Allsopp, "The future of macroeconomic policy in the European Union", speech to the Austrian Institute of Economic Research (WIFO), 23 January 2002 [www.bankofengland.co.uk].

[29] Ibid., pp. 21-29.

which has to cope with weak domestic demand and has difficulties in meeting the rules of the SPG, there will have to be a change in relative competitiveness, notably in relative wage levels, and associated changes in the current account balance (possibly a large surplus), if the economy is to achieve macroeconomic balance and grow at its potential rate. This may, however, be difficult to achieve (especially if other euro area countries are keen to maintain their competitiveness relative to Germany) and involve rising unemployment, the political costs of which may be difficult to sustain. This adjustment burden could be eased with a more flexible fiscal policy framework.

Changing the rule-based macroeconomic policy framework of EMU may be difficult however as any such suggestion will be interpreted by many as a threat to credibility. But ultimately, the credibility of EMU will depend on its ability to raise economic well-being for all its members and to avoid prolonged periods of anaemic growth and high unemployment. An appropriate moment to introduce changes to the macroeconomic policy framework might be during the preparations for EMU enlargement, not only to the east but also to other west European countries such as the United Kingdom. Enlargement might in fact be easier and more attractive to those currently outside if a less rigid approach were evident.

1.3 The transition economies

(i) Recent developments

Despite the negative repercussions of the global economic slowdown, 2001 turned out to be a relatively successful year for the ECE transition economies: with the exception of The former Yugoslav Republic of Macedonia, all of them posted positive rates of GDP growth and in some of them they were higher than in 2000. The transition economies' aggregate GDP increased by 5 per cent, making them one of the fastest growing regions in the world. The main factor behind this outcome was buoyant growth in the Commonwealth of Independent States where a strong recovery continued for a third consecutive year.

As in 2000, Russia remained the principal engine of growth for the CIS countries in 2001 (see chapter 3.1(iv)) with a 5 per cent increase in GDP. After the 1998 financial crisis, the Russian government introduced sweeping policy reforms, which have led to major structural adjustments in the economy and have moved it on to a path of strong growth. Considerable progress has been made in strengthening the Russian fiscal and judiciary systems, in rehabilitating the banking sector and the payments system in general, and in reducing administrative interference in the economy. The exchange rate realignment after the August 1998 financial collapse – equivalent to a competitive, real devaluation – provided an important stimulus to local producers, encouraging import substitution on a large scale. In addition, from mid-1999 until the fall of 2001, the Russian economy benefited substantially from the surge in world oil prices. Between 1999 and 2001, Russia's GDP increased by almost 21 per cent giving a much needed boost to popular support for the reforms. All the indications are that the Russian economy has crossed an important threshold in its systemic reforms, making the process of its transformation to a market economy now look irreversible.

Despite these positive developments, there are a number of uncertainties regarding Russia's economic prospects. Notwithstanding the recent progress in market reforms, Russia is far from the end of this process. Besides, it is not yet clear whether the institutional environment will be capable of implementing and enforcing efficiently all the newly adopted laws and regulations. In addition, the heavy dependence of the Russian economy on oil exports entails significant risks for macroeconomic performance in general due to the persistent volatility of international oil prices. Hence, some caution is needed in assessing the prospects for high and sustainable growth in Russia.

Another important development in the CIS region has been the continuing strong recovery of two of the larger economies, Kazakhstan and Ukraine, both of which had some of the highest rates of GDP growth in 2001. In the case of energy exporting Kazakhstan, the recent record rates of growth (13.2 per cent in 2001 after 9.8 per cent in 2000) reflect both the impact of a favourable external environment and balanced policies, which have helped to broaden the base of the recovery while maintaining macroeconomic stability. In Ukraine, strong domestic demand contributed to the 9.1 per cent GDP growth in 2001: the recent disinflation effort (which reduced the year-on-year inflation rate to single digits for the first time since independence), coupled with growing real incomes, has given a boost to consumer and investor confidence.

Although an important growth engine for the neighbouring CIS countries, Russia was not in fact the fastest growing economy in the region: 8 of the remaining 11 CIS member states in 2001 had annual rates of GDP growth higher than that of Russia (table 1.3.1). In most cases (Armenia, Kazakhstan, the Republic of Moldova, Turkmenistan, Ukraine and partly Tajikistan), strong growth was underpinned by the expansion of exports: commodity exporters benefited from favourable external market conditions while others were able to take advantage of rising import demand within the CIS itself. However, the surge in economic activity was mostly confined to the first half of 2001; in the second half of the year, there was a notable deceleration both in output and export performance throughout the CIS.

In 2001, strong rates of growth prevailed in most of the east European and Baltic states as well. In

TABLE 1.3.1

Annual changes in real GDP in eastern Europe, the Baltic states and the CIS, 1999-2002

(Per cent)

	1999	2000	2001 April forecast	2001 Actual outcome	2002 official forecast
Eastern Europe	1.7	3.7	4.2	3.0	2.7
Albania	7.3	7.8	5-7	7*	7
Bosnia and Herzegovina [a]	..	9.1	7-9	8*	6
Bulgaria	2.4	5.8	5	4.9	4
Croatia	-0.4	3.7	3-4	4.3*	4
Czech Republic	-0.4	2.9	3	3.6	2.4-3.4
Hungary	4.2	5.2	4.5-5	3.8	3-4
Poland	4.1	4.0	4.5	1.1	1
Romania	-1.2	1.8	4.1	5.3	4.5
Slovakia	1.9	2.2	3.2	3.3	3.6
Slovenia	5.2	4.6	4.5	3.0	2.9-3.6
The former Yugoslav Republic of Macedonia	4.3	4.6	6	-4.6	4
Yugoslavia [b]	-17.7	6.4	5	6.2	4
Baltic states	-1.7	5.4	4.7	6.2	4.2
Estonia	-0.7	6.9	6	5.3	3.5-4
Latvia	1.1	6.8	5-6	7.6	4.5-5.5
Lithuania	-3.9	3.9	3.7	5.7	4
CIS	4.5	8.3	4.2	6.2	4.8
Armenia	3.3	6.0	6.5	9.6	6
Azerbaijan	7.4	11.1	8.5	9.9	8.5
Belarus	3.4	5.8	3-4	4.1	4-5
Georgia	3.0	2.0	3-4	4.5	3.5
Kazakhstan	2.7	9.8	4	13.2	7
Kyrgyzstan	3.7	5.4	5	5.3	4.5
Republic of Moldova [c]	-3.4	2.1	5	6.1	6
Russian Federation	5.4	9.0	4	5.0	4.3
Tajikistan	3.7	8.3	6.7	10.2	8
Turkmenistan [d]	17.0	17.6	16	20.5	18
Ukraine	-0.2	5.9	3-4	9.1	6
Uzbekistan	4.4	4.0	4.4	4.5	5.1
Total above	3.3	6.5	4.2	5.0	4.0
Memorandum items:					
CETE-5	3.0	3.8	4.1	2.3	2.1
SETE-7	-1.7	3.5	4.5	4.9*	4.4

Source: National statistics, CIS Statistical Committee and direct communications from national statistical offices to UNECE secretariat.

Note: Aggregates are UNECE secretariat calculations, using PPPs obtained from the 1996 European Comparison Programme. Forecasts are those of national conjunctural institutes or government forecasts associated with the central budget formulation. Aggregates shown are: eastern Europe (the 12 countries below that line), with sub-aggregates CETE-5 (central European transition economies: Czech Republic, Hungary, Poland, Slovakia, Slovenia) and SETE-7 (south-east European transition economies: Albania, Bosnia and Herzegovina, Bulgaria, Croatia, Romania, The former Yugoslav Republic of Macedonia and Yugoslavia); Baltic states (Estonia, Latvia, Lithuania); and CIS (12 member countries of the Commonwealth of Independent States).

[a] Data reported by the Statistical Office of the Federation; these exclude the area of Republika Srpska.

[b] Data exclude Kosovo and Metohia.

[c] Excluding Transdniestria.

[d] Figures for Turkmenistan should be treated with caution. In particular, the deflation procedures that are used to compute officially reported growth rates are not well documented and the reliability of these figures is questionable.

the start of the year (table 1.3.1). Economic activity remained high, and in line with expectations in Albania, Bosnia and Herzegovina, Bulgaria and Estonia. In contrast, growth decelerated in Hungary and, especially, in Slovenia; in these two economies the effects of weakening west European import demand were probably most pronounced. Nevertheless, in both countries the annual rates of GDP growth were considerably higher than the west European average.

Two economies, Poland and The former Yugoslav Republic of Macedonia, have recently encountered serious economic difficulties. After nine years of uninterrupted and rapid expansion, the Polish economy came to a near standstill in 2001. The reasons for this are complex and deep-seated (see chapter 3.1(iii) for details) but they are indicative of the continuing fragility of the transition economies and the fact that even the more advanced reform countries are prone to unexpected setbacks. Some of the current problems in Poland stem from policy complacency: the reluctance by the authorities to undertake important but unpopular reforms during the boom period, when the excellent macroeconomic performance tended to mask some chronic economic problems. One of the lessons from the Polish case is that postponing policy reforms not only fails to resolve a pending issue but makes it more difficult to tackle later on when it may have escalated out of control, especially in a cyclical downturn, which may be exacerbated by the chronic structural weakness.

The former Yugoslav Republic of Macedonia was the only transition economy with falling GDP in 2001. This was not surprising against the background of widespread disruption caused by the internal military conflict; however, given the country's past record, it is likely that this can be regarded as a one-off setback. A relatively strong postwar recovery has continued in neighbouring Yugoslavia; however, this economy still faces formidable difficulties in implementing much needed but painful economic reforms.

In view of the increasing openness of the transition economies (chapter 3.2(iii)) and given the considerable weakening of global trade in 2001, their relatively strong performance in 2001 comes as a surprise. It is therefore instructive to compare the reaction of the transition economies to the negative external shock in 2001 with their handling of the consequences of a similar global recession at the beginning of the 1990s. Although in both cases the external impact was broadly similar, the outcomes could not have been more different. A decade earlier, the weakening of global and, especially, west European demand coincided with the initial shocks of the transition and at that time the external shock considerably amplified and prolonged the transformational recession. In contrast, in 2001 a shock of a similar or even stronger magnitude[30] has so

---

[30] At the beginning of the 1990s the cycles in the world's major economies diverged and there was no synchronous downturn in the global economy.

Croatia, the Czech Republic, Romania, Slovakia, Latvia and Lithuania the rate of GDP growth not only accelerated from 2000 but was also above expectations at

far had only a marginal impact on the transition economies. Although a detailed analysis of the factors behind this outcome remains a task for the future, it is worth drawing attention to the likely importance of two recent developments in the region.

First of all, thanks to the successful implementation of reforms which have bolstered consumer and investor confidence, domestic demand in the transition economies has generally been growing steadily in recent years. The recent global downturn has affected domestic demand (both private consumption and investment) in these economies to a lesser extent than in most of the industrialized countries. This relatively robust domestic demand helped to cushion the transition economies from the effects of the deteriorating external environment. As discussed in more detail in chapter 3.2 of this *Survey*, in many countries in 2001 there was a shift from external towards predominantly domestic sources of growth. An interesting aspect of this development – and a significant sign of the growing maturity of these economies – is the fact that, with a few exceptions, it was private domestic demand that played the key role in bringing about this shift; the fiscal stance of governments remained generally neutral in 2001 (chapter 3.1(ii)). Another sign of resilience was the fact that so far there were virtually no negative repercussions for the transition economies from the crisis in Argentina (for details see chapter 3.1(v)). Moreover, the flow of inward FDI to the transition economies was unabated, in many cases giving a further boost to final domestic demand.

Secondly, thanks to recent productivity gains, most east European transition economies have been able to improve their cost competitiveness vis-à-vis their main trading partners. Due to the fact that the productivity differential vis-à-vis western Europe was apparently retained in 2001, the transition economies held on to these gains or even enlarged them (chapter 3.1(i) and 3.3(ii)). The on-going improvement in their competitive position obviously helped east European exporters to perform better on west European markets in 2001 than some of their competitors. Thus, the negative repercussions on eastern Europe from the weakening in west European demand had a less than proportionate effect on east European exports (chapter 3.5(ii)). While the total volume of west European imports in 2001 increased by a little over 1 per cent, the volume of total exports from the central European and Baltic countries increased by some 11 per cent. The gains in competitiveness and the improved export performance also led to an increase in eastern Europe's share of the EU's extra-EU imports from 9.9 per cent in 2000 to 11.1 per cent in 2001.[31]

Although these are rather positive developments for the transition economies, their significance should not be overestimated. While being able to provide a temporary shield against a negative external shock, domestic demand has only limited potential as a leading factor of growth in the majority of the transition economies. The problem is that a number of these countries suffer from persistently large current account deficits (chapter 3.5(i)); excessive reliance on domestic absorption could drive these deficits out of control with consequent risks for their macroeconomic stability. As for their trade performance, it is not yet clear whether the negative repercussions from the weakening of global and west European demand will be confined to 2001; it may well be that due to lags in the economic system there will be negative carry-over effects in 2002 as well. In this regard, it should be emphasized once again that there is no room for policy complacency for the economies in transition; most of them still have a long way to go before they reach the stage of the mature market economies of western Europe.

(ii) The short-term outlook

It is generally expected that growth will moderate somewhat in the transition economies in 2002: according to the available official forecasts, aggregate GDP in the CIS will grow by close to 5 per cent, in the Baltic states by slightly more than 4 per cent and in eastern Europe by some 2.75 per cent (table 1.3.1). The average figures for the subregions are very much dominated by the expected developments in two of the largest economies in the region: Russia and Poland. In Russia, the 2002 budget assumes a 4.3 per cent rate of GDP growth. It should be noted though that the Russian Ministry of Economic Development drafted three possible growth scenarios for 2002, depending on the expected development of world oil prices, and the one that underlies the 2002 budget corresponds to the "optimistic" scenario in this set.[32] Private forecasters seem to be more conservative about Russia's growth prospects in 2002: according to a compendium of private forecasts collected and published by the World Bank, the rate of GDP growth in 2002 ranges between 1.6 and 3.8 per cent.[33]

---

[31] Based on the EU's imports from the 15 east European and Baltic states; full year data for 2000, and January-September data for 2001 (excluding Greece). UNECE secretariat computations on the basis of Eurostat, CD-ROM Theme 6: External Trade, *Intra- and Extra-EU Trade*, Monthly Data, No. 1, 2002.

[32] According to the report prepared by the Ministry of Economic Development, if the price for Russia's oil exports in 2002 averages $23.5 per barrel, GDP is expected to increase by 4.3 per cent; if oil fetches on average $18.5 per barrel, GDP growth in 2002 is estimated at 3.5 per cent; and if the price falls to $16.5 per barrel, GDP growth is expected to slow down to 3.1 per cent. *Interfax News Agency, Daily Financial Service*, 6 February 2002 as reported in *Reuters Business Briefing*, 12 February 2002. For more details on the impact of oil prices on Russia's economic performance see chap. 3.1(iv).

[33] *Interfax International, Weekly Business Report*, 19 February 2002 as reported by *Dow Jones Reuters Business Interactive* (Factiva).

In Poland, after prolonged consultations and policy debates, a revised 2002 budget reflecting the government's anti-crisis programme was finally voted in March. The budget contains a number of austerity measures aimed at reducing the fiscal deficit and these are expected to dampen further domestic demand. This adjustment is expected to have a negative impact on economic activity: the rate of GDP growth in 2002 incorporated in the budgetary framework is just 1 per cent, similar to the outcome in 2001. In contrast to Russia, however, some private forecasters seem to be more optimistic about the short-term outlook for the Polish economy.[34]

In the rest of eastern Europe, the governments in Bulgaria, Croatia, the Czech Republic, Hungary, Romania and Yugoslavia envisaged some deceleration of growth in 2002 as compared with 2001 (table 1.3.1). The most frequently quoted reason for this expected slowdown are the lagged effects of the global and west European slowdown. Nevertheless, the annual rates of GDP growth in most of these countries are expected to remain in the range of 3 to 4 per cent. In contrast, according to the official forecasts, GDP growth in Slovakia is expected to accelerate in 2002, consolidating the adjustment effort undertaken in 1999-2000, while The former Yugoslav Republic of Macedonia envisages a return to growth after the 2001 downturn.

After two years of robust economic growth, some slowdown is expected in the Baltic states in 2002. The deceleration is likely to be more pronounced (with GDP growing by some 4 per cent or even less) in Estonia and Lithuania, both of which are rather dependent on external factors (a very high degree of openness in the case of Estonia and a strong reliance on oil processing and exports of refined products in the case of Lithuania). In Latvia, where strong GDP growth in 2001 was mainly underpinned by buoyant domestic demand, aggregate output is likely to continue to grow at a high rate (around 5 per cent) in 2002.

Despite a certain slowdown, the CIS is likely to remain the fastest growing subregion within the ECE area in 2002. According to the official forecasts, Ukraine's GDP is expected to grow by 6 per cent in 2002, although some private forecasters are less optimistic.[35] In February, Kazakhstan's parliament approved a medium-term economic programme, which envisages GDP growing at an average annual rate of some 5 to 7 per cent in 2002-2004;[36] the country's 2002 budget is based on a 7 per cent growth rate assumption. In Georgia the budget projections envisage GDP growth of 3.5 per cent in 2002, although the Ministry of Economy, Industry and Trade is more optimistic and expects GDP growth to be in the range of 4.9 to 7.1 per cent.[37] In the majority of the other CIS countries, governments are expecting GDP growth rates in the range of 5 to 8 per cent in 2002.

---

[34] According to the forecast of the Centre for Socio-Economic Research (CASE), GDP in Poland may grow by close to 2 per cent in 2002. *Polish News Bulletin*, 15 March 2002 as reported by *Dow Jones Reuters Business Interactive* (Factiva).

[35] In February the International Center for Policy Studies forecast that the rate of growth of GDP would slow down to 4.5 per cent in 2002. *Ukrainian News*, 25 February 2002 as reported by *Dow Jones Reuters Business Interactive* (Factiva).

[36] *Interfax International, Daily Business Report*, 26 February 2002 as reported by *Dow Jones Reuters Business Interactive* (Factiva).

[37] *Black Sea Press*, 8 January 2002 as reported by *Reuters Business Briefing*, 18 January 2002.

CHAPTER 2

THE GLOBAL CONTEXT AND WESTERN EUROPE

The global economic environment for western Europe deteriorated significantly in the course of 2001 with all major regions of the world economy experiencing a marked slowdown in the rate of economic expansion. The United States and Japan moved into recession. The increased uncertainty about global short-term economic prospects was temporarily accentuated by the terrorist attacks of 11 September. In western Europe, against the background of rapidly weakening world trade, the annual growth of real GDP in 2001 fell to its lowest rate since 1993, due in large measure to a recession in Germany, the largest economy in the region. Annual economic growth in all countries in 2001 was less than in 2000. In the United States, there was a progressive and considerable easing of monetary policy in the course of 2001, which was complemented by an expansionary fiscal policy. In western Europe, the ECB and other central banks also reacted to the weakening growth forces by lowering interest rates, but this was less pronounced than in the United States. Fiscal policy in western Europe, on average, was broadly neutral in 2001.

2.1 The global context

(i) Overview

In early 2002, the world economy was still in a very weak state. The available short-term indicators are still ambiguous as to whether a cyclical recovery has begun. Industrial output in the industrialized countries continued to weaken in the final quarter of 2001 but activity appears to have stabilized in the early months of 2002. In the United States, real GDP edged up slightly in the final quarter, largely because of special factors, but the decline in manufacturing output appears to have bottomed out in January. In contrast, in the euro area, overall economic activity fell in the final quarter of 2001 while in Japan, the economic situation continued to deteriorate.

World output is estimated to have increased by only 2.25 per cent in 2001 compared with buoyant growth of 4.7 per cent in 2000. This was a much sharper deceleration than in the two preceding global recessions of 1991 and 1982. The weakening of economic growth spread to all the major economies and regions leading to an unusually synchronous international business cycle not seen since the early 1980s. China and the transition economies of eastern Europe and the CIS proved surprisingly resilient to the rapid weakening of the global economic environment (chart 2.1.1). The slowdown in global output growth was accentuated by a weakening of demand for foreign goods and services. The volume of world merchandise trade stagnated in 2001 in contrast to the preceding year when it increased by some 11.5 per cent.

The pronounced global downswing reflects a number of factors, the cumulative effect of which increasingly dampened the rate of economic expansion. A major factor was the abrupt end to the economic boom, especially the investment boom in information and communication technology (ICT) products, in the United States in the second half of 2000. The United States economy had been the main engine of global economic growth in the second half of the 1990s and when that engine failed, neither western Europe nor Japan were in a position to offset this adverse demand shock.[38]

The investment boom in ICT products collapsed against a background of a sharp increase of spare capacity in the manufacturing sector – a pointer to considerable overinvestment – and a pronounced fall in corporate profits. The end of the investment boom deflated the stock market bubble, which had led to excessively high valuations of companies, notably in the high-technology sector. The associated and considerable loss in financial wealth resulting from the sharp decline of equity prices dampened consumer spending. At the same time, lower equity prices raised financing costs in the corporate sector and was another factor restraining business investment. The sharp fall in demand for ICT products was not limited to the United States, but, rather, spread to other regions of the global economy, partly reflecting the intricate international supply chains for the production of these goods. This was a major factor behind the global industrial recession and the stagnation of world trade in 2001.

---

[38] Real imports of goods and services in the United States fell by 2.7 per cent in 2001 against an increase of about 13.4 per cent in 2000. At the same time, exports fell by 4.6 per cent in 2001, after an increase of 9.5 per cent in 2000.

CHART 2.1.1

Annual changes in real GDP in the world economy, 1980-2001
(Percentage change over preceding year)

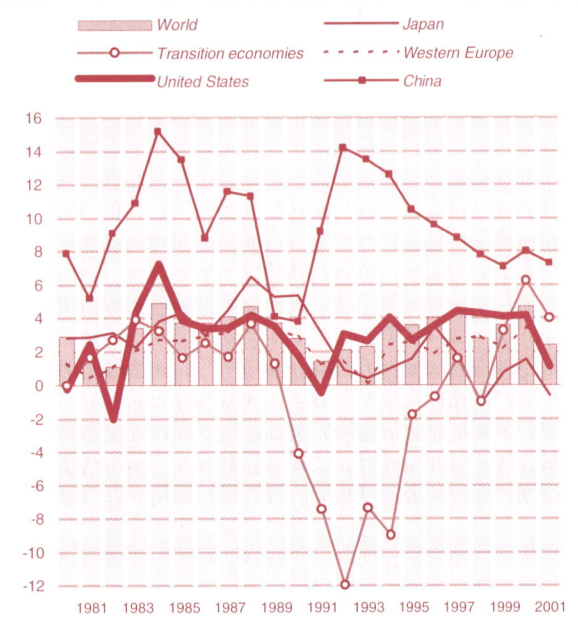

Source: IMF; Eurostat; national statistics.

Note: Net material product (NMP) for the former centrally planned economies for the period 1980-1990.

This demand shock followed on the heels of a significant supply shock, namely, the sharp rise in energy prices in 1999 and 2000 and was only reversed in the second half of 2001. Higher energy prices curbed households' real disposable incomes and, hence, the demand for non-energy goods and services with associated negative effects on domestic activity.[39] In the face of intense competition, moreover, enterprises had to absorb higher energy costs by reducing profit margins, which in turn affected business investment. Moreover, the lagged effects of tighter monetary policies in 1999 and the first half of 2000 in the major industrialized countries, intended to check inflationary pressures (chart 2.1.2), exacerbated these trends.

As a result, there was a progressive weakening of global economic growth in 2001, and for the industrialized countries as a whole real GDP fell after the first quarter of 2001 for three consecutive quarters (table 2.1.1). The United States Federal Reserve responded swiftly to the pronounced weakening of economic growth, followed by other major central banks. The European Central Bank (ECB) started to lower interest rates only in May 2001.

Prices in international equity markets fell strongly in the immediate aftermath of the terrorist attacks of 11 September in the United States. This reflected the general increase in uncertainty but also expectations of a more protracted cyclical slowdown. From October, however, equity prices started to rise again as it became apparent that the economic fallout from the attacks would be more limited than originally feared and as expectations about short-term economic prospects became more positive. At the end of 2001, share prices in the United States and western Europe had more or less recovered their mid-2001 levels, but were still between 10 and 20 per cent lower than in January 2001 (chart 2.1.3). The fall in share prices led to considerable falls in household financial wealth[40] and this is likely to have contributed to the slowing growth of household expenditures in the industrialized countries in 2001. In early 2002, global equity prices were being supported by expectations of an economic recovery in the United States and positive spillover effects in western Europe and other major regions. Despite the large fall in share prices since their peak in 2000, price-earnings ratios have remained at high levels, which largely reflect investors' expectations of a rapid return to a robust rate of growth of earnings. While equity markets in Europe have largely followed developments in the United States, in Japan they have been held back by the lingering deflationary tendencies and the need of banks to dispose of their equity holdings in other companies. In January 2002, Japanese equity prices came under renewed selling pressure, falling to their lowest level in 18 years. Since then, they have recovered somewhat partly as a result of government measures to prevent an excessive fall.

With hindsight it has become clearer that the United States and global economies were already in a rather fragile state at the time of the terrorist attacks on 11 September. Short-term economic indicators released before the attacks were already pointing to a more protracted slowdown than had previously been expected by many forecasters. Against this background, the attacks were a profound psychological shock that not only increased fears about security but further eroded business and consumer confidence in their short-term prospects.[41]

This more pessimistic outlook accentuated a wait-and-see attitude with regard to larger spending projects and added to the adverse effects of the attacks on activity in industries such as airlines and tourism. Given the size of the United States economy, the direct economic effects of the attacks have been fairly limited. The disruptions to financial markets were short-lived and the sharp fall in international stock market prices in the aftermath of the

---

[39] In western Europe, this was accentuated by the rise in food prices associated with animal diseases.

[40] In the United States, the fall in equity prices during the first three quarters of 2001 is estimated to have reduced household wealth by about $3.5 trillion or 8.25 per cent of total household net worth. These losses, however, were partly recovered by the rise in share prices in the final quarter of 2001. Board of Governors of the Federal Reserve System, *Monetary Policy Report to the Congress*, 27 February 2002, p. 25 [www.federalreserve.gov.].

[41] UNECE, *Economic Survey of Europe, 2001 No. 2*, p. 41.

CHART 2.1.2

Changes in key official interest rates in the major industrialized countries, January 1999-February 2002
(Per cent per annum)

Source: European Central Bank [www.ecb.int]; Bank of England [www.bankofengland.co.uk]; United States Federal Reserve Board [www.federalreserve.gov]; OECD, *Main Economic Indicators* (several issues); Bank of Canada [www.bank-banque-canada.ca].

TABLE 2.1.1

Quarterly changes in real GDP in the major seven economies, 2000QIV-2001
(Percentage change over preceding quarter)

	2000	2001			
	QIV	QI	QII	QIII	QIV
France	1.0	0.4	0.2	0.5	-0.1
Germany	0.2	0.4	–	-0.2	-0.3
Italy	0.8	0.8	–	0.1	-0.2
United Kingdom	0.6	0.7	0.5	0.5	0.2
Canada	0.4	0.3	0.2	-0.1	0.5
United States	0.5	0.3	0.1	-0.3	0.3
Japan	0.3	1.0	-1.2	-0.5	-1.2
Total above	0.5	0.5	-0.1	-0.2	-0.1
Memorandum items:					
Euro area	0.6	0.5	0.1	0.2	-0.2
European Union	0.6	0.5	0.1	0.2	-0.1
Western Europe [a]	0.6	0.5	0.1	0.2	-0.1
Total industrialized countries [b]	0.5	0.5	-0.1	-0.1	-0.1

Source: National statistics; Eurostat, New Cronos Database.

[a] The European Union plus Norway and Switzerland.

[b] Western Europe, North America and Japan.

attacks was more or less reversed by mid-October 2001. Also the quick response of the Federal Reserve and other major central banks in providing liquidity to ensure the functioning of financial markets in the days following the attacks helped to restore confidence. Further support was provided by an emergency fiscal package – totalling $40 billion – introduced by the United States administration to provide assistance and cope with other consequences of the attacks such as the need to increase airport security.

The pronounced cyclical downturn in the United States and other industrialized countries has had significant adverse effects on the economic performance of emerging markets and other developing countries. These effects were mainly transmitted via international trade. There was a sharp deceleration in the export growth of developing countries, reflecting the weakening demand for commodities and for ICT goods and services.

Private capital flows to emerging markets fell to some $115 billion in 2001, down from $169 billion in the preceding year.[42] This reflected, to a large extent, the capital outflows from Argentina and Turkey, which were caught up in deep economic and financial crises. Private flows to other emerging markets also fell somewhat. The increased risk aversion of international investors, notably reflected in the "flight to quality" in the wake of the terrorist attacks in September, led to rising credit risk premia in many emerging markets. The impact in the leading east European economies was subdued and temporary.[43]

The deepening economic, social and political crisis in *Argentina* in the second half of 2001 culminated in late 2001 and early 2002 in the government effectively defaulting on its $155 billion external debt (the largest default ever) and the end of the currency board, which had pegged the peso to the dollar since 1991. A dual exchange rate system was introduced but quickly abandoned and followed by the floating of the peso, which led to a sharp depreciation against the dollar by about 50 per cent in late February 2002.

The crisis in Argentina reflects the interplay of a host of factors. When the dollar appreciated against all the major currencies in the second half of the 1990s, the exchange rate peg to the dollar led to an increasing overvaluation of the peso. This reduced price competitiveness and held back export growth. The situation was aggravated when Brazil, a major trading partner, allowed its currency, the real, to float, which led to a sharp depreciation in early 1999. In addition, the continued slide in international prices of agricultural commodities since 1996 and the global cyclical downturn

---

[42] Institute of International Finance, Inc. (IIF), *Capital Flows to Emerging Market Economies* (Washington, D.C.), 30 January 2002 [www.iif.com].

[43] BIS, "International banking and financial market developments", *BIS Quarterly Review* (Basel), December 2001, p. 10.

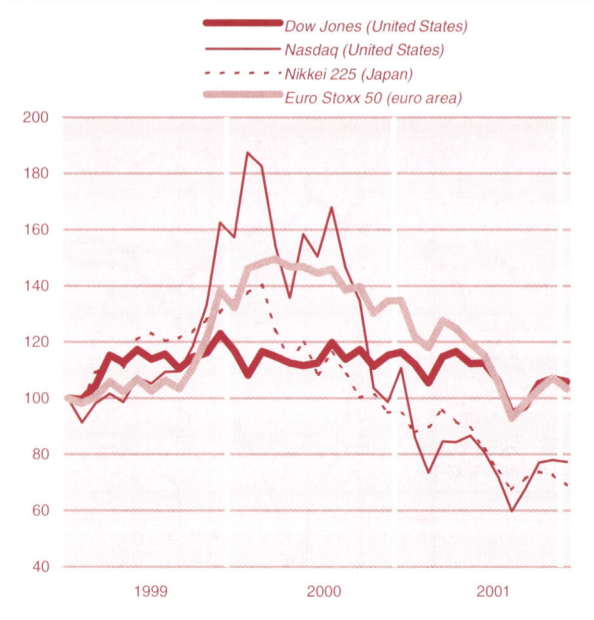

CHART 2.1.3

International share prices, January 1999-January 2002
(Indices, January 1999=100)

Source: The Financial Forecast Center [www.forecasts.org] and Stoxx [www.stoxx.com].

Note: Average monthly values.

in 2001 further depressed export revenues and domestic economic activity. Against this background real GDP, which had been declining since 1999, fell by at least 5 per cent in 2001. The decline in activity was accompanied by falling prices and wages but these were not sufficient to restore a competitive real exchange rate. The worsening domestic recession, in combination with deflation, eroded the government's tax base and led to a rise in public debt, which was financed by borrowing abroad. The failure to implement fiscal reforms progressively eroded financial market confidence and this led to capital outflows and a sharp rise in the country's risk premium,[44] which effectively cut off Argentina from the international capital markets in the second half of 2001. The deepening economic crisis undermined the credibility of the currency board and led to a run on the banks to which the government reacted by introducing capital controls, including a ceiling on monthly bank withdrawals, in December 2001. These measures only accentuated the political crisis and led to strong protests from the population.

The Argentine crisis has not led to significant contagion of other emerging markets.[45] Investors appear to have widely anticipated the outcome, adjusting their portfolios by reducing their exposure to Argentina and other markets that seemed to be risky. Another factor is that the overall exposure of investors to emerging markets in 2001 was relatively low compared with the levels prevailing at the time of the Asian crisis in 1997. In addition, the move away from rigid exchange rate regimes by many emerging markets allowed them to cope with the global cyclical downturn by letting the exchange rate adjust and thus avoiding the build-up of unsustainable external imbalances.[46] The risk of future contagion, however, cannot be excluded: much will depend on the future economic and financial developments in Argentina, and especially on how the default on the external debt is eventually resolved.

In *Latin America*, the main side effects of the protracted crisis in Argentina were felt in Brazil. Concerns at possible spillover effects led to temporary strong selling pressure on the national currency, the real, and a rise in the country's risk premium. A subsequent tightening of monetary policy accentuated the negative effects of a domestic energy crisis on domestic activity, but towards the end of 2001 the exchange rate recovered and the spread on borrowing rates declined. For the year as a whole there was a small increase of real GDP by 1.5 per cent. The decoupling from developments in Argentina was also apparent in the lack of any fall-out from the Argentine devaluation. In Mexico, the other major economy in the region, the stagnation of economic activity in 2001 was due to its close trade links with the United States. Overall, growth in Latin America is estimated at 0.5 per cent in 2001, a marked slowdown from the growth rate of 4.1 per cent in 2000.[47]

The *east Asian* emerging markets were particularly hard hit by the downturn in the global ICT industry due to the large share of these products in their total exports. Moreover, the lingering economic crisis in Japan depressed import demand from other countries in the region. Nevertheless, most of these countries continued to have a surplus on their current accounts in 2001. The continuation of strong economic growth in China stands in sharp relief to the marked weakening of activity in the other countries of the region.[48] Nevertheless, growth rates have been declining in recent years (chart 2.1.1), and the government has introduced a substantial fiscal stimulus to support economic activity.

The pronounced slowdown of the world economy in 2001 led to weaker demand in international commodity markets and a further pronounced decline of prices (chart 2.1.4). The fall in prices, however, appears to have

---

[44] The spread on Argentine international bonds was 5,500 basis points (55 percentage points) in December 2001, up from 700 basis points (7 percentage points) in January of the same year.

[45] See chap. 3.1(v).

[46] BIS, op. cit., p. 12.

[47] ECLAC, *Preliminary Overview of the Economies of Latin America and the Caribbean 2001* (Santiago), December 2001 [www.eclac.org].

[48] However, there are doubts about the accuracy of the official statistics. For a discussion of the current state of Chinese statistics see T. Rawski and W. Xiao, "Roundtable on Chinese economic statistics: introduction", *China Economic Review*, Vol. 12, No. 4, 2001. This issue contains 10 papers covering a wide range of statistics.

CHART 2.1.4

World commodity prices, January 2000-February 2002
(Indices, January 2000=100)

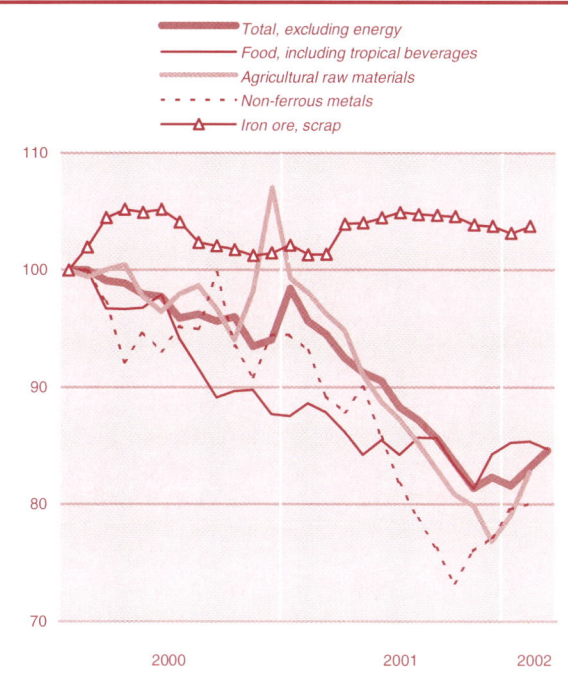

Source: Hamburg Institute for Economic Research (HWWA).
Note: Indices calculated on the basis of current dollar prices and weighted by commodity exports of the industrialized countries.

CHART 2.1.5

Crude petroleum prices, January 2000-February 2002
(Dollars/barrel)

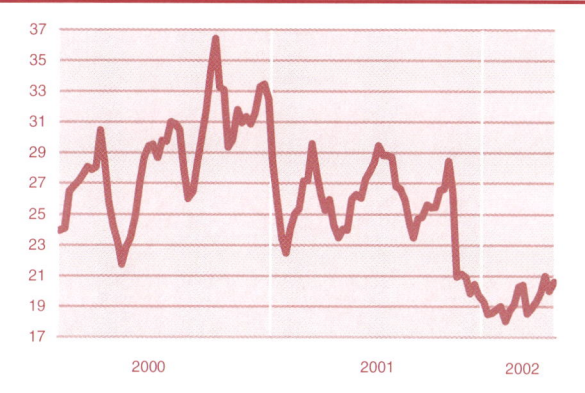

Source: United States Department of Energy, *Weekly Status Petroleum Report* (Washington, D.C.), various issues [www.eia.doe.gov].
Note: Brent spot price, weekly averages.

bottomed out in the final months of 2001 and, for some commodities, was partly reversed. Prices for non-energy commodities in dollars in January 2002 were some 13 per cent below their level of a year earlier.[49]

The spot price of crude oil (Brent crude) fluctuated around $26 per barrel during the first eight months of 2001, but fell sharply in the aftermath of 11 September and the more pessimistic assessment of short-term prospects for the world economy. The increase in world oil demand in 2001 was the smallest since 1985. Since autumn 2001 oil prices have fluctuated within a range of $18-$21 a barrel (chart 2.1.5), their lowest level since mid-1999 and below the target range of $22-$26 a barrel set by OPEC for its basket of crudes.[50] In the face of falling demand and weak prices, OPEC agreed to a further cut in output quotas from January 2002 and this was supported by cuts in exports or production of some oil producing countries outside OPEC. This was the fourth consecutive output cut agreed by OPEC members in less than a year, implying a cumulative fall in target output by nearly 20 per cent.

Outside the international oil markets, metal producers also reacted to severe imbalances between supply and demand by cutting production in mines and smelters, a move which helped to stabilize prices in the final quarter of 2001. Persistent excess supply continued to maintain downward pressure on the prices of agricultural commodities and industrial raw materials, with prices of rubber and cotton falling to their lowest level in two and three decades, respectively.

(ii) North America

In the *United States*, real GDP fell between the second and third quarters of 2001. Economic activity in this period was certainly influenced by the adverse impact of the terrorist attacks of 11 September on consumer and business spending, but the available statistics do not allow a quantitative estimate of that part of spending which occurred after the attacks in September.[51] In late November 2001, the Business Cycle Dating Committee of the National Bureau of Economic Research (NBER) announced that a peak in business activity had already occurred in March 2001, tantamount to the beginning of recession.[52] This meant that the economic expansion which started in March 1991, ended after exactly 10 years.[53] This is the longest expansion in the NBER business cycle chronology, which starts in 1854.

---

[49] As measured by the HWWA index in dollars. The corresponding index in euros fell by 8 per cent over the same period.

[50] The spot reference price for the OPEC basket is slightly lower than the spot price for Brent crude, but they move closely together.

[51] The Bureau of Economic Analysis (BEA) estimates the approximate value of the assets that were destroyed by the 11 September attacks at $15.5 billion, corresponding to somewhat less than 0.2 per cent of annual GDP in 2001. Note that these property losses had no direct immediate effect on real GDP because the latter is a measure of goods and services produced in a given period. BEA, *The Terrorist Attacks of September 11th as Reflected in the National Income and Product Accounts* (Washington, D.C.), November 2001 [www.bea.gov].

[52] NBER, *The Business Cycle Peak of March 2001*, 26 November 2001 [www.nber.org].

[53] The NBER bases its business cycle dating on a range of monthly indicators, not aggregate GDP, which is only available on a quarterly basis. The following are the four most important measures considered:

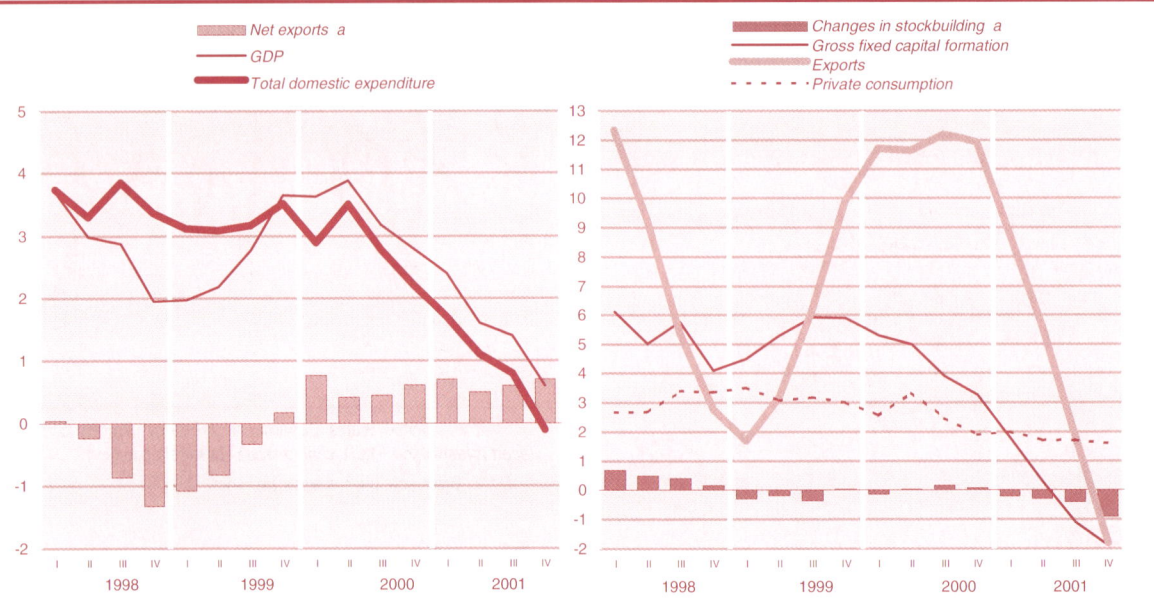

CHART 2.1.6

Quarterly changes in real GDP and major expenditure items in the United States, 1998-2001

(Percentage change over same period of previous year)

Source: Bureau of Economic Analysis (Washington, D.C.) [www.bea.gov].

a Growth contributions (percentage points).

Contrary to widely held expectations, real GDP did not decline further in the final quarter of 2001. Instead, economic activity edged up by 0.3 per cent just offsetting the fall in aggregate output in the third quarter (table 2.1.1). Year-on-year, GDP in the fourth quarter was 0.4 per cent above its level in the same quarter of 2000, the smallness of the increase largely reflecting the virtual stagnation of total domestic demand (chart 2.1.6).

The increase in real GDP in the final quarter of 2001 mainly reflects a sharp rise in private household expenditures on motor vehicles stimulated by very favourable discounts and financing schemes offered by the automobile producers. This probably led many households to bring forward their planned purchases from 2002 and so weaker demand for motor vehicles can therefore be expected in the first months of 2002. Household spending continued, moreover, to be underpinned by strong spending on furniture and household equipment, a reflection of the continuing high level of sales of new homes. This spending boom was supported by a surge in the growth of consumer credit and a sharp fall in the personal savings rate, which declined to 0.4 per cent of personal disposable income in the final quarter of 2001, down from 3.8 per cent in the preceding quarter. Household savings had surged in the third quarter in response to tax refunds and the reduction of various income tax rates introduced by the Economic Growth and Tax Reconciliation Act of 2001.[54] Households' real disposable incomes were also boosted by the fall in energy prices and the additional liquidity gained from refinancing mortgages at lower rates, the latter resulting from the easing of monetary policy in the wake of the terrorist attacks. These factors, however, were partly offset by the impact of declining employment on the growth of aggregate personal incomes. The sharp drop in consumer confidence in September and October bottomed out in November 2001 and has since been partly reversed. In January 2002, consumer confidence had regained its September 2001 level, but this was still the lowest level since March 1996 (chart 2.1.7). Also government expenditures surged in the final quarter of 2001, largely reflecting additional funds for the clean-up after the terrorist attacks in New York and for security and defence.

The considerable impact of total final consumption expenditures on domestic demand in the final quarter of 2001, however, was almost fully offset by a further weakening of private investment. Against a background of falling profits, large margins of spare capacity and

---

employment in the total economy; industrial production; the volume of sales in manufacturing and trade; and current real personal incomes less transfers. The NBER defines a recession as a significant decline in activity spread across the economy, lasting more than a few months. Note that another widely used definition of recession is a decline in real GDP in two consecutive quarters. According to the latter definition, the United States did not move into recession in 2001.

[54] It is estimated that the rebates and lower marginal tax rates reduced income tax liabilities by $44 billion (or 0.6 per cent of annual personal disposable incomes in 2001) and $52 billion in 2002. *Economic Report of the President, Transmitted to the Congress February 2002* (Washington, D.C.), 2002, p. 44 [www.whitehouse.gov/cea/pubs.html].

CHART 2.1.7

Consumer confidence and retail sales in the United States, January 1993-January 2002

(Index, percentage change)

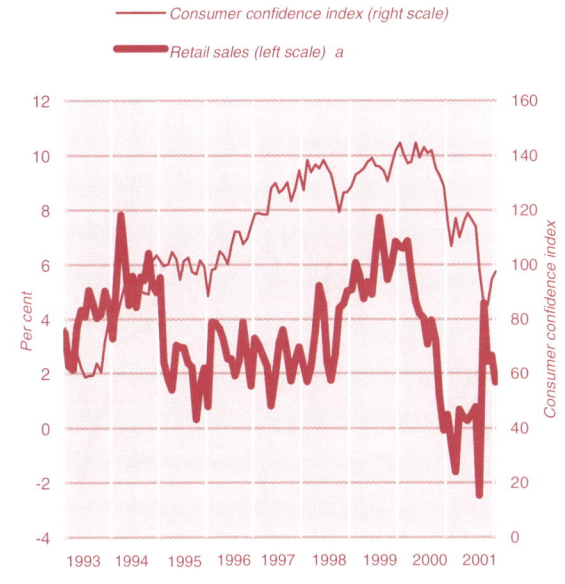

Source: The Conference Board (New York) [www.conference-board.org]; Federal Reserve Bank of St. Louis [www.stls.frb.org].

a Percentage change over same period of previous year (three-month moving average).

TABLE 2.1.2

Annual changes in real GDP and major expenditure items in the United States, 2000-2001

(Percentage change, percentage points)

	Percentage change over previous year		Growth contributions [a]	
	2000	2001	2000	2001
Private consumption	4.8	3.1	3.3	2.1
Government expenditures [b]	2.7	3.6	0.5	0.6
Fixed investment [c]	7.6	-1.9	1.4	-0.4
Stockbuilding	..	..	-0.1	-1.2
Total domestic expenditures	4.8	1.1	5.0	1.1
Net exports	..	..	-0.9	-0.1
Exports of goods and services	9.5	-4.6	1.1	-0.6
Imports of goods and services	13.4	-2.7	-2.0	0.4
GDP at market prices	4.1	1.2	4.1	1.2

Source: National statistics.

[a] Percentage points.

[b] Total government expenditure (consumption and investment).

[c] Private sector only.

lower equity prices, the business sector continued to cut back expenditures on equipment and buildings. But the steep fall in spending on information processing equipment and software during the first three quarters of 2001 appears to have bottomed out: real expenditures on computers rose between the third and last quarters of 2001, although they were still some 10 per cent below their level of a year earlier. Residential investment fell slightly in the final quarter, but expenditures remained at a relatively high level during 2001 as a result of favourable mortgage rates.

Inventory liquidation in the business sector was an important drag on output growth in the final quarter. This was the sixth consecutive quarter in which cutbacks of inventories had a negative effect on growth; their cumulative effect was to subtract some 1.2 percentage points from real GDP growth in 2001. This highlights the considerable progress made by the corporate sector in working off excess stocks of goods and suggests that the trough of the inventory cycle may have been reached or even passed. Real exports of goods and services continued to be depressed by weak foreign demand and the persistent strength of the dollar. Although weak domestic demand led to a further decline in real imports, the change in real net exports had a negative effect on overall economic growth in the final quarter of 2001.

For the year as a whole, real GDP rose by 1.2 per cent, the weakest growth rate since 1991 when there was a decline of 0.5 per cent. All of the average increase in real GDP in 2001 was due to the growth of personal and government consumption expenditures, although these were largely offset by falling investment and exports. The decline in the volume of exports of goods and services in 2001 was the first since 1983. The changes in real net exports, however, was less of a drag on domestic activity than in 2000 (table 2.1.2).

Manufacturing production fell in December 2001 for the fifteenth consecutive month, but stagnated in January 2002. For the year as a whole there was a decline of 4.4 per cent compared with 2000. Weak investment led to a sharp slowdown in the growth of manufacturing capacity in 2001, but the margins of excess capacity are still quite large. Capacity utilization fell throughout 2001 and in January 2002 was about 73 per cent, some 8 percentage points below the long-term average for 1967-2000 (chart 2.1.8). This hides a much lower capacity utilization rate of about 60 per cent in the high-technology industries.

Against such a background of sluggish economic activity, inflationary pressures have weakened markedly. The implicit GDP deflator fell slightly in the final quarter of 2001 (compared with the preceding quarter) and rose for the year as a whole by some 2.2 per cent. Consumer price inflation fell to only 1.6 per cent in December 2001, down from its previous peak of 3.6 per cent in May 2001. The December figure mainly reflected the large fall in the prices of final energy products triggered in turn by the weakness of oil and gas prices in the international commodity markets. Core inflation (which excludes food and energy prices) remained broadly steady throughout 2001, reaching 2.7 per cent in December. For the year as a whole, core inflation was 2.6 per cent, slightly up from 2.4 per cent in 2000 (chart 2.1.9).

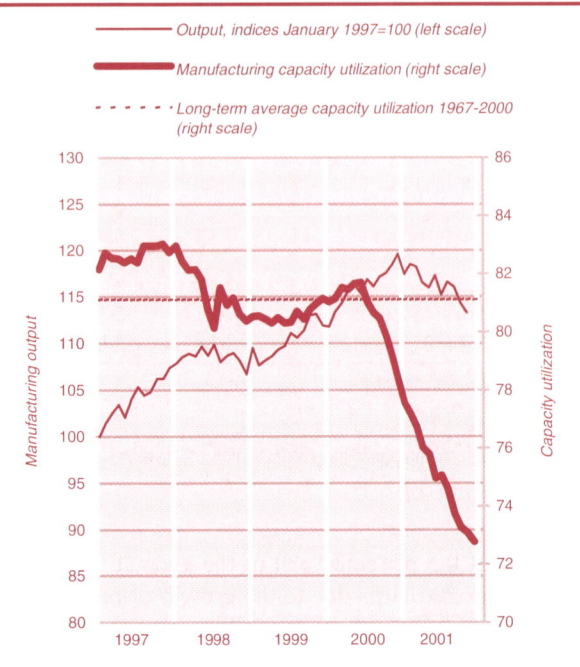

CHART 2.1.8

Output and capacity utilization in manufacturing industry in the United States, January 1997-January 2002
(Index, per cent)

Source: United States Federal Reserve Board [www.federalreserve.gov].

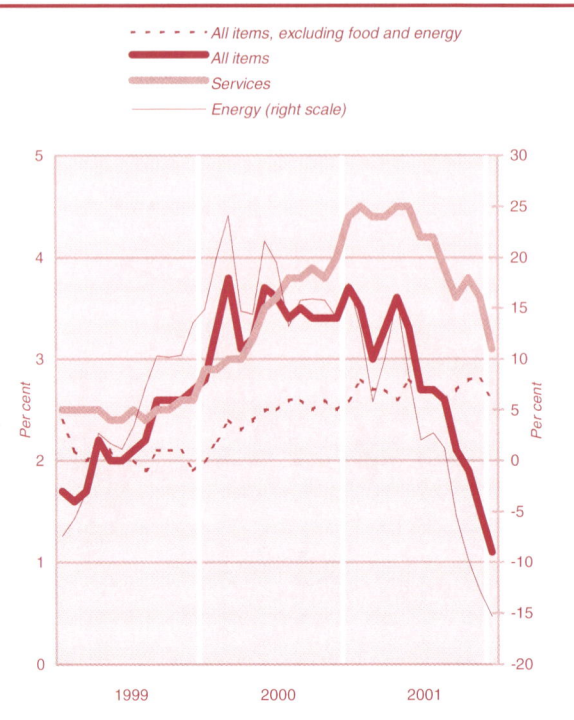

CHART 2.1.9

Consumer prices in the United States, January 1999-January 2002
(Percentage change over same month of preceding year)

Source: United States Bureau of Labor Statistics (Washington, D.C.) [www.bls.gov].

Note: United States city average. Data are seasonally adjusted.

Labour markets weakened substantially in 2001. Non-farm payroll employment fell by over 1.2 million in the 12 months to January 2002. Job losses were concentrated in manufacturing, while the gains in services steadily diminished. The unemployment rate rose throughout 2001; it fell slightly to 5.6 per cent in January 2002, but was still higher than in January 2001 (4.2 per cent) (chart 2.1.10).

The weaker demand for labour was reflected in a marked slowdown in the growth of labour costs (compensation per hour) during 2001. In addition there was a marked rebound in labour productivity growth (on account of significant labour shedding) in the final quarter of 2001, and this led to a fall in unit labour costs between the third and fourth quarters (chart 2.1.11). For the year as a whole, labour productivity growth decelerated to 1.8 per cent, down from 3.3 per cent in 2001. This reflects the lagged adjustment of hours worked to changes in output. Smaller productivity increases, combined with growth in compensation per hour by nearly 6 per cent, led to an increase in annual average unit labour costs by nearly 4 per cent in 2001 and to a squeeze on profit margins.

The external deficits have diminished only slightly. The merchandise trade deficit fell to $427 billion or 4.2 per cent of GDP in 2001, against 4.6 per cent in 2000. The values of exports and imports declined, reflecting to a large extent the flagging external trade in capital goods, in turn, a consequence of the cyclical sensitivity of investment in the United States. The current account deficit fell to some $415 billion or 4.1 per cent of GDP in 2001, down from some $445 billion or 4.5 per cent of GDP in 2000.

The progressive easing of monetary policy since the beginning of 2001, designed to arrest and reverse the cyclical downturn, continued in the aftermath of the 11 September attacks. Between mid-September and early December the target for the federal funds rate was reduced in four steps from 3.5 per cent to 1.75 per cent. There were altogether 11 interest rate reductions since early January 2001, when the target rate was 6.5 per cent. Nominal short-term interest rates have thus declined substantially and real short-term rates were close to zero in January 2002.

Changes in nominal long-term rates were rather volatile in 2001 and did not follow the downward trend of short-term rates. This was most pronounced in the first half of last year. In January 2002, the yield on 10-year treasury bills was 5.1 per cent, only slightly lower than 12 months earlier (chart 2.1.12). This may reflect the deteriorating outlook for government finances and expectations of an imminent economic upturn. A related concern is likely to be the persistent huge current account deficit. The maturity spread – the difference between yields on long-term government bonds and short-term interest rates – has widened significantly, largely because

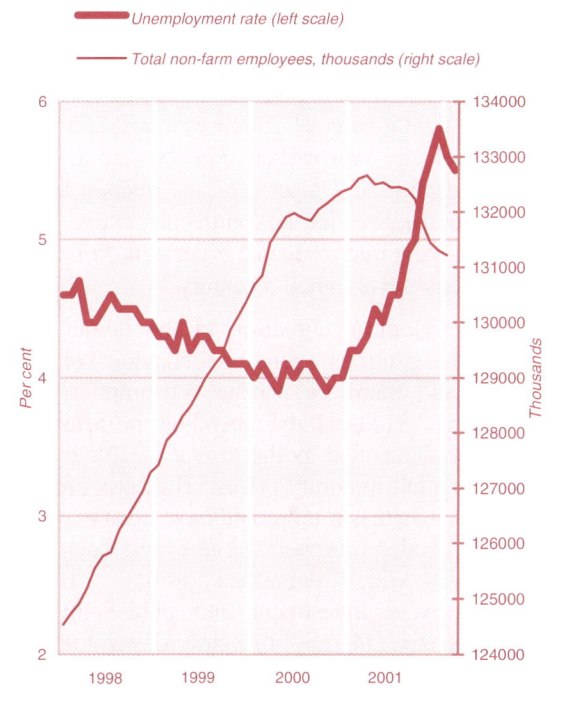

CHART 2.1.10

Employment and unemployment in the United States, January 1998-February 2002
(Thousands, per cent of civilian labour force, seasonally adjusted)

Source: United States Bureau of Labor Statistics (Washington, D.C.) [www.bls.gov].

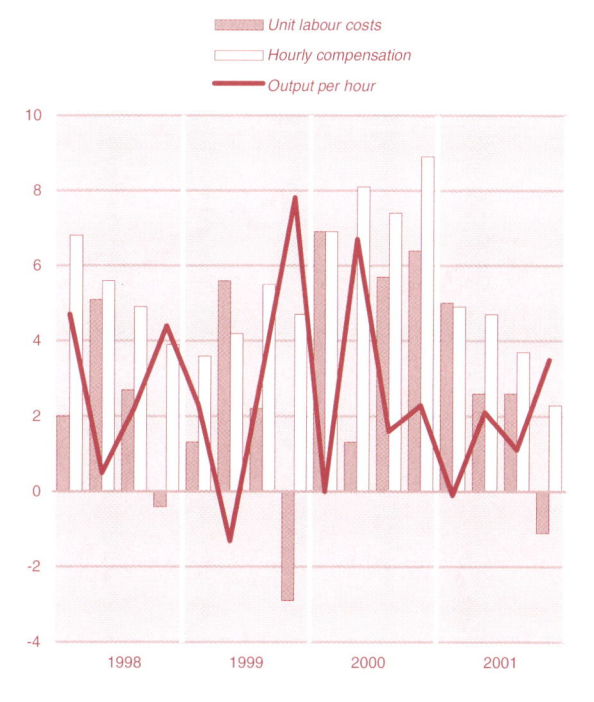

CHART 2.1.11

Unit labour costs in the United States, 1998-2001
(Percentage change over preceding quarter at annual rate)

Source: United States Bureau of Labor Statistics (Washington, D.C.) [www.bls.gov].
Note: Non-farm business sector. Seasonally adjusted at annual rates.

of the decline in the latter. Contemporaneous real long-term rates were 2 per cent in January 2002 up from 0.8 per cent in the same month of 2001. This increase reflects the decline in inflation, notably the fall in energy prices. If real interest rates are calculated using core inflation then they remained unchanged at some 2.5 per cent.

The dollar continued to appreciate against the major currencies in 2001, and continued to do so in early 2002. Its nominal effective exchange rate rose by 4.8 per cent over the 12 months to February 2002. In real effective terms the dollar rose by 5.4 per cent over the same period reaching its highest level in 16 years and partly offsetting the monetary stimulus of lower interest rates (chart 2.1.13).

The expansionary stance of fiscal policy in combination with the cyclical downturn resulted in a large fall in the federal budget surplus from 2.4 per cent of GDP in 2000 to 1.2 per cent ($127 billion) in 2001. The federal budget is currently forecast to move into deficit, to some 1 per cent of GDP, in fiscal 2002. This would be the first deficit since 1997, and deficits are projected to continue in each of the fiscal years up to 2004. The Administration's additional economic stimulus package (with proposed measures ranging from $69 billion to $89 billion) failed to win support in Congress in early February 2002. The general government financial balance was in deficit to the extent of 0.3 per cent of GDP in 2001, following a surplus of nearly 2 per cent of GDP in the previous year.

In *Canada*, the economy was adversely affected in 2001 by the sharp downturn in expenditure on capital goods in the United States and the significant slowdown in economic growth in other regions of the world economy. This led to a marked fall in exports, notably of information and telecommunication products and motor vehicles. Real GDP fell slightly in the third quarter of 2001 but edged up again in the final quarter mainly on the back of strong private consumption expenditures and an improvement in real net exports. For the year as a whole, real GDP rose by 1.5 per cent, down from 4.4 per cent in 2000. Manufacturing output, however, fell by almost 4.5 per cent. Annual private consumption growth weakened considerably in the wake of slowing real income growth and a deteriorating situation in the labour market. Business investment was held back by increasing margins of excess capacity and uncertain sales prospects. The unemployment rate rose to 8 per cent in December 2001, the highest in almost three years. Core consumer price inflation was below 2 per cent in 2001, well within the Bank of Canada's set target range of 1 to 3 per cent. In response to weakening domestic activity and low inflation, the Bank of Canada continued to loosen monetary policy in mid-January 2002. The target for the overnight interest rate has been lowered by a cumulative 3.75 percentage points since the beginning of 2001. This,

CHART 2.1.12

Average monthly nominal short-term and long-term interest rates in the United States, January 1998-January 2002
(Per cent per annum)

Source: United States Federal Reserve Board [www.federalreserve.gov].

a Yields on 10-year treasury bills less yields on three-month treasury bills.

CHART 2.1.13

Nominal and real effective exchange rates of the dollar, January 1995-February 2002
(Indices, January 1995=100)

Source: United States Federal Reserve Board [www.federalreserve.gov].
Note: Average monthly values.

in combination with a pronounced depreciation of the Canadian dollar, has led to a considerable easing of monetary conditions since mid-2000.

(iii) Japan

In *Japan*, the economic situation deteriorated markedly in 2001. The economy is in its third recession of the past 10 years. Real GDP fell by 0.5 per cent in 2001, with domestic demand stagnating and exports falling sharply (table 2.1.3). The decline in the global demand for information and communication technology products was the main factor in the slump of industrial production, which in the final quarter of 2001 was some 12.5 per cent below its level of a year earlier. For the year as a whole there was a decline of 7.2 per cent, the steepest fall since 1975. Employment fell for the fourth consecutive year and the unemployment rate reached 5.5 per cent in December 2001, its highest level in half a century.[55]

The deteriorating situation in the labour market reinforced the cautious spending behaviour of private households and therefore private consumption slowed down further. Households' spending propensity was probably also dampened by the loss of wealth associated with the sharp fall in equity prices. Business investment was stagnant, whereas a fall could have been expected in view of weak sales prospects, large excess capacity and falling prices, which depressed profits. However, residential private investment and public investment declined strongly, thereby bringing down total fixed capital formation. Weak export demand and growing import penetration, partly reflecting the continued relocation of manufacturing activities abroad, contributed to a sharp fall in the traditional merchandise trade surplus to its lowest level in 18 years.

Economic activity continued in a deflationary environment. Consumer prices in January 2002 were about 1.5 per cent lower than 12 months earlier. The implicit GDP deflator has fallen by 5 percentage points since 1994. The monetary policy measures introduced in 2001 to stimulate domestic demand and reverse deflation – targeting of the money stock and injection of liquidity by outright purchases of government bonds – have had no effect so far. Short-term interest rates have fallen to zero, but falling prices mean that real short-term interest rates are slightly positive. Nominal long-term interest rates were broadly stable at somewhat below 1.5 per cent in 2001. But the demand for credit has remained weak, notably because of the high debt burden in the corporate sector and the tightening of lending standards by banks in the face of their balance sheet problems caused by the huge volume of non-performing loans. The latter are estimated to be equivalent to at least 8 per cent of GDP, but the real figure could well be higher.[56]

The government's financial position, moreover, has become quite worrisome and greatly narrows the use of fiscal policy to support economic activity. Persistently high and rising budget deficits since 1995 – due to a

---

[55] In January 2002 the unemployment rate fell to 5.3 per cent, largely due to a decline in the labour force.

[56] "Japan's bad banks", *Financial Times*, 29 January 2002.

TABLE 2.1.3

Annual changes in real GDP and major expenditure items in Japan, 2000-2001

(Percentage change, percentage points)

	Percentage change over previous year		Growth contributions [a]	
	2000	2001	2000	2001
Private consumption	0.6	0.3	0.3	0.2
Government expenditures	4.6	3.1	0.7	0.5
Fixed investment	3.2	-1.7	0.9	-0.5
Stockbuilding	..	..	–	–
Total domestic expenditures	1.9	0.2	1.9	0.2
Net exports	..	..	0.5	-0.7
Exports of goods and services	12.4	-6.6	1.3	-0.7
Imports of goods and services	9.6	-0.6	-0.8	–
GDP at market prices	2.4	-0.5	2.4	-0.5

Source: National statistics.

[a] Percentage points.

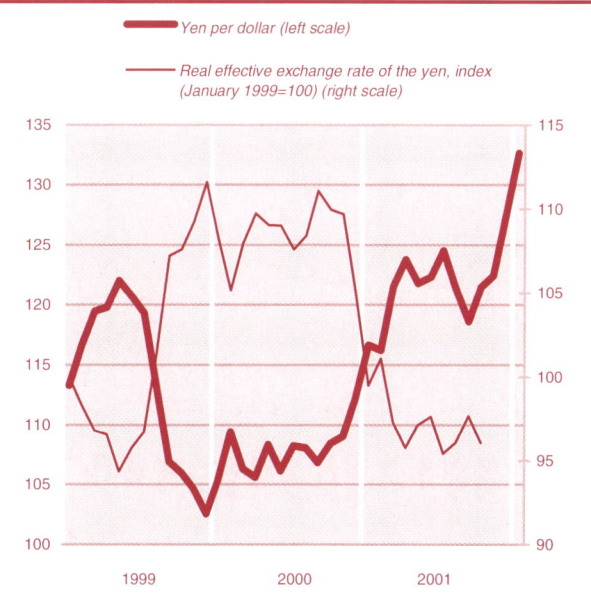

CHART 2.1.14

Exchange rate of the yen, January 1999-January 2002
(Yen per dollar, indices)

Source: IMF, *International Financial Statistics* (Washington, D.C.), various issues.

Note: Average monthly data. Real exchange rate based on consumer prices.

succession of fiscal stimuli and the impact of weak domestic activity and falling prices on government revenues – have led to a surge in gross government debt to more than 130 per cent of GDP in 2001, up from 85 per cent in 1997. In response to the deteriorating economic performance, the government adopted a supplementary budget in late 2001 to support economic activity. Concern at the state of the banking system and the high level of government debt has led to a downgrading of yen-denominated debt in international financial markets and a corresponding rise in risk premia.

Hopes for a recovery in Japan are pinned on an improvement in economic conditions in the United States and other parts of the world economy and on gains in price competitiveness from a depreciation of the exchange rate, which is also seen as a source of imported inflation. Interventions in the foreign exchange market by the Bank of Japan appear to have contributed to the marked depreciation of the yen against the dollar since late 2001; in February 2002, it had depreciated by 13 per cent compared with January 2001. The decline of the real effective exchange rate indicates a significant improvement in price competitiveness in 2001 (chart 2.1.14). However, there are concerns that a sharp depreciation of the yen could have adverse effects on Asian emerging markets, which rely strongly on shipments to Japan and in some cases compete with Japanese exports in third markets. Apart from a global recovery, the effective implementation of structural reforms (including, *inter alia,* the resolution of the bad loan problem, corporate sector restructuring and strengthening of competition policy) appears to be a *sine qua non* for a sustained recovery although the short-run adjustment costs are likely to be high.[57]

---

[57] It has been argued that at the root of the lingering economic crisis is the failure to radically reform the domestic institutions that helped rebuild Japan after the Second World War and led to its spectacular growth, but which are now stifling economic growth. W. Overholt, "Japan's economy at war with itself", *Foreign Affairs*, January/February 2002, pp. 134-147.

2.2 Western Europe

There was a marked slowdown in the rate of economic expansion in western Europe during 2001. For the year as a whole, real GDP rose by only 1.3 per cent, down from 3.5 per cent in 2000, and the smallest increase since 1993. This outcome partly reflects the slump in Turkey, where real GDP fell by more than 7 per cent in 2001, but in the European Union, real GDP rose by only 1.7 per cent in 2001, half the annual rate in 2000. Performance was broadly similar in the euro area (see table 1.2.1 above).

The cyclical downturn has partly reversed the improvements in the labour markets that had occurred in recent years. Labour cost pressures have remained moderate against the background of more uncertain employment prospects. Falling oil prices and weakening consumer demand contributed to a marked decline in rates of inflation in the second half of 2001. The short-term outlook is for a moderate recovery in the course of 2002, but this assumes a strengthening of cyclical growth forces in the United States and will need to be supported by an appropriate stance of macroeconomic policy in western Europe.

(i) Euro area

(a) Macroeconomic developments

In the *euro area*, the short cyclical upswing peaked in the first half of 2000 and the rate of economic growth has since steadily decelerated (chart 2.2.1). Between the third and final quarters of 2001, real GDP fell by 0.2 per

CHART 2.2.1

Quarterly changes in real GDP and major expenditure items in the euro area, 1998-2001

(Percentage change over same period of previous year)

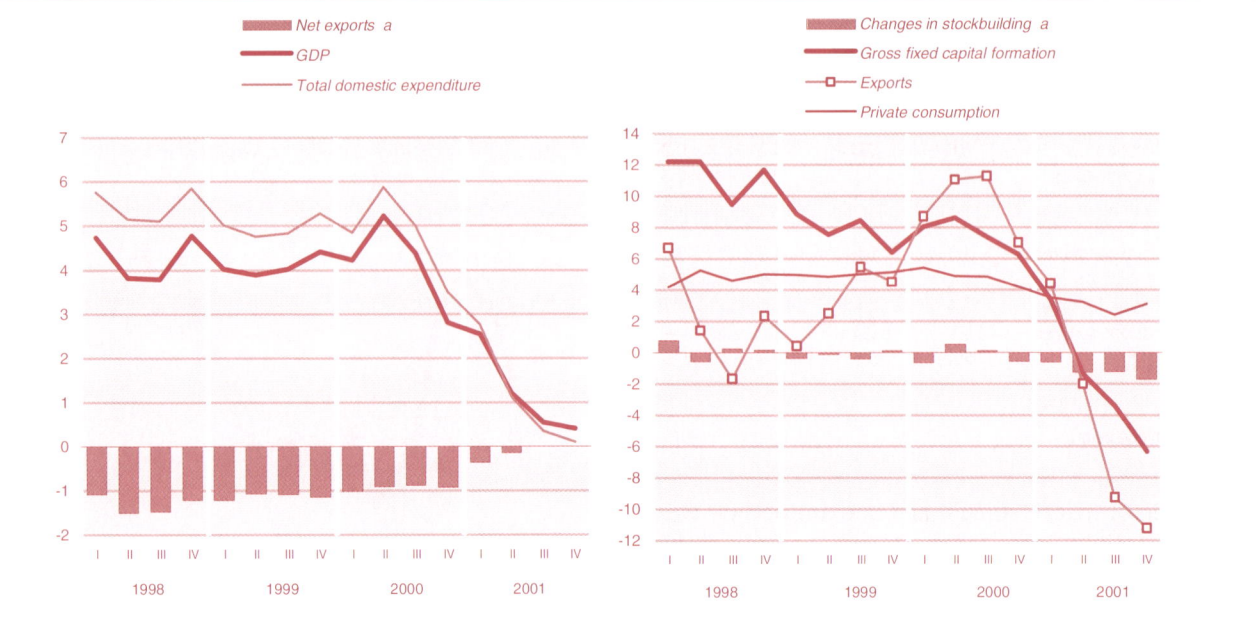

Source: Eurostat, New Cronos Database.

a Growth contribution (percentage points).

cent and was only about 0.75 of a percentage point above its level in the same period of 2000.

The tendency for rapidly weakening growth was general. The year 2001 was thus a cyclical turning point for countries such as Finland, France, Ireland, the Netherlands, Portugal and Spain where an upswing had been sustained for a number of years. The falls in the average annual growth rates of real GDP were particularly sharp (some 5 percentage points) in Finland and Ireland (table 1.2.1). In Ireland, the still very high growth rate of 6.5 per cent in 2001 is largely due to the substantial statistical carry-over effect from the buoyant growth in 2000. As in the rest of the euro area, economic activity stagnated in the second half of the year.[58]

The marked weakening in the overall growth rate in 2001 masks a severe recession in the manufacturing sector, where output fell throughout the year, leading to a pronounced decline in capacity utilization rates (chart 2.2.2). In December 2001, manufacturing output was some 5.5 per cent lower than 12 months earlier. Industrial confidence in November 2001 was at its lowest level since late 1993, but has edged up somewhat in early 2002 against a background of falling inventories and rising production expectations (chart 2.2.3).

The unexpectedly strong cyclical downturn reflects a marked weakening of both exports and total domestic demand, which mutually reinforced one another in the course of the year (chart 2.2.1). In the business sector,

firms responded quickly to the deteriorating prospects for sales and profits and to rising margins of spare capacity by cutting their expenditures on machinery and equipment. In addition, the growth of construction investment expenditures weakened. In total, gross fixed capital investment virtually stagnated in 2001. Changes in stockbuilding were also a drag on economic growth in 2001 (table 2.2.1).

There was a marked slowdown in the growth of real private consumption in the course of 2001. To a large extent this reflected the weaker growth of aggregate real disposable incomes which, in turn, was a result of the deteriorating situation in the labour markets, the effects of falling equity prices on net wealth, and sagging consumer confidence (chart 2.2.3). However, falling inflation rates supported real income growth in the second half of the year. In several countries (notably France, Germany and Italy), lower income tax rates increased disposable incomes. Government consumption held up quite well and contributed about half a percentage point to the overall rate of economic growth in 2001 (tables 2.2.1 and 2.2.2).

The deteriorating external environment had a considerable impact on export growth. In fact, the volume of exports of goods and services fell in each of the four successive quarters of 2001, but, given the significant statistical carry-over effect[59] from the end of

---

[58] ESRI, *Quarterly Economic Commentary* (Dublin), October 2001.

[59] Given the large rise of exports during 2000, the level at the end of the year was significantly above the average for the year as a whole. It is this statistical carry-over into 2001 which ensured positive annual growth in 2001, compared with 2000, despite declining trade *in the course of* 2001.

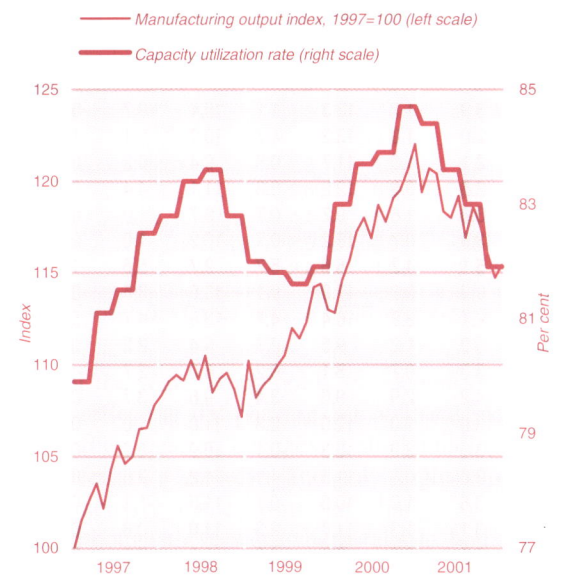

CHART 2.2.2

Manufacturing output and capacity utilization in the euro area, 1997-2001

(Indices, per cent)

Source: OECD, *Main Economic Indicators* (Paris), various issues; European Commission, *European Economy*, Supplement B (Luxembourg), monthly and direct communications.

Note: Capacity utilization rates are available only for January, April, July and October of each year. Data shown are approximations of quarterly data calculated as the arithmetic average of January and April data (first quarter), April and July (second quarter), etc.

CHART 2.2.3

Business and consumer confidence in the European Union, January 1995-February 2002

Source: European Commission, *European Economy*, Supplement B (Luxembourg), monthly and direct communications.

Note: Data show the net balance between the percentages of respondents giving positive and negative answers to specific questions. For details see any edition of the source.

2000 the annual increase was still some 2.8 per cent, albeit much lower than the 12 per cent or so in 2000. A similar development can be observed for the volume of imports.[60] Changes in real net exports continued to support economic growth to about the same extent as in 2000 (table 2.2.1).

Falling energy prices contributed to a marked fall in consumer price inflation from a peak at 3.4 per cent in May 2001 (chart 2.2.4). In addition, intense competition greatly reduced the scope of retailers to raise prices. By the end of the year, inflation was only slightly above the 2 per cent ceiling of the ECB's asymmetric inflation target. The average inflation rate for the euro area was 2.7 per cent in 2001, but there were large differences among countries, ranging from 1.6 per cent in France to 4.9 per cent in Ireland (table 2.2.3). The fact that the average inflation rate was above the target largely reflects the first-round effects of transitory supply shocks (i.e. the rise in oil prices and the effects of BSE and foot-and-mouth disease), which monetary policy should, in general, be prepared to accommodate. Some special factors (such as the effects of bad weather on fruit and vegetable prices, as well as indirect tax increases in a number of countries) led to a rise in the aggregate consumer price index by 2.7 per cent in January 2002 compared with the same month of the preceding year. (It has been surmised that the "innovative rounding" of prices during the euro currency changeover contributed to this rise in the January price level but so far there is only anecdotal evidence in support of this.) After increasing steadily since mid-2000, the core rate of inflation (which excludes energy and food products) has continued to hover at just over 2 per cent (year-on-year) since May 2001, broadly following the changes in the prices of services.

The cyclical downturn left its mark on the labour markets, bringing to a halt the improvements that occurred in 2000. Employment growth slowed to 1.2 per cent, down from 1.8 per cent in 2000. The marked fall in the unemployment rate, underway since 1997, bottomed out in the summer and in the final quarter of 2001 it rose slightly to 8.5 per cent, the first increase since the first quarter of 1997 when it stood at 11.6 per cent (chart 2.2.5). The average harmonized unemployment rate in the euro area in 2001 was 8.3 per cent, down from 8.8 per cent in 2000 (table 2.2.4). In the face of the adverse cyclical developments, labour cost growth remained moderate in 2001: unit labour costs in the total economy in the third quarter were 2.5 per cent higher than a year earlier (chart 2.2.6). Labour productivity stagnated, largely reflecting the lagged adjustment of labour input to changes in output over the business cycle.

In the balance of payments, the rise in the current account deficit to €70 billion in 2000 (corresponding to 1.1 per cent of aggregate GDP in the euro area) was more

---

[60] It should be recalled that these data include cross-border trade within the euro area.

CHART 2.2.4

Consumer prices in the euro area, January 2000-January 2002
(Percentage change over the same month of preceding year)

Source: Eurostat, New Cronos Database.
Note: Harmonized Index of Consumer Prices (HICP).

TABLE 2.2.3

Inflation in western Europe and North America, 1999-2001
(Percentage change over previous year)

	Consumer price index		
	1999	2000	2001
France	0.5	1.7	1.6
Germany	0.6	1.9	2.5
Italy	1.7	2.5	2.8
Austria	0.6	2.3	2.6
Belgium	1.1	2.5	2.5
Finland	1.2	3.4	2.6
Greece	2.6	3.2	3.4
Ireland	1.6	5.6	4.9
Luxembourg	1.0	3.2	2.7
Netherlands	2.2	2.5	4.5
Portugal	2.3	2.9	4.3
Spain	2.3	3.4	3.6
Euro area [a]	1.2	2.4	2.7
United Kingdom	1.6	2.9	1.8
Denmark	2.5	2.9	2.4
Sweden	0.3	1.3	2.6
European Union [b]	1.3	2.5	2.6
Cyprus	1.7	4.3	2.0
Iceland	3.2	5.1	6.4
Israel	5.2	1.1	1.1
Malta	2.1	2.3	2.9
Norway	2.3	3.1	3.0
Switzerland	0.8	1.6	1.0
Turkey	64.9	54.9	54.4
Western Europe	1.3	2.4	2.5
Canada	1.7	2.7	2.5
United States	2.2	3.4	2.8
North America	2.2	3.3	2.8
Japan	-0.3	-0.7	-0.5
Total above	1.5	2.4	2.3
Memorandum items:			
4 major west European economies [c]	1.0	2.2	2.2
Western Europe and North America	1.8	2.9	2.7

Source: OECD, *Main Economic Indicators* (Paris), various issues; national statistics.

Notes: All aggregates exclude Israel and Turkey.

[a] Twelve countries.
[b] Fifteen countries.
[c] France, Germany, Italy and the United Kingdom.

tax cut enacted in late 2000. Against a background of weakening domestic and external demand, the business sector reduced its spending on machinery and equipment. Investment expenditures were also probably held back in anticipation of tax incentives announced for the autumn of 2001, but these appear to have been offset by the deteriorating outlook for sales and profits. In contrast, construction investment held up quite well, and, in total, gross fixed investment rose by nearly 2.5 per cent, down from 6.5 per cent in 2000, which was the largest increase in more than a decade. Changes in real exports provided only slight support to economic activity in 2001.

In the manufacturing industry, output fell more or less continuously during 2001, and in the final quarter it was about 4.5 per cent lower than in the same period in 2000. Falling output and capacity utilization rates led to significant labour shedding in manufacturing, but this was more than offset by gains in other sectors, notably in services. Although total economy employment fell in the final quarter of 2001, the increase of 1.5 per cent for the year as a whole was still surprisingly strong.[62]

Among the other countries of the euro area, activity was also strongly affected by the downturn in the global business cycle with adverse consequences for exports and business fixed investment, although country-specific factors were also important. In *Finland* and *Ireland*, overall economic growth was especially hard-hit by the global downturn in the demand for ICT products, which account for a large share of their industrial output. In the *Netherlands*, tax increases significantly dampened the growth of private consumption spending and this was a major factor behind the marked slowdown in the rate of growth in 2001. In *Greece*, economic activity was supported by the large increase in fixed investment related to preparations for the 2004 Olympic games. In *Spain*, the slowdown was broadly based in all the major sectors of domestic demand. Changes in real net exports were neutral in their impact on overall economic growth in 2001, which remained above the EU average.

---

[62] Based on labour force survey data.

CHART 2.2.5

Employment and unemployment in the euro area, 1993-2001
(Index, 1995=100, per cent)

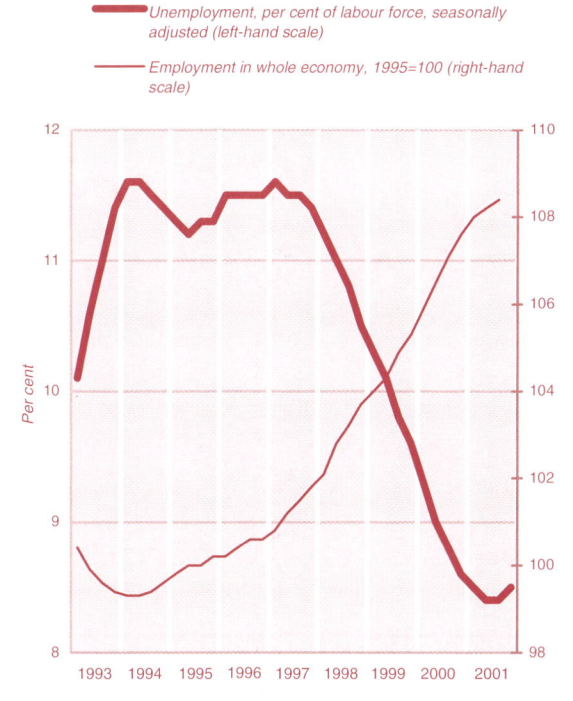

Source: Eurostat, New Cronos Database.

TABLE 2.2.4

Unemployment in western Europe and North America, 1999-2001
(Per cent of civilian labour force)

	Unemployment rate		
	1999	2000	2001
France	10.7	9.3	8.6
Germany	8.6	7.9	7.9
Italy	11.2	10.4	9.5
Austria	3.9	3.7	3.6
Belgium	8.6	6.9	6.6
Finland	10.2	9.8	9.1
Greece	11.6	10.9	10.2
Ireland	5.6	4.2	3.8
Luxembourg	2.4	2.4	2.4
Netherlands	3.2	2.8	2.4
Portugal	4.5	4.1	4.1
Spain	15.7	14.0	13.0
Euro area [a]	9.8	8.8	8.3
United Kingdom	5.9	5.4	5.1
Denmark	4.8	4.4	4.3
Sweden	7.2	5.9	5.1
European Union [b]	9.0	8.1	7.6
Cyprus [c]	3.6	3.5	3.5
Iceland	2.1	2.3	1.4
Israel [d]	8.9	8.8	9.2
Malta [e]	5.3	4.5	4.9
Norway	3.3	3.5	3.6
Switzerland	3.0	2.6	2.6
Turkey [d]	7.6	6.6	7.9
Western Europe	8.6	7.7	7.5
Canada	7.6	6.8	7.2
United States	4.2	4.0	4.8
North America	4.5	4.3	5.0
Japan	4.7	4.7	5.0
Total above	6.5	6.0	6.2
Memorandum items:			
4 major west European economies [f]	8.9	8.1	7.7
Western Europe and North America	6.9	6.2	6.4

Source: OECD, *Main Economic Indicators* and *Quarterly Labour Force Statistics* (Paris), various issues; Eurostat, New Cronos Database; national statistics.

Note: All aggregates exclude Israel.

[a] Twelve countries.

[b] Fifteen countries.

[c] Registered unemployment rate, average of monthly data.

[d] The definitions used comply with ILO guidelines but do not follow the Eurostat/OECD standards.

[e] Registered unemployment rate at the end of the year.

[f] France, Germany, Italy and the United Kingdom.

(b) Monetary policy

Against the background of a significant cyclical downturn and moderate inflationary expectations, the ECB continued to lower its key interest rate (minimum bid rate) by half a percentage point to reach 3.25 per cent in November 2001. This was the fourth interest rate reduction since May 2001, when the rate was at 4.75 per cent.

The fall in official rates is reflected in short-term interest rates, which were around 3.4 per cent in February 2002, compared with 4.8 per cent a year earlier (chart 2.2.7). Nominal long-term rates (yields on 10-year government bonds) displayed some limited volatility during 2001 but no sustained trend. There was some stronger downward pressure on yields in the wake of the terrorist attacks of 11 September, reflecting a flight-to-safety from stocks into government bonds, but this was fully reversed later in the year. By January 2002, average monthly yields were 5 per cent, just as they were 12 months earlier. The difference between yields on long-term United States treasury bills and government bonds in the euro area was close to zero in early 2002.

Average real short-term and long-term interest rates in the euro area have fallen to quite low levels, well below the long-term average rates in Germany, which traditionally provided the benchmark for other west European countries. In principle, financing conditions are relatively easy in early 2002, but what is missing is a perspective of sustained improvement in the overall economic outlook.

To some extent, the stimulus from lower interest rates may have been offset by the loss in financial wealth and the deteriorating financing conditions for the corporate sector caused by the sharp fall in equity prices. It is noteworthy that although the role of stock markets in the euro area's financial system is much less important than in the United States, there has been a significant

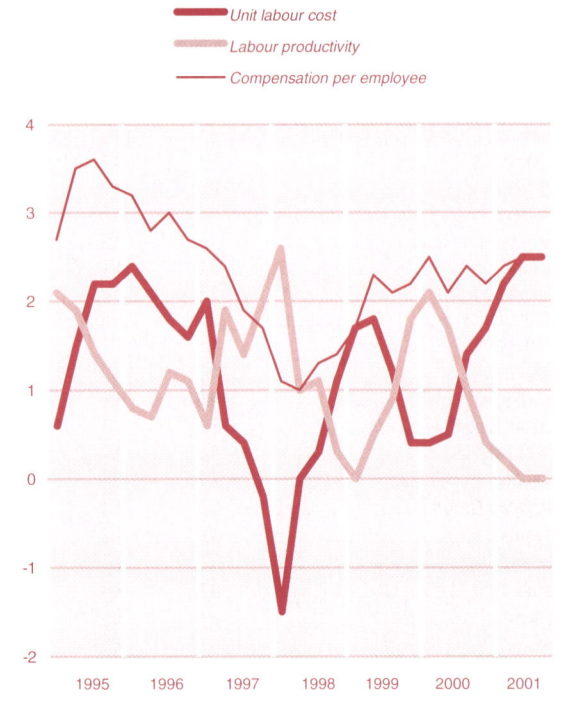

CHART 2.2.6

Unit labour costs in the euro area, 1995-2001
(Percentage change over same quarter of preceding year)

Source: Eurostat, New Cronos Database, European Central Bank [www.ecb.int].

Note: Total economy. Data are seasonally adjusted.

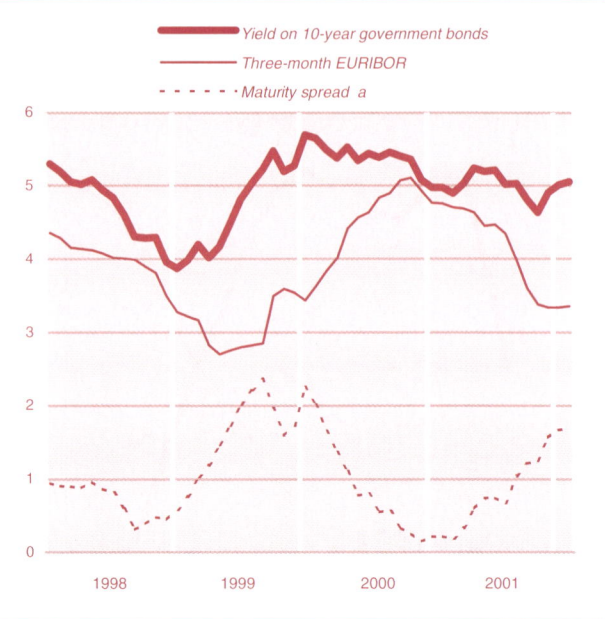

CHART 2.2.7

Average monthly nominal short-term and long-term interest rates in the euro area, January 1998-February 2002
(Per cent per annum)

Source: European Central Bank [www.ecb.int].

Note: Up to December 1998 interest rates for the euro area were aggregated using 1995 GDP weights of member countries.

a Yields on 10-year government bonds less three-month EURIBOR.

increase in the share of equity in private households' wealth over the past years.[63]

Some offset to lower interest rates also originated in the appreciation of the real effective exchange rate of the euro in the second half of 2001. In January 2001, it was some 3 per cent higher than its previous low point in June 2000 (chart 2.2.8). The euro-dollar exchange rate was quite volatile in 2001, with most of the ground gained in the third quarter of 2001 being lost in the following months. In the 12 months to February 2002, the euro had depreciated nearly 6 per cent against the dollar.

The combined impact of changes in real short-term interest rates and the real effective exchange rate on domestic economic activity (or aggregate demand) can be judged by using the monetary conditions index (MCI). The MCI is a weighted average of changes in the real effective exchange rate and the real short-term interest rate relative to a base period, the weights reflecting the relative impact that a given change in each of these variables is estimated to have on aggregate demand or output.[64] The rationale is that monetary policy affects demand not only via changes in interest rates but also via its effect on the exchange rate. Exogenous changes in the exchange rate also influence economic performance, although this channel is more important for small open economies.

The MCI fell throughout 1999, pointing to an easing of monetary conditions in the euro area in the first year of EMU (chart 2.2.9). This reflected to a large extent the depreciation of the euro. Since the spring of 2000, the index has been volatile and has not followed a clear trend. The decision of the ECB to increase its main refinancing rate in the second half of 2000 clearly led to a tightening of monetary conditions, but this was reversed with the easing of monetary policy and a weakening of the exchange rate in the second half of 2001.

The growth of money supply (M3) in the first half of 2001 fell below the reference value of 4.5 per cent fixed by the ECB. However, its growth accelerated strongly in the second half of 2001 and in December it rose (year-on-year) by 8 per cent. To a large extent, this acceleration reflects temporary portfolio adjustments in favour of short-term liquid assets triggered by the rise in uncertainty in the wake of the terrorist attacks of 11 September and the reduced opportunity costs of holding

---

[63] This is reflected, for example, in a pronounced increase in the ratio of market capitalization of domestic shares as a percentage of GDP by a factor of seven between 1990 and 2000. ECB, *Monthly Bulletin* (Frankfurt am Main), February 2002, p. 41.

[64] The MCI-1 variant uses weights estimated by the Bundesbank for Germany. MCI-2 uses weights employed by the European Commission for the euro area. Deutsche Bundesbank, "Taylor-Zins und Monetary Conditions Index", *Monatsbericht* (Frankfurt am Main), April 1999, pp. 47-63 and European Commission, *European Economy, Supplement A,* No. 10/11 (Luxembourg), October/November 2001.

CHART 2.2.8

The exchange rate of the euro, January 1999-January 2002
(Indices, January 1999=100)

Source: European Central Bank [www.ecb.int].
Note: Average monthly rates.

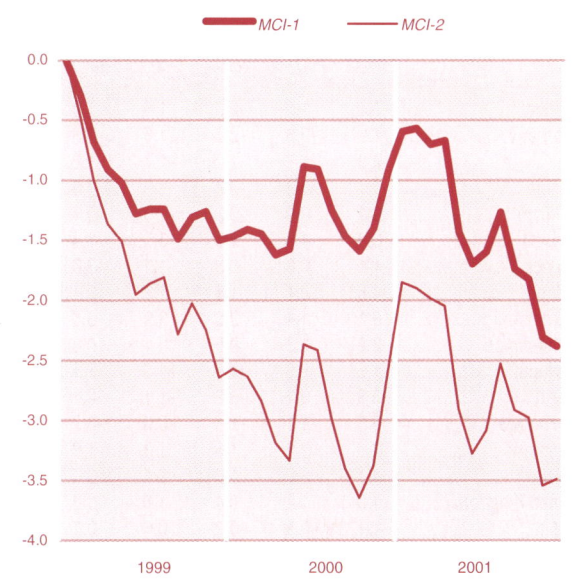

CHART 2.2.9

Monetary conditions index (MCI) for the euro area, January 1999-December 2001
(January 1999=0)

Source: UNECE/EAD calculations.
Note: The indices were calculated using different weights (w) for the real short-term interest rates (w1) and the real effective exchange rate (w2). The weights are as follows: MCI-1 (0.75/0.25); MCI-2 (0.85/0.15). The real short-term interest rate was calculated using the core inflation rate.

liquid assets given the fall in money market rates. This points to the need for caution when judging the relation between money supply and inflation in the short term (see also section 2.3).

(c) Fiscal policy

The cyclical downturn also led to a deterioration of government budgets in 2001 via the working of the automatic stabilizers (lower tax revenues and increased expenditure notably on unemployment benefits). In addition, cuts in income and other taxes enacted in a number of countries accentuated this development. As a result, the aggregate general government budget deficit in the euro area reached 1.1 per cent of GDP in 2001, up from 0.8 per cent in 2000.[65]

However, there was some variation in the levels and changes of actual budget balances in the 12 member countries of the euro area in 2001 (table 2.2.5). In view of the rules of the Stability and Growth Pact, which set a limit to the budget deficit of 3 per cent of GDP in normal times, fiscal developments were an increasing source of concern in Germany, where the deficit rose to 2.6 per cent of GDP in 2001, 1.3 percentage points higher than in 2000 (*excluding* the revenues from the Universal Mobile Telecommunication System (UMTS) licence sales in 2000). There was also a large increase in the budget deficit, to 2 per cent of GDP, in Portugal. Budget deficits in France and Italy were 1.5 per cent and 1.2 per cent of GDP, respectively, in 2001. These deficits contrast with more favourable budgetary positions in the smaller countries, which are either close to balance or in comfortable surplus.

Changes in the cyclically adjusted or structural budget deficits (again, excluding UMTS receipts) are a broad measure of the fiscal policy stance. At the aggregate level of the euro area, structural deficits have been stable (at 1.3 per cent of GDP) between 1999 and 2001, suggesting that fiscal policy has had a neutral impact on economic activity. At the individual country level, the impact of fiscal policy has been more varied with an expansionary stance in some countries (notably Finland, Germany, Ireland, Portugal but also France) offset by a neutral or restrictive orientation in the others (table 2.2.5).

Against a background of continuing cyclical weakness, actual budget deficits are forecast to increase further in 2002. A critical situation could arise in Germany, where the deficit is expected to remain at 2.7 per cent of GDP, i.e. very close to the 3 per cent ceiling imposed by the Stability and Growth Pact. In a number of countries (Belgium, Denmark, Finland, Greece, France, Ireland, Netherlands and Sweden) economic activity in 2002 will be supported by income and corporate tax cuts, but these will be offset by increases in other countries such as Germany. In the event, the actual

---

[65] This latter figure excludes the revenues from UMTS licence sales; including them yields a surplus of 0.3 per cent of GDP in 2000.

TABLE 2.2.5

General government budgetary positions in the European Union, 2000-2002

(Per cent of GDP)

	Actual balance			Cyclically adjusted balance		
	2000[a]	2001[b]	2002	2000	2001	2002
France	-1.4	-1.5	-2.0	-1.8	-1.9	-1.9
Germany	1.2	-2.5	-2.7	-1.6	-2.3	-2.0
Italy	-0.3	-1.2	-1.2	-1.7	-1.2	-1.0
Austria	-1.1	-0.2	-0.4	-1.9	-0.3	-0.2
Belgium	0.1	0.0	-0.2	-1.0	-0.7	0.1
Finland	6.9	4.8	2.9	3.7	3.4	2.4
Greece	-1.1	0.0	0.3	-1.2	-0.8	-0.1
Ireland	4.5	2.4	1.8	2.2	0.4	0.9
Luxembourg	6.1	4.4	2.8	3.9	3.1	2.8
Netherlands	2.2	1.3	0.5	0.1	0.8	0.8
Portugal	-1.5	-2.0	-1.6	-1.6	-2.5	-1.8
Spain	-0.3	0.1	-0.2	-1.1	-0.4	-0.3
Euro area	0.3	-1.1	-1.4	-0.6	-0.7	-0.7
Denmark	2.5	2.2	1.6	1.3	1.6	1.8
Sweden	4.1	3.9	1.6	2.4	3.0	1.5
United Kingdom	4.3	1.2	0.4	1.6	1.0	0.6
EU15	1.1	-0.5	-0.9	-0.6	-0.7	-0.7
Memorandum items:						
Japan	-7.6	-6.5	-5.9	..	..	..
United States	1.7	-0.3	-3.6	..	..	..

Source: European Commission, *European Economy*, Supplement A, No. 10/11 (Luxembourg), October/November 2001.

[a] Including UMTS receipts in 2000 for Austria, Germany, the Euro area, EU15, Italy, Netherlands, Portugal and Spain.

[b] Including UMTS receipts in 2001 for Belgium, Denmark, the Euro area, Finland and Greece.

budget deficit in the euro area is forecast to rise to some 1.5 per cent of GDP in 2002. On a cyclically adjusted basis, however, the deficit is expected to fall slightly, suggesting there will be no positive fiscal stimulus in 2002.

(ii) Other western Europe

Outside the euro area, real GDP in the *United Kingdom* rose by 2.3 per cent in 2001, significantly above the west European average but masking a steady slowdown in the course of 2001. As in 2000, buoyant private consumption was the main source of economic growth (table 2.2.2). Household spending was supported not only by the gains in income stemming from strong employment growth but also by the willingness of households to borrow, which was partly stimulated by rising house prices.[66] Government consumption and investment also continued to support domestic activity. In contrast, business investment weakened in the course of 2001, and growth in the volume of exports became increasingly sluggish as a result of the global downturn and the effects of the strong exchange rate of sterling.

---

[66] The link between house prices and borrowing is created by the use of houses as collateral. But rising house prices can also boost consumer confidence and thus the propensity to spend. K. Aoki et al., "Why house prices matter", *Bank of England Quarterly Bulletin*, Vol. 41, No. 4, Winter 2001, pp. 460-468.

These factors have led to a recession in the manufacturing industry, which contrasts with continued strong growth in the services sector. Changes in real net exports were broadly neutral in their impact on economic activity in 2001.

Given the differential strength of domestic and external demand, the merchandise trade deficit increased to a level corresponding to 3.3 per cent of GDP in 2001. The current account deficit is estimated at 1.5 per cent of GDP, reflecting the large surplus on invisibles. The labour market developments remained very favourable in 2001. The (standardized) unemployment rate edged up to 5.2 per cent in the final quarter of 2001, but this is still more than 3 percentage points below the average rate in the euro area. The average inflation rate, as measured by the retail price index excluding mortgage interest payments (RPIX), was 2.1 per cent in 2001, well below the government's inflation target of 2.5 per cent. Responding to the effects of the deteriorating global environment, the Bank of England reduced its base lending rate in several steps by 2 percentage points during 2001, but has kept rates unchanged since November. Fiscal policy supported economic activity in 2001. The government has launched a massive investment programme designed to improve the infrastructure (transport, health, education). The general government financial balance, however, remained in comfortable surplus to the extent of 1.1 per cent of GDP in 2001.

In *Denmark* and *Sweden*, a weakening in all the major expenditure items led to a pronounced deceleration of economic growth in 2001. Economic activity in Sweden was also strongly affected by the global crisis in the ICT sector, which led to a 30 per cent fall in Swedish exports in 2001.

Outside the EU, *Turkey* suffered a deep financial crisis in 2001 after the February devaluation led to turmoil in the financial sector and a deep recession. All the major components of domestic demand fell considerably. Exports were the only support to domestic activity but they were unable to prevent GDP from falling by over 7 per cent (as compared with 2000). The public sector's net debt-to-GNP ratio increased sharply to 95 per cent in 2001 from 57 per cent in the previous year, due mainly to the bail-out of the banking system. A major factor allowing the country to avoid default on its public debt was the support from the IMF, which provided loans totalling $31 billion during the year. (Turkey was the largest recipient of IMF loans in 2001.) In February 2002 financial market confidence in the country improved strongly after the announcement of a new package from the IMF comprising an additional credit of $12 billion. This is part of a three-year programme that foresees a continuation of the structural adjustment process already in place, a strengthening of the banking sector, the attainment of fiscal sustainability and the stabilization of the exchange rate and lower inflation. An annual primary budget surplus of 6.5 per cent of GNP should be achieved so as to reduce the public debt/GNP ratio and the debt service burden. To this end an austere fiscal policy must be adopted, including an increase in taxation as well as a restructuring of the public administration and state enterprises. The banking sector will receive additional public funding in order to deal with the problem of non-performing loans and to allow the consolidation and privatization processes to continue. The goal is to restructure the public banks and to recapitalize the private banks and ensure their compliance with the 8 per cent capital adequacy requirement. Efforts are also concentrated on strengthening the regulation and supervision of banks.

The devaluation and the adoption of a floating exchange rate spurred a steep rise in the inflation rate. The programme's goal is to reverse this and to reach an annual rate of 35 per cent by the end of 2002, down from 73 per cent in February 2002. To this end the backward indexation of wages will be scrapped and formal inflation targeting will be adopted in the course of 2002, to be pursued by an independent central bank. Disinflation should lead to lower interest rates, which in turn will reduce the costs of servicing the public debt and possibly allow banks to resume channelling funds to the private non-financial sector. Thus, a failure to lower the inflation rate would jeopardize the fiscal and economic growth targets of the adjustment programme.

2.3 The two pillars of the ECB monetary strategy

The ECB's monetary policy strategy defines the reference value for the growth rate of the money stock M3 as one of the two pillars of ECB's monetary policy strategy to achieve and maintain price stability. The second consists of a broadly based assessment of economic variables that allow inflationary pressures to be gauged in the short and medium term.

The theoretical case for a predefined monetary policy strategy arises from the idea that time inconsistency and uncertainty about policy decisions can be overcome by tying central bankers' hands with a policy rule. This is also reflected in the institutional arrangements for the conduct of monetary policy in the euro area: the ECB is an independent institution but it is constrained by its statute to achieve and maintain price stability. However, it is free not only to define price stability but also to determine the strategy of how to achieve it. The term "strategy" basically refers to the decision-making process, that is to say, the use of instruments to achieve the ultimate goals of monetary policy.

More generally, the need for a monetary policy strategy arises from the very complex environment in which a central bank has to pursue its ultimate goals.[67] As central banks have no direct control over their ultimate goal, i.e. mainly price stability and the creation of a conducive environment for economic growth and employment, they have to use intermediate (or operating) monetary policy targets to provide guidance for their interest rate decisions. Given the very limited knowledge of the so-called transmission mechanism of monetary policy, the central bank cannot rely on a single robust macroeconomic model for generating reliable forecasts of its operating targets. Instead, several models have to be used, and in addition a high degree of judgement is required.

In principle, an intermediate target can be seen as an attempt to reduce the complexity of the real world to one or a few relevant indicators that can signal the direction in which monetary policy should move. An important requirement, of course, is that the central bank has control over its operating target and that there is a close and stable relationship between the intermediate target and the ultimate target. In other words, given the ultimate goal of price stability, the intermediate target must be a reliable predictor of inflation. Basically, the choice of an intermediate target is tantamount to the adoption of a simple rule to guide the policy reactions of the central bank.

If such a rule can be identified, this will have the dual advantage of facilitating the internal decision-making of the competent body (here, the ECB's Governing Council) and helping to explain to the markets and the public at large the rationale for the central bank's interest rate decisions. In this respect, such a strategy can help to increase the transparency and thus credibility of monetary policy.

At the outset of its operation, the ECB had to choose essentially between two possible monetary policy strategies, inflation targeting or monetary targeting. Both strategies have price stability as their ultimate goal but differ with regard to the intermediate target. Within a framework of monetary targeting, the intermediate target is the growth of a broad monetary aggregate (such as M2 or M3), based on the quantity theory of money. In contrast, in the case of inflation targeting, it is the currently expected rate of future inflation, generally with a forecast horizon of two years, that becomes the intermediate target for the central bank. In practice, the differences between the two strategies are small, with monetary targeting putting a greater emphasis on the monetary aggregates in the formulation of monetary policy and in explaining monetary policy to the public.[68]

As a new institution, the ECB also faced the challenge of building its reputation and credibility. It appears that this was one of the reasons why it chose to follow the tradition of the Bundesbank, which had

---

[67] For a general treatment see P. Bofinger, *Monetary Policy. Goals, Institutions, Strategies, and Instruments* (Oxford, Oxford University Press, 2001).

[68] B. Bernanke, T. Laubach et al., *Inflation Targeting* (Princeton and Oxford, Princeton University Press, 1999).

announced money supply targets from 1974 until 1998, when the conduct of monetary policy was taken over by the ECB.[69] As the Bundesbank had effectively steered monetary developments in the EMS since its inception in 1979, presumably the rationale was that the emphasis on monetary targeting would help the ECB to inherit the Bundesbank's credibility from the outset of its operations. Moreover, an apparent advantage of the broad monetary aggregates is that they have always been easy to calculate and produce in a timely fashion.

However, given the uncertainties at the beginning of Stage Three of EMU about the stability of the money demand function in the euro area – and the related uncertainty concerning the existence of a strong and stable long-run relationship between money supply growth and income – the ECB decided to use a weaker form of reference value instead of announcing an outright monetary target.

It is noteworthy that the central banks in a number of major economies such as Canada, the United Kingdom and the United States had moved away from monetary targeting because the relationship between the broad monetary aggregates and overall economic activity (and, hence, inflation) had not proved to be sufficiently stable to justify their use in formulating monetary policy.[70] The reasons for this are mainly seen to lie in the development and internationalization of money and capital markets, financial innovations and the technology of the payments system.

Indeed, the difficulties involved in using money-supply targeting had already become clear from the experience of the Bundesbank which very often failed to meet its monetary target but managed to maintain its reputation as the guardian of price stability. This suggests that the Bundesbank had de facto followed a very pragmatic policy, which was not narrowly focused on changes in the monetary aggregates. Monetary policy decisions were rather based on consideration of a range of factors such as the state of the business cycle, the international economic environment, developments in the money and capital markets, and price and exchange rate developments. In addition, a considerable degree of interest-rate smoothing was observed, suggesting that changes in the policy instruments were only made when developments in a sufficient number of areas pointed in the same direction.[71] In fact, econometric estimates of the Bundesbank's monetary policy reaction function unanimously conclude that money supply growth had very little impact on its interest rate policy.[72]

In the case of the ECB, the experience with the first pillar is not very different from that of the Bundesbank. For most of the period since 1999, the growth of M3 has been above its reference value, which has been kept unchanged at 4.5 per cent since the start of monetary union.

In December 2001, the ECB adjusted the money stock M3 to allow for money market funds and short-term bonds held abroad. Preliminary estimates suggest that this reduced the annual growth rate of M3 by about three quarters of a percentage point so that the increase in money supply was for some months below its reference value. However, since June 2001, the adjusted M3 money stock has been on a rising trend and has increasingly exceeded the reference value. In December 2001, M3 rose by 8 per cent compared with the preceding year (chart 2.3.1).

These observations reinforce the view that the two pillar approach is widely perceived as not credible because of the possibility that they may not point in the same direction.[73] If, as announced by the ECB, money is given a "prominent role" in the its strategy, then the interest rate policy of the ECB, which is closely reflected in the overnight market rate, is difficult to reconcile with the directions in which the first pillar has been pointing (chart 2.3.1). When the ECB lowered its main refinancing rate in April 1999, M3 growth was above the reference value of 4.5 per cent and it held interest rates low until November 1999, M3 growth remaining above its reference value. The subsequent series of increases in interest rates that continued until October 2000 were compatible with the unrevised monetary statistics, but in retrospect this was not the case for the decisions taken in late August and October 2000. Furthermore, the three interest rate reductions between late August 2001 and early November 2001 occurred when monetary growth was growing by more than the reference value and even accelerating.[74]

This is not to say of course, that these interest rate decisions should not have been made. However, the upshot is that the first pillar cannot have played a dominant and consistent role in the ECB's policy considerations. Moreover, the statistical corrections to M3 compromise the prime purpose of the first pillar, namely its purported transparency.

The main reason for these shortcomings is the unstable demand for money in the short run. The monthly (year-over-year) growth rates of M3 have tended to be much more volatile than the corresponding inflation rate or changes in the other main macroeconomic variables. A comparison of actual M3 changes (even three-month

---

[69] In fact, the Bundesbank was the only major central bank to have set monetary targets over such a long period of time.

[70] This is reflected in the famous quip of one central banker: "We did not abandon the money supply targets, rather they abandoned us".

[71] J. von Hagen, "Geldpolitik auf neuen Wegen (1971-1978)", in Deutsche Bundesbank (ed.), Fünfzig Jahre Deutsche Mark: Notenbank und Währung in Deutschland seit 1948 (C.H. Beck, Munich, 1998), pp. 439-473.

[72] R. Clarida and M. Gertler, How the Bundesbank Conducts Monetary Policy, NBER Working Paper Series, No. 5581 (Cambridge, MA), May 1996.

[73] D. Gros, "The ECB's unsettling opaqueness: on transparency", paper presented to the Monetary Committee of the European Parliament (Brussels), 11 May 2001.

[74] There have been suggestions that the ECB did not loosen its monetary policy stance in response to the need for stronger economic growth, due to its concern at the weak exchange rate of the euro vis-à-vis the dollar.

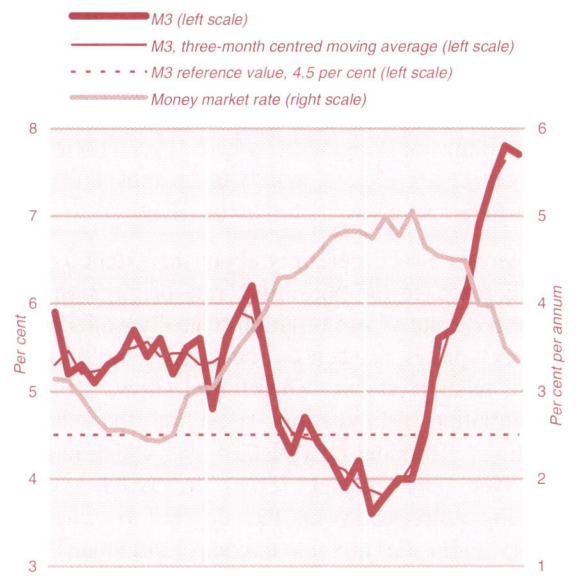

CHART 2.3.1

M3 growth and money market rates in the euro area, 1999-2001
(Percentage change, per cent per annum)

Source: European Central Bank [www.ecb.int].
Note: M3 data are seasonally and calendar effect adjusted. The money market rate is the overnight deposit rate.

moving averages) with the reference value is therefore bound to lead to erratic signals for interest rate policy.

In other words, the use of broad monetary aggregates such as M3 in the design and implementation of a monetary policy strategy only makes sense in a medium-term or even longer-term context. Such a medium-term perspective is reflected in the recent calculations of nominal and real money gaps by the ECB.[75] The nominal gap is defined as the percentage point deviation of the actual M3 stock from the level that would be observed if M3 had been increasing at a rate corresponding to its reference value using December 1998 as the base period. Similarly, the real gap is defined as the nominal gap less the deviation of consumer prices from the definition of price stability, again using December 1998 as the base period.

While such an exercise is better than focusing on monthly changes in money supply, it is, nevertheless, a rather ad hoc solution. First, the choice of December 1998 as a base period implies that money holdings were in equilibrium in that month. The ECB has so far not addressed this issue. Second, the calculation of a real money gap using the actual deviations of consumer prices from the definition of price stability neglects the relatively long lags (about 6 quarters) between monetary impulses and the price level.

Another important reason why changes in M3 are not a good guide for monetary policy in the short run concerns the ambiguous interest rate elasticity of demand for M3. The ECB has tried to justify the increase in official interest rates during 2000 by referring to the strong growth of M3. This would make sense only if an increase in short-term interest rates has a negative impact on the demand for M3. But such a negative interest rate elasticity has been found to exist only for the spread between nominal long-term and short-term interest rates.[76] In other words, an increase in short-term interest rates that does not feed through to interest rates at the longer end of the maturity spectrum will tend to have a *positive* impact on demand for M3 and vice versa. The ECB has recently implicitly acknowledged this by pointing to the "flat yield curve" as a factor behind the strong growth of M3.[77]

To sum up, the first pillar of the ECB's monetary policy strategy cannot be regarded as contributing to transparency and public understanding of the ECB's interest rate decisions. On the contrary, the first pillar is more of a barrier to effective communication with the public.[78] The repeated efforts to explain why special factors account for the overshooting of the reference value, the significant ad hoc adjustment of the underlying money supply statistics to allow for the effects holdings of money market paper and short-term debt securities by non-residents of the euro area, and the ambivalent impact of changes in short-term interest rates on the growth of M3, have all helped to create a continuing confusion about the role of the first pillar in the conduct of monetary policy in the euro area. This raises the wider issue of whether the relation between money and income in the long run really justifies the "prominent role" given to M3 in the ECB policy strategy, given its focus on the short and medium term. It might now be better to integrate the monitoring of changes in money supply into the second pillar, i.e. the broadly-based assessment of leading indicators of inflation in the short and medium term.

The second pillar of the ECB's strategy is de facto an inflation forecast, although the ECB has been using the term "assessment of the inflation outlook". It is not quite clear, however, why an "assessment of the inflation outlook" is conceptually different from an "inflation forecast". The inflation assessment is based, in principle, on the observation of all economic variables that are leading indicators of price developments. These include, *inter alia*, labour costs, the exchange rate of the euro, bond prices and the yield curve, measures of real sector activity, fiscal policy indicators, and price and cost indices as well as business and consumer surveys.[79]

---

[75] ECB, "Framework and tools of monetary analysis", *Monthly Bulletin* (Frankfurt am Main), May 2001 pp. 41-58 and June 2001, p. 9.

[76] G. Coenen and J.-L. Vega, *The Demand for M3 in the Euro Area*, ECB Working Papers, No. 6 (Frankfurt am Main), September 1999.

[77] ECB Press Conference, 30 August 2001 [www.ecb.int].

[78] See also A. Alesina et al., *Defining a Macroeconomic Framework for the Euro Area. Monitoring the European Central Bank 3* (CEPR, London, 2001), pp. 47-48.

[79] ECB, "The stability-oriented monetary policy strategy of the eurosystem", *Monthly Bulletin* (Frankfurt am Main), January 1999, pp. 39-50.

Given that the actions of the ECB cannot be successfully explained by the behaviour of the evolution of the first pillar, the ECB must de facto be pursuing a policy of inflation targeting.[80] The problem is that the ECB does not publish an explicit inflation forecast on which its inflation assessment is based although it has promised to do so.[81] It is therefore difficult to judge the extent of the role played by the second pillar in the internal decision making process, and this lack of transparency complicates communication with the public and weakens the accountability of the Bank. The ECB can, of course, always explain its interest rate decisions by referring to one or several of the variables included in inflation assessment, but this can hardly be labelled a "strategy".

In addition, the uncertainties about the monetary strategy of the ECB are amplified by its definition of price stability, which is a "year-on-year increase in the Harmonized Index of Consumer Prices (HICP) for the euro area of below 2 per cent". This is an asymmetric target, where only the upper bound is well defined. The lower bound was never explicit in the early statements of the ECB, but it has since made it clear that its definition excludes a *decline* of the price index. This would imply that the target range lies between 0 and 2 per cent, but there is no evidence to suggest that policy is focused in the middle of this range.

The reason given by the ECB for not specifying a lower bound for the band is that the size of the measurement bias in the HICP measure is not yet known. It is well known that price indices inadequately capture quality improvements and the introduction of new products and therefore tend to overstate the "true" inflation rate. This measurement bias will differ across countries but available estimates point to upward biases of around 1 percentage point. This would suggest that the actual lower bound of the inflation target range must be significantly above 1 per cent if the risk of a deflation is to be avoided. Some clues are provided by the way in which the ECB derived its reference value for the growth of money supply (M3), based on the quantity theory of money. This would imply an inflation rate of 1.5 per cent.[82] If this is the mid-point of the inflation range targeted by the ECB, it would imply a lower inflation bound of 1 per cent.[83]

This lack of clarity about the lower bound of the inflation range is damaging for the transparency and credibility of monetary policy. It increases the uncertainty about the policy reaction function, since whether and when the ECB is likely to act if inflation were to fall below 2 per cent for a sustained period of time becomes a matter of guesswork. This could be avoided with a symmetric inflation target (such as that set for the Bank of England), where overshooting and undershooting are seen as equally undesirable and in both cases require responsive action by the Bank.

There is also uncertainty about the extent to which the ECB reacts to the "headline" rate of inflation rather than the core rate. Core inflation typically excludes food and energy prices that tend to be more volatile than the prices of other products. Therefore it allows a focus on trend inflation, eliminating some of the short-run fluctuations. Eliminating variables over which monetary policy has little impact serves to avoid giving a misleading impression of the degree to which the monetary authorities are able to control inflation.[84] There is, however, a trade-off because transparency and communication become a problem when the inflation measure is too restrictive and less comprehensible to the public. Attempts to deduce a simple rule (policy reaction function) from the pattern of interest rate changes made by the ECB suggest that it has a hybrid rule in which both core inflation and the headline inflation forecast have equal weight.[85]

In addition, there is no consensus about the optimal rate of (low) inflation. It should be stressed that the downward stickiness of prices and wages leaves too little room for relative price changes to take place smoothly if the inflation target is too low (taking also into account the upward bias in the measurement of consumer price indices). This could have adverse effects on output and employment.[86] Without any explicit statement from the monetary authorities, it is up to the markets and other observers to make their own guess of what inflation rate (or range) the ECB regards as most desirable as well as what the consequent implications are for the setting of interest rates. Indeed, the above-mentioned "guestimate" about the actual target range and the perceived optimal mid-point inflation rate is based on simple plausibility considerations. It may well be, however, that the ECB's

---

[80] It has been claimed that the Bundesbank was also an inflation targeter in deed and only a monetary targeter in words. L. Svensson, "Price stability as a target for monetary policy: defining and maintaining price stability", paper presented at the Bundesbank Conference on *The Monetary Transmission Process: Recent Developments and Lessons for Europe*" (Frankfurt am Main), 26-27 March 1999.

[81] W. Duisenberg, *Presentation of the ECB's Annual Report 1998 to the European Parliament, Introductory Statement* (Strasbourg), 26 October 1999.

[82] This is derived as a residual from the change in money supply (4.5 per cent), the trend decline in money velocity (0.75 per cent) and an estimated potential rate of output growth of 2.25 per cent.

[83] L. Svensson, *The First Year of the Eurosystem: Inflation Targeting or Not?*, Institute for International Economic Studies (Stockholm University), February 2000.

[84] L. Svensson, "Price stability as a target for monetary policy ...", op. cit.

[85] A. Alesina et al., op. cit. It should be emphasized that the inflation forecasts embodied in these calculations are not from the ECB but based on opinion polls of consumer price expectations published in *The Economist*.

[86] C. Wyplosz, "Do we know how low inflation should be?", paper presented at the First Central Banking Conference on *Why Price Stability?*, organized by the ECB (Frankfurt am Main), 2-3 November 2000 (revised version 2001); G. Akerlof et al., "The macroeconomics of low inflation", *Brookings Papers on Economic Activity*, 1 (Washington, D.C.), 1996, pp. 1-76.

actual target does not lie at the centre of the band but is skewed to one end or the other of the range.[87]

The conclusion is that while the first pillar of the ECB monetary strategy is too narrowly focused on the money stock M3, the second pillar is much too broad to provide any guidance as to the ECB's internal decision-making process or to allow clear communication to markets and the public at large.

The lack of a transparent framework and strategy for the ECB's monetary policy was especially underlined in the course of 2001 when an increasing number of observers criticized the Bank for reducing its interest rates "too little, too late". The uncertainty that surrounds the monetary strategy of the ECB risks damaging the efficiency of its policy. Monetary policy effectiveness relies for a good part on the credibility of the central bank and the transparency of its policy actions. The central bank only has control over a very short-term interest rate (its main refinancing rate). However, the economy reacts mainly to longer-term interest rates, given the dominance of long-term financing in private and public sector spending. If the central bank is not credible, its actions will not necessarily be reflected in the longer-term interest rates set by the financial markets. This means that the monetary transmission mechanism will be hampered and the effectiveness of monetary policy will be correspondingly reduced.

2.4 Some reflections on the "weakness" of the euro

The positive economic performance of the euro area since 1999, especially until the first half of 2000, was accompanied by the continued depreciation of the euro vis-à-vis other major currencies, including the dollar, the yen and the pound sterling.

The movement in the euro-dollar exchange rate has been particularly remarkable, due not only to euro weakness but also to dollar strength against other currencies. In February 2002, the euro had lost some 25 per cent of its value vis-à-vis the dollar since its launch. Since the changeover to euro notes and coins in early 2002, the euro has failed consistently to breach the 90 cents level against the dollar. The euro also weakened considerably vis-à-vis the yen during the first two years since its launch, by almost 30 per cent, before recouping some of the lost value in 2001. In February 2002, it had depreciated by just over 11 per cent against the yen compared with January 1999.

Looking back further, however, using the ECU-dollar rate prior to 1999, it appears that the euro's decline vis-à-vis the dollar was the continuation of a longer-term trend since 1995. This puts both the euro's depreciation since 1999 and the short-lived 1998 appreciation of its predecessor currencies into perspective. During the period up to 1999, the ECU remained broadly stable against most other currencies.

The euro area's real effective exchange rate index – as calculated by Eurostat – declined by some 17 per cent between 1995 and January 2002, reaching a low in October 2000 when it had lost over 22 per cent of its value since 1995.

Most empirical studies of the equilibrium level of the "synthetic" euro suggest that at its current level and on the basis of macroeconomic fundamentals, it is substantially undervalued, possibly by as much as 20 per cent from a medium-term perspective.[88]

(i) Possible factors behind the euro's weakness

The reasons for the euro's weakness have been extensively debated. Some conjectures that have been offered appear less plausible than others. For example, Frenkel and Mussa[89] showed that, assuming Ricardian equivalence, fiscal tightening leads to a permanent increase in the net foreign assets position of a country. Provided that this fiscal consolidation is permanent, this will lead to an appreciation of the equilibrium exchange rate. While fiscal consolidation has been extensive throughout the euro area since the launch of the Stability and Growth Pact, this does not appear to have strengthened the euro.

Similarly, the idea that a country's real exchange rate will depreciate as the result of a negative terms of trade shock appears at first glance to correlate well with the decline in the effective value of the euro since its launch. This occurred simultaneously with the surge in global oil prices which – since the euro area is a net importer – represents a negative terms of trade shock. However, the net oil imports of the euro area were only slightly higher than those of the United States and therefore, one would expect this effect to be negligible.

Understanding the path of exchange rates and their determinants is complicated by the fact that they play a complex role in the economy and interact with a number of variables, including real interest rates and the trade and capital accounts. These variables all depend, to a large extent, on the expectations of market participants. There are also short-run effects, such as real interest rate differentials and psychological factors, which will cause the exchange rate to fluctuate temporarily around its long-term path.

---

[87] E. Clifton, *Inflation Targeting: What is the Meaning of the Bottom of the Band?*, IMF Policy Discussion Paper, PDP/99/8 (Washington, D.C.), December 1999.

[88] Surveys of such studies were conducted by V. Koen et al., *Tracking the Euro*, OECD Economics Department Working Paper No. 298 (ECO/WKP(2001)24) (Paris), June 2001; and ECB, "Economic fundamentals and the exchange rate of the euro", *Monthly Bulletin* (Frankfurt am Main), January 2002, p. 51, box 2.

[89] J. Frenkel and M. Mussa, "Exchange rates and the balance of payments", in R. Jones and P. Kenen (eds.), *Handbook of International Economics Volume 2* (Amsterdam, Elsevier, 1988).

(a) Monetary approach

One strand of economic theory that has attempted to shed some light on exchange rate behaviour is the theory of *relative* purchasing power parity (PPP), which states that the percentage change in the exchange rate over any period equals the difference between domestic and foreign inflation rates. The *monetary approach* to the exchange rate postulates that, in the long run (when prices have adjusted), the exchange rate is fully determined by the relative supplies and demands of money, i.e.

$$e_{\left(\frac{d}{f}\right)} = \frac{p_d}{p_f} = \frac{M_d^S}{M_f^S} \cdot \frac{M_f^D(i_f, Y_f)}{M_d^D(i_d, Y_d)}$$

where subscripts d and f denote domestic and foreign values, respectively, M^S and M^D are the supply of and demand for money, determined by the interest rate, i, and income, Y and e is the nominal exchange rate (national currency in terms of foreign currency).

This approach implies that, once price levels have adjusted, the equilibrium real exchange rate is constant. Adjustment towards equilibrium is assumed to be instantaneous. However, this approach has not satisfactorily explained the deviation of exchange rates from their equilibrium paths. In practice, real exchange rates have been found to follow a random walk. Moreover, panel data analysis has shown that the speed of adjustment to the equilibrium path is very slow.

Therefore, additional ideas have been put forward to explain the observed patterns in exchange rates, based on macroeconomic fundamentals, which include developments in domestic and overseas prices of goods and services, relative productivity growth, terms of trade shocks, fiscal variables and a country's international investment position.[90]

(b) Portfolio shifts

According to the portfolio balance theory, an increase in the relative net supply of domestic currency-denominated assets should lower their price relative to foreign assets, thus increasing their domestic currency yield and depreciating the exchange rate.

This effect works principally through a risk premium channel and a balance of payments channel. The first arises in the case of an increase in net foreign debt of a country, which has to be financed by international investors. However, this raises the risk premium which investors demand to keep their money invested in the country. At a given rate of interest, this higher risk premium requires a currency depreciation in the debtor country.

It has been shown that risk premium shocks are a possible source of exchange rate overshooting.[91] In Dornbusch's model,[92] a monetary policy change, given short-run rigidities, causes the exchange rate to overshoot, while in this extension, the shock occurs to the risk premium in the market for foreign exchange. Therefore, the author identifies risk premia shocks as an important source of deviations from the equilibrium exchange rate path.

The balance of payments channel operates if a trade surplus is needed to offset interest payments on foreign debt resulting from an excessive current account deficit. This can be brought about via a currency depreciation.

The introduction of the euro resulted in a sharp rise in the issuance (by domestic and foreign sources) of euro-denominated debt relative to that in predecessor currencies, at the expense of both dollar and yen-denominated securities. The share of new issues of euro-denominated debt securities has increased from 25-30 per cent of the total new issues prior to the euro's launch to some 40-45 per cent.[93]

This appears to have come about as resident issuers substituted euro-paper for foreign currency-denominated instruments in response to the increased homogeneity and liquidity of the domestic capital market, while non-resident issuers were motivated by the need to establish a presence in the market. Also, a wave of mergers and acquisitions vastly increased the financial needs of European corporations; low interest rates may have also played an important role.

This process would have put downward pressure on the euro, although modelling such an effect is complicated by the lack of data and by structural shifts. However, some commentators have cast doubt on the relevance of this channel, by arguing that this effect will depend on the maturity of the debt.[94] Longer maturities are less likely to demand an adjustment in the exchange rate, while in the case of short maturities the ECB might be expected to react to a drop in debt prices with an expansionary open market policy. However, this does not appear to be supported by the evidence.

---

[90] ECB, "Economic fundamentals ...", op. cit., p. 42; J. Coppel et al., *EMU, the Euro and the European Policy Mix*, OECD Economics Department Working Paper No. 232 (ECO/WKP(2000)5) (Paris), February 2000.

[91] See A. Isaac, "Risk premia and overshooting", *Economics Letters*, No. 61, 1998, pp. 359-364, who extended Dornbusch's seminal paper on exchange rate overshooting.

[92] R. Dornbusch, "Exchange rate expectations and monetary policy", *Journal of International Economics*, No. 6, August 1976, pp. 231-244; K. Rogoff, "Dornbusch's overshooting model after twenty-five years", Mundell-Fleming Lecture presented at the IMF's Second Annual Research Conference (Washington, D.C.), 30 November 2001.

[93] According to the Bank for International Settlements, the gross issuance of euro-denominated debt during the first half of 2001 was $408.5 billion out of a total value of new issues of $1,113.5 billion.

[94] European Economic Advisory Group at CESifo, *Report on the European Economy* (Munich), 2002.

(c) Fall in the demand for currency

Currency demand in the euro area has declined substantially since 1997. Calculations suggest that over the period 1997-2000, the demand for European currency dropped by €48 billion below trend, defined by GDP, interest rates and time.[95] An especially sharp drop during 2001 suggests that, between 1997 and October 2001, the overall demand for European currency declined by somewhere in the order of the equivalent of €90 billion.

Some reasons, partly anecdotal, have been advanced for this phenomenon. For example, it appears that for every three deutsche marks held in the euro area, one was held outside the euro area. These extra euro area currency holdings may have been converted into the local currencies of newly emerging states in eastern Europe as these regained public confidence toward the end of the 1990s. Equally, large cash holdings from illicit sources may have been converted into a variety of currencies other than the euro to minimize risks of detection. Another possible reason for this drop in currency demand could stem from a more widespread use of non-cash payments as a result of advances in technology.

All this would have had an impact in pushing down the euro's value. Indeed, it has been estimated[96] that each additional $1 billion of net dollar purchases increases the Euro price of a dollar by about 0.4 cents. This could explain a depreciation of the euro against the dollar by between 30 and 40 cents, if the observed fall in demand for euros translated fully into an increase in the demand for dollars. This would be enough to explain the actual decline in the foreign exchange value of the euro area currencies since 1997, which was about 40 cents. Whether or not the drop in demand for the euro and euro area currencies has actually fully translated into an increase in the demand for dollars is, however, difficult to gauge.

(d) Relative productivity growth

The well-known Balassa-Samuelson effect implies that increases in productivity in the traded goods sector, relative to the non-traded goods sector, are associated with a real appreciation of the currency of a country. Therefore, countries with lower productivity growth should see their real exchange rates depreciate in the long run vis-à-vis countries with higher productivity growth.

Meredith[97] reports a close correlation between the lagged difference in euro area and United States labour productivity and the change in the bilateral exchange rate since 1999. However, the magnitude of the effect appears to be small.[98]

Another channel through which productivity growth affects the exchange rate in the short run is the real interest rate channel. Higher productivity growth increases both investment demand and household consumption (via expectations of higher future income). In the medium term, overall productivity increases may support the exchange rate via higher real interest rates and better prospects for economic growth.

Indeed, there is evidence of a correlation between revisions to projected growth for the United States and the euro area and movements in the euro-dollar rate,[99] although this pattern loosened somewhat after the summer of 2000.[100] Moreover, there was evidence of a widening of long-term interest-rate differentials between the United States and the euro area, in favour of United States assets, until mid-2000 when this trend was reversed. This coincided with a surge in United States equity prices since the mid-1990s, which raised United States stock market capitalization relative to GDP to unprecedented levels, from 90 per cent of GDP in 1994 to 180 per cent in 1999, increasing the relative capitalization gap between the two economic areas.

These considerations suggest that the United States experienced a substantial positive aggregate demand shock relative to the euro area. According to standard open economy models with nominal rigidities, not only the dollar-euro appreciation and consequent worsening of the United States trade account but also aspects of the new economy appear to be consistent with this shock. This may have had an impact on the euro's weakness vis-à-vis the dollar during the first two years of its operation but is less potent in explaining the continued weakness in 2001 and early 2002.

(e) ECB policies and psychological impacts

A further argument cited for the explanation of the weak euro is related to the specific institutional and political environment under which the new European Central Bank is operating. The ECB has been criticized for lacking a clear-cut, transparent monetary policy, in the sense that changes in policy are difficult to anticipate. This was especially pronounced due to the lack of familiarity with the new currency. Clearly, given investors' risk aversion, such a factor has the potential to increase the volatility of financial variables,

---

[95] Ibid.

[96] M. Evans and R. Lyons, *Order Flow and Exchange Rate Dynamics*, NBER Working Paper, No. 7317 (Cambridge, MA), August 1999.

[97] G. Meredith, *Why Has the Euro Been So Weak?*, IMF Working Paper WP01/155 (Washington, D.C.), October 2001.

[98] S. Tille et al., "To what extent does productivity drive the dollar?", Federal Reserve Bank of New York, *Current Issues in Economics and Finance,* Vol. 7, No. 8, August 2001, pp. 1-6.

[99] G. Corsetti, "A perspective on the euro", paper presented at the CESifo Forum, Summer 2000, mimeo [www.econ.yale.edu/~corsetti/euro/ifo/pdf].

[100] European Economic Advisory Group at CESifo, op. cit.

thereby driving up risk premia and undermining confidence in the new currency.

In a similar vein some observers have argued that the European currency would suffer from the risks associated with the forthcoming enlargement of the European Union to countries in central and eastern Europe. This may have also contributed to the euro's weakness.

There is also some research, which points to market dynamics as part of the explanation for the euro's weakness. It has been argued[101] that due to the uncertainty surrounding the equilibrium levels of exchange rates, the short-run movement in the euro's exchange rate may have been driven by market participants extrapolating recent developments in the absence of well-formed views about the longer-run economic fundamentals of the euro area.

(ii) Conclusions

It is unlikely that any single cause can explain the euro's exchange rate path. A number of the factors discussed above are likely to have had an impact in weakening the euro. First, the global surge in equity prices since the mid-1990s may have created a demand shock that disproportionately hit the United States economy, causing the dollar to appreciate. Secondly, a mismatch in supply and demand of euro-denominated assets may have adversely affected the euro. Thirdly, a drop in the demand for European currencies will doubtlessly have contributed to its weakness. Finally, the euro was not helped by the opaque policy-making environment, which is likely to have exacerbated existing uncertainties about the euro's prospects. In the light of recent optimism voiced about the long-term growth prospects of the United States economy, it would not be implausible to expect the euro to remain at its current "weak" level for quite some time to come.

2.5 Overview of growth patterns in industrialized countries in the 1990s: the role of demand factors

(i) Growth performance in the 1990s

The industrialized countries[102] grew at an average annual rate of 2.5 per cent during the 1990s,[103] a performance that was boosted during the second half of the period, when growth accelerated significantly in most countries. Compared with previous decades, the 1990s were a period of slower expansion. While the average annual growth rate between 1990 and 2000 was close to that of the 1980s in the United States, it deteriorated somewhat in western Europe and sharply in Japan (table 2.5.1).

The variability of industrial country GDP growth rates decreased during the 1990s, as compared with the two previous decades.[104] This was the result of two processes: the cyclical desynchronization between North America, western Europe and Japan and changes in the conduct of macroeconomic policy.

(a) Cyclical desynchronization in the 1990s

A consequence of the cyclical desynchronization among the major industrialized regions during the decade was the smoothing of their average growth rates (whence lower variability). In the United States there had been a fall in annual GDP already in 1991, whereas this occurred in western Europe two years later. While a long cyclical expansion started in the United States in 1992, there was no upswing of comparable duration and magnitude in western Europe. On the other hand, growth in Japan slowed down sharply in 1991 and 1992, but it was only in 1998 that its annual GDP actually fell, although most of the decade was characterized by stagnation.

The main reason for cyclical desynchronization among the industrialized countries was that their economic developments were strongly influenced by country-specific (or region-specific) factors that did not spill over to other countries or regions[105] and, to a lesser extent, by the asymmetric effects of emerging market crises. The main positive shock affecting growth in the United States in the 1990s was the arrival of a new wave of technological innovation generated by the ICT sector during the second half of the decade. The same factor also played a role in the macroeconomic performance of the smaller west European countries that are more specialized in the ICT sector, particularly Finland, Ireland and Sweden. On the other hand the growth of western Europe during the first half of the 1990s was strongly influenced by the consequences of German reunification and the exchange rate crises of 1992-1993. During the following years, the macroeconomic performance of continental western Europe was restricted by the adjustments required to meet the convergence criteria for EMU. In Japan, in contrast, the main country-specific factor influencing macroeconomic performance was the protracted consequences of the bursting of the asset price bubble of the second half of the 1980s and the policy responses to it.

---

[101] P. De Grauwe, "Exchange rates in search of fundamentals: the case of the euro-dollar rate", *International Finance*, Vol. 3, Issue 3 (New York), 2000.

[102] In the present section "industrialized countries" refers to the 15 member states of the EU, Norway, Switzerland, Canada, United States and Japan.

[103] "1990s" refers to the last decade of the twentieth century, i.e. 1991 to 2000; reference is made analogously to previous decades. When average growth rates are mentioned, the basis year is included in the time range (e.g. growth rates for 1990-2000 cover 10 years of growth).

[104] The standard deviation of GDP growth rates during the 1990s was 0.9 percentage points, as compared to 1.4 in the 1980s and 2.1 in the 1970s (table 2.5.1).

[105] This is in contrast to the global shocks of the 1970s and 1980s, which affected all industrialized countries in a similar manner.

TABLE 2.5.1

Macroeconomic indicators for industrialized countries, 1970-2000
(Percentage, unless otherwise indicated)

	1970-1980	1980-1990	1990-2000	1990-1995	1995-2000
Average annual GDP growth rate					
Industrialized countries [a]	3.3	3.0	2.5	1.8	3.1
United States	3.3	3.2	3.3	2.4	4.1
Western Europe [b]	3.0	2.4	2.0	1.5	2.6
Japan	4.5	4.1	1.4	1.4	1.4
Standard deviation of annual GDP growth rate [c]					
Industrialized countries	2.1	1.4	0.9	0.9	0.3
United States	7.6	5.9	4.4	5.5	1.4
Western Europe	1.8	1.1	1.1	1.2	0.6
Japan	5.9	4.3	3.3	2.2	4.1
Inflation rate [d]					
Industrialized countries	8.8	5.0	2.5	3.1	1.9
United States	7.8	4.7	2.8	3.1	2.5
Western Europe	10.0	6.3	2.9	3.8	1.9
Japan	9.0	2.1	0.8	1.4	0.3
General government balance [e]					
Industrialized countries [f]	-3.1	-3.8	-2.9	-4.3	-1.6
United States	-1.9	-4.6	-2.5	-4.7	-0.3
Western Europe [g]	..	-4.3	-3.5	-5.4	-1.6
Japan	-2.3	-0.7	-3.0	-0.6	-5.3

Source: UNECE secretariat calculations, based on OECD, *National Accounts of OECD Countries* (Paris); various issues and IMF, World Economic Outlook (WEO) Databases [www.imf.org].

[a] Canada, Japan, the United States and western Europe.

[b] Fifteen member states of the EU plus Norway and Switzerland.

[c] Percentage points.

[d] Consumer prices.

[e] Percentage of GDP.

[f] Includes only Canada, France, Germany, Italy, Japan, the United Kingdom and the United States. First figure refers to 1977-1980.

[g] Includes only the EU.

(b) Changes in the conduct of macroeconomic policy

The second main cause of the fall in the variability of output of industrialized countries in the 1990s was the moderation of macroeconomic policy activism (as compared with the two previous decades), due to the enactment of medium-term fiscal consolidation programmes and the context of lower inflation. In order to reverse the growth of public debt accumulated during the 1980s, several countries started medium-term fiscal consolidation programmes at the end of the decade (e.g. Canada, United States). In western Europe such plans resulted from or were reinforced by the EMU fiscal convergence criteria.[106] In terms of monetary policy, the enactment of a more forward-looking approach to monetary policy (e.g. by the adoption of explicit or implicit inflation targeting in several industrialized countries during the decade) and the context of lower and more stable inflation (particularly during the second half of the 1990s) meant that monetary policy did not change as often as previously.[107] In turn, this contributed to a reduction in the variability of growth rates.

(ii) Demand factor contributions

The contributions of the various components of demand to total GDP growth in the 1990s mostly reflect the different shocks (negative and positive) that affected countries and regions, as well as the shape of macroeconomic policies during the decade.

(a) North America

North America was the region with the highest growth rate among the industrialized countries between 1990 and 2000 (table 2.5.2). The growth and pattern of the contributions of demand to growth were basically determined by developments in the *United States*, where there was a long cyclical upswing between 1992 and 2000 – at an average annual rate of growth of 3.7 per cent – with some particular features. Firstly, it was the longest expansion ever recorded in United States economic history.[108] Moreover, growth accelerated strongly towards the end of the upswing (i.e. between 1996 and 2000). Finally, the acceleration of economic growth was associated with a wave of technological innovation in the ICT sector and its spread to other sectors, which led to the acceleration of productivity growth.[109]

The rapid pace of technological innovation in the ICT sector led to a steep fall in the relative prices of its products and to a more rapid rate of technological obsolescence. The shortening of the life cycle of equipment required a faster rate of replacement of vintage capital, which in turn was facilitated by falling relative prices. This process spurred the rapid rise of investment in equipment and software, which grew by 12 per cent annually during the second half of the 1990s. In the context of robust economic expansion and household optimism, construction investment also expanded at a comparatively strong pace (4.7 per cent annually). On the financial side, this investment boom was stimulated by low capital costs to corporations (whether funding was

---

[106] The consequence of these policies can be seen in the large reduction of fiscal deficits (as a percentage of GDP) in North America and western Europe during the 1990s, as well as in the contrast between the first and second halves of the decade (table 2.5.1). The exception to this trend was Japan, where fiscal policy took the opposite direction, as mentioned below.

[107] IMF, "The business cycle, international linkages and exchange rates", *World Economic Outlook* (Washington, D.C.), May 1998, pp. 55-73.

[108] The present analysis relies on annual data. On the basis of monthly indicators the cyclical expansion is judged to have lasted 10 years (March 1991 to March 2001), NBER, *US Business Cycle Expansions and Contractions*, 2001 [www.nber.org/cycles.html].

[109] While there is a divergence of views on the effects of ICT-related innovations on economy-wide total factor productivity, there is consensus that it caused productivity to accelerate sharply, at least in the ICT manufacturing sector itself. IMF, "The information technology revolution", *World Economic Outlook* (Washington, D.C.), October 2001, pp. 103-144.

TABLE 2.5.2

Changes in expenditure items and their contribution to real GDP growth in industrialized countries, 1990-2000

(Per cent change, percentage points)

	1990-2000		1990-1995		1995-2000	
	Average annual growth rate	Contribution to GDP growth	Average annual growth rate	Contribution to GDP growth	Average annual growth rate	Contribution to GDP growth
Western Europe[a]						
Private consumption	2.1	1.2	1.4	0.8	2.7	1.6
Government consumption	1.6	0.3	1.6	0.3	1.6	0.3
Fixed investment	2.2	0.5	0.1	–	4.3	0.9
Changes in inventories	..	–	..	–	..	–
Total domestic expenditure	2.0	2.0	1.2	1.2	2.8	2.7
Net exports	..	0.1	..	0.3	..	-0.1
Exports	6.1	1.9	4.7	1.3	7.6	2.5
Imports	6.0	1.8	3.8	1.0	8.3	2.6
GDP	2.0	2.0	1.5	1.5	2.6	2.6
European Union (15)						
Private consumption	2.1	1.2	1.4	0.8	2.7	1.6
Government consumption	1.6	0.3	1.6	0.3	1.6	0.3
Fixed investment	2.2	0.5	0.1	–	4.3	0.9
Changes in inventories	..	–	..	–	..	–
Total domestic expenditure	2.0	2.0	1.2	1.2	2.8	2.8
Net exports	..	0.1	..	0.3	..	-0.1
Exports	6.2	1.9	4.8	1.3	7.7	2.5
Imports	6.1	1.8	3.9	1.0	8.3	2.6
GDP	2.0	2.0	1.5	1.5	2.6	2.6
Euro area (12)						
Private consumption	2.0	1.1	1.4	0.8	2.5	1.4
Government consumption	1.7	0.3	1.7	0.3	1.7	0.3
Fixed investment	2.1	0.5	0.2	0.1	4.0	0.8
Changes in inventories	..	–	..	–	..	–
Total domestic expenditure	1.9	1.9	1.2	1.2	2.6	2.6
Net exports	..	0.1	..	0.2	..	–
Exports	6.3	1.9	4.7	1.3	7.8	2.5
Imports	6.1	1.8	4.0	1.0	8.1	2.5
GDP	2.0	2.0	1.4	1.4	2.6	2.6
North America[b]						
Private consumption	3.3	2.2	2.5	1.6	4.2	2.8
Government consumption	0.8	0.1	-0.1	–	1.7	0.3
Fixed investment	6.0	1.1	3.6	0.6	8.5	1.6
Changes in inventories	..	0.1	..	0.1	..	0.1
Total domestic expenditure	3.5	3.6	2.3	2.3	4.7	4.8
Net exports	..	-0.4	..	–	..	-0.7
Exports	7.3	0.9	7.3	0.9	7.3	1.0
Imports	8.8	1.3	6.7	0.8	10.9	1.8
GDP	3.2	3.2	2.3	2.3	4.1	4.1
Industrialized countries[c]						
Private consumption	2.6	1.6	2.0	1.2	3.2	2.0
Government consumption	1.5	0.3	1.1	0.2	1.8	0.3
Fixed investment	3.2	0.7	1.1	0.2	5.4	1.1
Changes in inventories	..	–	..	0.1	..	–
Total domestic expenditure	2.5	2.6	1.7	1.7	3.4	3.4
Net exports	..	-0.1	..	0.1	..	-0.3
Exports	6.3	1.2	5.3	0.9	7.3	1.6
Imports	6.7	1.4	4.7	0.8	8.9	1.9
GDP	2.5	2.5	1.8	1.8	3.1	3.1

Source: UNECE calculations, based on OECD, *National Accounts of OECD Countries, Volume 1, Main Aggregates 1989-2000* (Paris), 2002.

[a] Fifteen member states of the EU plus Norway and Switzerland.

[b] United States and Canada.

[c] Canada, Japan, the United States and western Europe.

raised through equity markets or indebtedness) and rising asset prices. The fixed investment boom contributed 1.7 percentage points to the annual growth rate of 4.1 per cent during the second half of the 1990s (table 2.5.3).

Still, the main demand factor behind the strong cyclical expansion was private consumption, of which the contribution to total GDP growth rose from 1.7 percentage points to 2.9 percentage points between the first and the second halves of the decade. This acceleration was fuelled by a combination of growing employment and disposable household income, a falling personal savings rate, rising asset prices and growing household indebtedness.[110]

The robust expansion of domestic demand during the cyclical expansion in the United States coincided with slow growth in most of the other large industrialized economies. This combination weakened foreign demand for United States exports and at the same time strengthened United States demand for imports, leading to increasing current account deficits. Thus, the strong upswing of the second half of the 1990s was marked by the accumulation of domestic personal and corporate debt as well as foreign indebtedness, which constitute the major imbalances of the last business cycle of the twentieth century.[111]

During the first half of the decade the volume of exports and imports grew at approximately the same rate (7 per cent annually), but during the second half imports accelerated to 12 per cent per annum. Consequently, net exports made a negative contribution to GDP growth during most of the decade, and during the second half they subtracted 0.8 percentage points from the annual rate of GDP growth.

The financial crises originating in emerging markets in 1997 and 1998 eventually had a positive impact on the United States economy. Despite some temporary weakening of confidence, the country benefited via the trade and finance channels. In the first case, as the east Asian countries' current account balance turned from deficit to surplus in 1998, the United States deficit with them increased. Increased imports helped to offset the pressure of domestic demand expanding well ahead of domestic supply, which otherwise would have led to inflationary pressure and monetary tightening.[112] The crises in emerging markets also led to capital seeking safe havens (the "flight to safety"), particularly in United

---

[110] The growth in lending to households is partly explained by the financial deregulation enacted during the 1980s, which intensified competition among financial institutions and brought down lending costs to households.

[111] UNECE, "The ECE economies in autumn 2001", *Economic Survey of Europe, 2001 No. 2*, chap. 1, pp. 3-45.

[112] The patterns of trade between the United States and east Asia mean that the latter's exports were able to meet demand stemming from both private consumption and fixed investment.

TABLE 2.5.3

Changes in expenditure items and their contribution to real GDP growth in major economies, 1990-2000
(Per cent, percentage points)

	1990-2000		1990-1995		1995-2000			1990-2000		1990-1995		1995-2000	
	Average annual growth rate	Contribution to GDP growth	Average annual growth rate	Contribution to GDP growth	Average annual growth rate	Contribution to GDP growth		Average annual growth rate	Contribution to GDP growth	Average annual growth rate	Contribution to GDP growth	Average annual growth rate	Contribution to GDP growth
Canada							**Japan**						
Private consumption	2.4	1.4	1.3	0.8	3.5	1.9	Private consumption	1.6	0.9	2.2	1.2	1.0	0.5
Government consumption	0.6	0.1	0.4	0.1	0.9	0.2	Government consumption	3.1	0.5	3.2	0.5	3.0	0.5
Fixed investment	2.9	0.6	-1.1	-0.2	7.1	1.4	Fixed investment	0.1	0.0	-0.9	-0.3	1.2	0.3
Changes in inventories	..	0.2	..	0.4	..	–	Changes in inventories	..	-0.1	..	–	..	-0.2
Total domestic expenditure	2.3	2.3	1.0	1.0	3.6	3.5	Total domestic expenditure	1.3	1.3	1.4	1.4	1.2	1.2
Net exports	..	0.6	..	0.8	9.9	0.4	Net exports	..	0.1	..	0.0	..	0.3
Exports	8.1	2.9	8.2	2.6	8.1	3.3	Exports	4.4	0.4	3.1	0.3	5.7	0.6
Imports	6.8	2.3	5.7	1.7	7.9	3.0	Imports	3.6	0.3	3.3	0.2	3.8	0.3
GDP	2.8	2.8	1.7	1.7	3.9	3.9	GDP	1.4	1.4	1.4	1.4	1.4	1.4
France							**United Kingdom**						
Private consumption	1.4	0.8	0.7	0.4	2.0	1.1	Private consumption	2.7	1.8	1.5	0.9	3.9	2.6
Government consumption	2.0	0.5	2.3	0.5	1.7	0.4	Government consumption	1.3	0.3	1.2	0.2	1.4	0.3
Fixed investment	1.3	0.3	-1.2	-0.2	3.8	0.7	Fixed investment	2.8	0.5	-0.3	-0.1	6.1	1.1
Changes in inventories	..	–	..	–	..	–	Changes in inventories	..	0.1	..	0.2	..	–
Total domestic expenditure	1.5	1.5	0.7	0.7	2.3	2.3	Total domestic expenditure	2.5	2.6	1.3	1.3	3.8	3.9
Net exports	..	0.3	..	0.4	..	0.2	Net exports	..	-0.3	..	0.5	..	-1.1
Exports	6.6	1.5	5.3	1.1	8.0	2.0	Exports	6.1	1.8	5.3	1.4	7.0	2.2
Imports	5.5	1.2	3.4	0.7	7.7	1.8	Imports	6.4	2.1	3.2	0.9	9.7	3.2
GDP	1.8	1.8	1.1	1.1	2.5	2.5	GDP	2.3	2.3	1.8	1.8	2.8	2.8
Germany							**United States**						
Private consumption	1.9	1.1	2.3	1.3	1.6	0.9	Private consumption	3.4	2.3	2.6	1.7	4.3	2.9
Government consumption	1.6	0.3	2.0	0.4	1.2	0.2	Government consumption	0.8	0.1	-0.1	–	1.8	0.3
Fixed investment	2.2	0.5	2.5	0.5	1.8	0.4	Fixed investment	6.3	1.2	4.1	0.7	8.6	1.7
Changes in inventories	..	–	..	0.1	..	–	Changes in inventories	..	0.1	..	–	..	0.1
Total domestic expenditure	1.9	1.9	2.3	2.3	1.6	1.5	Total domestic expenditure	3.6	3.7	2.4	2.4	4.8	4.9
Net exports	..	-0.2	..	-0.6	..	0.3	Net exports	..	-0.5	..	-0.1	..	-0.8
Exports	4.5	1.3	0.7	0.2	8.4	2.3	Exports	7.0	0.7	7.0	0.7	7.0	0.8
Imports	5.6	1.5	3.6	0.8	7.7	2.1	Imports	9.3	1.2	7.0	0.8	11.6	1.6
GDP	1.7	1.7	1.6	1.6	1.8	1.8	GDP	3.3	3.3	2.4	2.4	4.1	4.1
Italy													
Private consumption	1.7	1.0	0.9	0.5	2.6	1.5							
Government consumption	0.3	0.1	-0.2	–	0.9	0.2							
Fixed investment	1.4	0.3	-1.2	-0.2	4.1	0.8							
Changes in inventories	..	-0.1	..	–	..	-0.2							
Total domestic expenditure	1.3	1.3	0.3	0.3	2.4	2.3							
Net exports	..	0.3	..	1.0	..	-0.4							
Exports	5.7	1.4	7.4	1.7	4.1	1.2							
Imports	4.7	1.1	3.0	0.7	6.3	1.6							
GDP	1.6	1.6	1.3	1.3	1.9	1.9							

Source: UNECE calculations, based on OECD, *National Accounts of OECD Countries, Volume 1, Main Aggregates 1989-2000* (Paris), 2002.

States bond and equity markets. In turn, this lowered the cost of capital in the United States and helped to prolong the investment boom.

Canada benefited from its deep economic integration with the United States economy. During the first half of the decade falling construction investment held back domestic demand, its contribution to total growth being only slightly higher than that of net exports (table 2.5.3). Between 1995 and 2000, however, the acceleration of economic activity in the United States led to increased optimism in Canada; household consumption accelerated and investment spending on equipment grew at the same strong pace as in the United States. Consequently, Canadian growth accelerated to 3.9 per cent a year, a rate close to that of its neighbour. Domestic demand accounted for most of this growth, with net exports making a positive (albeit smaller) contribution.

(b) Japan

The Japanese economy stagnated during most of the 1990s. Its average annual growth rate of just 1.4 per cent was the lowest among the large industrial economies and well below its own rates of over 4 per cent during the two previous decades (table 2.5.1). The investment and asset price boom of the late 1980s came to a halt in 1990, when asset prices started falling and investment spending was sharply reduced. The ensuing debt, capacity and stock overhang, combined with plummeting asset prices, resulted in large corporate and household portfolio imbalances.

The succession of corporate bankruptcies that followed and the decline of asset prices led to a deep financial crisis, which in turn severely restrained credit supply. The necessary portfolio adjustments and corporate restructuring caused considerable uncertainty and depressed fixed investment and private consumption. The policy response essentially consisted of a succession of fiscal stimulus packages and the easing of monetary policy; structural problems, particularly those of the financial sector, were not adequately addressed. The combination of enormous portfolio imbalances with poor policy responses led to protracted adjustment process that had not come to an end as the country entered the twenty-first century.

In the context of corporate restructuring, private expenditure on equipment fell by one fourth between 1991 and 1994 and, although it rose thereafter, by 2000 it had not recovered its level of 1991. During the second half of the decade, Japanese expenditure on equipment rose at the slowest rate among the industrialized countries (3.9 per cent per annum). Private construction expenditure fell during the 1990s and the building sector went into a deep crisis. The fall in private fixed investment was approximately offset by public fixed investment, which rose at an annual rate of 2.4 per cent. Consequently, total fixed capital formation provided virtually no support to GDP growth during the 1990s (table 2.5.3).[113]

The extended restructuring process caused employment to slow down and eventually fall from 1998. Together with falling asset prices and the stagnation of the economy, household pessimism increased throughout the 1990s. During the second half, household consumption grew at 1 per cent annually (the slowest rate among the industrial countries) and contributed just 0.5 percentage point to GDP growth (i.e. one third of the total).

As the Japanese government repeatedly enacted counter-cyclical fiscal policies, government consumption came to account for as much as one third of overall growth between 1990 and 2000. However, this fiscal activism was not enough to compensate for the contraction or stagnation of investment and consumption, so that the contribution of domestic demand to growth fell from 1.4 percentage points to 1.2 points between the two halves of the decade.

Between 1995 and 2000 Japan proved incapable of fully taking advantage of the strong expansion of world trade, in spite of its relative specialization in ICT-related goods. The volume of Japanese exports grew at 5.7 a year, one of the lowest rates among the industrial countries. The main reasons for this were the shift of manufacturing activity abroad and the east Asian turmoil of 1997-1998. The latter led to the demand for Japanese goods and services falling in the crisis-hit countries while at the same time the competitiveness of their exports improved in third markets.[114] Moreover, the Japanese currency appreciated by one fourth between early 1997 and late 2000 in real effective terms. Despite the sluggishness of domestic demand, net exports contributed 0.3 percentage points annually to total GDP growth between 1995 and 2000.

(c) Western Europe

Western Europe expanded by an average 2 per cent a year between 1990 and 2000, somewhat slower than during the previous decade (2.4 per cent per annum). Its growth rate was well below that of the United States, but higher than the Japanese (tables 2.5.2 and 2.5.3). Economic activity was particularly weak between 1990 and 1995, when the west European economies grew as slowly as Japan. This period included one year in which GDP actually fell (1993), as the growth forces of the late 1980s waned and economies were shaken by protracted exchange rate crises. During the first half of the decade growth was held back by weak domestic demand: private consumption expanded at a slow pace (1.4 per cent a year), while fixed investment stagnated. In this context, government consumption and net exports both contributed 0.3 percentage points to the expansion of GDP.

During the second half of the decade, stronger investment led to employment growth and improved consumer sentiment. Household consumption grew by 2.7 per cent annually. At the same time, investment in equipment and software grew at an annual rate of 7.2 per cent. Construction investment expanded more slowly, but still made a positive contribution to growth. Private consumption and fixed investment contributed 1.6 percentage points and 0.9 percentage point to total GDP growth, respectively. While real exports had grown at a stronger pace than imports during the first half of the decade, this was reversed during the second half, mainly because of the acceleration of domestic demand. Consequently, net exports were a slight drag on overall growth.

This overall picture of western Europe, however, masks striking differences among the 17 individual economies, both in terms of the overall growth rate and of the composition of GDP growth.[115]

---

[113] The neutral impact of fixed investment on growth between 1990 and 2000 stands in sharp contrast to the previous decade, when it had added 1.6 percentage points annually to GDP growth.

[114] Among other east Asian economies, Malaysia, the Republic of Korea, Singapore and Taiwan Province of China are the most direct competitors of Japanese exporters in third markets. OECD, "The influence of emerging market economies on OECD countries' international competitiveness", *OECD Economic Outlook*, No. 63 (Paris), June 1998, pp. 205-220.

[115] Thus, the acceleration of growth during the second half was mainly due to the stronger performance of the United Kingdom and some smaller economies (particularly Finland, Ireland, Spain and Sweden), which contrasts with the much lower growth rate in the large euro area economies.

In *Germany* the reunification boom prolonged the cyclical upswing of the second half of the 1980s until 1992. Private consumption rose strongly thanks to the rapid increase in household disposable income in the former German Democratic Republic, which unleashed a long repressed consumer demand. Construction investment was fuelled by the need to rebuild infrastructure and upgrade housing in the new *Länder*: its growth during the first half of the decade (4 per cent annually) was the highest among the industrialized countries and accounted for one third of the increase in GDP between 1990 and 1995. At the same time, investment in equipment and software stagnated. The steep rise in government spending associated with reunification was not only devoted to building but also to current consumption. Domestic demand grew faster than GDP and imports rose while exports stagnated. Consequently net exports subtracted 0.6 percentage points from total growth.

Once the reunification boom had petered out, growth was constrained by sluggish domestic demand. Fiscal spending was scaled back in order to rebalance the government finances, and to meet the EMU fiscal convergence criteria. Fixed capital formation slowed down sharply and its composition changed. Investment in equipment and software accelerated to 6.5 per cent annually between 1995 and 2000, but activity was scaled back in the construction sector, where there was considerable overcapacity after the reunification boom. The slow pace of economic activity led to rising unemployment and increasing consumers' pessimism, so that private consumption growth slowed from 2.3 per cent a year during the first half of the decade to 1.6 per cent. It contributed less than 1 percentage point to GDP growth, although this amounted to approximately half of the total increase.

Germany also suffered more than other west European economies from the emerging market crises of 1997-1998, due to its relative export specialization in machinery and transport goods, its greater trade exposure to east Asia, and its stronger financial links with the Russian Federation. Nevertheless, between 1995 and 2000 exports expanded to the United States and central and eastern Europe and net exports once again made a positive contribution (0.3 percentage point) to growth.

The combination of sluggish domestic demand with relatively weak export growth left Germany with the lowest growth rate in western Europe between 1995 and 2000 (1.8 per cent annually). Being the largest economy in western Europe, its slow rate of growth had a depressing effect on the other economies of the region, particularly those that are closest to it in economic and geographical terms.

Growth in *France* in the early 1990s was held back by the restrictive monetary policy[116] it had adopted in order to defend its exchange rate in the ERM. Fixed capital formation fell between 1990 and 1993, pulled down by construction. Employment fell between 1991 and 1994, contributing to the sluggish growth of private consumption (0.7 per cent annually). Consequently, the contribution of household consumption to growth during the first half of the decade (0.4 percentage points) was slightly less than that of government consumption (0.5 percentage points).[117] Given the weakness of domestic demand, net exports contributed 0.4 percentage points (one third) to total growth.

During the second half of the decade growth in France accelerated somewhat and was higher than that of Germany, thanks to a combination of fiscal and labour market measures that favoured employment creation,[118] and led to increased consumer optimism and stronger private consumption. Investment revived thanks to growth of 6 per cent a year in spending on equipment and software, while construction also rose, spurred by fiscal measures. While government consumption was restrained by the EMU convergence criteria, domestic demand accounted for the bulk of total growth and net exports played a less important role than during the first half of the decade.

France was less affected than other large European economies by the emerging market crises, thanks to its smaller trade and financial exposure to these countries. In spite of its relative recovery between 1995 and 2000, for the decade as a whole the French and German economies grew at virtually the same slow rate (1.7 per cent a year).

In the third largest economy of the euro area, *Italy*, growth was marginally slower during the decade (1.6 per cent per annum). It was the country where the struggle to meet the convergence criteria for EMU had the most restrictive impact on growth. Given Italy's situation of a high inflation rate and large fiscal deficits and levels of debt at the beginning of the 1990s, both monetary and fiscal policy had to be severely restricted in order to comply with inflation and fiscal convergence criteria. This generated a climate of uncertainty and pessimism that held back private consumption and investment. Fixed capital formation fell strongly at the beginning of the decade and it was only in 1998 that it regained its level of 1991. Consequently, during the first half of the decade, fixed investment subtracted 0.2 percentage points from GDP growth, while government consumption had a neutral impact. In this context of sluggish domestic demand, net exports were the main support of growth, contributing 1 percentage point to the average annual rate of 1.3 per cent.

---

[116] French inflation rates being lower than those of Germany, its real interest rates were higher.

[117] Although French private consumption recovered somewhat between 1995 and 2000, its contribution to GDP growth during the whole decade (0.8 percentage points annually) was one of the smallest among the industrialized economies.

[118] Deutsche Bundesbank, "The growth differential between Germany and France", *Monthly Bulletin* (Frankfurt am Main), August 2001, pp. 21-27.

Italy's economic performance improved towards the end of the 1990s, when its participation in the first phase of EMU was ensured and monetary policy was loosened. During the second half of the decade, private consumption and investment were the main supports of growth, with spending on equipment and software expanding at the same rate as in the other large continental European economies (6.1 per cent a year). In contrast, net exports were a drag on economic growth.

In contrast to the sluggish rates of expansion in France, Germany and Italy, the *United Kingdom* achieved the highest growth rate (2.3 per cent annually) among the large west European economies during the 1990s. Its macroeconomic developments resembled those of the United States, rather than those of continental Europe. The timing of monetary tightening and recession at the beginning of the decade was contemporaneous in the two "Anglo-Saxon" economies and the long British cyclical upswing (1993-2000) trailed that of the United States. The main factor behind this expansion in the United Kingdom was private consumption, which contributed 2.6 percentage points annually to growth during the second half of the decade. It was boosted by increases in employment, disposable household income and asset values and by a fall in the savings rate. Spending on equipment and software expanded at 8.5 per cent a year between 1995 and 2000, while investment in construction also grew strongly, in response to the steep rise in real estate prices. As in the United States, booming domestic demand was associated with the appreciation of the currency and growing current account deficits towards the end of the decade. While net exports had made a positive contribution to growth during the first half of the 1990s, during the second half they subtracted 1.1 percentage points from the growth rate.

Economic growth in the United Kingdom during the 1990s was highly correlated with that of the United States (with a correlation coefficient of 0.81) and very weakly with that of the other west European countries (the coefficient was just 0.26).[119] This was despite the fact that British trade links with other European economies are stronger than with the United States. In contrast, the financial links between the two Anglo-Saxon countries are more important than those between the United Kingdom and other west European countries, and these intensified during the 1990s.[120] Such links represent a strong channel of transmission of cyclical fluctuations and probably explain the close association of business cycles between the United Kingdom and the United States.

The smaller west European countries grew more rapidly than the large continental economies during the 1990s (except for Switzerland), although there were marked differences in their growth rates. The highest growth rates were in Ireland and Luxembourg. *Ireland* had by far the highest growth rate among the industrialized countries during the 1990s: 7.3 per cent a year, and accelerating to 9.9 per cent a year during the second half of the decade. It benefited from its strong specialization in the ICT sector, which was responsible for the rapid growth of exports. Thus, the external balance on goods and services contributed to 2.3 percentage points of GDP growth throughout the decade. Between 1995 and 2000 domestic demand strengthened thanks to the rapid expansion of fixed investment (by 14 per cent a year) and private consumption, which contributed 4.1 percentage points to annual growth. *Luxembourg* had the next highest growth rate among the industrialized countries during the 1990s: 5.9 per cent a year, the main source of growth being net exports, which were boosted by sales of high value added services (e.g. financial and business services).

In both *Finland* and *Sweden* asset bubbles burst in the early 1990s, leading to banking crises and recession. Between 1995 and 2000, however, these countries benefited from their relative specialization in the ICT sector, which was responsible for a rapid growth of exports from 1994. Net exports made a strong positive contribution to GDP growth during the second half of the 1990s, which in the Finnish case amounted to an annual rate of 1.5 per cent a year. The dynamism of the ICT sector stimulated strong investment in equipment, which in Sweden grew by 8 per cent annually. In turn, this led to increases in employment and household disposable income, so that private consumption was the main source of GDP growth in both countries. *Norway* maintained a strong growth rate throughout the decade (3.4 per cent a year). It was based on steady growth in private consumption, as well as investment in equipment, which expanded at 9.3 per cent annually between 1995 and 2000.

In the smaller south European economies – *Spain*, *Portugal* and *Greece* – economic performance had been sluggish between 1990 and 1995, but growth picked up vigorously between 1995 and 2000, reaching rates in excess of 3.4 per cent a year. This acceleration was based on domestic demand, with the large contribution of private consumption almost matched by that of fixed investment. Expenditure on both equipment and construction expanded at rates above 5 per cent, reflecting the need of these countries to upgrade their infrastructure and modernize industry.

---

[119] This is in contrast to the previous decade, when the British economy had moved more in tandem with the other west European countries (with a coefficient of correlation of 0.55) than with the United States (coefficient of 0.32).

[120] These financial links are established by cross-border short- and long-term capital flows (including financial assets such as equities and bonds as well as foreign direct investment). They are driven by the aims of international diversification of assets and liabilities and cross-border price arbitrage. IMF, "International linkages: three perspectives", *World Economic Outlook* (Washington, D.C.), October 2001, pp. 63-102.

During the first half of the 1990s, smaller countries close to Germany – the *Netherlands, Belgium* and *Austria* – benefited from the German reunification boom, as it greatly increased demand for their exports. Net exports provided a positive contribution to growth, except in Austria. Thereafter, however, the sluggish pace of activity in Germany dragged down the growth rate in Belgium and the Netherlands, which was further reduced by restrictive fiscal policies. The Netherlands, however, outperformed Germany by almost 2 percentage points annually between 1995 and 2000. The strong growth rate was based on domestic demand, mainly private consumption (boosted by the large increase in employment) and investment.

CHAPTER 3

THE TRANSITION ECONOMIES

In 2001 strong growth continued in most of the ECE transition economies. Rather unexpectedly, this part of the ECE zone turned out to be one of the fastest growing regions in the world in 2001, with aggregate GDP increasing by 5 per cent. This surprising buoyancy of output is a sign of the progress in transformation reforms, which has helped these economies build some resilience against adverse external shocks. Despite the favourable growth outcome, the situation in the labour markets of most transition economies remained tense and in a number of countries unemployment rates continued to grow. On the other hand, the disinflation efforts in these economies were facilitated by the falling international commodity and manufacturing prices. Notwithstanding the negative repercussions of the global and west European slowdown, the growth of exports from the transition economies remained relatively strong in 2001. This was partly due to gains in competitiveness, which allowed east European exporters to increase their share in EU markets and partly – especially in the case of the CIS – to increased trade among the transition economies. Although there was no notable improvement in the generally high current account deficits, in most cases these were easily financed, in part thanks to the continuing inward flows of FDI.

Against the background of a global economic downturn, the outcome for the year in terms of economic growth turned out to be surprisingly good for the ECE transition economies and, especially, for the CIS and the Baltic states. Total GDP in the transition economies as a whole increased by 5 per cent in 2001, a respectable outcome given the circumstances, albeit 1.5 percentage points lower than the rate achieved in 2000. With an average increase in GDP of 6.2 per cent, the CIS was one of the fastest growing regions in the world, with half of the CIS economies growing at annual rates above 9 per cent. The rate of GDP growth was the same (6.2 per cent) also in the Baltic region, with all three economies growing by more than 5 per cent. In most of the east European transition economies, GDP growth was above 3 per cent, considerably higher than the rates in western Europe, but the average for the region (3 per cent) was pulled down by the mediocre performance of the largest regional economy, Poland, where GDP increased by just over 1 per cent. The former Yugoslav Republic of Macedonia was the only transition economy where GDP fell in 2001, but this was mainly due to the negative impact of the internal conflict.

Given the ever closer trade links between the transition economies and the rest of the world, especially western Europe, the strong negative shock of the severe downturn in the global economy had a surprisingly muted effect on their economic performance. For the first time since the start of economic transformation, the transition economies as a whole have demonstrated considerable resilience to such a negative external disturbance. Although some of the repercussions of the global slowdown may still show up in the form of aftershocks, the outcomes in 2001 are encouraging signs of the progress that these countries have achieved in strengthening their institutions and reforming their economies.

3.1 Macroeconomic policy

(i) Monetary policy

In 2001, policy makers in the transition economies, and especially those in eastern Europe, were faced with the challenge of responding to the effects of the slowdown in the world economy which, *inter alia*, presented the commodity exporting CIS countries with faltering global demand and falling commodity prices. With hindsight, it is clear that governments in the transition economies were not prepared for such a rapid deterioration in the macroeconomic environment, especially against the background of the benign external conditions that prevailed in 2000. Yet, judging by the relatively favourable outcomes for 2001, policy makers in general appear to have done well: most transition economies not only avoided an outright recession but also maintained high (in some cases even accelerating) rates of growth.

The accession negotiations with the EU are now dominating the policy agenda in the 10 transition economies that are seeking EU membership. The process of intensive legislative and regulatory reforms in the candidate countries aimed at harmonizing their economic and business environment with that in the EU has already reached a fairly advanced stage. The European Council meeting in Laeken on 14 and 15 December 2001 acknowledged that these east European and Baltic transition economies have made considerable progress towards accession and reiterated that a large group of countries might be able to complete negotiations by the end of 2002, which would pave the way to full EU membership by 2004. These clear signals have been highly beneficial for business and investor confidence (in particular, for inward FDI flows) and were probably one of the factors that helped eastern Europe to weather the negative repercussions of the global slowdown.

Monetary conditions in the transition economies in 2001 and in early 2002 were dominated by two countervailing developments. On the one hand, the slump in world commodity and some manufacturing prices considerably alleviated supply-side cost pressures (section 3.3) thus facilitating the monetary authorities in the pursuit of their main objective, price stability. On the other hand, the major financial crisis in Argentina has renewed concerns among investors about macroeconomic and financial stability in emerging markets in general. Although, as discussed in section 3.1(v), there has been no direct fallout for the transition economies so far, the repercussions of the Argentine crisis are a potential source of macroeconomic and financial volatility for these economies and requires increased caution on the part of their monetary authorities. On balance, most central banks in the transition economies comfortably met (and in some cases even overshot) their monetary targets for 2001; however, the growing emerging market risks pose new challenges for the year 2002.

In the course of 2001 and early 2002 a number of important changes were made in the monetary regimes of the transition economies. In the case of the candidate countries these, together with other regulatory changes, were steps in their preparation for EU and EMU accession. In May 2001, the National Bank of Hungary widened the exchange-rate band within which the forint is permitted to fluctuate against the euro from ±2.25 to ±15 per cent. In June the Hungarian authorities lifted the remaining restrictions on short-term, cross-border capital movements, thus making the currency fully convertible. In a further move, the crawling devaluation of the forint was discontinued in October (before that the monthly devaluation rate was reduced from 0.3 to 0.2 per cent in April). The resultant euro peg with a 15 per cent fluctuation band in effect is equivalent to a unilateral commitment to adhere to a monetary regime, which is fully compatible with the EU's ERM-2.

On 1 February 2002, the Lithuanian litas was re-pegged from the dollar to the euro without any further ramifications for the country's currency board arrangement. This transition had been announced well in advance, in line with the monetary strategy for EU accession, and was carried out remarkably smoothly. As of January 2002, the National Bank of Romania, which follows the policy of a managed float, switched from targeting the dollar to targeting a 50:50 dollar-euro basket. This change was mainly aimed at reducing the impact of the dollar's appreciation on Romania's exporters (as the EU is Romania's main trading partner) but it also reflects the country's long-term monetary strategy.

Several east European countries changed their regulations relating to minimum reserve requirements. The central banks in Hungary, Poland and Slovakia reduced the required reserve ratios in line with their strategy of gradually harmonizing these regulations with those of the EU.[121] In addition, the Czech National Bank started paying interest on the mandatory reserves of commercial banks while the National Bank of Hungary increased the interest rate on these deposits.[122]

Within the CIS region, the National Bank of Belarus switched in January 2001 to targeting the Russian rouble,[123] which is an element of the joint Russian-Belarussian policy effort to establish a monetary union and adopt a single currency in 2005. The authorities in Uzbekistan intend to make the som convertible for current account transactions in 2002, which is one of the main preconditions for the resumption of IMF funding. The Russian authorities have further liberated the foreign exchange market: in particular, the share of mandatory sales of export revenues by local firms on the Moscow currency market was reduced from 75 to 50 per cent in mid-2001. In addition, in October, Russia's central bank eased the restrictions on borrowing from abroad by local residents and firms.

The general deceleration of inflation in 2001 – often faster than was expected – and, in some cases, an upward pressure on exchange rates caused by capital inflows, prompted the central banks in a number of east European transition economies to ease the stance of monetary policy by lowering interest rates.[124] This trend was especially pronounced in those central European countries with flexible exchange rate regimes and also (with the exception of Slovenia) to inflation targets.

---

[121] In Hungary the mandatory reserve ratio (the share of deposits that commercial banks have to redeposit with the central bank) was reduced from 7 to 6 per cent in July; in Poland the rate was cut from 5 to 4.5 per cent in December and the National Bank of Slovakia lowered its minimum reserve requirement from 5 to 4 per cent in January 2002. The mandatory reserve ratio was also reduced in Romania in 2001 (from 30 to 25 per cent) but it still remains considerably higher than those in the leading accession candidate countries.

[122] In Hungary the annual interest rate on mandatory reserves was increased from 3.25 to 4.25 per cent while the Czech National Bank undertook to pay interest equal to the two-week repo rate. Both changes were introduced in July.

[123] The policy target for 2001 was to prevent a monthly nominal depreciation of the Belarussian rouble against the Russian rouble greater than 3 per cent.

[124] There were some exceptions to this pattern. Thus, the Croatian National Bank tightened monetary policy in the autumn of 2001 aiming chiefly to strengthen the kuna. Among the policy changes were an increase in the Lombard rate from 9.5 to 10.5 per cent and a ruling that banks have to keep a portion (20 per cent as of October) of their mandatory reserves on foreign exchange deposits in kuna.

Thus, between January 2001 and February 2002 the Czech National Bank lowered its key interest rates by 1 to 2 percentage points;[125] during the same period the National Bank of Hungary reduced its base rates by more than 3 percentage points;[126] and the National Bank of Poland, in seven consecutive moves, cut its main interest rates by some 9 to 10 percentage points.[127]

Somewhat paradoxically, this monetary relaxation did not generally result in cheaper credit in real terms; on the contrary, in four of the five central European countries (Slovenia being the exception) real lending rates rose on average in 2001 (table 3.1.1). The main reason for this was the unexpectedly fast rate of disinflation, especially for producer prices (table 3.3.2). The extreme case was Poland, where despite a series of interest rate cuts by the central bank, real lending rates rose on average by more than 5 percentage points. The persistence of high interest rate differentials led to large capital inflows and a considerable real appreciation of the zloty in 2000-2001 (chart 3.1.1). Thus counter to the intention behind the easing of monetary policy, the monetary stance in Poland actually tightened in 2001 and this contributed to the weakening of economic activity (for further discussion see section 3.1(iii)). For similar reasons, real lending rates increased in 2001 in the other east European countries as well (table 3.1.1). Nevertheless, as shown in table 3.1.2, the volume of credit to the non-government sector kept growing at a high rate in Hungary, Slovenia, Estonia, Latvia and, to some extent, in Croatia and Poland (in the latter case it was mainly corporate credit). By contrast, credit to the non-government sector continued to shrink as a share of GDP in the Czech Republic and Slovakia, in both cases a consequence of the on-going restructuring of the banking sector.[128]

In the CIS region the situation was mixed, with very high real lending rates in some economies alongside negative rates in others (table 3.1.1). In the main this reflects a wide divergence in the stance of monetary policy: while some central Asian and Caucasian countries maintained a relatively tight stance of monetary policy, others such as Belarus have been persistently lax. A fast rate of disinflation allowed the National Bank of Ukraine to reduce substantially its refinancing rate in 2001[129] but, as in eastern Europe, real lending rates not only did not ease but climbed even further (table 3.1.1). It should be noted, however, that most CIS economies are still characterized by very low levels of monetization (table 3.1.2) and credit to the non-government sector has a smaller impact on economic activity than it does in eastern Europe.

In general real lending rates in the transition economies still remain relatively (in some cases excessively) high, which is an obstacle to business activity, investment and growth in these countries. On the other hand, real deposit rates in a number of economies were negative in 2001 (table 3.1.1), which may be a disincentive to private savings, especially as in some cases this has been chronic. Several key factors determine this relative positioning of real lending and real deposit rates. Firstly, in most transition economies the spread between nominal lending and deposit rates remains excessively high (table 3.1.1). This is due to the high risk premia demanded by commercial banks (which, in turn, reflects the perceived level of business risk) and, in some countries, to the poor average quality of bank assets (when bank portfolios are overburdened with substandard and non-performing loans, banks try to recover part of their losses by charging higher costs to their good clients). The large spreads between real lending and real deposit rates also reflect the fact that the consumer price index (CPI) (used to discount deposit rates) generally grows faster than the producer price index (PPI) (used to discount lending rates). This is due to both the continuing adjustment in relative prices (with prices of services which are an important component of CPI generally rising faster than the prices of goods) and to the Balassa-Samuelson effect, which implies that in a fast-growing, catching-up economy the prices of non-tradeable goods (with a large weight in the CPI) tend to rise faster than the prices of tradeable goods (closely correlated with the PPI).[130] Thus the large gap between real lending and deposit rates should be seen as a relatively persistent and structural phenomenon in the transition economies and policy makers need to take it into account when formulating their policy goals.

The exchange rate regimes in the transition economies were generally stable in 2001 with no major realignments. The above-mentioned regime changes were carried out smoothly and did not produce excessive volatility on the currency markets. Real effective

---

[125] Over this period the two-week repo rate fell from 5.25 to 4.25 per cent; the discount rate from 5 to 3.25 per cent, and the Lombard rate from 7.5 to 5.75 per cent.

[126] The key two-week deposit rate was cut from 11.75 to 8.5 per cent.

[127] During this period the 28-day intervention rate was cut from 19 to 10 per cent, the discount rate from 21.5 to 12 per cent and the Lombard rate from 23 to 13.5 per cent.

[128] In recent years, the governments of both countries undertook large-scale financial operations by restructuring some financially troubled banks with the aim to rehabilitate the banking systems as a whole. In Slovakia, as a result of the large one-off issue of loan portfolio restructuring bonds in January 2001, net credit to the government increased by 83.7 billion koruny while the stock of outstanding credit to enterprises and households dropped by 72.3 billion koruny (this corresponds to the volume of loans that was were written off the books of the restructured banks and transferred to the Slovak Consolidation Agency). National Bank of Slovakia, *Monetary Survey*, January 2001, p. 7. Consequently, the volume of credit to the non-government sector dropped as a proportion of GDP by more than 10 percentage points in 2001 (table 3.1.2).

[129] From 27 per cent in January to 12.5 per cent in December.

[130] For details see UNECE, "Economic transformation and real exchange rates in the 2000s: the Balassa-Samuelson connection", *Economic Survey of Europe*, 2000 No. 1, chap. 6, pp. 227-239.

TABLE 3.1.1

Short-term interest rates in selected east European, Baltic and CIS economies, 1999-2001

(Per cent)

	Short-term credits						Short-term deposits						Average yield on short-term government securities		
	Nominal			Real			Nominal			Real					
	1999	2000	2001	1999	2000	2001	1999	2000	2001	1999	2000	2001	1999	2000	2001
Albania	24.6	26.1	16.0	..	..	..	12.9	8.3	7.7	12.5	8.3	4.5	16.9	10.4	7.7
Bulgaria	13.6	12.2	11.7	9.9	-4.3	4.4	3.3	3.2	3.1	0.6	-6.5	4.0	5.5	4.3	4.7
Croatia	14.9	12.1	9.6	12.0	2.2	5.9	4.3	3.7	3.2	0.8	-1.5	-1.9	11.1	9.2	5.9
Czech Republic	8.7	7.2	7.0	7.6	2.1	4.0	4.5	3.4	3.0	2.3	-0.5	-1.6	..	..	..
Hungary	16.3	12.6	12.1	10.7	0.8	5.5	13.3	9.6	9.3	3.0	-0.2	0.1	14.7	10.9	10.8
Poland	17.1	20.0	18.5	10.7	11.3	16.4	11.2	13.9	9.9	3.6	3.5	4.2	13.1	16.6	15.4
Romania	65.5	53.8	45.4	16.4	1.5	2.2	46.0	32.9	26.6	0.1	-8.8	-5.9	74.6	52.3	42.2
Slovakia	19.6	13.6	11.2	15.2	3.5	4.4	10.5	7.2	5.1	-0.1	-4.2	-2.0	..	..	..
Slovenia	12.4	15.8	15.1	10.1	7.6	5.7	7.2	10.1	9.8	1.1	1.1	1.3	3.3	..	..
The former Yugoslav Republic of Macedonia	20.4	18.9	19.2	20.6	9.2	16.9	11.4	11.2	10.0	12.2	5.1	4.3	..	..	..
Estonia	11.1	7.8	7.8	12.5	2.8	3.2	4.2	3.8	4.1	0.9	-0.3	-1.6	..	..	..
Latvia	14.2	11.9	11.1	19.0	11.2	9.2	5.0	4.4	5.2	2.6	1.7	2.7	6.2	3.8	..
Lithuania	13.1	12.1	9.6	16.9	11.6	10.2	7.4	6.7	4.6	6.6	5.6	3.2	6.2	5.9	5.3[a]
Armenia	38.8	31.6	26.7	35.7	30.5	27.2	26.6	18.1	14.9	25.8	19.0	11.4	53.3	23.6	19.9
Azerbaijan	20.3	..	..	32.6	..	..	11.4	..	..	21.7	..	..	18.3	16.7	16.4
Belarus	51.0	67.7	47.2	-66.9	-41.3	-14.4	23.8	37.6	34.7[b]	-68.6	-48.8	-16.4[b]	..	..	..
Georgia	33.4	32.8	27.3	15.3	25.5	22.8	14.6	10.2	7.8	-3.7	5.9	2.9	..	..	..
Kazakhstan	23.7	19.4	..	4.1	-13.5	..	17.0	16.5	..	8.1	2.9	..	20.7	11.6	..
Kyrgyzstan	59.6	57.0	40.5	3.6	19.1	26.3	31.4	23.9	13.7	-3.3	4.4	6.4	47.2	32.3	19.0
Republic of Moldova	35.3	33.8	28.7	-6.1	4.1	14.6	27.5	24.9	21.2	-8.3	-4.7	10.6	28.5	22.3	..
Russian Federation	39.7	24.4	18.0[b]	-16.5	-5.4	-0.9[b]	13.7	6.5	4.8[b]	-16.7	-11.4	-13.8[b]	..	13.6	13.1
Ukraine	55.0	41.5	32.5	33.9	17.2	22.7	20.7	13.7	11.0	-1.6	-11.3	-0.9	..	..	..

Source: National statistics and direct communications from national statistical offices to the UNECE secretariat; IMF, *International Financial Statistics* (Washington, D.C.), various issues.

Note: Definition of interest rates:

Credits – Belarus: weighted average rate on short-term loans; Bulgaria: average rate on short-term credits; Croatia: weighted average rate on new credits; Czech Republic: average rate on total short-term loans; Estonia: weighted average rate on short-term loans; Hungary: weighted average rate on loans of less than one year; Latvia: average rates on short-term credits; Lithuania: average rates on loans of one to three months; Poland: weighted average rate on low-risk short-term loans; Romania: average short-term lending rate; Kazakhstan: weighted average interest rates (for new credits); Kyrgyzstan: weighted average rate on loans in sums for one- to three-month maturities; Russian Federation: weighted average rate on loans of up to one-year maturity; Slovakia: average rate on new short-term loans; Slovenia: average rate on short-term working capital loans; The former Yugoslav Republic of Macedonia: median rates for short-term loans to all sectors; Ukraine: weighted average rate on short-term loans. The real lending rates are the nominal rates discounted by the average rate of PPI for the corresponding period.

Deposits – Belarus: weighted average rate on short-term deposits; Bulgaria: average rates on one-month time deposits; Croatia: weighted average rate on new deposits; Czech Republic: average rate on short-term time deposits; Estonia: weighted average rate on short-term deposits; Hungary: weighted average rate on deposits fixed for more than one month, but less than one year; Latvia: average rates on short-term deposits; Lithuania: average rates on deposits of one to three months; Poland: weighted average rate (according to information collected from 15 biggest commercial banks) on short-term household deposits in domestic currency; Romania: average short-term deposit rate; Kazakhstan: weighted average interest rates (for new deposits); Kyrgyzstan: weighted average rate offered on time deposits of three-month maturities; Russian Federation: prevailing rate for time deposits with maturity of less than one year; Slovakia: average rate on time deposits; Slovenia: average rate on time deposits of 31-90 days; The former Yugoslav Republic of Macedonia: lowest reported interest rate on household deposits with maturities of three to six months; Ukraine: weighted average rate on short-term deposits. The real deposit rates are the nominal rates discounted by the average rate of CPI for the corresponding period.

Yields of government securities – Bulgaria: yield on government securities is computed as the average weighted yield of all issues during the calendar month; Croatia: interest rate on NBC bills, due in 91 days; Hungary: weighted average yield on 90-day treasury bills sold at auction; Poland: yield on bills purchased, weighted average, 13 weeks; Romania: rate on 91-day treasury bills; Slovenia: BS tolar bills, 14 days overall nominal rate; Latvia: weighted average auction rate on 91-day treasury bills; Lithuania: average auction rate on treasury bills with maturity of 91-days; Kazakhstan: yield based on treasury bill prices established at the last auction of the month; Kyrgyzstan: weighted average rate on three-month treasury bills sold in the primary market; Russian Federation: weighted average rate on government short-term obligations (GKO) with remaining maturity of up to 90 days.

[a] January-September.

[b] January-November.

exchange rates in eastern Europe followed divergent trends in 2001 (chart 3.1.1), reflecting both the realignment of the major world currencies among themselves and the main macroeconomic trends in the transition economies. In most cases (with the exception of Slovenia), real effective exchange rates appreciated in 2001, in line with the general trend of recent years. In a medium-term perspective, real exchange rate appreciation in the transition economies is an equilibrium phenomenon, reflecting the Balassa-Samuelson effect.[131] In some cases, especially among the central European countries, this process was reinforced in 2001 by a continuing surge in capital inflows; the attempts of central banks (particularly that of the Czech Republic) to intervene in order to prevent further appreciation were not sufficient to arrest this process.

---

[131] Ibid.

The Transition Economies

CHART 3.1.1
Real effective exchange rates in selected east European and Baltic economies, 1995-2001
(Indices, first quarter 1995=100)

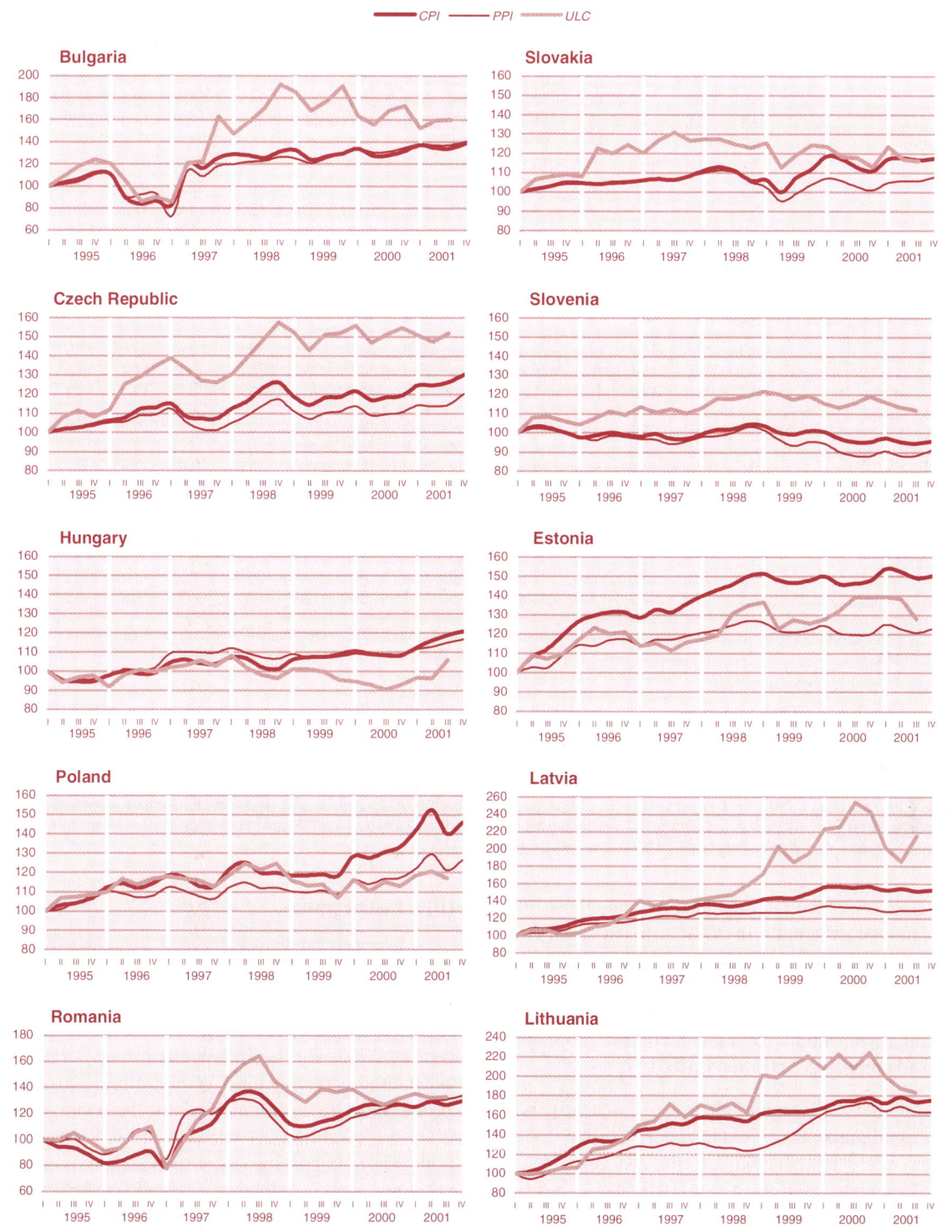

Source: National statistics; UNECE Common Database; OECD, *Main Economic Indicators* (Paris), various issues.

Note: The real effective exchange rates were computed from the nominal exchange rates against the deutsch mark and the dollar, deflated respectively by the domestic and German or United States consumer and producer price indices, and by indices of estimated unit labour costs in industry, while the shares of the EU and the rest of the world in total exports of individual transition economies were used to determine the deutsch mark and the dollar trade weights, respectively. Indices of unit labour costs in industry are computed from seasonally adjusted rates of average quarterly gross wages in industry and of estimated labour productivity. Unit labour costs for Germany and the United States are from the OECD Database.

TABLE 3.1.2

Monetization in selected east European, Baltic and CIS economies: share of monetary aggregates [a] in GDP, 1997-2001

(Per cent)

	M1 [b]					Total broad money [c]					Total credit [d]				
	1997	1998	1999	2000	2001 [e]	1997	1998	1999	2000	2001 [e]	1997	1998	1999	2000	2001 [e]
Albania	27.7	18.0	17.8	20.4	21.3	52.3	47.8	53.5	56.7	60.1	4.4	3.8	3.7	4.3	4.8
Bulgaria	6.5	10.1	10.7	12.1	13.5	24.9	28.1	28.3	31.9	34.7	17.4	16.4	16.8	16.6	14.8
Croatia	9.9	9.6	9.2	9.7	11.2	35.7	39.0	38.7	40.4	48.6	32.9	39.9	40.8	36.4	39.5
Czech Republic	25.2	21.8	23.5	26.3	27.6	68.1	66.3	69.6	72.4	74.2	64.0	61.4	57.3	52.2	44.5
Hungary	14.7	15.2	15.9	16.1	16.1	35.6	41.2	42.9	42.7	42.9	20.8	22.3	23.0	26.1	29.7
Poland	13.8	13.4	14.5	13.5	12.8	41.0	43.2	48.0	48.7	50.7	20.3	22.7	25.9	28.5	29.7
Romania	5.0	4.7	4.1	4.1	4.1	18.3	19.2	20.0	19.0	19.2	14.8	13.3	12.2	11.8	11.1
Slovakia	23.0	20.3	17.5	18.0	19.9	61.2	60.3	60.2	63.6	64.4	53.3	50.4	49.5	46.3	33.8
Slovenia	7.9	8.9	9.9	9.6	9.5	42.5	47.5	49.2	50.9	55.0	26.3	28.5	32.3	35.9	38.0
The former Yugoslav Republic of Macedonia	6.4	7.0	8.2	8.3	9.0	12.0	13.0	15.3	17.1	19.0	25.6	22.1	18.3	17.7	18.0
Yugoslavia	6.4	6.1	6.8	5.7	..	31.4	35.6	34.6	22.2	..	..	..	..	..	..
Estonia	19.3	17.8	19.4	22.4	23.3	26.7	27.9	30.6	35.6	38.6	25.0	33.0	31.9	35.4	41.1
Latvia	14.1	16.0	15.1	15.5	16.2	22.5	25.4	24.1	26.3	28.2	8.9	14.4	15.8	18.0	23.1
Lithuania	10.7	11.9	12.4	11.4	11.8	16.3	17.6	20.3	20.9	23.4	10.4	12.1	14.9	13.5	12.7
Armenia	5.1	4.8	4.8	5.2	5.7	7.8	8.6	10.1	11.9	13.4	5.5	6.6	8.8	9.9	8.5
Azerbaijan	7.7	7.6	5.9	..	5.4	11.2	11.6	12.9	9.6	9.9	12.1	12.4	12.6	8.0	7.4
Belarus	6.5	7.1	5.0	4.0	4.1	11.7	13.2	11.4	11.6	12.8	6.3	7.4	6.4	4.8	7.1
Georgia	5.0	5.0	4.6	5.0	5.5	6.6	7.3	7.6	8.7	10.0	3.7	5.3	6.9	7.9	8.9
Kazakhstan	6.3	7.2	6.5	8.0	8.0	..	8.9	9.5	12.4	14.2	4.5	5.9	7.5	9.2	12.5
Kyrgyzstan	9.6	9.2	7.9	6.5	6.0	12.1	13.7	12.4	10.7	9.4	2.8	4.8	4.8	3.9	3.6
Republic of Moldova	12.5	12.1	11.0	10.4	11.2	18.7	19.7	19.0	18.7	21.5	17.6	18.4	13.8	12.1	13.9
Russian Federation	9.5	10.2	8.6	9.4	10.9	17.0	17.4	16.3	17.9	20.0	11.1	11.1	9.5	10.3	13.0
Ukraine	8.5	9.0	9.2	9.9	11.5	11.8	13.1	13.9	15.4	17.4	7.5	8.2	8.5	10.0	12.5

Source: National statistics and direct communications from national statistical offices to the UNECE secretariat; IMF, *International Financial Statistics* (Washington, D.C.), various issues.

[a] Averages of monthly or quarterly figures.

[b] Currency in circulation plus demand deposits.

[c] M1 plus time deposits in domestic currency and foreign currency deposits.

[d] Total outstanding claims on firms and households (except claims on government).

[e] January-June for Albania, The former Yugoslav Republic of Macedonia and Republic of Moldova; January-September for Armenia, Azerbaijan, Belarus, Czech Republic, Georgia, Kazakhstan, Kyrgyzstan and Ukraine; full year for the rest of the countries. GDP data for 2001 are based on preliminary reports by national statistical offices, wherever available, otherwise they are based on UNECE secretariat estimates.

Real effective exchange rates deflated by unit labour costs generally moved differently in 2001: in most cases (with the possible exception of Hungary) they declined, suggesting beneficial shifts in productivity and cost structures in the transition economies. On balance, there was an improvement in the east European economies' international competitive position in 2001.

Within the CIS region, there have recently been important realignments of exchange rates due to the continuing real appreciation of the Russian rouble (discussed in more detail in section 3.1(iv)). As a result, most of the CIS exchange rates have been depreciating in real terms against the Russian rouble. As shown in chart 3.1.2, the cumulative real depreciation during the last three years ranges between 20 and 45 per cent. This boost to competitiveness (attributable to the real depreciation of national currencies), together with buoyant domestic (and import) demand in Russia, have been important factors for the continuation and strengthening of the recovery throughout the rest of the CIS region.

(ii) Fiscal policy

In 2001 the transition economies made further progress towards fiscal consolidation. The unweighted average of the general government deficits[132] in the region as a whole was 2.7 per cent of GDP against 3.3 per cent in 2000.[133] The improvement was largely due to a substantial strengthening of the fiscal position in the Baltic and CIS economies where the average deficits in 2001 were 1.4 and 2.1 per cent of GDP, respectively, in both cases improving by 1 percentage point from the previous year. However, fiscal performance across subregions and countries was uneven. Thus, the average

---

[132] Throughout this section the fiscal position in the transition economies is measured by the balance of the consolidated general government, which includes the central government, the regional governments and any extrabudgetary public funds. For more details, see box 3.1.1, the note to table 3.1.3 and UNECE, *Economic Survey of Europe, 2000 No.1*, p. 51, box 3.2.1.

[133] Calculated on the basis of the data shown in table 3.1.3. The data for 2001 reported in this table are preliminary and incomplete, and subject to revision.

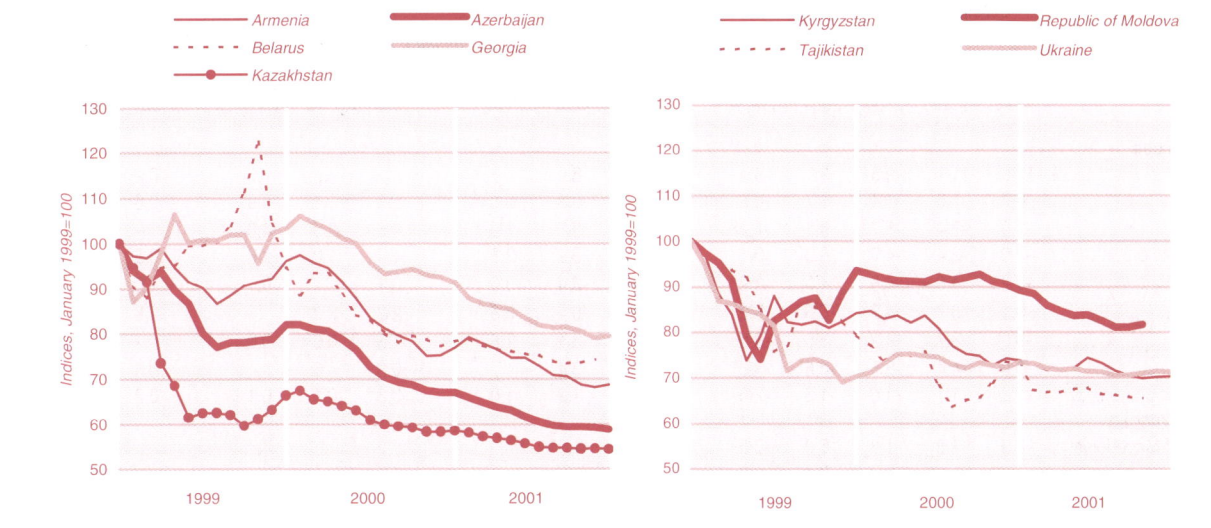

CHART 3.1.2

Real exchange rates against the Russian rouble in selected CIS economies, 1999-2001

(Indices in per cent)

Source: UNECE secretariat calculations, based on national statistics.

fiscal balance in eastern Europe widened in 2001 to -4.5 per cent of GDP from -4.2 per cent in 2000 mostly due to a deterioration in The former Yugoslav Republic of Macedonia and Yugoslavia but also in some central European countries (the Czech Republic and Poland).

Chart 3.1.3 indicates that the number of countries where the fiscal position improved in 2001 (those that lie above the x-axis) was about the same as that where it worsened (those below the x-axis).[134] In this sense the improvement was not as widespread as in 2000. On the other hand, given the negative impact of the global slowdown and the generally unfavourable external environment, the fiscal outcomes on the whole can be regarded as relatively successful. It should also be noted that due to the existing macroeconomic constraints, policy makers in the transition economies have limited room for manoeuvre to counteract negative external shocks with fiscal measures. As discussed below, Hungary and to some extent the Czech Republic were the only transition economies to adopt an openly counter-cyclical fiscal policy in 2001.

The recent stance of fiscal policy in the transition economies is illustrated in chart 3.1.4. The very weak correlation between the changes in fiscal position and in domestic absorption in 2001 (as suggested by the scatter diagrams) and the fact that changes in the absorption ratio were almost evenly spread between positive and negative suggest that fiscal policy in the transition economies generally maintained a neutral stance in 2001 (with some exceptions discussed below). This stance was one of the factors that allowed for more flexibility in the conduct of monetary policy in 2001. This was in contrast to the situation in 2000 when fiscal policy was more active (and on the whole restrictive) in a number of transition economies while monetary policy remained broadly neutral.

The recent mediocre fiscal performance in some of the most advanced reforming economies (such as the Czech Republic, Poland, Slovakia[135] and, to some extent, Hungary and Latvia) calls for some comment. On the one hand, it should be borne in mind, that a number of the reforms already implemented in these economies (such as the comprehensive pension and health care reforms and the overhaul of other institutional systems, a process which is well advanced in the leading reform countries) are costly and sometimes involve large one-off claims on public funds. The same is true for some of the measures that the accession candidates have to undertake in the process of policy and institutional harmonization with the EU, in particular, complying with EU regulatory norms and standards. The new NATO members, as well as the candidates for membership of the alliance, are also required to incur additional public spending in order to upgrade their defence systems to NATO standards. All these demands put a considerable strain on the public finances despite the fact that some of these countries (particularly the Czech Republic, Hungary and Slovenia) are able to raise relatively large amounts of public revenue as a proportion of GDP (table 3.1.4).

---

[134] Due to the absence of data on the structural budget deficits in the transition economies, cash deficits are used in charts 3.1.3 and 3.1.4 as proxies for the fiscal policy stance.

[135] It should be noted that in the case of Slovakia the figures for 2001 refer to the non-consolidated central budget only, and thus are not comparable with previous years. The large consolidated budget deficit for 2000 (-8.5 per cent) includes one-off extrabudgetary expenditures, which were covered by privatization revenues in the same year (14.1 billion koruny from the sale of a state owned commercial bank and 26.3 billion koruny from the sale of Slovak Telecom).

TABLE 3.1.3

Consolidated general government deficits and their sources of financing in eastern Europe, the Baltic states and the CIS, 1998-2002
(Per cent of GDP)

	Consolidated general government deficit/surplus[a]					Financing of consolidated general government deficit by components								
					2002	Borrowing			Privatization receipts			Other capital receipts		
	1998	1999	2000	2001*	target	1999	2000	2001	1999	2000	2001	1999	2000	2001
Eastern Europe														
Albania	-10.6	-11.1	-9.8	-9.1	-8.4	10.9	8.0	..	..	..	..	0.2	1.8	..
Bulgaria	0.7	-1.6	-1.4	-1.7	-0.8	-1.4	-0.4	0.6	2.3	1.4	0.8	0.6	0.5	0.2
Croatia	-3.0	-7.4	-5.7	-5.2	-4.2[b]	2.9	3.7	1.2	4.5	2.0	4.0	–	–	–
Czech Republic	-2.9	-2.4	-4.7	-6.0	-9.1	0.6	3.2	2.8	1.4	1.0	2.8	0.5	0.5	0.4
Hungary	-7.2	-4.6	-4.5	-4.0	-3.2[b]	3.6	3.4	3.0	0.1	–	0.1	1.0	1.0	0.9
Poland	-3.1	-3.7	-4.3	-5.5[b]	-5.2[b]	0.8	-0.2	3.5	2.2	3.9	..	0.8	0.5	..
Romania	-5.5	-3.7	-4.2	-3.5	-3.0[b]	2.4	3.4	2.8	1.3	0.6	0.6	–	–	–
Slovakia	-6.0	-4.4	-8.5	-4.5[b]	-3.5[b]	3.6	3.4	..	0.3	4.6	..	0.5	0.6	..
Slovenia	-1.2	-1.1	-1.6	-1.3	-2.6	0.7	1.2	1.0	0.3	0.1	0.1	0.2	0.2	0.2
The former Yugoslav Republic of Macedonia	-1.9	-1.6	-1.3	-6.0	-3.5[b]	-1.6	1.3	..	..	..	..	..	..	..
Yugoslavia	..	..	-0.3	-2.4	-6.2	..	..	..	..	..	..	..	..	..
Baltic states														
Estonia	-1.6	-5.0	-1.3	-0.5	–[b]	4.6	0.7	-0.4	–	–	–	0.5	0.5	–
Latvia	0.1	-4.0	-5.2	-4.3[b]	-2.7[b]	4.0	2.8	1.8	–	2.4[c]	2.5[c]	–	..	..
Lithuania	-5.5	-8.4	-3.1	-1.5	-1.5	7.4	1.4	..	1.0	1.7	..	–	0.1	..
CIS														
Armenia	-5.9	-10.1	-7.8	-4.6[b]	-3.4[b]	9.0	5.8	..	1.1	2.1	..	–	–	..
Azerbaijan	-1.9	-4.5	-2.2	-1.6[b]	-1.2[b]	4.5	2.2	..	..	..	..	..	..	..
Belarus	-1.0	-2.1	-0.1	-1.4	-1.5[b]	1.7	-0.3	1.1	0.1	0.1	0.1	0.2	0.3	0.3
Georgia	-6.2	-6.7	-4.1	-1.2	-3.1[b]	5.8	3.7	1.1	0.9	0.3	0.2	–	–	–
Kazakhstan	-7.9	-5.3	0.8	2.9	1.5	3.5	-1.8	..	..	..	..	1.8	1.0	..
Kyrgyzstan[d]	-2.7	-2.5	-9.3	-5.0	-4.8	1.6	9.5	5.3	0.5	-0.3	-0.4	0.3	–	..
Republic of Moldova	-4.1	-4.2	-1.9	-2.6[b]	-1.4[b]	3.0	1.0	..	1.1	0.9	..	..	..	..
Russian Federation	-6.1[e]	-0.8[e]	1.9[e]	2.9[f]	1.6[g]	0.3	..	..	0.6[c]	0.7[c]	..	..	..	..
Tajikistan	-1.0	-3.1	-0.6	-0.5	..	2.3	-0.4	0.1	0.8	1.0	0.4	–	..	..
Turkmenistan	-6.0	..	..	-1.2[b]	-0.8[b]	..	..	..	..	..	..	..	..	..
Ukraine	-2.7	-2.4	-1.3	-1.7	-1.8	1.7	-0.1	–	0.6	1.3	1.7	–	–	..
Uzbekistan	-3.4	-2.2	..	-2.5[b]	-3.5[b]	..	..	..	..	..	..	..	..	..

Source: UNECE secretariat estimates and calculations, based on direct communications from national Ministries of Finance and IMF data.

Note: The consolidated general government deficit, or financing requirement, is defined here as (current revenue and grants) − (current and capital expenditure plus net lending for policy purposes). A deficit is negative, a surplus is positive. With this definition of the deficit, it follows that privatization and other capital receipts are components of financing, not of revenue. The three components of financing (borrowing, privatization and other capital receipts) sum to the general government deficit (or surplus) with the opposite sign. Where the borrowing item is negative, this indicates net repayment of government debt. The "IMF" method of the IMF Fiscal Affairs Division is generally to treat only privatization receipts, but not other capital receipts, as financing. Thus the "IMF methodology deficit" is normally equal to the general government deficit plus other capital receipts. The general government deficit here is closest to the present definition of the "Maastricht criterion", as presently interpreted by Eurostat. The "IMF-GFS" method, frequently cited by national sources, defines the general government deficit as in the first panel of this table but national practices may differ. Deficits projected at the start of 2002 are official budget deficits, forecast in the initial budget proposals, necessarily involving GDP and inflation projections as well as fiscal data. The definitions of the projected deficits as well as some of the preliminary estimates of the deficits in 2000 may differ from the above definition. Sources are national Ministries of Finance, official press releases from Reuters, IMF publications and country information (www.imf.org/external/country) and official websites of Ministries of Finance.

[a] For some countries the deficit shown in this table may not equal the difference between the revenue and expenditure shown in tables 3.1.4 and 3.1.5. The most current aggregate information was used for table 3.1.3, while the more detailed breakdown by type of revenue or expenditure in tables 3.1.4 and 3.1.5 in some cases had to be based on more dated information.

[b] Central government deficit/surplus.

[c] Including other capital receipts.

[d] The officially reported deficit for Kyrgyzstan in 1998 and 1999 does not include the national public investment programme. According to IMF estimates, if expenditure under this programme were included in the fiscal accounts, the consolidated government deficit would be as follows, 1998: -9.5 per cent; and 1999: -12.0 per cent.

[e] Consolidated central government (including social security and extrabudgetary funds) plus (without consolidation) regional and local government.

[f] Without final adjustments.

[g] Federal government.

The Transition Economies

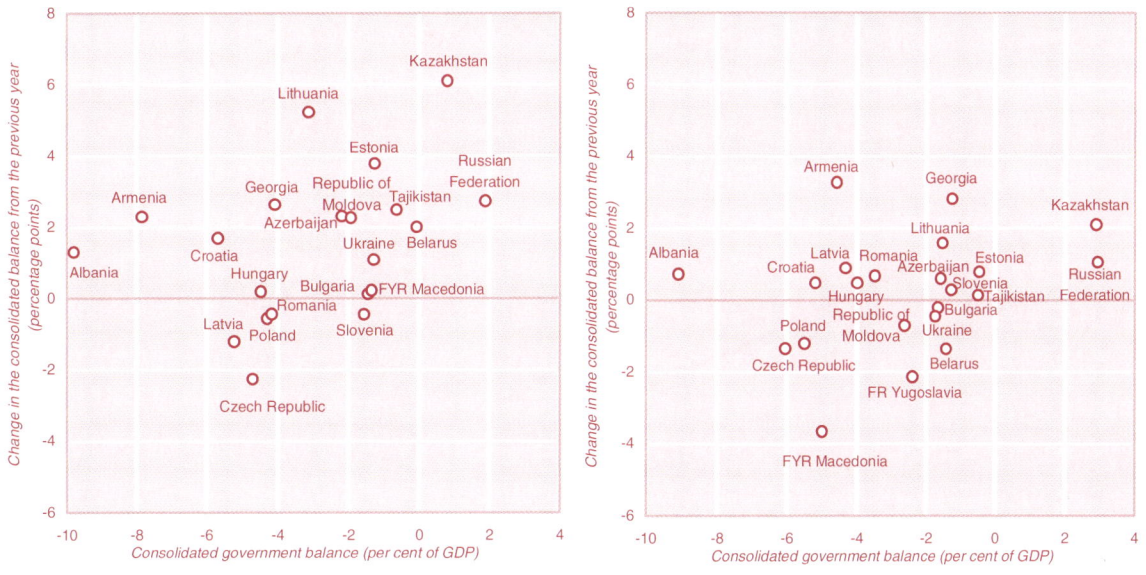

CHART 3.1.3

Fiscal balance and its change from the previous year in eastern Europe, the Baltic states and the CIS, 2000-2001
(Per cent of GDP and percentage points)

Source: UNECE secretariat calculations, based on direct communications from national Ministries of Finance and IMF data.

On the other hand, however, the current deterioration in the fiscal position of some countries merely brings into the open the negative consequences of past policies and measures whose fiscal implications do not appear to have been properly assessed at the time when they were undertaken (this issue is also related to the accurate reporting of fiscal deficits – see box 3.1.1). Thus, the absence of strict supervision over the banking system, and the reluctance of the authorities to engage in more radical financial reforms in the past, has led to the escalation of bad loans in many transition economies. The recent large-scale and costly financial rehabilitation of the banking sector in the Czech Republic and Slovakia[136] (which is partly responsible for their current fiscal deterioration) merely reflects the fiscal cost of policies pursued in the past. In Poland, the present fiscal problems partly reflect the fact that some of the policies undertaken by the authorities in the past were launched without a proper assessment of their fiscal implications, an omission which subsequently resulted in large-scale claims on public funds.[137]

Apart from the specific factors mentioned above, fiscal performance in some of these economies was also affected by the general weakening of economic activity in 2001. The unexpectedly sharp downturn in Poland (for details see section 3.1(iii)) brought about a very large shortfall in government revenues, raising serious concerns about the sustainability of the government's policies. In the course of 2001 the budget for the year was revised twice to reduce expenditures and to allow for a larger than expected deficit. The budget adopted for 2002 also contains a series of austerity measures to curb the growth of public spending. Despite this major fiscal adjustment, the general government deficit is likely to remain high in the short run (table 3.1.3) and further efforts to reduce it may be needed if Poland is to stay on track for EU and EMU accession.

In 2001, the governments of Hungary and the Czech Republic introduced large fiscal packages to stimulate their economies and boost economic growth. Although the two programmes are quite different in their goals,[138] scale and scope, conceptually they are very similar and amount to large-scale public investment programmes.[139]

---

[136] In both cases the state intervened to bail out state owned commercial banks by taking over their non-performing loans and placing their management in rehabilitation agencies. Special government bonds were then issued to cover the losses of the latter.

[137] For more details see UNECE, "Fiscal turmoil in Poland", *Economic Survey of Europe, 2001 No. 2*, box 1.2.1 and sect. 3.1(iii), pp. 20-21 and 78-79.

[138] The Czech Republic's stimulus programme was designed in the context of a longer-term effort to revitalize the economy which was slow to recover after the 1997 financial crisis; in Hungary the package was more a policy response to the global slowdown in 2001 aimed at giving an additional boost to growth.

[139] The Czech "Big Bang Plan" is estimated at some 86 billion koruny (4 per cent of GDP), almost equally distributed between 2001 and 2002, and includes aid to the business sector and funding for job creation, housing and infrastructure. *Interfax Czech Republic* and *Slovakia Weekly Business Report*, 27 July 2001, quoted by *Reuters Business Briefing*, 30 July 2001. The Hungarian package (known as the "Széchenyi Plan") involves public spending of 55 billion forint in 2001 (0.4 per cent of GDP) in grants for infrastructure development and interest subsidies on housing loans. *Interfax Hungary Weekly Business Report*, 25 October 2001, quoted by *Reuters Business Briefing*, 30 October 2001. A further 110 billion forint are earmarked to be spent under the Széchenyi Plan in 2002. *EIU Viewswire*, 12 February 2002.

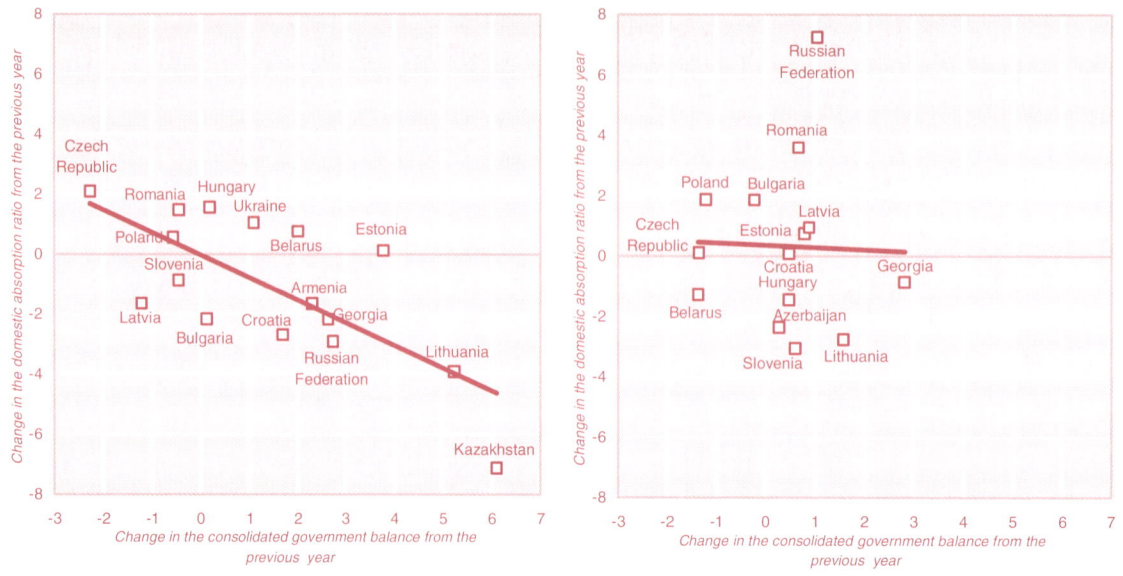

CHART 3.1.4

Changes in the fiscal balance and domestic absorption in eastern Europe, the Baltic states and the CIS, 2000-2001
(Percentage points of GDP)

Source: UNECE secretariat calculations based on direct communications from national Ministries of Finance and IMF data.

In both cases the adoption of these plans was accompanied by heated public debates, questioning both the rationale of the measures involved and their fiscal implications. While the authorities in both countries are optimistic about the positive results of these measures, it will take time before their effects can be assessed. In the meantime they undoubtedly contributed to the widening of the fiscal deficit in the two countries in 2001 and are likely to do so again in 2002.[140] The fact that the timing of these measures correlates closely with the political cycles in the two countries (with parliamentary elections due in 2002) has also raised some doubts about their rationale.

Although all three Baltic countries managed to reduce their budget deficits in 2001, the targets were not always met. In June, the Latvian government reduced its deficit target for 2001, bowing to pressure from interest groups to increase spending. The subsequent failure to reach an agreement with the IMF over the target for 2002[141] led to a temporary suspension by the IMF of the 18-month stand-by agreement and delays in the disbursement of a structural adjustment credit by the World Bank. In the context of the 2002 budgetary process, the Lithuanian parliament voted to reduce corporate taxation (in particular, the corporate profits tax was lowered from 24 to 15 per cent from 1 January 2002). While this measure may stimulate business activity, it will also make it considerably more difficult to meet the 2002 fiscal targets.

In south-east Europe, Romania made some progress in reducing its fiscal deficit in 2001 (to 3.5 per cent of GDP from 4.2 per cent in 2000) and the government has pledged to make further reductions in 2002. Despite some fiscal loosening in the first half of the year, the authorities in Croatia managed to keep the deficit in line with the target agreed with the IMF (5.3 per cent) and committed themselves to further fiscal tightening in 2002. In recent years Croatia has made considerable progress in reducing the size of the budget in proportion to GDP (which was one of the goals of the fiscal adjustment): the share of public expenditure in GDP has fallen from nearly 50 per cent in 1999 to 42.6 per cent in 2001 (table 3.1.5). On the eve of the June parliamentary elections, there was also some fiscal easing in Bulgaria; however, in the context of negotiating a new IMF stand-by agreement, fiscal policy was tightened in the second half of the year. The Bulgarian authorities are committed to keeping the fiscal deficit in 2002 below 1.0 per cent of GDP despite the lowering of some corporate taxes. The internal military conflict in The former Yugoslav Republic of Macedonia led to a considerable widening of the fiscal deficit in 2001 as a result of both increased military spending and the weakening of economic activity. The fiscal deficit also increased in Yugoslavia in 2001 (with further widening expected in 2002), reflecting the mounting costs of the extensive reforms initiated by the government.

---

[140] In both cases the IFIs and the European Commission have been critical of the programmes, arguing that they amounted to a de facto loosening of the fiscal stance in the two countries.

[141] The parliament voted the 2002 fiscal programme with a central government budget deficit of 2.45 per cent of GDP while 1 per cent had been agreed with the IMF. *Baltic Business Daily*, 30 November 2001 quoted in *Reuters Business Briefing*, 3 December 2001.

TABLE 3.1.4

Consolidated general government current revenue in eastern Europe, the Baltic states and the CIS, 1999-2001

(Per cent of GDP)

	Total current revenue and grants			Taxes on income, profits and capital gains			Indirect taxes and customs duties			Non-tax revenue			Social security contributions		
	1999	2000	2001	1999	2000	2001	1999	2000	2001	1999	2000	2001	1999	2000	2001
Albania	21.2	22.9	..	1.8	1.8	..	9.6	11.6	..	4.8	3.8	..	3.6	3.7	..
Bulgaria	39.0	42.1	..	7.9	6.9	..	12.6	13.8	..	7.8	8.2	..	7.9	11.2	..
Croatia	42.9	39.9	37.9	5.0	3.8	3.3	21.5	20.8	20.2	2.2	1.9	1.6	13.8	13.0	12.5
Czech Republic	39.0	39.4	39.5	8.8	8.9	9.2	12.7	12.6	12.0	2.5	2.5	3.1	14.4	14.7	14.7
Hungary	46.8	45.0	45.0	9.1	9.4	9.8	15.7	15.7	15.1	7.4	5.6	5.9	13.1	12.8	12.6
Poland	40.3	..	..	7.9	..	..	13.2	..	..	6.5	..	..	11.3	..	..
Romania	32.2	31.4	32.0	6.5	5.9	5.5	11.7	11.4	11.6	1.7	1.9	2.0	10.7	10.9	11.6
Slovakia	38.5	36.2	..	8.5	7.6	..	12.2	13.0	..	6.7	3.7	..	10.5	11.2	..
Slovenia	43.4	42.5	43.0	7.5	7.7	7.9	17.7	15.9	15.4	2.2	2.4	3.1	13.6	13.7	13.7
The former Yugoslav Republic of Macedonia	36.2	43.2	..	6.6	7.3	..	15.1	18.1	..	2.4	3.1	..	11.6	14.1	..
Estonia	36.0	35.4	35.7	10.7	8.7	8.2	12.1	13.0	12.9	2.9	3.3	3.5	9.5	9.9	10.0
Latvia	40.8	35.0	33.6	8.5	7.7	8.1	12.5	11.9	11.1	5.9	3.0	2.8	..	10.7	10.1
Lithuania	31.5	..	..	9.3	..	..	12.5	..	..	1.7	..	..	6.8	..	..
Armenia	22.6	19.5	..	4.6	3.8	..	10.1	10.0	..	1.8	1.2	..	..	..	..
Azerbaijan	19.6	15.7	..	5.3	4.2	..	10.1	7.7	..	1.6	1.5	..	..	..	..
Belarus	43.7	44.2	43.6	7.8	7.8	7.8	21.3	21.4	20.5	3.0	3.3	3.3	8.7	8.9	9.1
Georgia	15.4	15.3	16.7	2.8	2.9	..	8.8	8.8	9.5	0.8	0.8	1.0	..	..	..
Kazakhstan	17.8	..	..	1.8	..	..	6.8	..	..	1.3	..	..	..	..	..
Kyrgyzstan	15.7	18.6	21.1	1.7	2.1	2.8	6.6	8.9	8.8	2.7	2.4	3.3	..	3.5	3.9
Republic of Moldova	30.4	30.7	..	3.7	3.4	..	13.1	13.8	..	3.5	4.5	..	6.4	6.2	..
Russian Federation [a]	35.1	40.1	..	7.3	8.1	..	12.3	14.0	..	4.3	6.4	..	6.8	7.9	..
Tajikistan	13.5	13.6	13.9	2.3	1.8	1.8	8.4	8.7	8.6	0.5	0.6	0.9	..	..	..
Ukraine	33.8	35.1	32.4	8.9	8.7	9.0	11.9	10.8	9.9	1.7	5.4	4.6	..	..	..

Source: UNECE secretariat estimates and calculations, based on direct communications from national Ministries of Finance and IMF data. Data not available for Yugoslavia, Turkmenistan or Uzbekistan.

[a] Consolidated central government (including social security and extrabudgetary funds) plus (without consolidation) regional and local government.

There was a general improvement in the fiscal position of the CIS countries in 2001. Russia ended the year with a large surplus of some 3 per cent of GDP, according to the preliminary estimates. Although strong output growth and the favourable world prices for oil provided a strong boost to budget revenues in 2001, the comprehensive tax reforms introduced in 2000-2001 also contributed to the ongoing fiscal consolidation in Russia. In particular, the unification of the personal income tax and the simplification of corporate taxation,[142] together with a marked improvement in tax collection, played a major role in strengthening the public revenue. The large fiscal surpluses in 2000 and 2001 allowed the Russian authorities to repay in advance some of the outstanding public debt (in particular, to the IMF) and to increase fiscal reserves.[143] Likewise, in Kazakhstan, the fiscal position improved considerably in 2001 and the public finances remained in large surplus (table 3.1.3). The factors behind this improvement were basically the same as in Russia: strong output growth, increased oil revenue, tax reforms,[144] and an improvement in tax collection. In 2000 Kazakhstan established a national stabilization fund which, so far, has benefited from the large increase in oil revenue.[145] However it remains to be seen whether and how the fiscal authorities will manage to smooth out an eventual future downturn.

Despite the general improvement in fiscal positions during the last two years, a number of CIS countries have encountered problems in meeting the fiscal targets agreed with the international financial institutions. In several cases (Armenia, Georgia and Ukraine) these have led to temporary suspensions in the disbursement of funds,[146] while in Kyrgyzstan they caused a delay in the negotiations

---

[142] In 2000, the previous progressive personal income tax scale was replaced by a flat, single 13 per cent rate and a unified social security tax was introduced. On the changes in corporate taxation in 2001, see sect. 3.1(iv).

[143] However, the long-debated stabilization fund (which is expected to accumulate the extrabudgetary revenue in periods of high oil prices) has not yet been formally established. P. Kadochnikov, "Establishing a stabilization fund in Russia", *Russian Economic Trends*, Vol. 11, No. 1, 2001, pp. 8-17.

[144] A new tax code was adopted in 2001, which unified all tax payments and removed all previous exemptions.

[145] By the end of 2001, the National Fund for the Republic of Kazakhstan had accumulated assets amounting to $1.24 billion or 5.5 per cent of GDP. *Dow Jones International News*, 5 February 2002, reported in *Reuters Business Briefing*, 6 February 2002.

[146] In the case of Armenia and Georgia the reason was the failure to meet revenue targets, which led to an escalation of the fiscal deficit; in Ukraine, the main stumbling block was a large accumulation of overdue VAT refunds to exporters. IMF funding to Georgia was resumed in November after the parliament adopted a revised budget for 2001 which included a 15 per cent cut in public spending.

BOX 3.1.1

Why are fiscal deficits not measured accurately?

Despite the gradual adoption of internationally accepted standards (in most cases the IMF's GFS methodology), there still exist important differences in the methodology of fiscal statistics even in some of the more advanced reformers among the transition economies. The most important differences are related to the treatment of various extrabudgetary public funds (all of which should in principle be included in the consolidated general government budget, but is not always the case in practice), the treatment of transactions in public assets (whether they should count as current revenue or as sources of deficit financing) as well as the proper accounting of various government guarantees. Due to the existence of such problems, this *Survey* has adopted a uniform definition of the general government deficit (explained in the note to table 3.1.3) and, to the extent possible, reports the fiscal position of all the transition economies in accordance with this definition. Consequently, in a number of cases, the numbers reported in table 3.1.3 differ considerably from the officially reported figures. However, due to the lags in fiscal reporting, the relevant data required to compile the deficits in accordance with the uniform definition, especially for the most recent period, are not always available for all countries at the moment of writing. Hence, for some countries, the deficits for 2001 are shown in table 3.1.3 as reported officially (and do not necessarily meet the uniform definition). The same is true for most of the target figures for 2002.

The deviations from standard practice are a major problem both for the accurate measurement of the fiscal position in the transition economies and for cross-country comparisons of fiscal policy and performance. The different methodological practices and, sometimes, "creative accounting", lead to widely differing estimates of the fiscal position of a country. For example, the discrepancy between the IMF's estimates of Kyrgyzstan's fiscal deficit and that officially reported in some years falls within a range of 8-10 percentage points;[1] the situation is similar in the case of the Republic of Moldova. However this problem exists also in the more advanced reform countries. For example, the consolidated fiscal deficit in the Czech Republic in 2001 would have been considerably higher than the one reported in this *Survey* (6 per cent of GDP) if all the losses of the Consolidation Agency had been covered by public funds (as originally envisaged in the 2001 budget); instead, some of the payments were deferred to 2002 which helped to reduce the 2001 deficit but will show up in the 2002 figures.[2] According to some analysts, if all the off-budgetary fiscal items in Hungary were properly accounted for, the country's 2001 deficit would have been 5.5 per cent of GDP rather than 4 per cent.[3]

In 2001, in the context of preparing its annual reports on progress towards EU accession,[4] the European Commission requested for the first time that candidate countries compile their national fiscal statistics in accordance with the European System of Accounts (ESA95), which is the standard methodology in the EU and is close to the definition used in this *Survey*. One of the revealing outcomes of this exercise – which is indicative of the existing problems – was that most of the discrepancies in the measurement of the deficits were on the negative side (i.e. they enlarged rather than reduced the estimated deficits). In some countries, including some among the more advanced candidate countries, fiscal deficits turned out to be much higher than was usually believed.

The failure to apply rigorous methodological standards in fiscal statistics is not just the result of inertia in statistical procedures; some of the existing practices (for example, the extrabudgetary public funds whose operations are not always adequately reported or included in the general government balance) are deeply entrenched in the policy process of these countries, often (but not always) dating back to the communist past. Their negative impact is widely recognized: they make fiscal policy non-transparent, create coordination problems in the domain of public finance and preclude proper accountability, and encourage fiscal abuse, all of which may potentially endanger long-term fiscal sustainability. Considerable efforts for further reform will be needed in this area in order to improve the transparency of fiscal policy. The political elites may not always be in favour of such changes because, if rigorous accounting standards are thoroughly and consistently adopted, they might lead to a reassessment of fiscal positions and past policies. Besides, changes in established budgetary practices often involve institutional changes which may trigger strong resistance by interest groups. Notwithstanding these difficulties, such reforms are a key step not only towards establishing the institutional foundations of a mature market economy but also towards a more democratic and transparent society.

---

[1] One of the causes for the discrepancy was the treatment of a large extrabudgetary public investment programme (amounting to several percentage points of GDP – see table 3.1.5, last panel), which was apparently not included in officially reported public deficits prior to 2000 (table 3.1.3).

[2] *Interfax Czech Republic* and *Slovakia Weekly Business Report*, 22 January 2002, quoted by *Reuters Business Briefing*, 23 January 2002.

[3] Statement by Roger Nord, local representative of the IMF, reported in *Interfax Hungary Weekly Business Report*, 8 November 2001 and quoted by *Reuters Business Briefing*, 9 November 2001.

[4] The annual progress reports are available at [http://europa.eu.int/comm/enlargement/report2001].

TABLE 3.1.5

Consolidated general government expenditure in eastern Europe, the Baltic states and the CIS, 1999-2001
(Per cent of GDP)

	Total expenditure including capital outlays			Current expenditure on goods and services			Subsidies and other current transfers			Interest payments			Capital outlays		
	1999	2000	2001	1999	2000	2001	1999	2000	2001	1999	2000	2001	1999	2000	2001
Albania	31.0	35.8	..	10.8	17.4	..	7.4	5.5	..	6.9	4.4	..	5.7	6.7	..
Bulgaria	40.8	43.1	..	20.3	20.2	..	13.2	14.5	..	3.9	4.1	..	4.6	3.9	..
Croatia	49.2	44.8	42.6	22.1	21.0	17.5	19.8	18.6	19.2	1.6	1.8	2.1	5.6	3.3	3.8
Czech Republic	42.2	44.3	45.7	8.5	8.7	8.9	27.2	28.6	30.0	1.0	1.1	1.2	5.5	5.9	5.6
Hungary	50.1	50.0	49.6	17.7	17.4	18.2	19.4	19.4	19.8	7.4	6.1	5.4	5.5	7.1	6.2
Poland	43.7	..	..	17.0	..	..	20.5	..	..	3.1	..	..	3.1	..	..
Romania	35.8	35.1	35.3	12.8	12.6	12.4	14.9	14.6	15.2	5.3	4.9	4.1	2.8	3.1	3.5
Slovakia	39.6	41.8	..	10.1	10.1	..	22.3	22.9	..	3.2	2.9	..	4.0	5.8	..
Slovenia	44.2	44.1	44.6	17.7	17.9	18.5	20.2	20.2	20.0	1.4	1.5	1.6	4.9	4.5	4.5
The former Yugoslav Republic of Macedonia	35.2	40.3	..	13.6	12.7	..	20.0	23.8	..	1.6	1.8	..	2.6	4.2	..
Estonia	41.0	36.8	36.2	20.5	17.9	17.3	15.8	15.4	15.3	0.4	0.3	0.3	4.4	3.2	3.3
Latvia	44.5	40.2	37.9	18.0	16.1	14.6	21.2	18.9	18.4	0.8	1.1	1.1	4.6	4.1	3.8
Lithuania	39.0	..	..	19.1	..	..	12.7	..	..	1.5	..	..	2.2	..	..
Armenia	30.1	25.9	..	11.9	11.6	..	8.8	7.2	..	2.0	1.7	..	4.7	3.9	..
Azerbaijan	24.1	17.9	..	15.0	12.1	..	4.4	3.9	..	0.5	0.4	..	6.3	5.1	..
Belarus	45.7	43.1	44.5	15.7	15.6	17.3	18.7	18.8	21.6	0.6	0.8	0.7	10.7	8.3	4.8
Georgia	25.5	19.4	17.8	18.8	14.5	13.1	2.5	2.8	1.9	2.8	3.0	2.2	2.1	1.0	1.4
Kazakhstan	21.5	..	..	11.1	..	..	8.1	..	..	1.1	..	..	1.5	..	..
Kyrgyzstan	19.8	28.9	27.6	13.9	15.5	16.0	2.9	2.5	3.0	3.4	2.8	1.9	..	8.1	6.8
Republic of Moldova	34.6	32.7	..	12.9	9.3	..	11.2	14.0	..	7.4	6.5	..	2.6	3.1	..
Russian Federation [a]	35.3	37.3	..	13.1	15.0	..	14.7	14.4	..	4.0	3.8	..	3.4	4.1	..
Tajikistan	16.6	14.2	14.3	9.9	8.2	9.1	2.7	2.9	2.7	0.6	0.4	0.9	3.4	2.6	1.5
Ukraine	34.2	29.8	..	..	..	..	..	..	..	..	..	..	..	..	..

Source: UNECE secretariat estimates and calculations, based on direct communications from national Ministries of Finance and IMF data. Data not available for Yugoslavia, Turkmenistan or Uzbekistan.

[a] Consolidated central government (including social security and extrabudgetary funds) plus (without consolidation) regional and local government.

for a new funding agreement with the IMF.[147] It should be noted however that in many cases the available fiscal statistics for the CIS countries are still of dubious quality (box 3.1.1) which is an impediment to a more accurate assessment of their fiscal position and long-run fiscal sustainability.

(iii) Economic distress in Poland

Over 10 years ago, Poland received much attention and praise for its radical programme of economic transformation. While the country's move "from plan to market" began in an atmosphere of pessimism given the presence of many seemingly insurmountable problems inherited from the communist regime, the successful macrostabilization policies brought inflation down from near hyperinflation, maintained current account convertibility and increased domestic confidence in the zloty.[148] Simultaneously with macroeconomic stabilization, the Polish economy began to undergo extensive restructuring that has led to the largest productivity gains in the region and considerable inflows of foreign direct investment.[149] The reforms designed to put the country on the path towards a market economy also laid the basis for a sharp rebound in economic activity and, in 1992, Poland became the first transition economy to return to growth. Its high growth rate continued uninterrupted throughout the 1990s, earning the post-1989 economy such popular labels as the "soaring eagle" and "European tiger".[150]

Recently however the Polish economy – and its dynamic image – has begun to falter. The rate of GDP

---

[147] The main reason for the delay was the adoption of a controversial reduction in income and profit taxes; after an amendment to the tax code, an agreement with the IMF was finally reached in December.

[148] The country's advantages such as its proximity to west European markets, the existence of a modicum of a pre-communist institutional framework, and a pent-up entrepreneurial spirit played significant roles in the transition. In terms of the latter, about two million businesses have been created and they have become a powerful engine of change in Poland by contributing to output growth, job creation, and the formation of a new class of consumers.

[149] Compared with 1991, the average industrial worker in Poland produced 91 per cent more in 2000. His Hungarian counterpart produced 85 per cent more, while all other eastern and central European workers were more productive by a third or less during the same period. In addition to FDI, these impressive comparative productivity gains reflected the much higher degree of labour hoarding or hidden unemployment in pre-1989 Poland.

[150] Since 1989, Polish GDP has increased by almost a third while the aggregate GDP of eastern Europe as a whole has yet to reach the 1989 level. In 2000, the GDP of the Baltic states was about two thirds of the 1989 levels while the CIS countries had reached about 60 per cent of their 1989 level (appendix table B.1).

growth gradually declined from 5 per cent in the second quarter of 2000 to 0.3 per cent in the fourth quarter of 2001, triggering fears of future economic hardship and possible negative implications for its efforts to gain EU membership.[151] Along with the rapid reduction in output growth, other macroeconomic variables also point to a pronounced economic slowdown: private consumption is weakening, fixed investment and employment are falling, and the real interest and unemployment rates remain high.

Although there is no recession at present, the many weaknesses in all aspects of the economy have set off an urgent domestic debate on the causes of the slowdown. The debate intensified when the Minister of Finance warned (subsequently dismissed for not saying so earlier) that without urgent austerity measures Poland's 2002 budget deficit could rise to over 10 per cent of GDP from an estimated 4 per cent. The discovery of what was called a "budgetary hole" several weeks before the parliamentary elections prompted an active search for those responsible. The politicians blamed one another but most of them held the Monetary Policy Council and its restrictive monetary policy accountable.[152] The central bank, in turn, has consistently put the blame on expansive fiscal policies as well as inadequate reform of the labour and product markets.[153]

(a) Recent economic performance

After the initial sharp contraction in output, following the introduction of the macroeconomic stabilization programme, the Polish economy gathered speed in 1992-1993 and enjoyed a very fast rate of growth throughout the decade. In national accounting terms, growth in the 1990s was determined by domestic aggregate demand, which almost always outpaced GDP growth (chart 3.1.5). During the early stages, industrial output underpinned this expansion while improvements in corporate profitability, tax concessions, the start of considerable inflows of foreign direct investment and the greater use of bank credit led to an investment boom (chart 3.1.6).

In 1996-1997, domestic demand accelerated – increasing by almost 10 per cent a year. The investment boom accounted for almost half of the increase – driven

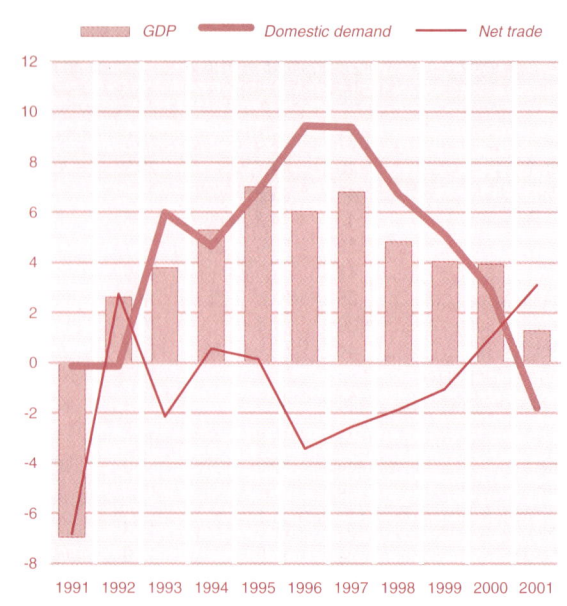

CHART 3.1.5

Contributions to Poland's GDP growth, 1991-2001QIII

Source: UNECE secretariat, based on national statistics.

by FDI flows, high corporate earnings, buoyant residential construction and increased bank lending to businesses. Consumption – particularly of consumer durables – also rose dramatically fuelled by high real wage growth and the availability of consumer loans.[154] At the same time, export growth slowed down owing to, among other factors, sluggish demand in the European Union, where about two thirds of Polish exports are destined. Imports, driven by overheated domestic demand, FDI and import-inducing trade policy changes, continued to grow rapidly. As a result, the current account shifted from a surplus to a sizeable deficit triggering concerns about the economy's external vulnerability (chart 3.1.7).

Fears of a possible currency crisis intensified in the wake of the Russian financial crisis and the economic slowdown in Europe at the end of 1998. Both events adversely affected exports and, more importantly, business and consumer confidence. These external shocks weakened output growth, caused a lasting loss of food exports to countries of the former Soviet Union and reduced the border shuttle trade. In the second half of 1999 with confidence returning in Poland and elsewhere, the economy bounced back and the current account deficit improved, but after four quarters of robust expansion the economy began to lose steam. Its descent started in the second half of 2000 when the rate of GDP growth halved from the high rates seen in previous years.

---

[151] Poland's accession to the EU will be the subject of a national referendum, probably in 2003. In October 2001, fewer than 50 per cent of Poles supported membership, more than 10 percentage points less than in 2000.

[152] See, for example, *Raport Otwarcia* (The situation in Poland at the beginning of the new Parliament), The Chancellery of the Prime Minister, 3 January 2002 [www.kprm.gov.pl].

[153] National Bank of Poland, *Monetary Policy Guidelines for the Year 2002* (Warsaw), September 2001 [www.nbp.pl]. The public appears to associate the weak economic performance with incompetence, self-interest and the corruption of public officials. This is reflected in opinion polls as well as the results of the September 2001 elections when two populist parties got 20 per cent of the seats in parliament and the previous governing coalition got none.

[154] Commercial bank credit to households more than tripled between 1995 and 1997, from 5.6 billion to 18.4 billion zloty.

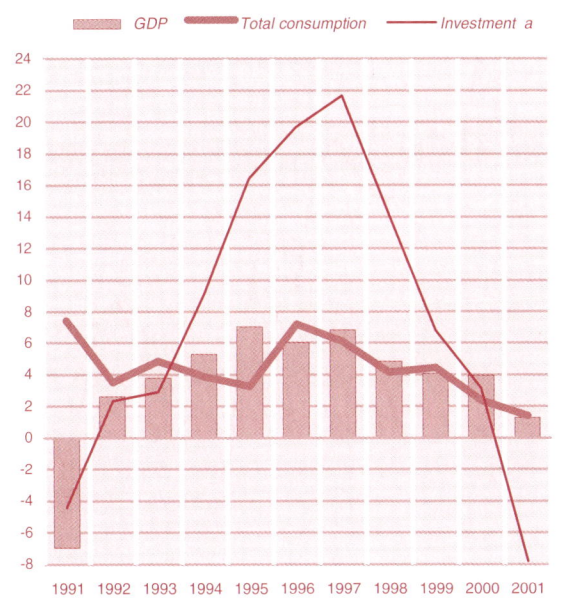

CHART 3.1.6

Consumption and investment growth in Poland, 1991-2001QIII
(Percentage change over the same period of previous year)

Source: UNECE secretariat, based on national statistics.

a   Gross fixed capital formation.

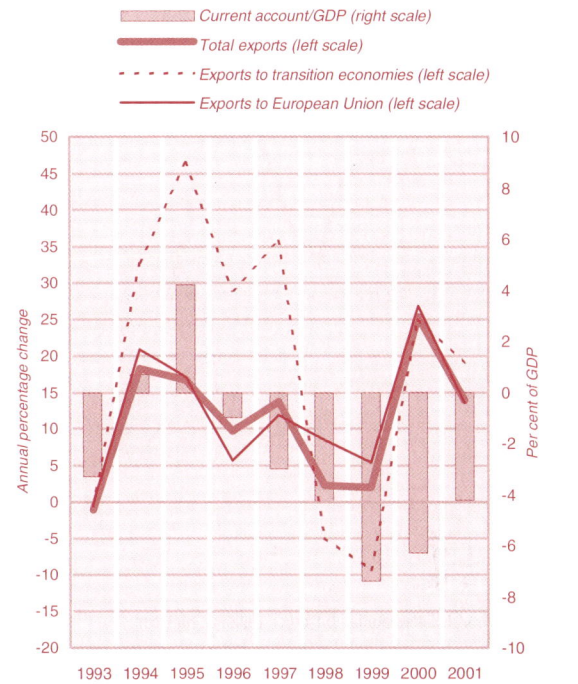

CHART 3.1.7

The current account and the volume of exports in Poland, 1993-2001QIII
(Percentage change over the same period of previous year)

Source: UNECE secretariat, based on national statistics.

For the first three quarters of 2001, the Polish economy continued to grow weakly at 1.3 per cent, year-on-year, mainly because of plummeting investment expenditures. Individual consumption expenditure continued to increase slowly (by 1.8 per cent) supported by still increasing household borrowing and a reduced rate of savings.[155] The dramatic decline in the growth of retail sales (chart 3.1.8) partly reflects smaller real disposable incomes in 2000, but it may also anticipate the crisis of consumer confidence and possibly the end to the euphoric post-transformational spending boom on consumer durables.[156] The current sharp slowdown in car sales (down by a third in 2001, year-on-year) and an excess supply of residential housing suggest the end of the consumer boom is highly plausible.[157] High real interest rates will have had an impact on consumers demand for credit and their spending, as will the slowing of income growth, and lower levels of employment (see below). Consumption in 2001 was increasing at two to three percentage points below its rate in previous years (chart 3.1.6).

Given weak yet still positive consumption growth, the decline in domestic demand was due to declining gross fixed capital formation (down by 7.8 per cent, year-on-year). With excess productive capacity increasing and bleak prospects for growth in sales, companies – accounting for roughly two thirds of all investment outlays – held back their investment plans.[158] The demand for capital usually falls when expected output weakens, but with about 60 per cent of corporate investment in Poland financed with internally generated funds (i.e. retained earnings), current profitability is a major factor in investment decisions. Corporate profits have been declining, thus reinforcing the incentive to postpone investment (chart 3.1.9).[159] The cost of financing also plays a role, but the influence of real interest rates on domestic investment is not straightforward. With roughly 15 per cent of investment expenditures financed through domestic bank credit, the direct impact of real interest rates on investment in

---

[155] Preliminary data indicate that loans to households in 2001 increased by 15 per cent and zloty deposits by 12 per cent. In both cases, the rates of growth were about half those in 2000.

[156] The growth of annual disposable incomes has been slowing from 5.5 per cent in 1998 to 2.7 per cent in 1999 to the reduction of 0.2 per cent in 2000. National Bank of Poland, *Inflation Report*, Second Quarter 2001, Monetary Policy Council (Warsaw), August 2001.

[157] In view of the relatively low per capita GDP in Poland, spending on, say, cars has to stabilize at a commensurate level.

[158] A survey of enterprises employing more than 49 employees indicated that investment outlays were down 12 per cent, expenditures on buildings declined by 17 per cent, while spending on machinery and equipment was down by 9 per cent in the first three quarters of 2001. Central Statistical Office, *Poland Quarterly Statistics* (Warsaw), December 2001, p. 13.

[159] The increase in profitability in 1999-2000 was the result of increases in producer prices and unit export values concurrent with relatively smaller unit labour cost increases.

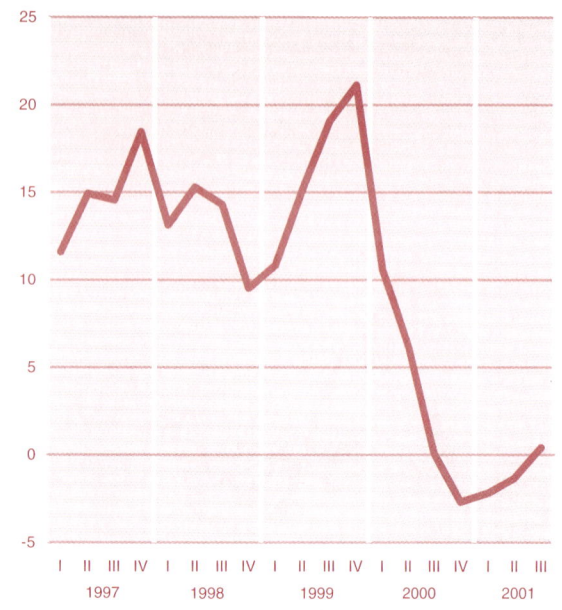

CHART 3.1.8

The volume of retail sales in Poland, 1997-2001
(Percentage change over the same period of previous year)

Source: UNECE secretariat, based on national statistics.

CHART 3.1.9

The average profitability of Polish enterprises, 1995-2001
(Per cent of sales)

Source: Poland Quarterly Statistics (Warsaw), various issues.

Poland is relatively weak.[160] Nevertheless, high real interest rates raise the cost of capital, lower the incentive to invest and contribute to the strength of the zloty. While there appears to be a wide consensus that the Polish currency is overvalued, the impact of the strong zloty on exports is inconclusive. Historically, the level of Polish exports has been closely related to the level of external demand, thus the recent gradual slowdown in export growth appears to be due more to the slowdown in the EU than to the zloty's appreciation relative to the euro.[161] The profitability of exports, however, is linked to the value of the zloty and an increasing number of export sales (particularly of industry) was reported to be unprofitable in the first three quarters of 2001.

On the production side, in January-December 2001, industrial output remained flat after a 7 per cent increase in 2000. The recession in the construction sector, which began in the second half of 2000, continued into 2001 with output falling by almost 10 per cent through the year. The construction sector is likely to remain stagnant until capital spending recovers and the residential market absorbs the existing oversupply.[162]

The deteriorating financial position of firms has spilled over to the labour market. In the first three quarters of 2001, employment declined by over 3 per cent, year-on-year, and the unemployment rate rose to about 17 per cent. This outcome however is only partly the result of slowing GDP growth. Despite the very high rate of economic growth throughout the 1990s, the (annual) unemployment rate in Poland never fell below 10 per cent.[163] While the pre-1989 legacy of labour hoarding and the ongoing process of corporate restructuring are the main culprits, the economy has not created many new jobs.[164] With one in four Poles unemployed or underemployed,[165] more supply-side

---

[160] Periodic corporate surveys appear to confirm the weakness of this relationship. Only one in five Polish firms admits to being affected by the level of interest rates. Those that do, however, do not consider it as important as aggregate demand or the value of the exchange rate. National Bank of Poland, *Wstepna informacja o kondycji sektora przedsiebiorstw ze szczegolnym uwzglednieniem stanu koniunktury w trzecim kwartale 2001*, Department of Statistics (Warsaw), p. 23 [www.nbp.pl].

[161] The euro area economy has shown increasing signs of cyclical weakness since the beginning of 2001. After growth of 3.4 per cent in 2000, second quarter figures for euro area GDP in 2001 showed year-on-year growth down to 1.7 per cent, with a third quarter estimate of 1.3 per cent. GDP in the euro area's largest market for Polish exports, Germany, fell in the third quarter of 2001 following zero growth in the previous three months.

[162] Residential housing prices have fallen by as much as 10 per cent and the number of construction permits issued by 11 per cent in the first half of 2001.

[163] Changes in aggregate demand account for about a quarter of the unemployment level above 10-12 per cent. M. Gora, "Konsekwencje uczestnictwa w Unii Gospodarczej i Walutowej dla polskiego rynku pracy" (The consequences of EU membership: the Polish Labour Market), paper presented at the Polska Droga do euro Seminar (Falenty), 22-23 October 2001 [www.nbp.pl].

[164] About 40 per cent of the 1.5 million (net) new jobs created during the boom of 1993-1998 were in agriculture, and "non-market services" such as public administration, education or health.

[165] "Hidden unemployment" in rural areas was estimated to be 900,000 individuals in 1999. Ministry of Finance, *The Strategy of Public Finance and Economic Development, Poland 2000-2010*, Council of Ministers (Warsaw), June 1999, p. 13.

reforms are necessary, but lowering the unemployment rate is complicated by demography, while job creation is made difficult by institutional factors.[166] A costly tax wedge – comprising personal and corporate income taxes, pension contributions and social assistance – increases the costs of employment, discourages hiring and reduces workers' earnings.[167] It also forces legitimate economic activities outside the formal economy. The higher the price of labour, the greater the incentive to employ labour-saving capital and considering that two thirds of the Polish unemployed have very low educational attainments (at most secondary vocational), they represent the easiest type of labour to substitute.[168]

(b) Macroeconomic policies and structural changes

In view of slowing output growth, weak export prospects and a rapidly deteriorating labour market, the Polish government is searching for ways to revive the economy. With binding constraints on the public finances, which are now stabilized but in a large deficit, the government's short-term strategy is focused on monetary policy. The approach appears to be based on the understanding that the only way to get a quick recovery is through a weaker zloty, and this requires the central bank to lower interest rates. Additionally, it is also assumed that cheaper credit will stimulate aggregate demand.

The conduct of monetary policy, however, is the sole responsibility of the central bank and, in particular, of the independent Monetary Policy Council (MPC).[169] The Council has consistently argued that lower interest rates cannot be provided "on demand" and that, considering the circumstances of the Polish economy, they must be preceded or accompanied by structural changes in fiscal management and other areas under the responsibility of government. The MPC has argued that lower interest rates cannot substitute for these measures which, in fact, are indispensable for lasting reductions in interest rates. These divergent views on the causes of the crisis and the ways out of it have spurred an open debate between the government and the Council. The government argues that lower interest rates are justified given the current low inflation rate. In response, the Council regards the current low inflation rate in Poland as ephemeral owing to the existence of many factors influencing prices that are clearly beyond its control.[170]

Members of the government argue that the MPC is misreading the economy as it has in the past. In fact, the Council has never achieved its own annual inflation targets which are supposed to pave the way to the medium-term objective (4 per cent inflation by the end of 2003) leading to claims, supported by the former chairman of the central bank, that direct inflation targeting was adopted prematurely.[171] Moreover, the government has stressed that while the Council's constitutional and legislative mandate is to safeguard the value of the domestic currency, it is the government's role to ensure economic growth and macroeconomic stability, and to bear the responsibility for the costs related to structural impediments and the risks associated with the external imbalances.

The government's claim that restrictive monetary policy is stifling growth does not appear to rest on completely solid foundations. First, in the absence of structural reforms, lower rates may not necessarily result in faster and sustainable growth. Second, it is difficult to assess the impact of interest rate cuts given the relative ineffectiveness of monetary policy in Poland.[172] The magnitude of the impact aside, the transmission process might involve a lengthy lag. By some estimates, it may take up to nine months before a financial improvement is detected at the firm level.[173] Admittedly, lower interest rates are likely to weaken the zloty (although the ongoing political pressures on the central bank have also put short-term upward pressure on the zloty), but given weak external demand, a lower zloty may not necessarily translate in higher export volumes.[174] Overall, the net result is difficult to estimate.[175]

---

[166] Due to the baby boom of the early 1980s, it is estimated that the Polish economy must create 250,000 non-agricultural jobs a year until 2010 just to absorb new labour force entrants. Ministry of Finance, op. cit.

[167] In the 1990s, it was estimated that payroll taxes almost doubled the cost of net pay. "Zawinily przepisy i biurokracja", *Rzeczpospolita* (Warsaw), 1-2 December 2001.

[168] The government of 1997-2001 did little to lower the cost of labour but it raised the cost of capital by limiting investment write-offs and reducing depreciation allowances. In contrast, the current government appears to favour preferential tax treatment of investment.

[169] The central bank's independence is enshrined in the Constitution, while the bank's fundamental goal of price stability is spelled out in the relevant legislation. Recently there have been attempts to pass legislation aimed at reducing this independence.

[170] Differences of opinion over macroeconomic stability are illustrated by two recent contributions. A member of the MPC claims that the movement from a current account surplus to a deficit in the mid-1990s "destabilized the economy". B. Grabowski, "Wzrost popytu nie wystarczy", *Rzeczpospolita*, 5 December 2001. A prominent government consultant argues that the same movement represented the "strengthening of the economy" because the shift in the current account balance was a necessary equilibrating macro-adjustment. S. Gomulka, "Recepta na recesje", *Rzeczpospolita*, 14 December 2001.

[171] H. Waltz-Gronkiewicz, "Jaka powinna byc polityka pieniezna", *Rzeczpospolita*, 10 October 2001. More importantly, the price level reacts to changes in interest rates with a lag of between 9-24 months making annual targets largely unattainable.

[172] Among the factors that make the monetary policy in Poland relatively ineffective are the low monetization, the relatively undeveloped financial sector and the availability of large (relative to the size of the economy) capital inflows from abroad. See, for example, K. Rybinski, "Wplyw polityki pieniezjnej na process dezinflacji w Polsce", *Bank i Kredyt*, July-August 2000, pp. 56-79.

[173] "Za wysokie stopy procentowe", *Rzeczpospolita*, 9 January 2002.

[174] The high real interest rates in Poland attract foreign portfolio investment in bonds and treasury bills. However, expectations of an imminent loosening of monetary policy play a dominant role given the

The absence of policy coordination between the government and the MPC is rather damaging to the policy process in Poland. Stretching their constitutional responsibilities, the fiscal authorities are attempting to gain more influence on the conduct of the monetary policy, while the MPC is seeking influence on fiscal policy through its own pressures on the government. Being unable to forecast inflation accurately, the MPC has, by necessity, been looking backwards.[176] In doing so, it has also been seen, to a large degree, to be inconsistent, non-transparent and internally divided. Facing such substantive criticism, the MPC has tried to (re)gain credibility by conducting an overly restrictive monetary policy. As a result, the Council appears to have successfully stamped out inflationary expectations, but at the cost of contributing to the economic slowdown. In general, the differences between the government and the MPC illustrate the practical difficulties of conducting monetary policy in a transition economy and have highlighted the government's reluctance to undertake necessary structural reforms.

This insufficient resolve on the part of government to accelerate the needed structural reforms in order to establish the supply-side foundations for sustained growth has also contributed to the slowdown. Admittedly, the transition in Poland has involved massive changes to the country's institutional foundations, but much remains to be done. The unfinished privatization process is one of the sources of inefficiency and politicization of the economy. While most of GDP and employment originate in the private sector, a majority share of fixed productive assets in Poland is still in state hands (the state still owns up to 800 companies). Governments which are unwilling to sell them, allow them to go bankrupt or are unable to find buyers for them, have been forced to shift the associated liabilities onto the taxpayers. Moreover, some 2,300 companies with partial state ownership continue to blur the line between the public and private spheres, thereby potentially contributing to corruption.[177] As a result, taxpayers carry the burdens of redundant employment and excessively high wages in the state owned companies as well as the costs of state owned firms defaulting on their tax and social security obligations.[178] The prospects for accelerating the sale or closure of state owned companies are not very good. The current government – judging by the first three months in office – appears to prefer slow and measured progress in restructuring and consolidation as opposed to rapid sales or dissolution, which in many cases would mean massive job losses. Many of the government's parliamentary opposition members argue that this approach represents a step backwards, claiming it is aimed more at increasing the state control over the Polish economy than an industrial policy. While it may be so, if restructuring means significant job losses, it is the government's role to confront it as a real and tangible social problem.[179]

In addition, some policies benefiting single interest groups suggest the government's preference is for job protection over job creation. First, the labour code negotiated by trade unions in the large state companies and the communist government in 1989 has not been fundamentally modified to reflect the interests of the now dominant employers from small- and medium-size enterprises. As a result, the labour market has not been flexible enough to accommodate the inevitable reallocation of labour in the transition economy. Moreover, high payroll taxes – related to, in particular, social transfers for pensions, early retirement and disability benefits – increase the cost of labour. Second, the lingering education reform still apparently does not fully reflect the needs of students to prepare for employment in a market economy. Third, protectionist policies have blocked the modernization of agriculture.[180] Food prices are kept artificially high and during poor harvests they climb even higher contributing to inflation.[181] Consumers also face non-competitive prices in other sectors. Some sectors such as telecommunications,

---

inverse relationship between bond yields and their prices. Political pressures on the central bank to lower the interest rates signal future capital gains in fixed income securities, and thus result in the appreciation of the zloty. After the 7.5 point decrease in the interest rates by the central bank, the zloty continued to appreciate in 2001 as foreigners continued to bet on future reductions.

[175] A lower zloty would also increase the costs of foreign loan repayments – often incurred to avoid high domestic real interest rates – for both private and public borrowers.

[176] This runs against the basic principle of direct inflation targeting. P. Christofferson, T. Slok and R. Wescott, "Is inflation targeting feasible in Poland?", *The Economics of Transition*, Vol. 9, Issue 1, 2001, pp. 153-174.

[177] In *The Strategy of Public Finance and Economic Development, Poland 2000-2010*, the Council of Ministers agreed: "the search continues, by way of political pressure, for subsidies to keep inefficient firms operational, and Poland's privatization rate in the 1990s cannot be considered satisfactory", p. 18.

[178] In 2000, employees in state owned firms earned almost 50 per cent more than those in the private sector (excluding foreign owned companies), Central Statistical Office, *Statistical Yearbook of the Republic of Poland* (Warsaw), 2001, p. 164. It should be noted that earnings in the private sector are likely to be underestimated to some extent due to tax evasion. For a more detailed breakdown of wages see, Central Statistical Office, *Poland, Quarterly Statistics* (Warsaw), December 2001, pp. 38-43.

[179] The difficulty of adequately addressing the nexus of "restructuring-unemployment" was apparent during the 1990s when previous governments did not take advantage of the booming economy to move more rapidly with restructuring.

[180] The agricultural sector is characterized by low productivity, high unemployment and underemployment, and limited access to education and health facilities. Subsistence farming continues to dominate in Poland with over one million farms (55 per cent) with revenues of less than $50 a year and only 5 per cent with incomes comparable to average incomes in the non-farm sector.

[181] The Polish government in its "Government plenipotentiary for negotiations of Poland's EU membership" (undated, available at www.minrol.gov.pl) claims that: "Incorporation of Polish agriculture into the CAP instruments is not going to destabilize the common agricultural markets (because) ... the purchase prices of Polish products such as cereals, pig meat and poultry are comparable to EU prices, and sometimes even higher".

transportation, banking, fuels and energy are still dominated by monopolistic or oligopolistic structures, which increase the cost of living and of doing business in Poland. These sectors are also the key areas where state ownership is prevalent. Breaking up the current monopolies and sheltered areas could lead to lower prices for consumers and lower costs for business thereby stimulating both consumption and entrepreneurship.

(c) The way ahead

Economic slowdowns rarely have a singular cause and the Polish economy attests to that. Expansive fiscal policies – with budget deficits incurred even at peak rates of economic growth – sustained a "post-transformational" consumption boom but they discouraged domestic savings necessary to establish the foundations for sustainable growth. Neither has the structure of public finances – with transfers to households amounting to 20 per cent of GDP – been conducive to growth.[182] In response to loose fiscal policies, an inappropriate structure of public expenditure and a slow pace of reform, the central bank has pursued a restrictive monetary policy primarily aimed at offsetting the expansionary fiscal policy. Its policy has successfully restored external balance and delivered disinflation, but the ensuing high real interest rates have helped to increase the value of the zloty, reduce investment and consumption spending and thus contribute to lower growth.

In January 2001, the government published its economic strategy called *Entrepreneurship, Growth, Work*, in which it anticipates economic growth rates of 1, 3 and 5 per cent, respectively, for the next three years. While the programme assumes less restrictive monetary policy and some effort to rationalize public finances, in the absence of fundamental reforms in this area, the predicted growth rate in 2004 may turn out to be optimistic.

From the macroeconomic perspective, a comparable and sustainable growth rate requires a favourable external environment, solid consumer confidence and an investment-GDP ratio of about 30-35 per cent (that is 10 to 15 percentage points above the current levels).[183] Given the low level of domestic savings, the savings-investment gap must be closed by foreign capital. While EU membership could bring transfers worth about 2 to 3 per cent of GDP in 2004, foreign capital must be encouraged by macroeconomic stability and solid structural foundations that include the rule of law, an attractive tax system and a competitive, well-educated labour force. Aside from the feasibility of attracting so much foreign capital (and the risks related to the magnitude of current account deficits), the authorities must also focus on policies that aim to mobilize domestic resources. In view of the stability of private savings in Poland, the government needs to reduce its own borrowing by spending less and by spending more efficiently. This implies a restructuring of public revenues and expenditures as well as a series of comprehensive reforms including some that may carry high political risks.

From the microeconomic perspective, more attention should be paid to the private sector. In response to a high tax burden, overregulation, frequently changing and opaque rules and discriminatory treatment – small and medium enterprises, for the most part the engine of growth in Poland, have stopped growing.[184] The private sector, to be made more competitive, has to be less taxed. In this regard, the government economic package announced in January 2001 moves in the right direction as one of its elements – *Above All: Entrepreneurship* – aims at reducing the costs of doing business, creating more flexibility in the labour market and providing a regulatory framework that would give greater encouragement to entrepreneurship.

(iv) Can Russia be a regional growth engine?

Large economies can and do serve as growth engines for their smaller trading partners. The best example is the United States economy which has been pulling (directly and indirectly) the whole world economy during most of the 1990s and up until the present slowdown. During a boom, the trading partners of a large economy can benefit markedly from the above average growth of domestic demand in the latter. The extra "pull" effect is generated by the size differential: even if the increment in the large economy's import demand is small relative to its own size, it can be much larger relative to the size of its smaller trading partners. As with any other external disturbance, however, the situation can be reversed: during the downward phase of the large economy's cycle, the shrinking of import demand can exert an extra negative "push" onto its smaller trading partners.

The recent strong recovery in the Russian economy raises the question as to whether and to what extent it can serve as a growth engine for the other economies in the region: in the first place, for the neighbouring CIS countries, but also (although to a lesser extent) for some of the other transition economies. An economy's potential as a regional growth engine is determined by two main factors: its relative size and the intensity of its trade links with the other economies in the region. In terms of its size, as reflected in the absolute level of GDP measured at purchasing power parities, Russia's economy

---

[182] The Polish government agreed with this characterization in 1998. "The (Polish) authorities view budgetary savings as insufficient, investment too limited, the tax burden too heavy and the budgetary expenditure too skewed towards social transfers". *OECD Economic Surveys: Poland* (Paris), January 2000, p. 54.

[183] B. Grabowski, "Wzrost popytu nie wystarczy", *Rzeczpospolita*, 5 December 2001 and S. Gomulka, "Recepta na wzrost", *Rzeczpospolita*, 21 December 2001.

[184] For more analysis see, J. Winiecki, "The Polish generic private sector in transition: developments and characteristics", *Europe-Asia Studies*, Vol. 54, No. 1, January 2002.

is about double the size of the rest of the CIS taken as a whole and about equal to the aggregate of eastern Europe (including the Baltic states). In a pan-European perspective, however, it is smaller than any of the four largest west European economies (France, Germany, Italy and the United Kingdom).[185] In any case, the size of Russia's economy is sufficiently large for it to exert a perceptible external effect on the rest of the transition economies.

Apart from size, it is the intensity of trade and other economic links that determine the strength of the external "pull" or "push" that a given economy can potentially exercise on others. In this regard, despite some weakening of Russia's economic links with the rest of the CIS these are still quite intensive and for some of them (Belarus, the Republic of Moldova and Ukraine) Russia is still the largest trading partner. Hence the recovery in Russia's domestic demand in 2000-2001 has been mirrored by an even stronger (in relative terms) recovery in intra-CIS trade (table 3.5.5), Russia accounting for the bulk of the latter. In turn, the surge in exports to Russia, partly driven by the real depreciation of other CIS currencies against the rouble (section 3.1(i)), has been an important factor for the acceleration of growth in many of its neighbouring countries. As for eastern Europe, although trade links with Russia have weakened considerably in recent years (table 3.5.3),[186] Russia still exerts a considerable influence on some of these economies, as demonstrated by the strong aftershocks of the 1998 crisis.

Being an effective growth engine, on the other hand, implies high and sustained rates of growth of the large economy's domestic demand. Thus the main question that needs to be considered in this respect is whether and to what extent the present high rates of economic growth in Russia can be sustained over the medium term. In addition, this growth must be balanced in the sense that the expansion of domestic and import demand should not endanger macroeconomic stability.

In less than four years after one of the most devastating financial crises in recent history, Russia's economy has changed dramatically for the better. Three years of strong growth have improved the welfare of the population and boosted public confidence – both domestically and internationally – and expectations about the future. This positive change in public perception has been underpinned not only by the strength of the economy but also by the government's effort to break with the stop-go policies of the past and to accelerate systemic transformation and market reforms.

Probably more sweeping and comprehensive legislative and regulatory reforms were introduced in Russia in 2001 than in any other year since the start of economic transformation. Among the most important measures were the changes in the tax code, which aim at more transparency and uniform treatment of taxpayers. The tax reform involves a reduction in the rate of corporate profits tax to a unified 24 per cent (previously it was 35 per cent for enterprises and 43 per cent for banks) and the elimination of all previous exemptions (which were numerous). It also includes further steps to strengthen fiscal federalism in Russia by clearly defining the powers and responsibilities of the central and the local governments. Among the other important legislative measures is the major pension reform (introduced in a series of laws and due to start in 2002), which will transform the present pay-as-you-go into a three-pillar system. The Duma also finally approved the long-awaited Land Code, which for the first time allows the free sale of land in residential areas (the sale of farmland is due to be addressed by a separate law). An important step towards economic liberalization was the move to simplify administrative procedures and reduce bureaucratic intervention in the economy.[187] In preparation for WTO accession the government has drafted a new customs code (simplifying considerably the previous one in line with WTO norms), which is to be applied in 2002. A new labour code approved by the Duma in the final months of 2001 envisages further liberalization of the labour market coupled with measures for employee protection in line with European labour legislation. The government has also made considerable progress in enterprise and bank reforms.[188] The new bankruptcy law streamlines procedures, providing better protection of creditors' rights, while the so-called "corporate conduct code" (which is voluntary rather than mandatory) prescribes norms of behaviour aimed at improving corporate governance. The government, together with Russia's central bank, launched a three-year programme of bank restructuring which, *inter alia*, envisages further recapitalization of the smaller banks, the introduction of international accounting standards and the reorganization of three of the largest banks (Sberbank, Vneshekonombank and Vneshtorgbank). A special law to combat money laundering was also voted by the legislature.

The list of important legislative and regulatory changes is much longer but even this small sample indicates the government's ambition to widen and deepen

---

[185] According to the 1996 round of the European Comparison Programme. UNECE, *International Comparisons of Gross Domestic Product in Europe, 1996* (United Nations publication, Sales No. E.99.II.E.13).

[186] Russia accounts for the major share of these countries' trade flows to and from the CIS.

[187] More than 1,000 regulatory norms were scrapped in the first half of 2001. *PlanEcon Report*, Vol. 17, No. 13 (Washington, D.C.), July 2001, p. 13. In addition, the new law on licensing, which entered into force in February 2002, reduced the number of economic activities subject to licensing from 2000 to 120. N. Savvina, "Novyi zakon o litsenzirovanii", *StranaRu*, 8 February 2002 [http://www.strana.ru/].

[188] Although these have not been completely successful: despite the government's efforts, little progress was made in reforming the monopolistic structures operating the public utilities.

systemic reform. More importantly, most of the reforms are marked by the spirit of economic liberalization: they are aimed at fostering entrepreneurship and developing the infrastructure of the market economy in Russia. They are expected to contribute to a better business and investment climate and thus should contribute to sustaining growth in the future.

The ongoing process of economic liberalization will benefit the economy but it will take some time before its positive effects are directly reflected in higher levels of economic activity. The notable improvement in the business environment thus cannot alone explain Russia's recent remarkable growth (in 1999-2001 GDP grew at an average annual rate of 6.5 per cent). Most analysts tend to agree that this period of growth was largely driven by two main factors: the sharp real depreciation of the rouble after the August 1998 financial collapse and the windfall gain in export revenue due to the upturn in world commodity prices and, especially, of oil. What follows is a closer look at these two factors and their impact on Russia's economic performance.

Crude oil is Russia's major export item:[189] in recent years its share in the value of total exports has ranged between 17 and 24 per cent, depending on the volumes sold and the level of oil prices.[190] Between 1998 and 2000, Russia's monthly revenue (in dollars) from oil exports roughly tripled, largely thanks to the upturn in oil prices but also to a rise in export volumes (chart 3.1.10).[191] Increasing the volume of exports was an important element in Russia's strategy to profit as much as possible from the favourable external environment in this period. The continuing increase in export volumes in 2001 helped to offset to a large extent the negative impact on export earnings of the fall in oil prices (chart 3.1.10).

In order to be able to assess more accurately the effect of oil revenue on economic performance, it is useful to eliminate the effect of price changes on nominal values. Chart 3.1.11 shows the development of "real revenue from oil exports"[192] (or simply, real oil revenue)

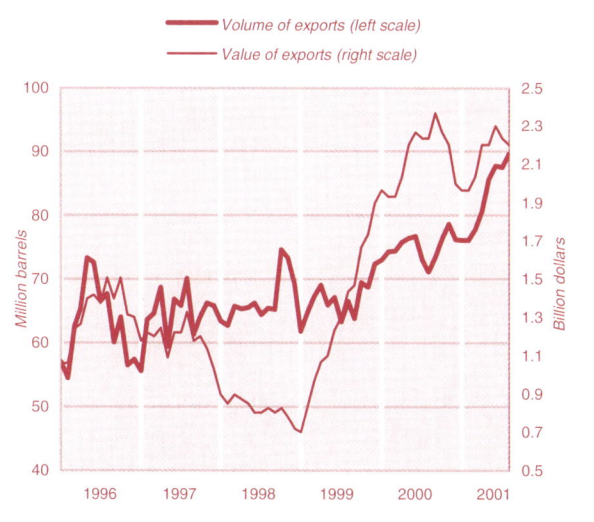

CHART 3.1.10

Monthly exports of crude oil from Russia, 1996-2001
(Volume in barrels, value in dollars, three-month moving averages)

Source: UNECE secretariat calculations, based on national statistics.

in the period 1996-2001 expressed in constant domestic prices of January 1995 (deflated by CPI). Even a simple visual inspection of chart 3.1.11 reveals the significant gains for Russia in the period after 1998. It is instructive to compare the change in real revenue from oil exports in two consecutive time periods, namely, 1996-1998 and 1999-2001. If the real revenue from oil exports in each of the years from the second subperiod are compared with the average annual real oil revenue during the first subperiod, the extra real oil revenue in 1999 amounted to approximately 2.7 per cent of GDP; in 2000 it was roughly 6.3 per cent of GDP; and in 2001 around 4 per cent of GDP.[193] These estimates are admittedly approximate and tentative in character; nevertheless, they indicate the strong positive external support that the Russian economy enjoyed in the period after 1998.

Going one step further, the real revenue from oil exports is determined by the dynamics of three main variables: the volume of exports (chart 3.1.10), the international price of oil[194] and the real exchange rate[195] (both shown on chart 3.1.11). Chart 3.1.12 presents a

---

[189] It should be noted that natural gas occupies a share similar to oil in Russia's exports. Hence, most of the conclusions concerning the impact of oil exports on Russia's economic performance are also relevant with respect to gas exports; moreover, the world prices of natural gas usually follow the dynamics of oil prices with a lag of several months. The main difference in the recent export trends for the two commodities was that in the case of gas exports volumes were relatively stable, while the volumes of oil exports increased substantially after 1999.

[190] In 2000 and 2001, 15 of Russia's top 50 exporters were firms in the oil and gas industry; another 13 top exporters belong to the metal processing industry. K. Liuhto and J. Jumpponen, "International activities of Russian corporations – where does Russian business expansion lead?", *Russian Economic Trends*, No. 3-4, 2001, pp. 19-29.

[191] In 2000 the volume of Russia's exports of crude oil rose by 13 per cent and in 2001 by a further 10 per cent. UNECE secretariat estimations based on national statistics.

[192] The real revenue from oil exports is determined by first converting the dollar value of oil revenue into current roubles (on the basis of the current rouble exchange rate) and then by deflating it with the CPI index (in other words, by deflating the dollar revenue with the CPI-exchange rate differential).

[193] For these estimates, the current value of annual GDP has also been expressed in January 1995 prices.

[194] The average international price of Russian export oil (the so-called "Urals crude") in 2000-2001 was some 3 per cent lower than that of Brent crude, which is shown on chart 3.1.11. Russian-European Centre for Economic Policy, *Russian Economic Trends*, monthly edition (Moscow), February 2002. The price of Brent crude is used as a proxy for Urals crude in all subsequent calculations due to the absence of a sufficiently long series for the latter.

[195] The term "real exchange rate" is used here for the simplicity of presentation. Actually, what is denoted by the real exchange rate in charts 3.1.11 and 3.1.12 is the above-mentioned differential between changes in the nominal exchange rate and in the CPI (changes in the United States price level, which should also be part of the proper definition of the real exchange rate are not included).

CHART 3.1.11

Brent crude price, real exchange rate of the rouble and Russia's real revenue from oil exports, 1996-2001

(Price and exchange rate as indices in per cent, revenue in billion roubles in constant January 1995 prices)

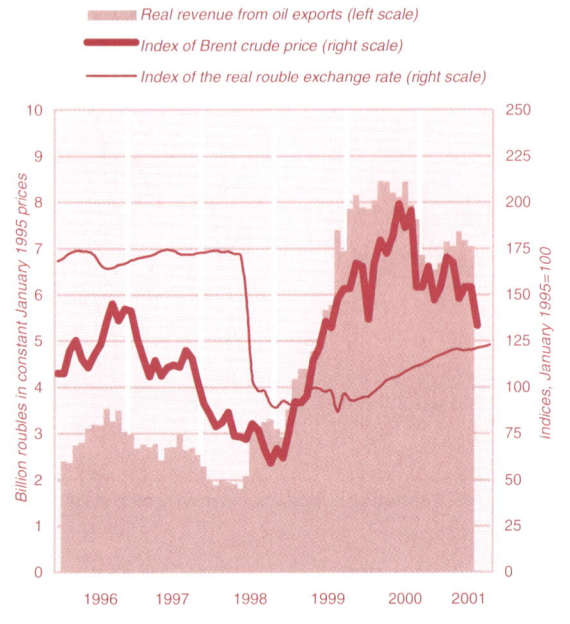

Source: UNECE secretariat calculations, based on national statistics; oil prices from United States Department of Energy, *Weekly Status Petroleum Report* (Washington, D.C.), various issues [www.eia.doe.gov].

CHART 3.1.12

Annual change in Russia's real revenue from oil exports and its components, 1997-2001

(Billion roubles in constant January 1995 prices)

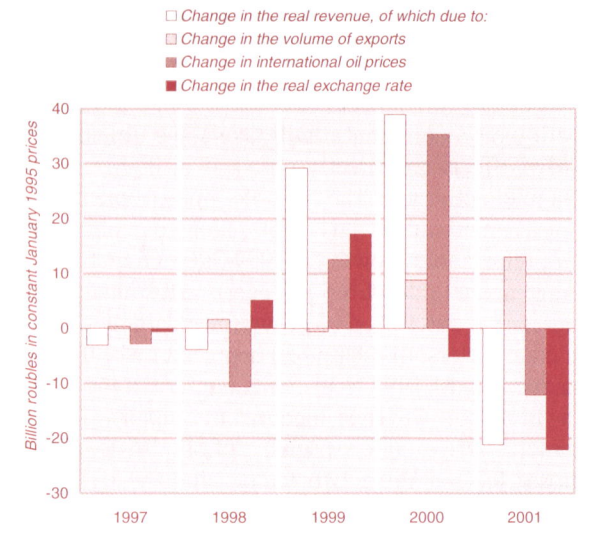

Source: UNECE secretariat calculations, based on national statistics.

rough estimate of the contribution of each of these variables to the changes in real oil revenue in the period 1997-2001. In 1999, the increase in revenue was due in almost equal proportions to the rise in oil prices and the change (depreciation) in the real exchange rate. In 2000, the strongest boost to real oil revenue came from the surge in oil prices; the rise in the volume of exports also contributed positively while the impact of the real exchange rate was negative (due to its appreciation). The situation changed radically in 2001: oil revenue decreased in real terms from 2000 and it was only the increase in export volume that made a positive contribution to the outcome; both oil prices and the real exchange rate had a negative effect on real oil revenue. It should be noted though that despite its decline, the real revenue from oil exports in 2001 still remained high in absolute terms and much above the average level in 1996-1998 (chart 3.1.11).

The fact that real oil revenue has a strong impact on Russia's macroeconomic performance is illustrated in charts 3.1.13 and 3.1.14, which indicate a strong statistical association between this variable, on the one hand, and the growth of GDP and the fiscal balance, on the other. Naturally, the real effect of oil-related revenue is not the only factor of growth in Russia; however, its present-day importance should not be underestimated either.[196] The mechanism through which a change in real oil revenue is transmitted to the macroeconomy is complex and combines the effects of each of the three primary variables. Thus the increase in oil production feeds directly into higher aggregate output, while higher export prices are equivalent to a terms of trade gain. In addition, the change in the real exchange rate has a multiplicative effect: it amplifies the terms of trade effect, which accrues to exporters in the case of a real depreciation and vice versa. In turn, the higher real revenue allows exporters to increase investment and to pay higher wages, both of which add to domestic demand and, consequently, to domestic output. In addition, there is a fiscal transmission channel: directly, due to higher profit taxes paid by oil exporters' and, indirectly, due to rising profits and incomes in the rest of the economy.

The heavy reliance on oil exports, however, is a mixed blessing for the Russian economy. In times of boom, as seen during the past three years, it may provide a welcome boost to the economy; however, things may go into reverse when both prices and demand weaken. Moreover, even long booms carry risks due to the dangers of the "Dutch" disease. If Russia is to follow the path of industrial modernization, it will have to aim to gradually reduce its reliance on oil exports. However, this can only be a medium- or even long-term goal; current judgements about Russia's future growth prospects cannot ignore the expected impact of oil

---

[196] Russian analysts have estimated that, *ceteris paribus*, a change in world oil prices by one dollar is likely to be associated with a 0.4 to 0.6 percentage point change in Russia's GDP and with a change in fiscal revenue amounting to $0.8-$0.9 billion. Center for Macroeconomic Analysis and Short-Term Forecasting, *The Macroeconomic Consequences of the Falling World Oil Prices,* December 2001 (in Russian) [http://www.forecast.ru/]. However, as noted above, apart from oil prices, for a proper account one should also take into consideration the volume of exports and the dynamics of the real exchange rate.

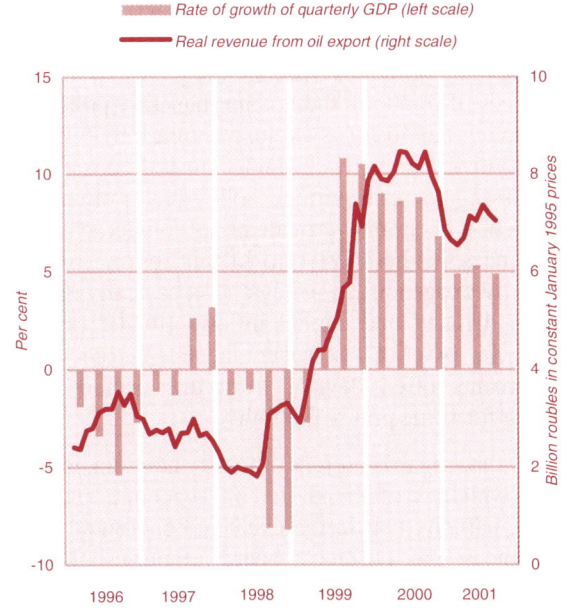

CHART 3.1.13

Russia's real revenue from oil exports and the rate of GDP growth, 1996-2001

(Revenue in billion roubles in constant January 1995 prices, growth rates over the same period of the previous year, per cent)

Source: UNECE secretariat calculations, based on national statistics.

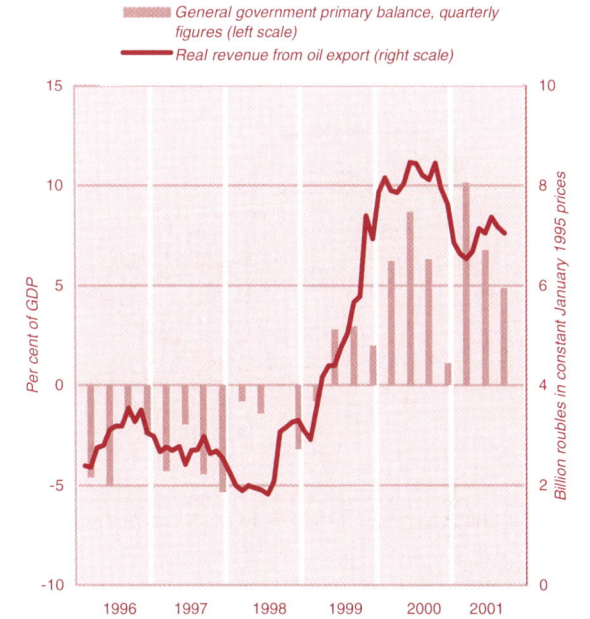

CHART 3.1.14

Russia's real revenue from oil exports and the general government fiscal balance, 1996-2001

(Revenue in billion roubles in constant January 1995 prices, fiscal balance as percentage of GDP)

Source: UNECE secretariat calculations, based on national statistics.

exports and this has to be incorporated into any assessment of its role as an engine of growth for the surrounding region.

As argued above, real oil revenue has three main components and thus a review of Russia's short-term prospects involves looking at the outlook for each of them. After the sharp downturn towards the end of 2001, international oil prices started to recover in March-April 2002, partly due to the political uncertainties in the Middle East. In general it seems likely that in the short run they will remain within a range that is favourable for Russian exporters. Raising the volume of oil exports was the main avenue followed in 2001 in an attempt to offset the drop in prices. However, it is not quite clear whether Russia would be in a position to follow a similar strategy in the short run. On the one hand, in view of the continuing weakness in the world economy, global demand is also likely to remain weak. On the other hand, the potential to export larger quantities in the short run depends on the existing extraction capacities which, according to many analysts, are limited. Moreover, in December 2000, Russia made a unilateral commitment to cut oil exports by 150,000 barrels a day as part of the OPEC-led effort to reduce global supply in an attempt to stop oil prices from falling further.[197] This is only a voluntary restraint, though, which is unlikely to be binding in a situation when Russia's vital interests might be at stake. In any case, there are some uncertainties, both on the supply and on the demand side, regarding the prospects of increasing the volumes of Russia's oil exports.

The real exchange rate is an important instrument on which the Russian authorities can rely in order to influence the level of real oil revenue. The impact of the real exchange rate on economic performance is much broader than its effect on real oil revenue; however, as noted above, this latter aspect should not be underestimated either. During recent years, the real exchange rate of the rouble has been quite volatile. Between January 1995 and December 2001 it has actually gone full circle: from a period of excessive real appreciation (between September 1995 and September 1998) to a period of excessive real depreciation (between September 1998 and mid-2001) and returning close to its initial level in the second half of 2001 (chart 3.1.15). The Russian authorities are concerned at the potentially damaging effect of another bout of real appreciation of the rouble and appear to be inclined to keep it close to its present level.[198] This might be a reasonable strategy to follow, given the fact that the current level of the real exchange rate still seems to be quite favourable for the economy as a whole and apparently does not endanger

---

[197] RFE/RL, *Newsline*, Vol. 5, No. 229, Part I, 5 December 2001.

[198] Recently the Russian central bank openly admitted that it was using the real exchange rate as the main indicator of the efficiency of its exchange rate policy. Central Bank of the Russian Federation, *Vnutrennyi valyutnyi rynok v oktyabre 2001 goda* [http://www.cbr.ru/analytics/money_market/].

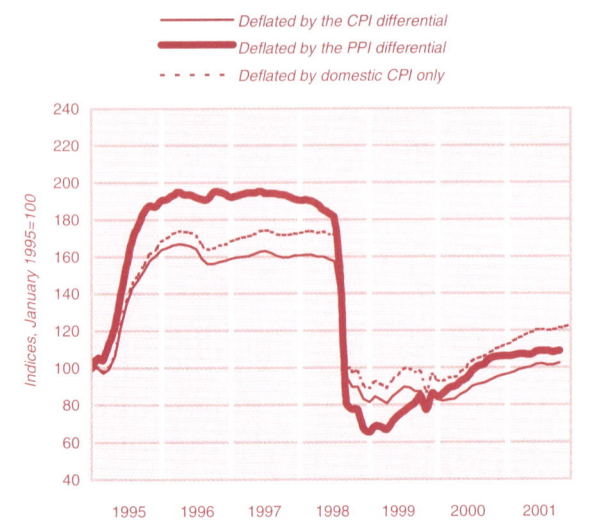

CHART 3.1.15

Changes in the real exchange rate of the Russian rouble, 1995-2001
(Indices in per cent)

Source: UNECE secretariat calculations, based on national statistics.

the competitiveness of local manufacturers. On balance, real oil revenue in 2002 is likely to remain sufficiently high to continue to provide a positive impulse to the economy.

In conclusion, Russia's growth prospects in the short to medium run appear to be moderately favourable, provided that it follows prudent macroeconomic policies and continues with its programme of reform. Given the large current account surplus (table 3.5.1), the external balance does not seem to be a constraint even if domestic absorption grows faster than output for some time to come. Thus, there are good prospects for Russia continuing to be a regional engine of growth despite a likely slowdown in 2002. As for the longer run, the prospects for growth will depend on the successful restructuring of the Russian economy, and especially on the success or failure of efforts to transform it into a knowledge-based economy whose exports are not dominated by primary commodities but by technology intensive, high value added manufactured goods.

(v) Is there a risk of "currency board contagion" for the transition economies?

The recent collapse of the currency board in Argentina has given rise to new concerns about the stability and sustainability of currency boards. For almost a decade the currency board in this country was considered a great success and its sustainability was rarely questioned. Nonetheless, Argentina's forced exit from the currency board arrangement brought to the foreground the question as to whether there are implications – and lessons – from this painful experience for the ECE transition economies and, especially, for those with similar monetary regimes.

A country that adopts a currency board arrangement basically gives up the privilege of conducting an independent monetary policy. This regime, in its orthodox form, implies a hard peg to a reserve currency, full backing by foreign reserves of the central bank's liabilities denominated in national currency (so that, accordingly, changes in reserves are reflected in proportionate changes in the monetary base), no central bank intervention on the financial markets and no central bank lending to the government or to the banking sector. Under such a regime macroeconomic policy basically has no degrees of freedom: there is no monetary policy while fiscal policy has to focus exclusively on long-term fiscal sustainability. As a result, under an orthodox currency board arrangement, economic policy has no instruments to influence economic growth or to counteract external disturbances in the short run: in this regime growth is essentially exogenous to policy and is basically driven by external demand and capital inflows (in the first place FDI). In practice, however, actual currency board arrangements often deviate from this orthodox form, allowing for some policy flexibility.

In this regard, currency boards have an important feature which is often ignored in academic and policy debates, namely, that in the event of concurrent external disturbances the currency board amplifies their real domestic effect as compared with other monetary regimes. The reason for this is that, under this arrangement, all external disturbances – real as well as nominal – have to be absorbed by the real economy since, in the absence of monetary policy, there are no nominal buffers to cushion them. Hence in the event of simultaneous real and/or nominal shocks, their real domestic effects are mutually reinforcing. It is important to note that this mechanism works in both directions: with respect to positive as well as negative disturbances. To understand this feature it is necessary to take a closer look at the mechanisms through which external disturbances are transmitted under this regime.

Typical external shocks to a currency board regime are changes in external demand (real disturbance), appreciation or depreciation of the reserve currency against other major world currencies, or changes in the direction of capital flows (nominal disturbances). Consider a situation of rising external demand (positive real disturbance) coupled with a surge in capital inflows (positive nominal disturbance). Higher real demand implies that domestic output has to increase in order to meet the increased demand, resulting in higher output growth. In turn, the extra capital inflow adds to the foreign reserves, which has the effect of a proportionate expansion of the monetary base. As a result of this superfluous monetary injection, money supply may exceed current money demand, leading to a fall in domestic interest rates.[199] In turn, the latter may

---

[199] Depending on the nature of the capital inflows, the change in interest rates may act as an automatic adjustment mechanism. Thus, in the case of short-term financial inflows, when interest rates fall below a certain level, they may halt or even reverse the direction of the flows due to the change in interest differentials. In this regard, FDI usually exhibits a smaller degree of short-term volatility.

provide an extra stimulus to domestic producers, giving a further boost to economic activity. Hence, in the event of a combination of positive external disturbances (nominal and real), the currency board amplifies their positive effect on the real economy. Other combinations of positive external disturbances may have a similar overall effect.[200] Under a symmetric mix of negative external shocks (for example, falling external demand coupled with capital outflow), the situation will be reversed and the same mechanism of the currency board will amplify their negative real impact resulting in a further lowering of domestic economic activity.

It is instructive to look at Argentina's experience in light of the different combinations of external factors that affected its economy in the course of the 1990s. After the introduction of the currency board in 1991, not only was monetary stabilization achieved but also growth strengthened considerably, stimulated by a combination of capital inflows and a relatively weak currency (pegged to the dollar). However things started to change after 1997 when the economy was hit by a series of external shocks starting with the Asian crisis and continuing to the present global slowdown. The emerging market crises had a negative impact on the inflows of capital, which gradually dried up and moved into reverse. In addition, in recent years the economy was adversely affected by the substantial appreciation of the dollar (and hence the peso), which resulted in deteriorating competitiveness. The government then started to experience problems with the servicing of the public debt (denominated in dollars and dominated by short-term maturities); these problems were aggravated by the fact that a large share of the public debt was contracted by local governments, which complicated the task of fiscal coordination.

In principle, under such circumstances, the mechanisms of the currency board arrangement should be capable of triggering the needed adjustment: a rise in money demand due to higher public borrowing should push up interest rates and, when they reach a certain level, this should help to reverse the outflow of capital due to the emergence of a high enough interest differential. However, in the final months of Argentina's currency board this mechanism failed to attract external funds despite the skyrocketing of interest rates, while the economy fell into a deep depression. The main reason for this was that the aggravation of the situation changed investors' perceptions and started to erode their confidence. The sharp increase in interest rates implied a simultaneous increase in the investors' risk premium; in turn, the latter was equivalent to a lowering of the (perceived) risk-free level of public and foreign debt. So although Argentina's public and foreign debt was not high by international standards (relative to GDP) this change in perception started to deter foreign investors. It was this rapid erosion of investor and public confidence that, in the end, led to the collapse of the regime. In retrospect, the authorities in Argentina failed to recognize early enough the severity of their existing fiscal problems and the looming debt crisis. The absence of timely policy responses was another major factor in the crisis.

Does Argentina's experience raise new concerns about macroeconomic stability in the transition economies that adhere to similar monetary regimes? Currency boards are in operation in four of the ECE transition economies: Bulgaria, Bosnia and Herzegovina, Estonia and Lithuania.[201] However, compared with Argentina, there are differences in some of the specific arrangements. While the Argentine peso was pegged to the dollar, the transition economies have chosen to peg to the euro.[202] This difference is important in assessing the risks of a loss of competitiveness due to an exchange rate misalignment (which was an important factor in Argentina's failure). In principle, this risk should be much lower in the case of the transition economies, since the EU is now their main trading partner and thus the major share of their trade is not subject to exchange rate risks. Another difference is that in the transition economies there are some deviations from the orthodox currency board to which Argentina was sticking. Thus, both in Estonia and Lithuania, the central banks have some residual, although limited, powers of monetary discretion, mostly in the form of open market operations.[203] The specificity of Bulgaria's currency board is that it requires a much larger backing of the domestic currency by foreign reserves: in addition to the monetary base, it requires the foreign exchange reserves to back the so-called "fiscal reserve account" of the government (which is a buffer fund designed to cushion short-term liquidity problems of the latter)[204] and a special fund designed to provide a quasi "lender-of-last-resort" facility vis-à-vis the commercial banks in case of systemic risks to the banking system as a whole.[205] These deviations from the orthodox form of the currency board create some additional room for manoeuvre for the monetary authorities in the case of threats to macroeconomic or financial stability.

Another distinctive difference is the sheer size of the economies: the four transition economies are much smaller than Argentina. Experience has shown that in small economies the above noted adjustments are easier to make; a larger economy is subject to greater frictions

---

[200] Thus, a depreciation of the reserve currency vis-à-vis other major currencies may also have a stimulating real domestic impact as it transforms into competitive gains.

[201] Latvia adheres to a hard currency peg regime whose policy implications are similar to those of a currency board.

[202] Lithuania's litas used to be pegged to the dollar until 2 February 2002, when it was re-pegged to the euro.

[203] Such as the sales of certificates of deposits by the Bank of Estonia or the repo auctions of the Bank of Lithuania.

[204] Estonia has also established a similar "stabilization reserve fund" but in relative terms it is smaller in size.

[205] As a result Bulgaria's foreign exchange reserves are nearly twice as large as the monetary base.

and, in general, has more to lose from the loss of sovereign monetary policy. Finally, the absence of a time consistent exit rule or a strategy of orderly exit from the currency board arrangement played a major role in the case of Argentina, further increasing financial instability and reinforcing the economic turmoil. In the case of the transition economies that aspire to EU membership, an exit strategy from the currency board is already in place as all of them have declared that this regime will be preserved until their accession to ERM-2.

At this point in time the available evidence does not raise immediate concerns about the stability of the currency board arrangements in the transition economies. In the design of their monetary regimes, the latter have created some policy flexibility to cope with short-term macroeconomic and financial volatility while the risks of an exchange rate shock are relatively small. Even more important, these countries have been following quite prudent fiscal policies (section 3.1(ii)) and, by and large, their long-term fiscal sustainability does not appear to be under threat. As fiscal prudence is the key to a balanced macroeconomic performance under a currency board arrangement, there do not seem to be any visible risks to the four transition economies from this side.[206] The levels of foreign debt do not appear to be a reason for immediate concern either (table 3.5.12); although Bulgaria's foreign debt is somewhat high, it has been declining while its maturity structure is quite balanced (with a large share of long-term debt).[207] Although these economies have been affected by external shocks, none of them has plunged into a prolonged recession; in fact they have weathered relatively successfully two disturbances in recent years: the Russian crisis of 1998 and the present global economic slowdown.

This does not mean, however, that the risks to the monetary regimes in these transition economies are non-existent. The greatest risk is indeed that of a prolonged recession caused by a combination of adverse external shocks.[208] The main issue then would be how these economies would absorb such a prolonged negative shock. The latter ultimately implies a downward adjustment in output and employment but, as seen in the case of Argentina, such a painful adjustment may be extremely difficult and may turn out to be politically unacceptable. Such an adjustment also requires a high level of labour and wage flexibility and appropriate institutional arrangements, something that has not been tested under critical circumstances in the transition economies. There is also the question of how the adjustment mechanisms of the currency boards in the transition economies would operate in the face of a prolonged external shock. In an orthodox arrangement (when only the monetary base is backed by foreign reserves) a capital outflow leads to an equiproportionate shrinking of the money supply; this, in turn, should lead to higher interest rates which, eventually, may help to halt the outflow. However, as shown by Argentina's experience, when confidence is low such an automatic adjustment may fail to materialize. Besides, in some of the transition economies, such a mechanism may not be operational due to the specific design of the currency board arrangement. This is especially the case in Bulgaria, where the level of reserves is much higher than the monetary base as a result of which capital flows do not cause immediate changes in the money supply.

To conclude, at present there do not appear to be any immediate risks of "currency board contagion" for the transition economies that adhere to this particular monetary regime. However, due to the loss of degrees of policy freedom, the countries that operate it need to be very cautions in their policies in order to prevent a dangerous macroeconomic destabilization, especially as they are prone to external shocks. Among the key factors of macroeconomic stability and growth in these countries are the pursuit of prudent and sustainable (in the long run) fiscal policies and the creation of a favourable investor climate to attract FDI. However, the important lessons from Argentina are that a society has to make a judgment about the potential long-term risks associated with the currency board as a monetary regime and that it also has to assess whether the long-term benefits of this arrangement justify the implied long-term costs, including the extra costs of safeguarding its stability. These are issues that do not have a universal, "one-fits-all" answer; they have to be addressed by each country individually.

3.2 Output and demand

(i) Patterns of output and demand in 2001

After impressive growth in 2000, the transition economies slowed down during 2001. However – on the basis of preliminary data – they were not greatly affected by the global downturn and their growth in 2001 is expected to be significantly higher than in 1999 (table 3.2.1). Regional differences were marked: the Baltic economies accelerated on average, while both eastern Europe and the CIS slowed down, hampered by industrial weakness. The CIS, however, remained one of the fastest growing regions in the world, and performance improved in Armenia, Kazakhstan, the Republic of Moldova and Ukraine, countries that have only recently recovered from their long transformational recession and the effects of the Russian crisis in 1998. In assessing regional divergences, the impact of the largest economies (Poland and Russia)

---

[206] However, there are still problems with the accurate measurement of the fiscal deficits of the transition economies (box 3.1.1). Consequently, the assessment of their present policy course and long-term fiscal sustainability should be regarded as tentative.

[207] As a word of caution, however, the case of Argentina suggests that even at relatively low levels foreign debt may become a serious risk factor if public confidence falters.

[208] One hypothetical combination of such negative external shocks might be a prolonged slowdown in their main trading partner (the EU) coupled with a persistent appreciation of the euro. Such a slowdown could not only reduce external demand but might also cause a reduction or even reversal in capital flows.

TABLE 3.2.1

GDP and industrial output in eastern Europe, the Baltic states and the CIS, 1999-2001
(Percentage change over the same period of the preceding year)

	GDP		2001					Industrial output		2001				
	1999	2000	QI	QII	QIII	QIV	2001	1999	2000	QI	QII	QIII	QIV	2001
Eastern Europe	1.7	3.7	..	..	..	..	3.0	0.3	8.4	..	..	..	..	3.2
Albania	7.3	7.8	..	..	..	..	7.0*	16.0	12.0	..	..	..	..	-20.0*
Bosnia and Herzegovina	..	9.1	..	..	..	..	8.0*	10.6	8.8	14.5	12.6	13.9	6.7	12.2
Bulgaria	2.4	5.8	4.5	5.1	4.5	..	4.9	-9.3	5.8	5.6	-2.1	3.2	-3.8	0.7
Croatia	-0.4	3.7	4.2	4.7	4.1	4.3*	4.3*	-1.4	1.7	5.6	6.2	6.0	6.1	6.0
Czech Republic	-0.4	2.9	3.8	3.8	3.5	3.2	3.6	-3.1	5.4	10.0	7.2	4.2	5.8	6.8
Hungary	4.2	5.2	4.4	4.0	3.7	3.3	3.8	10.4	18.7	10.6	6.2	-0.7	0.9	4.1
Poland	4.1	4.0	2.3	0.9	0.8	0.3	1.1	3.6	6.7	4.5	-0.7	-0.8	-2.6	–
Romania	-1.2	1.8	4.8	5.1	6.8	..	5.3	-2.2	8.2	10.8	10.0	4.2	7.8	8.2
Slovakia	1.9	2.2	3.0	2.8	3.5	3.9	3.3	-3.0	9.3	5.2	5.8	5.1	6.1	5.6
Slovenia	5.2	4.6	3.2	2.7	3.3	2.7*	3.0*	-0.5	6.2	4.7	1.8	2.7	2.6	2.9
The former Yugoslav Republic of Macedonia	4.3	4.6	-6.4	-5.0	..	..	-4.6	-2.6	3.5	-8.7	-8.9	-14.1	-8.8	-10.1
Yugoslavia	-17.7	6.4	..	..	..	..	6.2	-23.1	11.2	-0.5	-4.1	-6.0	10.1	–
Baltic states	-1.7	5.4	..	..	..	..	6.2	-8.0	7.1	..	..	..	..	12.5
Estonia	-0.7	6.9	5.8	5.0	5.0	5.5	5.3	-3.4	13.1	7.0	5.8	7.7	9.3	7.5
Latvia	1.1	6.8	8.3	9.3	6.4	6.3	7.6	-5.4	4.7	8.5	9.7	8.8	6.9	8.4
Lithuania	-3.9	3.9	4.4	5.7	5.0	7.9	5.7	-11.2	5.3	12.4	19.0	10.3	25.8	16.9
CIS	4.5	8.3	..	..	..	..	6.2	9.2	11.6	..	..	..	..	6.7
Armenia	3.3	6.0	12.3	3.1	13.5	8.9	9.6	5.3	6.4	9.2	-3.2	5.9	3.8	3.8
Azerbaijan	7.4	11.1	8.1	8.7	11.1	11.6	9.9	3.6	6.9	6.1	4.1	6.0	4.2	5.1
Belarus	3.4	5.8	2.4	5.1	4.5	4.0	4.1	10.3	7.8	2.1	5.9	5.3	7.9	5.4
Georgia	3.0	2.0	1.6	8.6	3.0	4.8	4.5	7.4	6.1	-10.6	5.1	4.3	-2.3	-1.1
Kazakhstan	2.7	9.8	11.2	16.6	13.2	11.8	13.2	2.7	15.5	11.2	15.9	13.7	12.8	13.5
Kyrgyzstan	3.7	5.4	5.2	4.9	8.3	1.9	5.3	-4.3	6.0	14.2	-0.1	5.8	-0.8	5.4
Republic of Moldova	-3.4	2.1	2.6	3.6	5.1	13.0	6.1	-11.6	7.7	3.1	21.3	12.4	19.8	14.2
Russian Federation	5.4	9.0	4.9	5.3	4.9	..	5.0	11.0	11.9	5.2	5.9	4.5	4.1	4.9
Tajikistan	3.7	8.3	7.6	13.0	15.7	4.9	10.2	5.6	10.3	13.9	12.1	22.0	11.5	14.8
Turkmenistan	17.0	17.6	10.0	20.0	24.2	27.4	20.5	15.0	30.0	4.5	12.6	10.9	16.0	11.0
Ukraine	-0.2	5.9	7.8	10.8	10.7	..	9.1	4.0	12.4	15.1	18.4	12.5	7.6	14.2
Uzbekistan	4.4	4.0	2.8	5.6	5.1	4.5	4.5	6.1	6.4	7.0	8.0	7.8	0.6	5.8
Total above	3.3	6.5	..	..	..	..	5.0	4.9	10.1	..	..	..	..	5.3
Memorandum items:														
CETE-5	3.0	3.8	..	..	..	..	2.3	2.4	8.7	..	..	..	..	2.8
SETE-7	-1.7	3.5	..	..	..	..	4.9	-6.6	7.2	..	..	..	..	4.4
Former GDR	1.4	2.1	..	..	..	..	-0.7	7.4	10.5	7.3	4.2	3.9	-0.1	3.7

Source: National statistics; CIS Statistical Committee; direct communications from national statistical offices to UNECE secretariat.

Note: Quarterly industrial output figures for 2001 in table 3.2.1 are based on monthly data. Various types of monthly and quarterly indices of industrial output (fixed-base index, month over previous month, month over corresponding month of previous year, cumulative months over cumulative months of previous year, quarter over previous quarter, quarter over corresponding quarter of previous year) published by transition countries often contradict each other. The reasons for that are perhaps related to problems in source data and to inadequate revision and dissemination practices. In the above table, officially reported quarterly indices were used for Poland and Slovenia. Quarterly indices were calculated from fixed-base monthly indices, either reported by countries or constructed by the UNECE secretariat from other types of monthly indices, for Bosnia and Herzegovina, Croatia, the Czech Republic, Hungary, Romania, Slovakia, The former Yugoslav Republic of Macedonia, Yugoslavia, Estonia, Latvia, Lithuania, Belarus, Kazakhstan, Kyrgyzstan, the Russian Federation and Ukraine. For Bulgaria, Armenia, Azerbaijan, Georgia, the Republic of Moldova, Tajikistan, Turkmenistan and Uzbekistan fixed-base indices are not available, but indices of cumulative months over cumulative months of the preceding year are published. For these countries a special technique was used to determine a unique decumulated index level based on a criterion of minimum seasonality. This technique was also used to estimate quarterly GDP growth rates for Azerbaijan, the Republic of Moldova, Tajikistan, Turkmenistan and Uzbekistan, countries which do not produce national accounts figures on the basis of discrete quarters. On regional aggregates and other information see the notes to table 1.3.1.

on the aggregate performance of eastern Europe and the CIS should be borne in mind. Given its weight within eastern Europe, the decline in Polish GDP growth from 4 per cent year-on-year in 2000 to 1.1 per cent in 2001 cost the region about 1 percentage point in growth, more or less the full amount of the deterioration between 2000 and 2001. Similarly, the deceleration of growth in Russia more than accounted for the decline of about 2 percentage points in the average CIS growth rate between 2000 and 2001.

Apart from the effect of the two largest economies, where activity slowed down for quite different reasons, there were also significant differences both over time and across countries. While growth decelerated in slightly less than half of the transition countries during 2001, the majority were able to sustain or even increase their growth rates. Overall, industrial production, the largest and most outward oriented component of output in most countries, decelerated much more sharply than GDP (table 3.2.1), which suggests that external demand

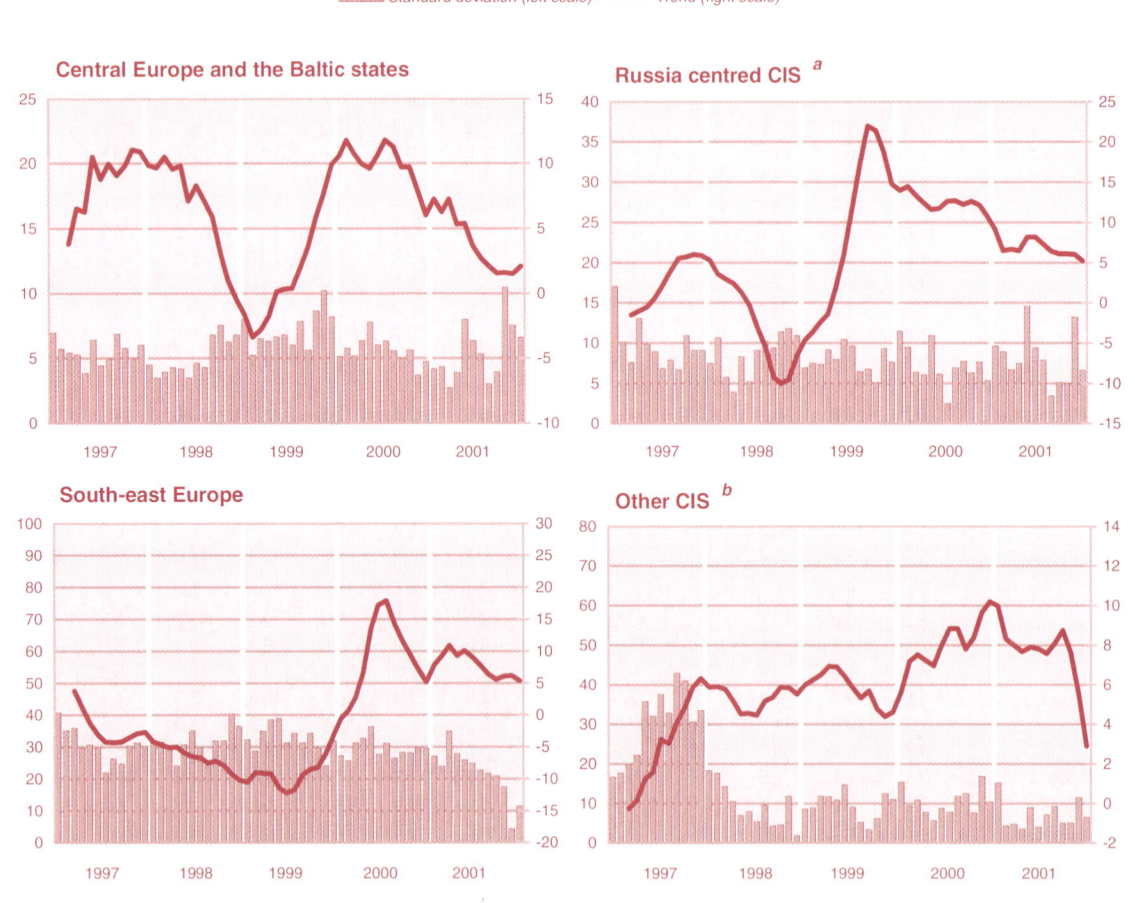

CHART 3.2.1

Trend and dispersion of monthly changes in industrial production in transition economies by subregions, January 1997-December 2001

(Per cent)

Source: UNECE Common Database.

Note: Trend is the three-month moving average of cross-country weighted averages of monthly year-on-year rates of change of industrial output, using constant 2000 industrial production weights. Dispersion is measured by the unweighted standard deviation on the same rates of change. South-east Europe excludes Bosnia and Herzegovina.

[a] Armenia, Belarus, Georgia, Kazakhstan, Republic of Moldova, Russian Federation and Ukraine.

[b] Azerbaijan, Kyrgyzstan, Tajikistan, Turkmenistan and Uzbekistan.

explains much of the decline in the area's growth during 2001. Chart 3.2.1 traces the dynamics of industrial production, usually a good leading indicator of aggregate GDP developments, and shows decreasing industrial output growth in all subregions in the second half of 2001. Apart from Poland, *central Europe* and the *Baltic states* essentially constitute a group of small open economies, to different degrees integrated with western Europe (chart 3.2.2), and the variations of industrial production across these countries often signal turning points in their aggregate output. Industrial output growth in this group of countries has been declining since mid-2000 in the wake of west European demand, and industrial growth in 2001 was significantly lower than in 2000. The deceleration of GDP growth, however, was weaker than for industrial output. This was due to stronger agricultural output and to the response of the services sectors to increased domestic demand, which in part was a lagged reaction to higher real wages and profits during the earlier period of trade-led growth. In *south-east Europe,* industrial production in 2001 followed a cycle that was a milder version of that in 2000: the slowdown, which began in mid-2000, was interrupted in early 2001. The decline in industrial production growth was therefore less marked than in central Europe, while GDP growth even accelerated.

As Russia clearly dominates the aggregate performance of its most integrated neighbours (Armenia, Belarus, Georgia, Kazakhstan, the Republic of Moldova and Ukraine),[209] the dispersion of industrial production across the Russia centred CIS cannot necessarily be taken as a leading indicator for the regional aggregate. This aggregate – along with Russian industry – has been on a

---

[209] Industrial production in all of these countries is well correlated with Russian industrial output except for neighbouring Belarus, which, however, enjoys close trade links with Russia.

The Transition Economies 79

CHART 3.2.2

Dynamics of monthly industrial production in transition economies and the European Union, centred three-month moving averages, January 1998-December 2001

(Year-on-year percentage change)

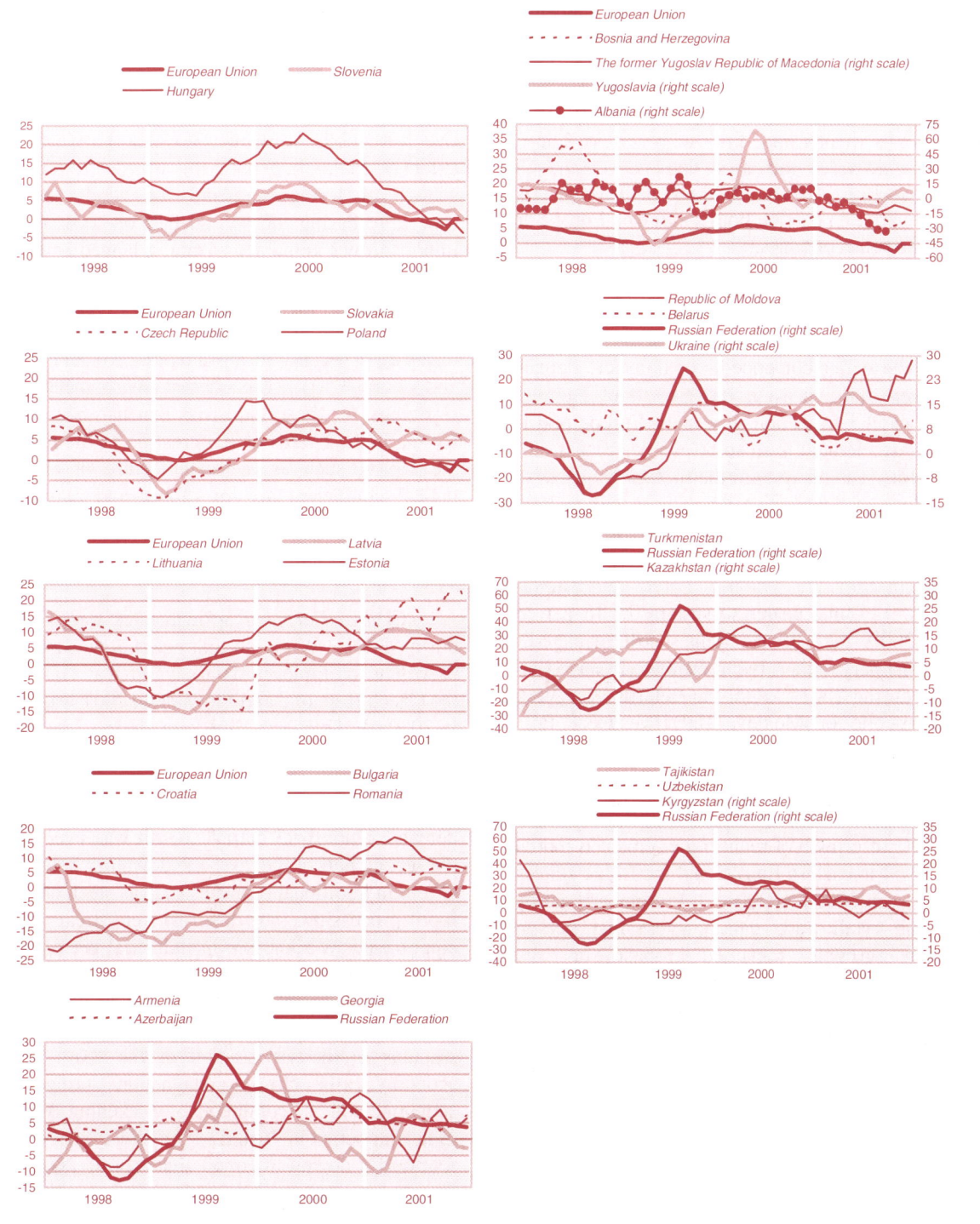

Source: UNECE Common Database.

declining growth path since the fall of 1999, when Russian industrial output growth peaked at an annual rate of more than 25 per cent. The other CIS economies comprise energy exporters (Azerbaijan and Turkmenistan) and agricultural and non-agricultural commodity producers (Kyrgyzstan, Tajikistan and Uzbekistan) significantly less integrated with Russia. Recent developments in this group – dominated by the gradual decline in Uzbek industrial output growth – indicate that the upturn of 2000 was reversed during 2001.

These general developments were related to various supply- and demand-side impacts. Supply-side influences in eastern Europe and the Baltic states included an ongoing, often FDI-induced, industrial restructuring, and the emergence of more favourable energy prices since mid-2001, but also disruptions due to regional conflicts, such as that in The former Yugoslav Republic of Macedonia. While there were similar disruptions in the Caucasian rim and in central Asia, supply-side influences in the CIS were dominated by the improving business climate in Russia. The impact of external demand on eastern Europe and the Baltic states was mainly due to the west European slowdown (chart 3.5.1). The relatively favourable performance of the CIS reflected to a great extent its weaker trade links with western Europe, while Russia provided increased demand for the CIS economies (section 3.1(iv)). The recent Turkish crisis has had a negative impact on both south-east Europe and the Caucasian rim economies.

As a result, the general pattern in 2001 was one of strengthening domestic demand versus a weakening net external demand throughout the whole area (table 3.2.2). Changes in the major components of domestic demand, however, were quite uneven: the growth of aggregate gross fixed capital formation in eastern Europe and the Baltic states decelerated from some 3.6 per cent in 2001 to 1.8 per cent (year-on-year) in the first three quarters of 2001.[210] Much of this decline was due to falling Polish demand. Investment also declined in Slovenia, grew moderately in Hungary and more strongly elsewhere, particularly in Slovakia, Estonia, Latvia and Bulgaria (table 3.2.3).[211] While unable to sustain the 18 per cent growth of 2000, the CIS economies still managed to increase investment outlays by an estimated year-on-year 10 per cent during 2001.[212] Only in Belarus, Kyrgyzstan and the Republic of Moldova did investment fall,[213] suggesting that their industrial output growth may have been achieved by greater utilization of existing fixed assets rather than by their modernization (table 3.2.4).

The rate of growth of retail trade remained more or less unchanged in eastern Europe and the Baltic states as a whole between 2000 and 2001.[214] Preliminary national accounts and retail trade data suggest that personal consumption growth remained robust or even strengthened in Croatia, the Czech Republic, Hungary, Slovakia, Latvia and especially in Romania. Retail trade accelerated in the CIS driven by higher rates of real wage growth than in eastern Europe and the Baltic states (table 3.2.5).

---

[210] UNECE secretariat calculations, based on PPP-adjusted GDP-weighted data from the UNECE Common Database.

[211] All of these countries enjoyed relatively large FDI inflows during 2001, ranging from 4.4 per cent of GDP (in Croatia) to 11 per cent (in Estonia). See sect. 3.5(iii).

[212] Interstate Statistical Committee of the CIS, "Economic situation of the countries of the Commonwealth of Independent States in 2001", *Press Release* (Moscow), 4 February 2002.

[213] Kyrgyzstan had to scale down public investment in an IMF-assisted attempt to reduce public deficits.

[214] According to UNECE secretariat calculations, based on PPP-adjusted GDP-weighted data from the UNECE Common Database.

(ii) Eastern Europe and the Baltic states

Preliminary information and national accounts data for the first three quarters of 2001 point to a switch from net external demand to domestic demand as the main support of real GDP growth. The exceptions were Poland and Slovenia, where domestic demand was weak, and Hungary, where import growth declined ahead of a deceleration in exports (chart 3.5.3). However, the switch was quite pronounced in Bulgaria, Croatia, the Czech Republic, Romania, Slovakia, Latvia and, to some extent, Estonia. In general this was due to the weakening growth of foreign demand for goods and services, especially during the second half of 2001. In contrast, import growth remained strong, a lagged response to the earlier phase of trade-led growth and some real appreciation of the region's major currencies vis-à-vis the euro,[215] which in turn led to a worsening of external balances, especially in Slovakia and Romania (table 3.2.2 and section 3.5(i)). The external environment deteriorated especially for those industries that had been responsible for the highest rates of job creation in the region during the 1990s, such as the manufacture of electrical and optical equipment and of rubber and plastics (section 3.4(iii) and chart 3.2.3). Poland, Slovenia and Estonia were especially affected by waning growth in both of these important branches.[216]

Central European growth rates were quite diverse during 2001. While an aggregate downward trend was prevalent, preliminary data suggest that GDP growth in the Czech Republic and Slovakia in 2001 was higher than in 2000. In Hungary, GDP growth decelerated gradually through the year, and industrial output, on a downward trend since mid-2000, increased by less than one quarter of the previous year's rate due to waning external demand. Industrial output growth was even negative in the second half of the year, the downturn being especially severe in traditional branches such as steel, chemicals and food processing, but also for electrical and optical equipment. Weaknesses in industrial performance were partly offset by stronger growth of retail trade and a boom in housing and public construction fuelled by strong wage growth, especially in the public sector, and a large fiscal stimulus in the second half of the year. The Polish economy came close to stagnation during 2001 with economic activity continuously weakening over the year. Industrial output in Poland – with minor interruptions – has followed a downward trend since spring 2000, and there was no year-on-year growth at all in 2001. While consumption provided a modest support to GDP growth, a

---

[215] Except for the Slovenian and Baltic currencies in 2001.

[216] The Slovak rubber and plastics industry suffered especially from the weakening of EU demand for cars. By localizing its supply chain, the country's leading exporter, Volkswagen Bratislava, had earlier provided a strong stimulus to that branch. On recent production losses in the electrical equipment sector due to Finnish electronic parts maker Elcoteq's loss of markets in the mobile phone business, see UNECE, *Economic Survey of Europe, 2001 No. 2*, p. 22, box 1.2.2.

TABLE 3.2.2

Contribution of final demand components to real GDP growth in selected east European, Baltic and CIS economies, 1999-2001
(Percentage points)

	1999	2000	2001 [a]		1999	2000	2001 [a]
Bulgaria				**Slovakia**			
Consumption	4.1	4.1	0.2	Consumption	-1.6	-1.9	3.0
Fixed investment	3.3	1.3	2.7	Fixed investment	-7.0	-0.2	3.3
Changes in stocks	-0.2	0.1	3.3	Changes in stocks	3.4	0.8	1.1
Domestic demand, total	7.2	2.6	6.2	Domestic demand, total	-5.1	-1.3	7.4
Net trade	-5.1	3.1	-2.6	Net trade	7.1	3.6	-4.1
Exports	-2.5	10.7	7.7	Exports	2.3	10.9	5.0
Imports	-2.6	-7.6	-10.2	Imports	4.8	-7.3	-9.1
GDP	2.4	5.8	4.7	GDP	1.9	2.2	3.3
Croatia				**Slovenia**			
Consumption	-1.4	2.2	1.6	Consumption	4.3	1.1	1.6
Fixed investment	-0.3	-0.8	1.7	Fixed investment	4.9	0.1	-1.1
Changes in stocks	-0.4	0.8	3.2	Changes in stocks	0.2	0.1	-0.6
Domestic demand, total	-2.1	2.2	6.5	Domestic demand, total	9.4	1.2	-0.1
Net trade	1.7	1.5	-2.2	Net trade	-4.2	3.4	3.2
Exports	0.3	3.7	3.9	Exports	1.1	7.4	4.9
Imports	1.4	-2.2	-6.1	Imports	-5.3	-4.0	-1.7
GDP	-0.4	3.7	4.3	GDP	5.2	4.6	3.1
Czech Republic				**The former Yugoslav Republic of Macedonia**			
Consumption	1.0	0.8	1.9	Consumption	3.4	..	..
Fixed investment	-0.2	1.4	2.3	Fixed investment	-0.3	..	..
Changes in stocks	-1.4	1.7	1.8	Changes in stocks	-1.6	..	..
Domestic demand, total	-0.6	3.9	6.0	Domestic demand, total	1.5	..	..
Net trade	0.2	-1.0	-2.4	Net trade	2.8	..	..
Exports	4.2	12.4	9.9	Exports	2.8	..	..
Imports	-4.0	-13.4	-12.3	Imports	-0.1	..	..
GDP	-0.4	2.9	3.6	GDP	4.3	..	..
Hungary				**Estonia**			
Consumption	3.1	2.9	2.5	Consumption	0.5	5.0	1.7
Fixed investment	1.4	1.8	0.6	Fixed investment	-4.6	0.5	3.5
Changes in stocks	-0.4	0.5	-0.8	Changes in stocks	-0.2	2.0	2.1
Domestic demand, total	4.1	5.2	2.4	Domestic demand, total	-4.2	7.5	7.3
Net trade	0.1	–	1.6	Net trade	5.5	-1.2	-4.3
Exports	6.6	12.0	8.1	Exports	-0.5	27.5	2.0
Imports	-6.5	-12.0	-6.5	Imports	6.0	-28.7	-6.3
GDP	4.2	5.2	4.0	GDP	-0.7	6.9	5.3
Poland				**Latvia**			
Consumption	3.5	1.9	1.2	Consumption	3.3	3.4	5.6
Fixed investment	1.7	0.7	-1.7	Fixed investment	-1.1	5.2	2.6
Changes in stocks	-0.1	0.3	-1.3	Changes in stocks	-1.1	-5.0	0.5
Domestic demand, total	5.1	2.9	-1.8	Domestic demand, total	1.1	3.6	8.7
Net trade	-1.0	1.0	3.1	Net trade	-0.1	3.2	-0.8
Exports	-0.7	6.1	..	Exports	-3.6	6.4	4.7
Imports	-0.3	-5.1	..	Imports	3.6	-3.2	-5.4
GDP	4.1	4.0	1.3	GDP	1.1	6.8	7.9
Romania				**Lithuania**			
Consumption	-2.3	1.0	4.9	Consumption	-2.2	3.3	2.6
Fixed investment	-0.9	0.8	1.2	Fixed investment	-1.9	-1.2	1.6
Changes in stocks	-0.9	2.5	2.4	Changes in stocks	-1.3	-1.9	0.6
Domestic demand, total	-4.0	4.3	8.6	Domestic demand, total	-5.4	0.3	4.7
Net trade	2.8	-2.5	3.3	Net trade	1.5	3.5	0.3
Exports	2.4	6.8	3.5	Exports	-10.3	7.2	11.2
Imports	0.5	-9.3	-6.8	Imports	11.8	-3.6	-10.9
GDP	-1.2	1.8	5.3	GDP	-3.9	3.9	5.1

(For source and notes see end of table.)

TABLE 3.2.2 (concluded)

Contribution of final demand components to real GDP growth in selected east European, Baltic and CIS economies, 1999-2001
(Percentage points)

	1999	2000	2001 [a]		1999	2000	2001 [a]
Armenia				**Kyrgyzstan**			
Consumption	0.4	6.8	7.9	Consumption	1.0	-2.8	..
Fixed investment	0.1	1.9	1.4	Fixed investment	3.6	4.2	..
Changes in stocks	0.1	-0.3	-0.6	Changes in stocks	–	-0.2	..
Domestic demand, total	0.6	8.4	8.6	Domestic demand, total	4.6	1.2	..
Net trade	3.3	-0.9	5.8	Net trade	-0.9	4.2	..
Exports	3.3	4.8	..	Exports	-3.8	4.4	..
Imports	–	-5.7	..	Imports	2.9	-0.2	..
GDP	3.3	6.0	10.0	GDP	3.7	5.4	..
Azerbaijan				**Republic of Moldova**			
Consumption	9.1	9.1	..	Consumption	-15.9	15.4	9.2
Fixed investment	-0.7	0.7	..	Fixed investment	-5.1	-1.6	-0.5
Changes in stocks	–	-0.1	..	Changes in stocks	–	4.3	-1.6
Domestic demand, total	8.5	9.8	..	Domestic demand, total	-21.0	18.2	7.1
Net trade	11.6	-2.9	..	Net trade	17.6	-16.1	-1.1
Exports	11.4	4.3	..	Exports	1.4	5.0	7.3
Imports	0.2	-7.2	..	Imports	16.2	-21.1	-8.4
GDP	7.4	11.1	..	GDP	-3.4	2.1	6.1
Belarus				**Russian Federation**			
Consumption	6.5	5.8	8.1	Consumption	-1.8	5.1	3.8
Fixed investment	-1.0	0.6	-0.4	Fixed investment	0.9	2.1	1.3
Changes in stocks	-3.2	2.9	-0.6	Changes in stocks	0.5	2.6	1.9
Domestic demand, total	2.2	9.2	7.1	Domestic demand, total	-0.5	9.8	7.0
Net trade	4.3	-0.8	1.8	Net trade	5.7	-0.9	-2.0
Exports	4.1	7.5	..	Exports	3.5	4.4	1.3
Imports	0.1	-8.3	..	Imports	2.3	-5.2	-3.3
GDP	3.4	5.8	4.1	GDP	5.4	9.0	5.0
Kazakhstan				**Ukraine**			
Consumption	1.2	5.9	8.8	Consumption	-3.0	1.5	6.8
Fixed investment	0.1	2.3	4.1	Fixed investment	–	2.4	1.6
Changes in stocks	1.5	-0.7	3.4	Changes in stocks	-3.5	1.9	0.2
Domestic demand, total	2.7	7.5	16.3	Domestic demand, total	-6.6	5.8	8.6
Net trade	0.7	2.5	-4.4	Net trade	6.4	0.1	0.6
Exports	0.9	14.0	-0.6	Exports	-0.9	11.7	1.8
Imports	-0.2	-11.5	-3.8	Imports	7.4	-11.6	-1.3
GDP	2.7	9.8	11.2	GDP	-0.2	5.9	9.1

Source: National statistics; CIS Statistical Committee; direct communications from national statistical offices to UNECE secretariat.

Note: The sum of component changes does not add up to the GDP change for Bulgaria, Estonia, Armenia, Azerbaijan, Belarus, Kazakhstan, the Russian Federation and Ukraine because of a reported statistical discrepancy in each case.

[a] QI-III for Bulgaria, Croatia, Hungary, Poland, Slovenia, Estonia, Latvia, Lithuania, Armenia, Belarus and Kazakhstan.

sharp plunge in investment led to a negative contribution of domestic demand to GDP growth in 2001 (table 3.2.2). The small increase in GDP was in fact due to imports growing more modestly than exports in the face of subdued domestic demand. A milder version of the pattern of Polish domestic demand can also be seen in Slovenia, where GDP growth, however, showed no clearly decelerating trend during 2001. The downward trend in industrial production, following the EU pattern since mid-2000, was interrupted in the spring of 2001, but on average output growth in 2001 was half its rate in 2000.[217] Construction was in recession throughout the year, as investment was cut back in response to relatively high real interest rates (table 3.1.1) and there was only little support from FDI inflows (section 3.5(iv)). Slow consumption growth was unable to offset the negative impacts of fixed investment and stock changes on GDP growth, but the competitiveness of the tolar proved sufficient to withstand the slowdown in European activity, although the positive contribution of net trade to GDP growth diminished over the year.

In contrast to other countries in the region, manufacturing output in the Czech Republic and Slovakia, although decelerating over the course of the year, remained the engine of growth on the supply side. Growth in personal consumption, supported by moderate increases in real wages,[218] and above all FDI-fuelled capital formation,

---

[217] Sales of manufactured goods, however, proved unable to keep up with production, and there was a significant increase of stocks during the first three quarters of the year. Institute of Macroeconomic Analysis and Development, *Slovenian Economic Mirror* (Ljubljana), October 2001, p. 8.

[218] Consumption reacted positively to the repayment of National Property Fund bonds issued as compensation for the cancellation of the 1995 mass privatization.

TABLE 3.2.3

Real domestic demand components in selected east European, Baltic and CIS economies, 1999-2001

(Annual percentage change)

	Private consumption expenditure [a]			Government consumption expenditure [b]			Gross fixed capital formation		
	1999	2000	2001 [c]	1999	2000	2001 [c]	1999	2000	2001 [c]
Eastern Europe									
Bulgaria	5.2	3.4	-0.2	2.0	9.8	1.8	25.3	8.2	18.2
Croatia	-2.7	4.1	4.8	0.8	-0.7	-4.5	-1.1	-3.5	7.4
Czech Republic	1.9	1.9	3.7	-0.1	-1.3	-1.0	-0.6	4.2	7.0
Hungary	5.4	4.3	4.5	1.5	3.2	1.0	5.9	7.7	3.0
Poland	5.2	2.7	1.8	1.0	1.3	-0.1	6.8	2.7	-7.8
Romania	-2.1	-0.6	6.5	-4.5	10.0	1.8	-4.8	4.6	6.6
Slovakia	-0.2	-3.4	4.0	-6.9	-0.9	5.2	-18.8	-0.7	11.6
Slovenia	6.0	0.8	1.6	4.6	3.1	3.3	19.1	0.2	-3.8
The former Yugoslav Republic of Macedonia	3.6	..	..	4.4	..	..	-1.4	..	..
Baltic states									
Estonia	-0.6	8.2	3.0	3.8	0.1	-0.6	-14.6	2.0	13.7
Latvia	5.1	5.6	8.1	–	-1.9	1.0	-4.0	20.0	10.1
Lithuania	2.1	4.6	3.6	-17.5	-0.7	-0.6	-6.3	-3.9	6.2
CIS									
Armenia	0.4	7.1	6.0	-0.6	-0.3	17.8	0.5	11.8	7.8
Azerbaijan	11.6	11.6	..	-1.3	2.3	..	-2.0	2.6	..
Belarus	9.3	7.8	13.6	5.8	6.0	2.7	-4.0	2.3	-1.9
Kazakhstan	0.5	2.4	10.2	7.6	36.1	18.7	0.5	14.0	25.5
Kyrgyzstan	0.3	-5.0	..	4.1	5.9	..	28.1	26.9	..
Republic of Moldova	-8.5	20.9	11.7	-38.3	-1.2	-7.7	-23.1	-8.7	-3.3
Russian Federation	-4.1	9.1	8.5	3.0	1.4	-1.0	5.1	14.6	8.1
Ukraine	-2.2	2.3	8.7	-7.9	1.0	9.6	0.1	12.4	8.3

Source: National statistics; CIS Statistical Committee; direct communications from national statistical offices to UNECE secretariat.

[a] Expenditures incurred by households and non-profit institutions serving households.

[b] Expenditures incurred by the general government on both individual consumption of goods and services and collective consumption of services.

[c] QI-III for Bulgaria, Croatia, Hungary, Poland, Slovenia, Estonia, Latvia, Lithuania, Armenia and Belarus.

backed by high government consumption[219] and the resumption of the highway construction programme, supported Slovak domestic demand after two years of contraction. The high import content of both domestic demand components and the removal of the import surcharge led to a significant widening of the current account, signalling a high external risk in 2002. While overall economic performance has been improving for two and a half years in the Czech Republic, this disguises a fall in industrial production since early spring 2001, which however was reversed in the final quarter of the year. While large inflows of FDI have been beneficial for the level of investment, they have also exerted upward pressure on the koruna reinforcing the influence of weakening external demand and boosting the growth of imports. As a result, GDP growth was driven mainly by strong investment and consumption, and by an expansionary fiscal policy (section 3.1(ii)).

Rather surprisingly for a group of very small, open economies, preliminary data suggest that the *Baltic states*, in aggregate, were able to accelerate their growth during 2001 despite the weakening of global and, especially, west European economic activity. The slowdown in external demand was partly offset by some reorientation of trade away from the EU, as growth in Russia and the real appreciation of the Russian rouble boosted their competitiveness in that market. However, this could only slightly compensate for decelerating EU demand,[220] implying some deceleration for the Estonian economy, which perhaps has the closest links of all the east European and Baltic economies with the European Union (chart 3.2.2). In contrast, Latvia and Lithuania – significantly involved in the transit (and, in the Lithuanian case, refining) of Russian oil – were able to grow strongly during 2001. The very high rate of Latvian GDP growth for the second year running was based to a large extent on the transport sector, which profited considerably from the transit shipment of Russian energy products.[221] Manufacturing output also accelerated sharply. Lithuania's economy also grew more rapidly during 2001, although agricultural output declined by 8.5 per cent.[222]

---

[219] Government consumption contributed 1 percentage point to real 2001 GDP growth in Slovakia.

[220] In the Estonian case, this holds especially for the two industries noted in chart 3.2.3.

[221] Cargo turnover in Latvian ports increased by 9.8 per cent, year-on-year, during 2001; 58 per cent of outgoing cargo consisted of exports of crude and oil products. RFE/RL, *Newsline*, Vol. 6, No. 7, Part II, 11 January 2002.

[222] Statistics Lithuania, *Economic and Social Development in Lithuania* (Vilnius), January 2002, p. 15.

TABLE 3.2.4

Investment outlays in selected east European, Baltic and CIS economies, 1999-2001

(Annual percentage change)

	1999	2000	2001 [a]
Eastern Europe			
Czech Republic	-6.7	3.8	..
Hungary	5.3	7.4	3.5
Poland	5.9	1.4	-18.5
Romania	-5.9	5.0	5.1
Slovakia	-18.9	4.3	..
The former Yugoslav Republic of Macedonia	30.7	52.1	-11.3
Baltic states			
Latvia	-5.2	22.4	10.0
Lithuania	-6.8	-2.0	11.7
CIS			
Armenia	-2.0	25.9	14.0
Azerbaijan	-2.0	3.0	17.1
Belarus	-8.0	2.0	-6.1
Georgia	-51.4	2.0	3.0*
Kazakhstan	22.4	38.2	21.0
Kyrgyzstan	-1.8	4.2	-16.0
Republic of Moldova	-22.0	-15.0	-2.0
Russian Federation	5.3	17.4	8.7
Turkmenistan	14.0	8.0	..
Ukraine	0.4	14.4	17.2
Uzbekistan	2.0	1.0	3.0

Source: National and CIS Statistical Committee data; direct communications from national statistical offices to UNECE secretariat.

Note: "Gross capital formation" and "gross fixed capital formation" are standard categories of the United Nations 1993 SNA (System of National Accounts) and the European Union's 1995 ESA (European System of Accounts). Gross capital formation includes gross fixed capital formation plus changes in inventories and acquisitions less disposal of valuables. "Investment outlays" (also called "capital investment" in some transition economies) mainly refers to expenditure on construction and installation works, machinery and equipment. Gross fixed capital formation is usually estimated by adding the following components to "capital investment": net changes in productive livestock, computer software, art originals, the cost of mineral exploration and the value of major renovations and enlargements of buildings and machinery and equipment (which increase the productive capacity or extend the service life of existing fixed assets).

[a] QI-III for Latvia and Uzbekistan; QI for The former Yugoslav Republic of Macedonia.

TABLE 3.2.5

Retail trade in the east European, Baltic and CIS economies, 1999-2001

(Percentage change over the same period of previous year)

	1999	2000	2001
Eastern Europe			
Albania	..	..	..
Bosnia and Herzegovina	..	59.4	..
Bulgaria	12.3	0.7	4.3
Croatia	-4.7	8.0	10.0
Czech Republic	3.3	5.3	4.3
Hungary	7.9	2.0	5.4
Poland	4.0	1.0	0.7
Romania	-5.0	-3.8	1.3
Slovakia	9.8	2.3	4.2
Slovenia	2.9	7.4	7.8
The former Yugoslav Republic of Macedonia	17.6	11.1	0.3
Yugoslavia	-13.5	10.3	16.0
Baltic states			
Estonia	4.4	12.9	13.0
Latvia	12.0	9.0	9.0
Lithuania	-5.0	10.9	7.4
CIS			
Armenia	11.0	8.5	15.5
Azerbaijan	13.3	9.8	9.9
Belarus	10.7	12.0	21.2
Georgia	4.6	11.4	5.6
Kazakhstan	2.3	4.8	14.2
Kyrgyzstan	0.8	7.7	5.9
Republic of Moldova [a]	-27.4	4.0	18.2
Russian Federation	-7.7	8.9	10.8
Tajikistan	4.0	-21.2	1.2
Turkmenistan [b]	52.0	28.0	32.0
Ukraine	-4.1	6.4	11.6
Uzbekistan	10.5	7.8	9.5

Source: National statistics, CIS Statistical Committee and direct communications from national statistical offices to UNECE secretariat.

Note: Retail trade covers mainly goods in eastern Europe, the Baltic states, Kazakhstan and Russia; it comprises goods and catering in other CIS countries. The most recent data for The former Yugoslav Republic of Macedonia are subject to regular and considerable revisions.

[a] Registered enterprises through 2000.

[b] On growth rates for Turkmenistan, see notes to table 1.3.1.

The basis of this acceleration was a 16.9 per cent increase in industrial output, much of which was due to a 47.3 per cent increase in refined oil products.[223] The sources of growth on the demand side differed to some extent between the Baltic neighbours: in both Estonia and Latvia fixed capital formation rose at two-digit growth rates, making Lithuania's 6.2 per cent increase look comparatively modest.[224] Large increases in retail sales also indicate strong domestic consumption in Estonia and Latvia, and to a lesser extent in Lithuania, where net foreign trade still provided some slight support to GDP growth during the first three quarters of 2001.

[223] Statistics Lithuania, *Business Activity Statistics*: [http://www.std.lt/STATISTIKA/Verslas/Pramone/pard_prod_ind3e.htm].

[224] All year-on-year increases during the first three quarters of 2001 (table 3.2.3).

In the *south-east European* economies aggregate performance was also much better than in central Europe, as growth accelerated in the largest economy in the region, Romania, and remained strong in Bulgaria. Supported by a broadly-based expansion of industrial output Romania's economy grew by 5.3 per cent in 2001, following a tentative recovery in 2000 from the long 1997-1999 recession. Export-oriented branches, including machinery and equipment, electrical machinery and appliances, and clothing, grew particularly fast. Support for the recovery also came from agriculture, following a severe drought in 2000. The demand side reflected strong consumption,[225] bolstered by higher real net wages and lower unemployment, and FDI-led investment demand,

[225] Despite strong consumption growth, as reported in the national accounts, retail trade remained subdued (table 3.2.5).

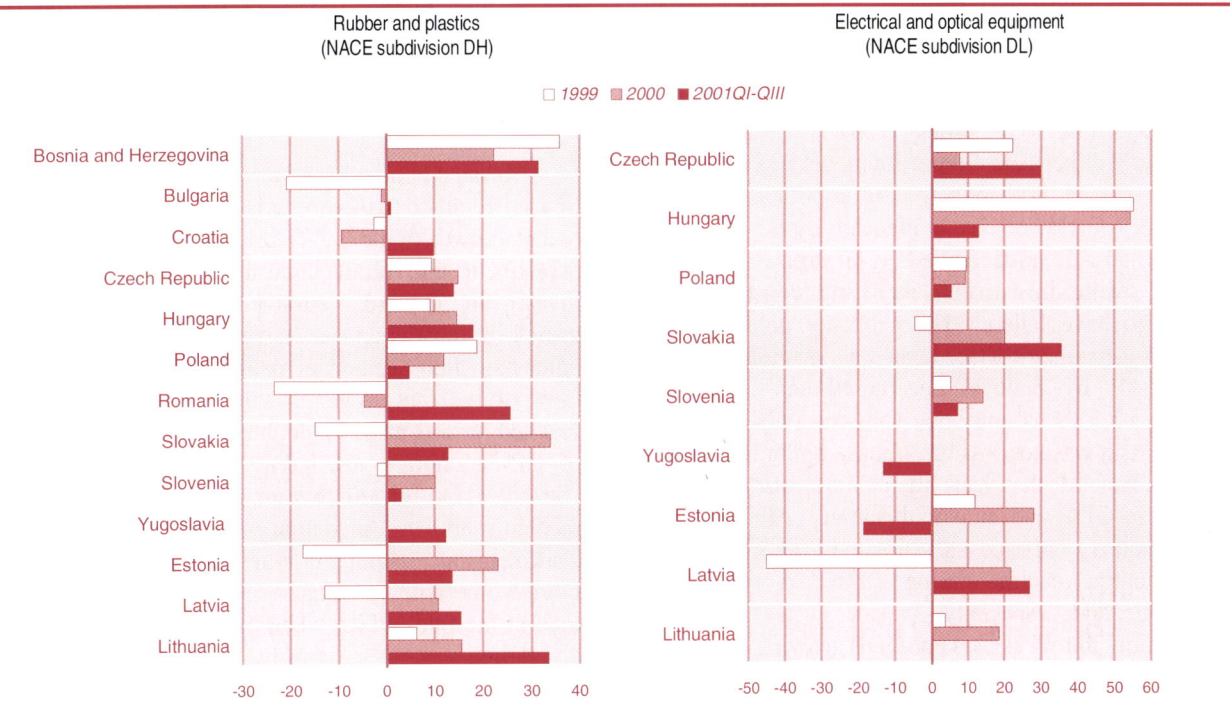

CHART 3.2.3

Annual rates of change of manufacturing output by NACE industries in selected east European and Baltic economies, 1999-2001
(Per cent)

Source: UNECE Common Database.

Note: Data for 2001 for Bosnia and Herzegovina, Croatia and Slovakia cover the full year. For Yugoslavia data for 1999 and 2000 are not available. Lithuania registered no change in output of NACE DL during 2001QI-QIII.

both supported by relatively low real interest rates (table 3.1.1). However, strong export growth was surpassed by an even larger increase in imports, raising concerns about the external balance of the economy.

Although Bulgaria was affected in 2001 both by the EU downturn and the recent Turkish crisis, it proved possible to maintain a high rate of growth. Domestic demand strengthened, almost exclusively due to buoyant investment, while neither private or public consumption nor net trade generated significant support to demand. Under these conditions, industrial output remained almost stagnant, significantly down from 2000.[226] The performance of the traditional export branches reflected the changes in external demand: while the textile industry grew 12.9 per cent year-on-year over the first nine months, basic metals manufacturing remained unchanged, and wood and paper output fell 6 per cent over the same period. Overall, GDP growth was mostly due to construction and services.

The Croatian economy also grew faster in 2001, due to a strong increase in domestic demand, buoyant industrial output, and increased tourist revenues, the latter continuing its upward trend from 2000. Both total industrial output and manufacturing production rose markedly and developed more or less in line. Double-digit increases in the output of machinery and equipment, transport equipment, and pulp and paper are encouraging signs of strengthening industrial export capacities. Due to the strong response of imports to the recent recovery, growth in 2001 was supported by domestic rather than net external demand, in contrast to the previous year. Croatia is currently taking measures to stabilize its hitherto fragile fiscal position; these have encouraged buoyant FDI-supported investment demand after two years of contraction.

The situation in the other south-east European transition economies generally remained rather difficult. Domestic demand in these countries still relies heavily on foreign aid and on remittances from their citizens working abroad, while the shadow economy dominates the supply side. Revitalization requires a change from aid-driven to investment-driven economies, but the integration of large groups of displaced people remains a pressing problem.[227] Given the need to create a genuine single economic space, the situation in Bosnia and Herzegovina remains especially challenging, despite

---

[226] The monthly industrial output data provided by the National Statistical Institute must be treated with some care. The three indices provided – each using a different base period – are not always consistent.

[227] More than 50 per cent of the local population (2.2 million people) have left Bosnia and Herzegovina since 1992. W. Petritsch, "Bosnia and Herzegovina: economic consolidation by reform," Oesterreichische Nationalbank, *Focus on Transition* (Vienna), January 2001, p. 127.

substantial recent achievements that include the creation of a stable currency, notable improvements in the banking sector and steps towards reform of the pension system. The industrial recovery strengthened in 2001 and GDP growth for the whole year is expected to reach about 8 per cent. While there was no increase of industrial output in Yugoslavia, relatively high GDP growth was due to a strong recovery in agricultural output by 23.2 per cent from a low level in 2000.[228] On the demand side, both private consumption and imports continued to recover, as evidenced by a very large increase in retail trade fuelled by a strong rise in real wages,[229] while investment, including construction, is reported to have fallen. The economy of The former Yugoslav Republic of Macedonia was seriously disrupted during 2001 by political uncertainty and by military conflict within the country that subsided only during late August. As it was not possible to make up for the losses in the first half of the year, there were marked annual declines in GDP and industrial output. Albania's state owned industry suffered an estimated 20 per cent year-on-year output decline during 2001. A recovery of agriculture and, especially, private sector growth in construction and services, kept GDP growing strongly in 2001. The prospects for a more sustainable economic development, however, remain seriously compromised by chronic energy crises and the low level of investment in industry.

(iii) Commonwealth of Independent States

The boom in Russia and other CIS energy exporters (Kazakhstan, Azerbaijan, Turkmenistan), coupled with the real appreciation of the Russian rouble vis-à-vis other regional currencies, contributed to a high rate of regional growth in 2001. The aggregate GDP of the CIS increased by 6.2 per cent. In general, growth began to slow down during the second half of 2001 in comparison with the corresponding period of 2000. With outstanding double-digit GDP growth rates in Kazakhstan, Tajikistan and Turkmenistan, growth of 4 to 5 per cent (for Belarus, Georgia and Uzbekistan) appears relatively modest in this high growth region. As in eastern Europe, however, the deceleration in more export-oriented industry was stronger: the aggregate rate of growth of industrial production in the CIS as a whole almost halved from 11.6 per cent in 2000 to 6.7 per cent in 2001. The dispersion of industrial output growth rates across the CIS was more pronounced than for GDP, although the pattern did not quite correspond to a divide between energy exporters and the rest. Georgia was the only country where there was a decline in industrial production, while there was double-digit growth in gas-exporting Turkmenistan, but also in Tajikistan, Ukraine and the Republic of Moldova. In oil-exporting Azerbaijan, there was a comparatively modest increase of 5.1 per cent in industrial output (table 3.2.1). Due to better weather, especially in the Caucasian rim and central Asia, total agricultural production in the CIS increased by 8 per cent in 2001, against 7 per cent in 2000.[230]

Russian national accounts suggest the dominance of domestic rather than external sources in the economy's recent growth (table 3.2.2), and preliminary CIS data on investment and retail trade indicate some shift from investment towards stronger personal consumption growth in most countries (tables 3.2.4 and 3.2.5).[231] The latter has often been accompanied by comparatively high rates of real net wage growth.[232] Retail trade growth reached, or was close to, double-digit rates in the majority of CIS economies, including Kazakhstan, Russia and Ukraine. Oil-related investment has been booming in recent years in Azerbaijan, Kazakhstan, Russia and Turkmenistan,[233] and all four countries are now much more vulnerable to volatile international energy prices than in the mid-1990s, when their economies were more diversified.

There were some genuine economic reform initiatives in Russia in 2001, concerning a customs reform, corporate tax reform, simpler business registration and licensing, and stronger property rights in general. The tax burden has been alleviated, with further reductions to follow in 2002, with a consequent decline in real government consumption expenditure, all of which will bear fruit over the medium to long term. But the impetus for reform has now collided with a weakening of external demand, falling prices for oil and metals, and a rapidly appreciating rouble. The economy grew by 5 per cent in 2001, a significant drop from 2000, with a more pronounced slowdown in the second half of the year. While growth decelerated in four of the five key sectors (agriculture, industry, transport and the still-booming construction industry), retail trade was the only sector to feature an increasing growth rate. Industrial output, on a declining growth trend since the fall of 1999, grew at less than half its rate in 2000. With the falling oil price

---

[228] Yugoslav Federal Statistical Office, *Basic Data on Socio-Economic Trends in the FR Yugoslavia, 2001* (Belgrade), 7 February 2002, pp. 1-4.

[229] Strong wage growth may also have been the reason for the decline in textile production from mid-2001 and could endanger the ability of the industry to exploit the 20 per cent increase in its export quota granted by the EU from 1 January 2002.

[230] Interstate Statistical Committee of the CIS, op. cit.

[231] Double-digit rates of growth of investment outlays are reported for Armenia, Azerbaijan, Kazakhstan and Ukraine.

[232] According to the UNECE Common Database, the exception is Tajikistan, the poorest CIS economy. Nevertheless, real wage increases may not have led to significant changes in living standards everywhere, given the prevailing poverty especially in Armenia, Kyrgyzstan and Tajikistan. World Bank, *Transition – The First Ten Years: Analysis and Lessons for Eastern Europe and the Former Soviet Union* (Washington, D.C.), 2002, p. xiii.

[233] The new Caspian Pipeline considerably increased Kazakhstan's oil export capacity, while Azerbaijan's will remain more constrained until 2005, when the Baku-Tbilisi-Ceyhan pipeline is scheduled to open according to current investment plans. Turkmenistan remains dependent on Russia both as a market and as a transit country for gas.

depressing real export revenue, growth in 2001 was due to domestic sources (table 3.2.2), especially private consumption which was fed by real wage increases in excess of productivity gains, as the investment boom moderated.[234] Nevertheless, the Russian economy still provided enough steam to remain the growth engine for the CIS as a whole (section 3.1(iv)) and, in particular, for those economies for which it remains the largest trading partner, namely, Belarus, the Republic of Moldova and Ukraine.

While there is no obvious single driving force behind the recent recovery of the Ukrainian economy, the combination of a number of factors help to explain its strong performance. These include a low base after a prolonged transformational recession, the remonetization of the economy, a monetary policy leading to some real depreciation against the Russian rouble, and strong external demand, especially from Russia. GDP grew by 9.1 per cent in 2001, although decelerating in the course of the year. In contrast to 2000, when the recovery relied predominantly on booming, export-driven steel production, the expansion in 2001 was more broadly-based, and reflected growth in agriculture, construction, and in manufacturing industries other than steel. The metal processing sector, accounting for more than 20 per cent of industrial output, slowed considerably, but growth was taken up by oil refining, wood products, machine building and food processing. The recent economic acceleration in 2001 has benefited more from domestic demand, especially consumption. Weakening external demand was not only due to the global slowdown, but also to protectionist measures introduced against a background of worldwide overcapacity of heavy metal production, especially steel.[235]

In sharp contrast to Ukraine, the economy of Belarus showed clear signs of supply-side weakening in 2001, as indicated by rising levels of non-cash transactions, rising inventories and shrinking investments (table 3.2.4). While expansionary wage policies in the wake of last year's presidential election campaign led to sharply rising real wages and buoyant personal consumption,[236] this is likely to have had an adverse impact on industrial competitiveness. There were still large increases in production in the food and metal processing industries, driven mostly by domestic demand. Similarly, rising wages and pensions in the Republic of Moldova boosted retail trade, thereby giving some support to domestic demand in the face of shrinking investment. Building on the previous year's recovery, output was fuelled by a record increase in industrial production led by food processing, a major export industry.

The performance of the economies of the *Caucasian rim* varied across countries during 2001, reflecting differences in endowments, exposure to conflicts and resulting supply-side disruptions in output,[237] as well as the strength of their trade links with crisis-ridden Turkey. In all of them, however, there was a strong recovery of agricultural output during 2001 after a serious drought in 2000. This was especially true for Georgia, where industrial output actually declined in 2001. Supply-side constraints, especially irregular power supplies, seem to have been largely responsible for this, and were only partially offset by electricity imports from Armenia. Weaker demand from neighbouring Turkey was coupled with a surge in imports during the first three quarters of 2001, which was influenced by real exchange rate appreciation against the weakened Turkish lira. The strong performance of the Armenian economy in 2001 was also due more to agriculture and services than to the slow-growing industry. The high rate of GDP growth was driven by consumption and investment, both boosted by remittances and other transfers from abroad that are reported to have doubled between 2000 and 2001.[238] Azerbaijan's GDP growth in 2001 continued a long expansion underway since the second half of 1996. Since then, growth has been driven mainly – directly or indirectly – by the oil sector; in 2001 there was a construction boom, led by investments in oil extraction and related transport infrastructure. The uneven pace of growth is illustrated by a 5.9 per cent increase in the output of the extractive industries, but only 2.8 per cent in manufacturing.[239]

The *central Asian* CIS economies also profited from better harvests, as the persistent drought in parts of the region proved less devastating for irrigation-dependent cotton production than a year previously. Industrial output remained dominated by a small range of commodities, although most countries maintained high rates of growth during 2001. The Kazakh economy has improved since the end of 1999 owing to favourable international conditions and prudent policies, including the adoption of a new tax code and a well-balanced monetary policy, which brought about a welcome

---

[234] With real unit labor costs in industry rising sharply, decelerating investment growth has been explained by the declining profits of large and medium enterprises, "Russlands Wirtschaft auf riskantem Kurs", *DIW Wochenbericht*, No. 6 (Berlin), 7 February 2002. However, this explanation contrasts with the narrowing but still sizeable gap between savings and investment, a reflection of chronic capital flight.

[235] For more details see sect. 3.5(ii).

[236] IMF, "IMF concludes 2001 Article IV Consultation with the Republic of Belarus", *Public Information Notice*, No. 02/10 (Washington, D.C.), 19 February 2002.

[237] This relates to energy disruptions, but also to labor force changes in response to poverty and conflicts in the region. For example, according to the preliminary national census conducted in October 2001, some 950,000 people have left Georgia since the last Soviet census in early 1989. RFE/RL, *Newsline*, Vol. 6, No. 33, Part I, 20 February 2002.

[238] *BBC Monitoring Central Asia*, quoted by *Reuters Business Briefing*, 12 February 2002.

[239] *Interfax News Agency*, quoted by *Reuters Business Briefing*, 21 January 2002.

financial deepening of the economy and a steady real depreciation of the currency against the Russian rouble. While the acceleration of growth remained primarily based on oil and gas extraction, which together rose by 13.9 per cent in 2001, and a related construction boom, the economic surge was broadly based. This is shown by a 16.9 per cent rise in agricultural output,[240] as well as growth in non-energy industrial branches such as machinery and equipment and chemicals. On the domestic demand side, robust personal consumption based on large increases in real net wages has reinforced the effects of the continuing investment boom, which is now in its fifth consecutive year. Growth was somewhat narrowly based in mainly gas-exporting Turkmenistan, with the qualification that GDP and industrial output growth figures should be treated with caution.[241] Reports of food production as the fastest growing industrial branch (up by 24 per cent in 2001), however, are consistent with better harvests.[242] Kyrgyzstan's economy, relying heavily on gold mining, posted GDP growth of 5.3 per cent in 2001, to which agriculture and gold mining together contributed about 3.7 percentage points. However, industrial activity apart from gold mining declined by 1.6 per cent, year-on-year.[243] Data for the first three quarters of 2001 indicate a 6.2 per cent increase in agricultural output in Uzbekistan,[244] the most populous country in the region, but the monthly data show a marked deceleration of industrial activity in the course of the year. The economy was faced with falling world prices for its main export commodities, cotton and gold, and limited access to foreign exchange. The system of multiple exchange rates, which has been in place since independence, has so far deterred foreign investment in its rich oil, gas and metal endowments.[245] In Tajikistan, retail trade turnover was almost flat during 2001, pointing to the lack of support of personal consumption for GDP growth, which accelerated mainly because of an increase in agricultural output by 11 per cent[246] and stronger industrial activity.

(iv) Openness and the cyclical behaviour of selected east European and Baltic economies

This note investigates the relationship between trade and macroeconomic fluctuations, focusing on the national income accounts concept of trade in goods and services for 11 east European and Baltic economies (i.e. the 10 current EU candidate countries plus Croatia), hereafter referred to as the EE-11. The first section documents the increasing international openness of these economies and is followed by an account of the role of international trade as a catalyst for recent recoveries. After a brief review of trade and output variables over the cycle, an attempt is made to identify the relative influence of exchange rate changes and foreign activity on exports over the short run.

(a) Reform, openness and integration

Transition and liberalization, unprecedented in scale and speed, have facilitated an increasing degree of openness (measured as the sum of real exports and imports of goods and services relative to real GDP) in most transition economies with tangible effects on the geographic orientation and commodity composition of foreign trade. As a result, external liberalization *ceteris paribus* has significantly enhanced the rates of economic growth of the transition economies.[247] Chart 3.2.4 provides evidence for this growing role of international trade. With the exception of Slovenia,[248] EE-11 countries have increased their openness since the early 1990s, some of them to a considerable degree.

Openness to international trade does not necessarily imply integration into one particular economic area with an identifiable business cycle. In fact, the short-term gains from openness derive from diversifying trade to a range of countries, all with different cycles, while an increasing synchronization of business cycles across countries reduces the likelihood of trade having a significant influence on the domestic cycle. In the case of the EU candidate countries, increasing openness has naturally been mainly with the European Union.[249] However, "... trade links alone do not ensure the convergence of business cycles if countries are not

---

[240] State Committee of Kazakhstan on Statistics, "Itogi sotsial'no-ekonomichskogo razvitia respubliki Kazakhstan za janvar'–dekabr' 2001 goda", *Press Release* (Almaty), 18 January 2002.

[241] Officially reported growth rates are based on deflation procedures that are not documented, see note to table 1.3.1.

[242] *Interfax New Agency*, quoted by *Reuters Business Briefing*, 1 February 2002.

[243] National Statistical Committee of Kyrgyz Republic, *Sotsial'no-ekonomicheskoe polozhenie Kyrgyskoi respubliki 2001* (Bishkek), 2002, pp. 4-6.

[244] Interstate Statistical Committee of the CIS, op. cit.

[245] The authorities are committed to unifying the exchange rate regime that has so far transferred an estimated 16 per cent of GDP from gold and cotton exporters to importers, with high welfare losses on import and export markets. C. Rosenberg and M. de Zeeuw, "Welfare effects of Uzbekistan's foreign exchange regime", *IMF Staff Papers*, Vol. 48, No. 1 (Washington, D.C.), December 2001, pp. 160-178.

[246] Interstate Statistical Committee of the CIS, op. cit.

---

[247] The growth effects of external liberalization have been estimated to be of the same order of magnitude as those of privatization and institutional reforms (i.e. price liberalization and competition). S. Fischer, R. Sahay and C. Végh, *From Transition to Market – Evidence and Growth Prospects*, IMF Working Paper, No. WP/98/52 (Washington, D.C.), April 1998.

[248] This notable exception might be taken as a hint to the formidable role of FDI in facilitating export capacity and openness, as this country has so far featured one of the lowest cumulative FDI shares in GDP in the region. UNECE, "Economic growth and foreign direct investment in the transition economies", *Economic Survey of Europe, 2001 No. 1*, chap. 5, especially table 5.2.2, p. 190.

[249] Already at the end of the 1990s the trade intensity of some of these countries with the EU was almost matching intra-EU trade. UNECE, *Economic Survey of Europe, 1998 No. 1*, p. 134.

CHART 3.2.4

The openness of 11 east European and Baltic economies, 1991-2000

(Ratio)

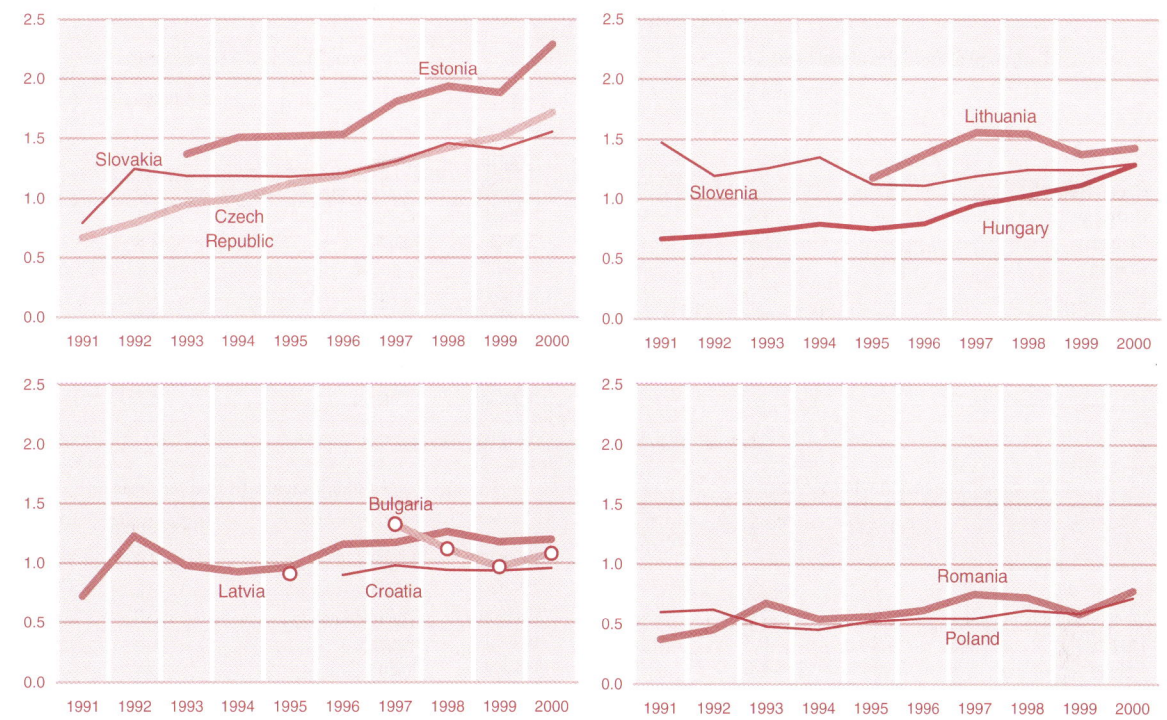

Source: UNECE Common Database.
Note: Openness is defined as the ratio of the sum of real exports and imports to real GDP.

sufficiently similar".[250] As a result, although the east European countries enjoy considerably close trade links with the EU, their respective synchronization with the European business cycle differs greatly: most findings[251] conclude that the Hungarian cycle is quite well correlated, as is the Slovenian and perhaps the Estonian. This holds to a much lesser extent for the others, possibly indicating that a significant asymmetry of shocks between the EU and accession countries still prevails. Recent work on this has arrived at some rather clear results: the correlation of supply shocks with the euro area differs greatly from country to country,[252] with Hungary and Estonia exhibiting the highest correlation, as high as for many current EMU members. Hungary also has a high correlation of demand shocks, which is much lower for the others, even for advanced reformers such as Slovenia and Estonia,[253] or the Czech Republic, while demand shocks are even negatively correlated for Latvia and Lithuania. In addition, still perceived as "emerging markets," east European economies also suffer from different exchange-rate shocks than the EMU. This suggests that foreign trade, especially with the EU, may have a significant influence on short-term output fluctuations in the east European economies, both via foreign activity and exchange rate channels.

(b) Two recent recoveries

The increased role of foreign trade in the east European economies has thus improved their prospects both for long-term growth (via FDI, technological spillover, etc.) and for short-term recoveries via trade-led expansions. One would therefore expect to find – assuming some kind of adequate normalization – that trade had a larger and faster impact on the transition economies' recovery from the Russian crisis, which hit

---

[250] J. Fidrmuc, *The Endogeneity of Optimum Currency Area Criteria, Intra-industry Trade and EMU Enlargement*, Bank of Finland Institute for Economics in Transition (BOFIT) Discussion Papers, No. 8 (Helsinki), 2001, p. 23. Integration of dissimilar countries will simply enforce specialization according to comparative advantage resulting in susceptibility to quite dissimilar shocks. Intra-industry trade (IIT), however, has been demonstrated to be an important measure of structural similarity and thus a factor for inducing harmonization of business cycles within the OECD. Within the logic of theoretical IIT models, however, it should be *horizontal* rather than *vertical* intra-industry trade that qualifies as a proxy for similarity. For evidence on horizontal versus vertical IIT between EE-11 and the EU see sect. 3.5(iii).

[251] Summarized in I. Korhonen, *Some Empirical Tests on the Integration of Economic Activity Between the Euro Area and the Accession Countries*, BOFIT Discussion Papers, No. 9 (Helsinki), 2001.

[252] J. Fidrmuc and I. Korhonen, *Similarity of Supply and Demand Shocks Between the Euro Area and the CEECs*, BOFIT Discussion Papers, No. 14 (Helsinki), 2001.

[253] These two countries' similarity with EU cycles might therefore be related to the *absence* of major shocks, rather than to their correlation with shocks to EU output, especially when considering the short time horizon of these studies.

TABLE 3.2.6

Contributions of exports and net exports to cumulative real GDP growth during two recent recoveries in eastern Europe and the Baltic states
(Percentage points)

	Cumulative growth contributions of exports							
	T = trough of transformational recession				t = trough during the Russian crisis			
T		T+1 year	T+2 years	T+3 years	t		t+4 quarters	t+8 quarters
					1998QIII	Bulgaria	2.0	11.4
					1998QIV	Croatia	1.5	3.2
1992	Czech Republic	6.7	7.5	15.8	1998QIV	Czech Republic	9.3	26.7
1993	Hungary	4.3	7.4	10.6	1999QI	Hungary	11.1	22.4
1991	Poland	3.3	-5.2	-3.8	1999QI	Poland	7.5	6.7
1992	Romania	11.0	8.0	10.9	1999QI	Romania	7.2	19.2
1993	Slovakia	8.1	10.0	10.5	1998QIV	Slovakia	1.6	13.8
1992	Slovenia	0.4	8.2	-0.9	1999QI	Slovenia	7.7	15.0
1994	Estonia	3.9	5.6	28.9	1999QI	Estonia	28.7	47.1
1995	Latvia	9.5	16.9	20.0	1998QIV	Latvia	1.9	6.8
					1999QIII	Lithuania	8.7	20.1

	Cumulative growth contributions of net exports							
	T = trough of transformational recession				t = trough during the Russian crisis			
T		T+1 year	T+2 years	T+3 years	t		t+4 quarters	t+8 quarters
					1998QIII	Bulgaria	3.6	8.9
					1998QIV	Croatia	-2.0	-2.3
1992	Czech Republic	-2.0	-7.9	-10.7	1998QIV	Czech Republic	-0.5	-2.4
1993	Hungary	0.5	9.7	10.4	1999QI	Hungary	0.6	0.7
1991	Poland	2.8	-1.0	2.1	1999QI	Poland	0.4	1.6
1992	Romania	-1.4	4.1	3.2	1999QI	Romania	1.2	-2.1
1993	Slovakia	10.2	6.6	-4.1	1998QIV	Slovakia	8.3	6.1
1992	Slovenia	-9.5	-10.3	-9.2	1999QI	Slovenia	0.5	5.4
1994	Estonia	-0.4	-5.0	-8.4	1999QI	Estonia	2.4	0.9
1995	Latvia	-4.6	-1.5	-11.2	1998QIV	Latvia	2.4	3.1
					1999QIII	Lithuania	3.9	5.7

Source: UNECE Common Database.

Note: Troughs T (years) and t (quarters) were identified as the largest negative deviations from national GDP trends during the relevant time periods. The time elapsed since the Russian crisis does not allow for more than eight quarters of data, while for the transformational recession quarterly data are not available. Data availability restricts analysis of the recovery following the transformational recession to eight countries.

---

the region during the summer and fall of 1998,[254] than from the transformational recession during the early 1990s. For the purpose of this comparison, the contribution of exports and net exports to cumulative real GDP growth over a period of 4 to 12 quarters from the business cycle trough is measured by holding everything else constant. Over a j-period horizon starting with t, this measure is defined as:

$$CG(X_{t,j}) = (X_{t+j} - X_t)/GDP_t = (1 + w_{GDP(t+j,j)})x_{t+j} - x_t,$$

where w_{GDP} denotes the cumulative GDP growth rate, X stands for exports or net exports, and x is the ratio of exports or net exports to GDP.

When interpreting table 3.2.6, it should be borne in mind that the causes of and the recoveries from the two crises differed a lot: the transformation recession was caused by a number of massive real shocks on the supply and demand side, i.e. the dismantling of the central planning system, the disorganization of production and the collapse of CMEA trade, with implications far beyond those of a cyclical downswing. The beginning and duration of that recession varied greatly across countries, and so did the external environment at the time of each recovery: during the Polish recovery (1992-1994), the GDP of the European Union grew by 3.6 per cent cumulatively, with aggregate east European GDP falling slightly by 1 per cent. In contrast, during the Hungarian recovery (1994-1996), the EU grew by 7.1 per cent, with east European GDP increasing by 14.2 per cent. The Russian crisis led to contagion effects and resembled much more a cyclical downturn. During the recovery external demand conditions were favourable for all countries, and real exchange rate developments were less dramatic (with CPI-based real effective exchange rate appreciation within the first eight quarters of each recovery averaging 4.6 per cent) than during the recovery from the transformational recession (when CPI-based real effective exchange rate appreciation during the first two years of recovery averaged almost 14 per cent for the same 11 countries).[255]

---

[254] While Hungary, Poland, Slovakia and Slovenia did not in fact suffer an actual output loss as a result of the Russian crisis, all of them had to cope with a perceptible negative deviation from their trend rates of growth.

[255] Aggregate east European growth figures and real exchange rate data are taken from the UNECE Common Database and national banks of ECE member countries.

The Transition Economies

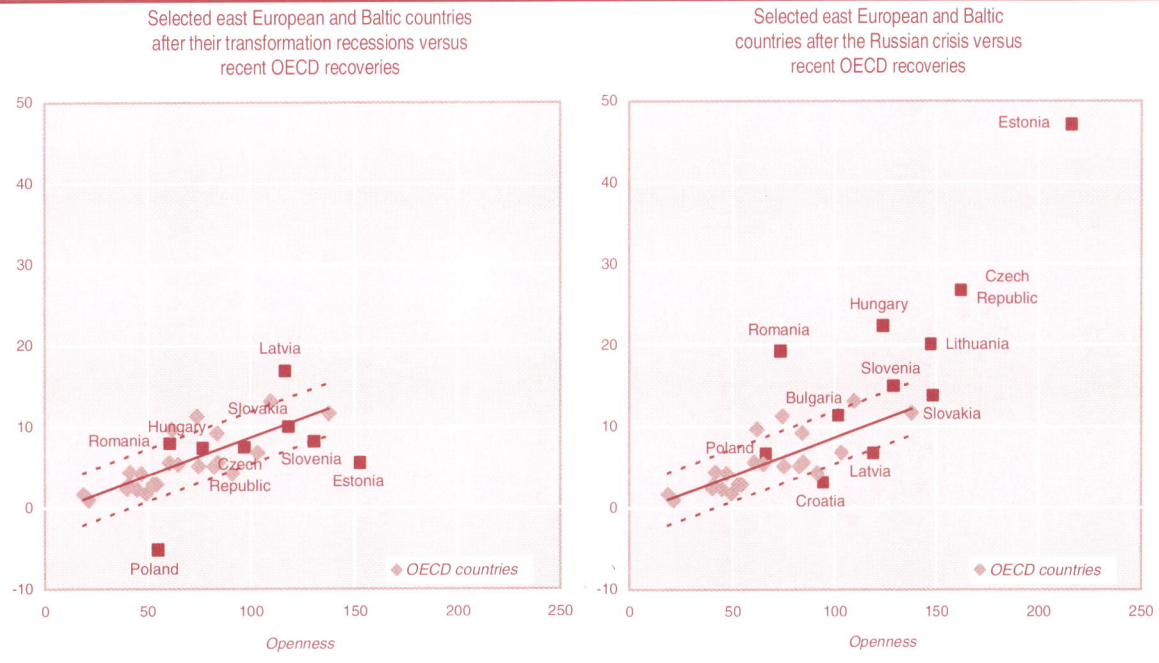

CHART 3.2.5

The contribution of exports to cumulative real GDP growth and openness in eastern Europe and the Baltic states
(After two years, percentage points)

Source: UNECE Common Database. OECD data supplied by the IMF directly to UNECE secretariat.

Note: For the east European and Baltic countries, the chart compares horizons of two years after their transformational recessions to eight quarters after the Russian crisis. For each OECD country the recovery denotes the first eight quarters after its most recent cyclical trough; the OECD sample covers 22 countries that have been members since 1970; accordingly, recently admitted east European countries are excluded from the OECD sample, in which Japan has the lowest (20 per cent) and Belgium the highest (140 per cent) degree of international openness. Openness is defined as the average percentage ratio of the sum of real exports and imports to real GDP over the respective time period. The regression lines and ±1.5 standard deviation lines in the figures are based only on the OECD sample.

Furthermore, in the light of liberalization, it might be expected that strong positive impulses from the external sector for east European countries recovering from their recent recession would significantly exceed the experience of "normal" economies recovering from "normal" business cycle troughs. To make such a comparison, OECD countries' most recent recoveries up to the mid-1990s, again over a recovery period of eight quarters,[256] can be used as a benchmark. Charts 3.2.5 and 3.2.6 show the cumulative real growth contributions of exports and net exports of goods and services for OECD and east European countries against their respective degrees of international openness.

Charts 3.2.5 and 3.2.6 first of all confirm that the – mostly very small – EE-11 economies are now more open than during their recovery from the transformational recession, and on average are significantly more open than the OECD economies. Table 3.2.6 and chart 3.2.5 indicate that exports have made sizeable and significant contributions to the cyclical recoveries of GDP growth, although their extent differs across countries. The average magnitude of these contributions is clearly linked to the openness of the economy. Accordingly, the contributions of exports to east European countries' economic growth, although varying across countries, was significantly larger during the post-Russian crisis recovery both in comparison with their recovery from the transformational recession and with the OECD countries' recent experience.[257]

As for the recovery from transformational recession, it might be thought a priori that external sector liberalization would necessarily result in a rapid and large contribution of the export sector to real economic growth. However, as already indicated, the transformational recession varied greatly across countries in timing and in the external environment.[258]

As is evident from chart 3.2.6, all this does not hold to the same extent for net exports: for both the OECD and the transition economies, there is no convincing evidence that net trade has on average contributed to recovery within eight quarters of their

---

[256] E. Prasad and J. Gable, "International evidence on the determinants of trade dynamics", *IMF Staff Papers*, Vol. 45, No. 3 (Washington, D.C.), September 1998, pp. 401-439.

[257] Including the EE-11 economies in their post-transformational recession recovery does not significantly alter the slope of the OECD regression line in chart 3.2.5, but reduces the constant. Including the EE-11 economies in their recovery from the Russian crisis both reduces the constant and significantly increases the slope of the same regression line.

[258] R. Frensch, "Internal liberalization as a barrier to export-led recovery in central European countries preparing for EU accession", *Comparative Economic Studies*, Vol. 42, No. 3, Fall 2000, pp. 31-47.

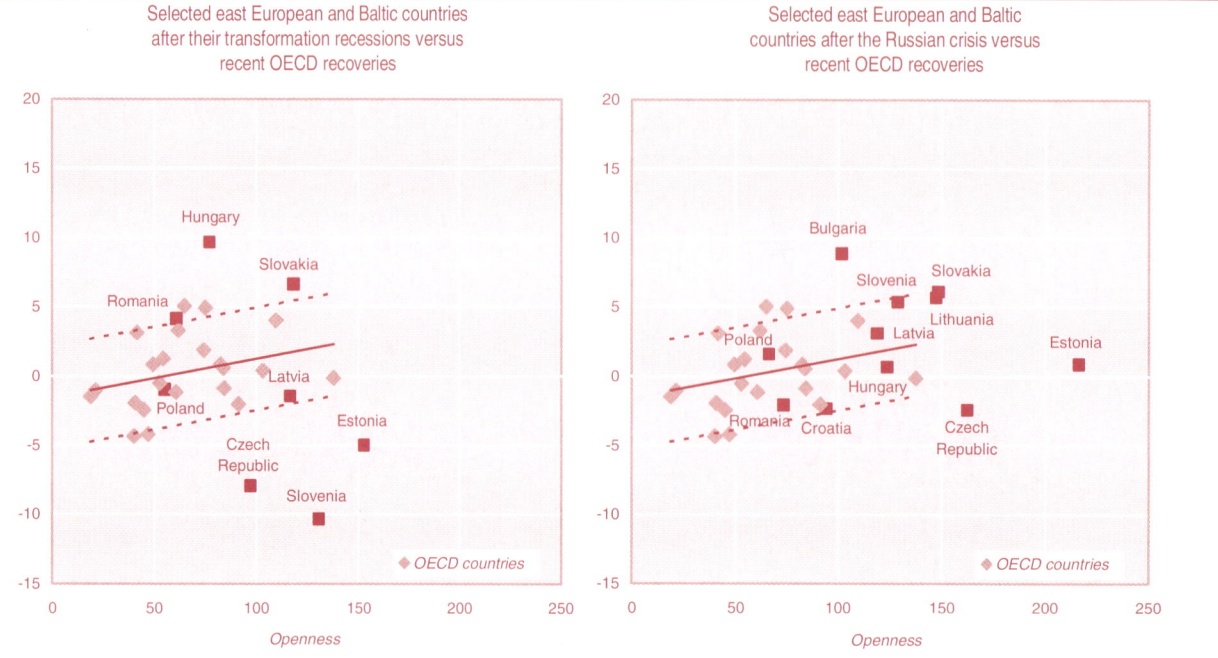

CHART 3.2.6

The contribution of net exports to cumulative real GDP growth and openness in eastern Europe and the Baltic states
(After two years, percentage points)

Source: As for chart 3.2.5.
Note: As for chart 3.2.5.

respective troughs.[259] However, at least the variance of EE-11 economies' deviations from average OECD experience has decreased between the recoveries from the transformational and Russian-crisis recessions, probably due to more neutral external conditions during the latter. As a corollary of the above, there is no correlation between cumulative export and net export contributions to real GDP growth: very open economies (such as the Czech Republic and Estonia) were exceptional in terms of the export contributions to GDP growth after the Russian crisis, but the net export contributions to growth during both recent recoveries were small or even negative.

Accordingly, examination of charts 3.2.5 and 3.2.6 raises the question as to whether exports or net exports are the "true" international trade catalyst for cyclical recoveries. While net trade appears to be more appropriate within the conceptual framework of the national income accounts, the logic of recovery rather points to exports: increasing exports trigger the recovery by stimulating domestic demand including import demand.[260] An examination of trade and output variables over the cycle might help to answer this question.

(c) Trade and the business cycle

A recent study of OECD experience points to the following general conclusions:[261] net exports behave anti-cyclically, largely driven by the strong procyclical behaviour of imports, while export behaviour in this respect varies widely across countries, due to different levels of foreign activity and exchange rate developments. The results in table 3.2.7 are not quite in line with this conclusion: there are significant differences in export as well as import behaviour among the transition countries.[262] Contemporaneous imports are anti-cyclical in Latvia, and above all in both the Czech Republic and in Slovakia, and procyclical only with a one quarter lag. Also, the contemporaneous import response generally appears to be weaker than in the OECD countries. Accordingly, net exports are not anti-cyclical in Poland, Latvia, Lithuania or, again especially, in the Czech Republic.

These differences from OECD experience in the behaviour of the trade variables – together with a higher correlation of contemporaneous export and import

---

[259] The regression line in chart 3.2.6 describes a very weak (and not significant) cross-country relationship, which in fact disappears altogether for the OECD countries after 12 quarters. Accordingly, Prasad and Gable (loc. cit., p. 402) find that quite contrary to exports "... surprisingly ... the trade balance ... has in fact played only a limited role in business cycle recoveries in the OECD economies".

[260] Prasad and Gable, loc. cit., p. 411.

[261] Ibid., p. 410.

[262] Analyzing trade ratios to control for size effects has the advantage that they can be analyzed like "normal" trade variables, while at the same time coming close to the "growth contribution" concept discussed above. The correlation between changes in trade variables and trade ratios is very high over the sample period for each country, typically above 0.9, as trade variables are much more volatile than output.

TABLE 3.2.7

Simple correlations between trade ratios and GDP in eastern Europe and the Baltic states, 1995QI-2001QII

	Exports/GDP, GDP						
	-4	-2	-1	0	1	2	4
Bulgaria	0.17	-0.11	0.03	-0.42	-0.02	0.15	0.47
Croatia	-0.12	-0.45	0.34	-0.61	0.24	-0.13	0.33
Czech Republic	-0.24	0.06	0.15	-0.08	0.15	-0.35	-0.12
Hungary	-0.27	0.12	0.21	-0.39	0.20	0.12	-0.25
Poland	-0.43	0.08	-0.24	0.23	-0.11	0.11	0.31
Romania	0.69	-0.24	0.20	0.40	-0.34	-0.69	0.08
Slovakia	-0.26	0.23	-0.26	-0.49	0.36	-0.29	-0.09
Slovenia	-0.44	-0.58	0.40	-0.07	–	0.16	0.76
Estonia	-0.09	0.20	0.43	0.35	0.14	0.16	-0.27
Latvia	0.13	0.09	-0.05	0.21	0.32	0.16	-0.07
Lithuania	0.57	0.42	-0.18	0.44	0.09	-0.22	-0.41

	Imports/GDP, GDP						
	-4	-2	-1	0	1	2	4
Bulgaria	-0.44	-0.25	-0.05	0.20	-0.40	0.47	0.07
Croatia	-0.14	-0.15	-0.02	0.33	0.10	-0.33	-0.38
Czech Republic	-0.32	-0.38	0.60	-0.40	0.43	-0.31	–
Hungary	0.30	0.11	0.39	0.08	0.12	0.15	-0.04
Poland	-0.20	0.10	-0.15	0.12	-0.13	0.29	0.09
Romania	0.21	-0.07	0.48	0.42	-0.38	-0.65	-0.01
Slovakia	0.25	0.14	-0.06	-0.30	0.50	-0.07	-0.13
Slovenia	0.27	0.23	-0.65	0.88	-0.91	0.87	-0.07
Estonia	-0.21	-0.03	0.61	0.42	0.45	0.10	-0.42
Latvia	0.03	-0.18	-0.16	-0.06	0.34	-0.16	0.04
Lithuania	0.06	0.24	-0.06	0.26	0.08	-0.12	-0.10

	Net exports/GDP, GDP						
	-4	-2	-1	0	1	2	4
Bulgaria	0.53	0.12	0.08	-0.59	0.36	-0.30	0.43
Croatia	-0.03	-0.29	0.27	-0.66	0.12	0.09	0.53
Czech Republic	0.16	0.58	-0.65	0.47	-0.43	-0.02	-0.18
Hungary	-0.52	0.04	-0.16	-0.57	0.11	-0.02	-0.29
Poland	-0.32	-0.02	-0.17	0.19	0.02	-0.23	0.36
Romania	0.29	-0.10	-0.57	-0.27	0.34	0.40	0.18
Slovakia	-0.57	0.02	-0.16	-0.05	-0.35	-0.18	0.10
Slovenia	-0.30	-0.29	0.67	-0.85	0.88	-0.77	0.55
Estonia	0.16	0.24	-0.31	-0.11	-0.42	0.08	0.18
Latvia	0.03	0.24	0.14	0.18	-0.17	0.26	-0.08
Lithuania	0.37	0.08	-0.08	0.08	–	-0.06	-0.21

Source: UNECE Common Database.

Note: Correlations are based on quarterly rates of change with different leads (–) and lags (+) of trade variables. All EE-11 countries (except Romania, starting only in 1996) have been publishing quarterly real GDP since at least 1995, although mostly they are not seasonally adjusted. These data, however, are not always accompanied by a breakdown in expenditure, thus not allowing a truly consistent analysis of GDP and trade variables since 1995 for all countries: the respective series start only with 1996 (Bulgaria), 1997 (Croatia), 1998 (Romania) or even 1999 (Slovenia). If not published as such, all data have been seasonally adjusted by applying a X-12 filter.

changes[263] than in the OECD – suggest that the import response in the east European countries is less a reflection of a "normal" consumption and investment demand reaction to output changes (e.g. induced by exports), typical of the OECD countries, than of the particular nature of their trade which favours outward processing, subcontracting and other "import for export" arrangements resulting in large import content of exports. On the other hand, according to table 3.2.7, exports signal GDP changes much more consistently than net exports over the short run (of one to two quarters). Thus, the evidence from tables 3.2.7 and 3.2.8 is insufficient to allow for much more than the conclusion that the import content of exports in EE-11 is larger than in the OECD economies. In particular, the evidence does not suggest a qualitatively different pattern of trade behaviour over the cycle. This justifies treating exports, rather than net exports, as the true catalyst through which trade influences GDP in the short run. The "effectiveness" with which export stimuli are translated into GDP growth, however, are determined by the degree to which the import response is export related, i.e. by the import content of exports, which differs across countries.

(d) Foreign activity versus exchange rate channel

Long-run developments in the trade shares of GDP follow trends in trading costs, which are influenced, *inter alia*, by the relative size of the economy, liberalization policies and the intertemporal aspects of consumption smoothing. The conclusion to be drawn from the previous section, however, is that to describe short-run trade behaviour over the cycle appropriately the effects of both external activity and real exchange rate behaviour have to be included.

The available data[264] do not allow the estimation of rigorous multivariate econometric relationships between these variables across countries in a meaningful and comparable way: the availability of quarterly GDP data by expenditure varies greatly among the east European countries so that there is a trade-off between coverage and length of series, which in turn largely determines the methodology used to analyse the series. The choice made here has been to favour breadth of coverage and comparability across countries at the obvious cost of analytical depth.

A first attempt at assessing the influence of foreign, especially EU, activity on EE-11 export fluctuations involves some basic principal components analysis. For 9 of the 11 countries (except the two with the weakest trade links with the EU, i.e. Bulgaria and Lithuania), there exists a common set of factors, expressed as the first principal components vector, accounting for 43 per cent of the variance of their export ratios (in terms of quarterly rates of change). This vector, in turn exhibits a statistically significant correlation coefficient of 0.47 with the quarterly rates of change in European Union GDP, which can be taken as evidence for a relationship between east European export fluctuations and European Union activity.[265]

---

[263] On average, 0.46 for all EE-11 countries over the whole sample, but perceptibly increasing over time for some countries, especially for Bulgaria, the Czech Republic, Hungary and Estonia; see table 3.2.8.

[264] See note to table 3.2.7.

[265] To create a common sample size starting with the first quarter of 1996, some annual data for Croatia, Romania and Slovenia had to be re-

TABLE 3.2.8

Correlations between quarterly rates of change of exports and imports in eastern Europe and the Baltic states, 1995QI-2001QII

	Full sample	1998QI – 2001QII
Bulgaria	0.47	0.59
Croatia	-0.04	0.04
Czech Republic	0.64	0.88
Hungary	0.50	0.79
Poland	0.81	0.77
Romania	0.84	0.84
Slovakia	0.73	0.77
Slovenia	0.60	0.60
Estonia	0.71	0.88
Latvia	0.44	0.19
Lithuania	0.51	0.47

Source: UNECE Common Database.

Note: For data constraints, see note to table 3.2.7.

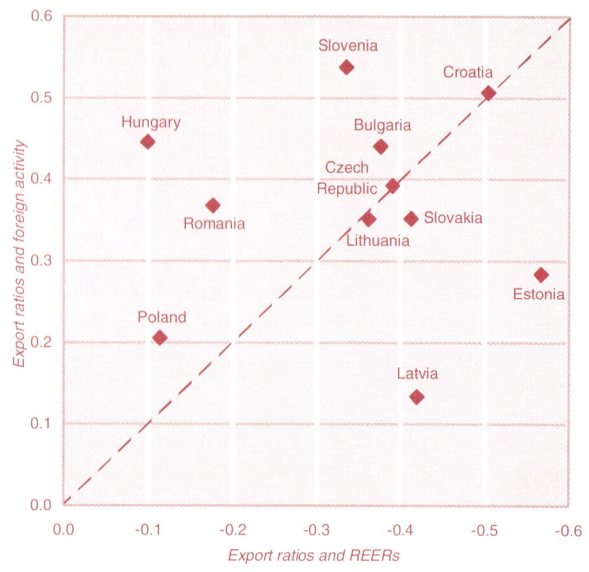

CHART 3.2.7

Partial correlation coefficients between export ratios and real effective exchange rates (REERs) versus the partial correlations between export ratios and foreign (EU) activity, 1996QI-2001QII

Source: UNECE Common Database and national banks of ECE member countries.

Note: The correlations are between quarterly seasonally adjusted rates of change. The partial correlation coefficients have been adjusted for bivariate correlations between foreign activity and real exchange rate developments. For data constraints, see note to table 3.2.7. Foreign activity is proxied by EU GDP except for Bulgaria and Lithuania, for which an index of trade-weighted GDP of EU, eastern Europe, the Baltic states and Russia is used. Real effective exchange rates are producer-price based and trade weighted.

In addition, as a rough assessment, chart 3.2.7 presents the partial correlations between export ratios, real exchange rates and foreign activity variables over the short run. The a priori expectations are that exchange rate changes are negatively, and foreign activity positively, correlated with changes in exports.[266] Chart 3.2.7 shows that across countries foreign activity (especially EU GDP) and real exchange rate effects vary in their importance as determinants of short-run changes in export ratios. Among the higher correlations, changes in foreign activity do not in general appear to outweigh real exchange rate effects, except for Hungary, Romania and Slovenia, while real exchange rate effects seem to dominate for Estonia and Latvia (in terms of deviations from the 45° line in the chart).

However, while there seems to exist a perceptible effect of foreign – especially EU – activity on export activity in the short run for most countries, the high import content of exports in some of them presumably dampens the influence of foreign activity upon output fluctuations, more so than the real exchange rate influence, which has an additional impact on direct imports over and above the impact on imports for exports. This is probably especially true for those countries with a sizeable contemporaneous correlation between changes in exports and imports (table 3.2.8), i.e. mainly the Czech Republic, Hungary, Poland, Romania, Slovakia and Estonia. Of the countries considered in chart 3.2.7, Hungary, Slovenia and perhaps Romania appear to be the countries where foreign (and especially EU) activity may have a more significant influence than the real exchange rate not only on short-term exports but also on short-term variations in output.[267]

(e) Conclusions

A rapidly increasing degree of international openness, on average now much higher than that of the OECD countries, has led to significant contributions of exports to GDP growth during the east European countries' recovery from the regional recession that followed the Russian crisis. At the same time, short-run fluctuations in exports appear to be influenced by both real exchange rate developments and foreign, especially European Union, levels of activity. Considering the close links between exports and imports for many east European economies, a cautious interpretation of the evidence suggests that in general – with the possible exceptions of Hungary, Slovenia and perhaps Romania – the impact of foreign activity is probably weaker than the influence of the real exchange rate on short-run output fluctuations.

---

estimated on a quarterly basis; see note to table 3.2.7. Leaving out the data for these countries increases the share of the variance accounted for by the first principal components vector.

[266] Lag selection for the real effective exchange rate and activity indicators was made by preliminary country-by-country ordinary least squares regression of export ratios on differently lagged real exchange rates and activity variables. For example, chart 3.2.7 reproduces the effects of contemporaneous real exchange rate changes for Slovakia and Lithuania, and of lagged changes for the rest except for Hungary, Poland and Romania (lagged twice).

[267] The presence of Romania in this group is noteworthy, as it has one of the lowest degrees of openness in the region (chart 3.2.4).

3.3 Costs and prices

Disinflation resumed its downward trend in early 2001 and by mid-year overall price inflation remained an immediate policy concern in only a few transition economies. Year-end rates in some east European economies even overshot the original forecast/targets, which a year ago were thought to be rather optimistic. This favourable inflation performance mainly reflects the significant decline of imported inflation which was the key cause of the setback in the disinflationary process from mid-1999 through 2000. The fall in world commodity prices, particularly of crude oil and some industrial raw materials,[268] together with the global weakening of prices of manufactures,[269] have greatly alleviated the external supply-side cost pressures. The domestic effect of weaker import prices was further amplified in some countries by appreciating exchange rates. On the domestic side, however, wage inflation remained sticky or even accelerated and rose faster than measured labour productivity, particularly in those countries where the slowdown in industrial output accelerated in the second half of the year. Consequently, the downward trend in unit labour costs was checked and with few exceptions they picked up strongly.[270] Nevertheless, real unit labour costs in eastern Europe and the Baltics generally continued to fall in 2001, albeit at a slower rate than in 2000. In addition, given the dampening effect of imported disinflation on the relative prices of material inputs, there was little or no pressure on unit operating profits, which probably increased in some manufacturing branches in eastern Europe and the Baltics. In contrast, in most of the CIS economies real unit labour costs rose sharply in the course of the year, reaching double-digit annual average rates in many countries. Given the significant deceleration in producer price inflation in most of them, some of this intensified labour cost pressure must have been absorbed by unit operating profits.

On the demand side, the pressures on prices were less homogenous across the countries. There was a general increase in real wages, which was combined in some cases with increased income from tourism and/or fiscal stimuli. Hence, demand-pull factors probably intensified in countries, such as in Croatia, the Czech Republic, Hungary, Romania, Slovakia and Latvia, as suggested by the preliminary national accounts data for household consumption.[271] In the other east European and Baltic economies, private consumption growth continued to be weak or lose steam during 2001, reflecting in the main unfavourable labour market developments.[272] Rising real wages and some improvement in the labour markets boosted private consumption throughout the CIS except in Tajikistan and probably Georgia.[273]

In short, the lower rates of inflation in 2001 were not the result of more moderate domestic cost-push and/or demand-pull factors in general, but rather the consequence of a negative external price shock assisted by good harvests in some countries. Domestic macroeconomic policies cannot claim much credit for either of these factors, with the exception of exchange rate policy in some cases.[274] Thus, although 2001 was a favourable year in terms of overall price performance, achieving sustainable price stability through the moderation of domestic demand-pull and cost-push pressures still remains a challenge at the start of 2002. This underlines once again the necessity of fiscal prudence, especially during election periods, in order to support monetary policies, which have been usually restrictive, and, in some countries, even overly restrictive and counterproductive. In addition, further, and in some cases faster, progress is needed with industrial restructuring and in implementing other reforms, *inter alia*, those that will improve the functioning of labour markets (such as incomes and employment policies, social safety nets, etc.).

In the short run, a further slowdown in headline inflation rates can be expected. This assumes no external shock that would raise input costs,[275] and that adjustments of administrated prices (many of which were postponed in 2000 and some also in 2001), are made gradually and not bunched together, particularly in those countries which are candidates for joining the EU. The major source of domestic cost pressure in 2002 is likely to come from unit labour costs due to a possible further slowdown in export-led growth in output and productivity. However, this may be partly offset by the deterioration in the labour markets, which is likely to ease wage pressures. Furthermore, the expected moderation in household income growth and the sense of insecurity created by rising unemployment may check consumption demand in the near future. Consequently, the inflation forecasts and targets incorporated in the budgets for 2002 are expected to be reached, especially in those countries where these forecasts were made before the fourth quarter of 2001, i.e. before the current disinflation gained momentum.

---

[268] Over the 12 months of 2001 world market prices for energy commodities (in dollars) decreased by 27 per cent while those for industrial raw materials fell by nearly 22 per cent; see chap. 2.1 above.

[269] Export unit values in western Europe, in dollar terms, for manufactured goods declined by 2.4 per cent, year-on-year, in the first three quarters of 2001; see sect. 3.5(ii)(a) below.

[270] In 2001, however, unit labour costs in western Europe grew even faster (chart 3.1.1), which helped the transition economies to maintain or even improve their competitive position.

[271] Tables 3.2.2 and 3.2.3.

[272] Table 3.4.1.

[273] Table 3.2.5. In the absence of quarterly national accounts data for the CIS economies, except Armenia, the Republic of Moldova and Russia, retail trade data are used as a proxy for developments in private consumption in 2001. Retail trade data in the CIS not only cover goods but also catering.

[274] This has been the case in general throughout the transition except for the period immediately following the initial price shock; see sect. 3.3(iii) below.

[275] See chap. 2.1(i) for prices in the international commodity markets during early 2002 and expectations for the rest of the year.

TABLE 3.3.1

Consumer prices in eastern Europe, the Baltic states and the CIS, 2000-2001

(Percentage change)

	Annual average		December over previous December		2001, year-on-year			
	2000	2001	2000	2001	QI	QII	QIII	QIV
Albania	–	3.1	4.2	3.5	2.2	3.2	4.4	2.7
Bosnia and Herzegovina	1.7	1.8	3.4	1.5	2.4	3.0	1.0	1.0
Bulgaria	10.2	7.3	11.2	4.8	8.8	9.5	6.2	4.9
Croatia [a]	6.4	5.0	7.5	2.5	6.7	6.4	4.2	2.8
Czech Republic	3.9	4.7	4.1	4.2	4.2	5.1	5.4	4.3
Hungary	9.9	9.2	10.1	6.9	10.4	10.6	8.7	7.2
Poland	10.2	5.5	8.6	3.6	6.8	6.6	4.8	3.8
Romania	45.7	34.5	40.7	30.2	40.1	36.9	31.8	30.5
Slovakia	12.0	7.3	8.3	6.5	7.1	7.7	7.7	6.6
Slovenia	9.0	8.6	9.0	7.1	8.9	9.6	8.5	7.4
The former Yugoslav Republic of Macedonia	6.6	5.2	6.1	3.7	6.0	4.7	6.0	4.1
Yugoslavia	77.5	90.4	115.1	40.5	117.1	126.3	106.0	46.1
Estonia	3.9	5.8	5.0	4.3	5.9	6.8	6.1	4.5
Latvia	2.8	2.4	1.9	3.0	1.2	2.4	3.2	3.0
Lithuania	1.0	1.5	1.5	2.1	0.2	1.5	2.0	2.2
Armenia	-0.8	3.2	0.4	2.8	2.5	4.1	3.2	2.7
Azerbaijan	1.8	1.5	2.1	1.5	1.5	1.8	1.5	1.3
Belarus	168.9	61.4	108.0	46.3	83.3	70.6	55.0	46.3
Georgia	4.2	4.6	4.6	3.4	5.5	6.4	4.1	2.5
Kazakhstan	13.4	8.5	10.0	6.6	9.1	9.7	8.3	7.0
Kyrgyzstan	18.7	7.0	9.5	3.8	8.3	9.3	6.6	3.8
Republic of Moldova	31.3	9.8	18.5	6.4	15.9	13.0	5.4	5.6
Russian Federation	20.8	21.6	20.1	18.8	22.3	24.5	21.1	18.9
Tajikistan	32.9	38.6	60.6	12.5	61.7	53.6	38.6	12.6
Turkmenistan	..	..	..	..	..	..	..	..
Ukraine	28.2	12.0	25.8	6.1	19.4	14.5	8.9	6.1
Uzbekistan	24.9	27.3 [b]	28.2	22.2 [c]	25.4	28.4	28.1	..

Source: UNECE secretariat estimates, based on national statistics.

[a] Retail price index.

[b] First three quarters.

[c] November 2001 over December 2000.

(i) Consumer prices in 2001

The downward trend in *consumer price inflation* in the transition economies as a whole resumed in early-2001 and gained momentum in the second half of the year (table 3.3.1). This disinflation largely followed the sharp reduction in import price pressures on non-food consumer goods (chart 3.3.1). These were accompanied in many countries by considerably weaker food prices during the summer, thanks mainly to better than average harvests. Given the important share of food in the average household's consumption expenditure in these economies, particularly in those of south-east Europe and the CIS, the fall in the overall rate of inflation accelerated in the summer. Nevertheless food prices still remained an important inflationary factor in 2001 as a whole in Croatia, Hungary, Slovenia, The former Yugoslav Republic of Macedonia, the Baltics, Armenia, Kazakhstan and Uzbekistan. In Russia, despite some deceleration in the last two quarters, the average rate of increase in food prices still remained high, at around 20 per cent, in 2001. Nevertheless, within the consumer price index, the major source of upward pressure in 2001 generally came from service prices, particularly in Croatia, the Czech Republic, Yugoslavia, Lithuania, Kyrgyzstan and Russia. In these countries they not only rose faster than the other two components of the index but also faster than in 2000.

At the end of 2001, the year-on-year rate of consumer price inflation was higher than in 2000 only in Latvia, Lithuania and Armenia, albeit their rates were among the lowest (2-3 per cent) in the region (table 3.3.1). In contrast, although the rate fell in Romania, it still remained at about 30 per cent, the third highest among all the transition economies after Yugoslavia (40.5 per cent) and Belarus (46.3 per cent). In Poland a further slowdown in real wage growth and soaring unemployment dampened household demand in 2001. Subdued consumer demand, combined with a strong zloty and considerably weaker food prices during the second half of the year, lowered the 12-month inflation rate at the end of the year by 5 percentage points to 3.6 per cent, one of the lowest rates in the region and well below the expectations of the authorities at the beginning of the year. In contrast, the strong recovery in private consumption demand continued

The Transition Economies

CHART 3.3.1

Components of consumer prices in eastern Europe, the Baltic states and the CIS, 1998-2001

(Year-on-year, monthly percentage change)

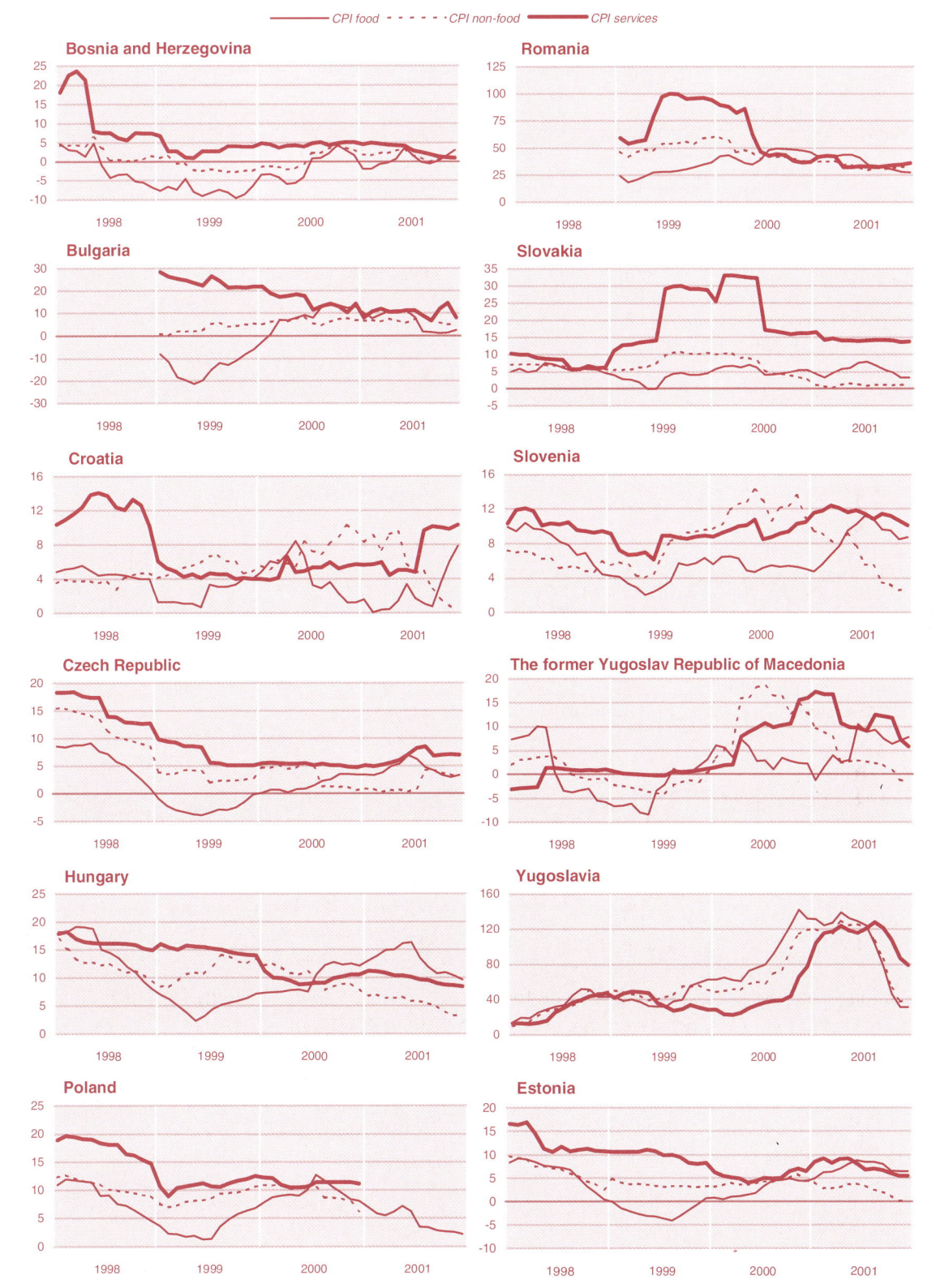

(For source see end of chart.)

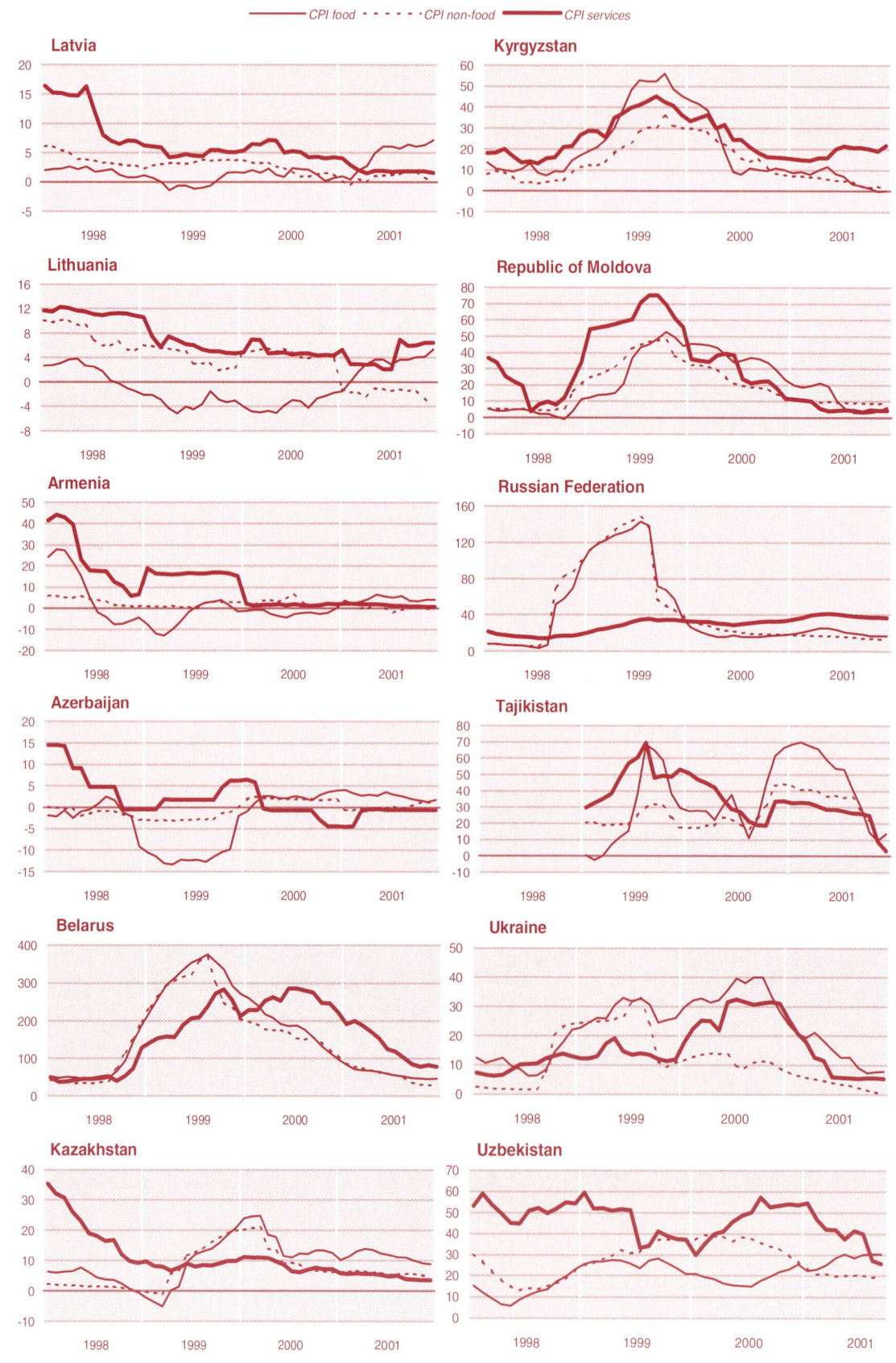

CHART 3.3.1 (concluded)

Components of consumer prices in eastern Europe, the Baltic states and the CIS, 1998-2001
(Year-on-year, monthly percentage change)

Source: National statistics and UNECE secretariat estimates.

TABLE 3.3.2

Producer prices, wages and unit labour costs in industry [a] in eastern Europe, the Baltic states and the CIS, 2000-2001
(Annual average percentage change)

	Producer prices		Nominal wages [b]		Real product wages [c]		Labour productivity [d]		Unit labour costs [e]		Real unit labour costs [f]	
	2000	2001	2000	2001 [g]	2000	2001 [g]	2000	2001 [g]	2000	2001 [g]	2000	2001 [g]
Albania	..	..	..	..	..	..	57.0	..	..	..	..	..
Bosnia and Herzegovina	0.9	2.3	12.3	7.4[h]	11.3	6.0[h]	8.8	20.3	3.2	-10.6[h]	2.3	11.8[h]
Bulgaria	16.9	6.7	10.9	6.7	-5.1	-2.1	18.4	6.4	-6.3	0.3	-19.9	-8.0
Croatia	9.6	3.4	6.0	8.4	-3.2	3.2	4.4	8.4	1.6	–	-7.3	-4.8
Czech Republic	5.1	3.0	7.1	6.7	2.0	2.9	8.2	6.9	-1.0	-0.3	-5.7	-3.8
Hungary	11.4	5.7	15.0	14.4	3.2	8.3	20.2	2.3	-4.3	11.8	-14.1	5.9
Poland	7.8	1.8	10.9	7.0	2.9	4.4	13.5	6.4	-2.3	0.6	-9.4	-1.9
Romania	53.4	41.0	41.7	54.6	-7.6	6.7	15.5	10.9	22.7	39.5	-20.0	-3.7
Slovakia	9.8	6.7	9.1	9.7	-0.6	1.8	12.8	4.1	-3.3	5.3	-11.9	-2.2
Slovenia	7.7	9.0	11.7	11.4	3.8	1.7	6.9	2.1	4.5	9.1	-2.9	-0.4
The former Yugoslav Republic of Macedonia	9.0	2.6	5.5	3.9	-3.2	-0.2	8.4	-4.8	-2.7	9.1	-10.7	4.8
Yugoslavia	105.4	84.4	97.8	129.2	-3.7	10.3	15.1	0.2[h]	71.8	127.6[h]	-16.3	4.3[h]
Estonia	4.9	4.5	10.6	13.5	5.4	7.8	9.2	6.1	1.2	7.0	-3.5	1.6
Latvia	0.8	1.6	15.1	-0.6	14.2	-2.0	2.5	9.2	12.4	-8.9	11.4	-10.2
Lithuania	17.8	-1.5	1.2	-2.0	-14.1	-1.8	7.5	18.2	-5.8	-17.1	-20.1	-16.9
Armenia	-0.4	1.1	14.1	11.0	14.5	9.9	15.6	3.4	-1.3	7.4	-0.9	6.3
Azerbaijan	9.4	-2.5	15.0	26.7	5.1	30.0	11.0	4.4	3.6	21.4	-5.3	24.6
Belarus	185.6	71.8	201.9	110.3	5.7	22.4	8.2	5.7	179.1	89.9	-2.3	15.8
Georgia	5.8	3.6	..	..	..	..	..	..	..	..	..	..
Kazakhstan	38.0	0.4	25.9	25.7	-8.8	20.8	22.2	24.4	3.0	1.1	-25.4	-2.8
Kyrgyzstan	31.7	12.1	21.7	22.1	-7.6	7.1	14.2	8.8	6.6	12.2	-19.1	-1.6
Republic of Moldova	33.6	12.6	32.3	25.8	-1.0	11.7	7.0	11.4	23.6	12.9	-7.5	0.3
Russian Federation	46.5	19.1	42.5	45.7	-2.7	22.4	9.5	3.4	30.2	40.9	-11.1	18.3
Tajikistan	39.0	25.1	30.7	50.0	-6.0	13.6	20.9	16.0	8.0	29.3	-22.3	-2.1
Turkmenistan	..	..	..	..	..	..	..	..	..	..	..	..
Ukraine	20.8	8.6	30.2	34.9	7.8	24.2	19.1	14.9	9.3	17.4	-9.5	8.1
Uzbekistan	61.1	43.4[i]	39.0	51.1	-13.8	5.4	..	..	..	..	..	..

Source: UNECE secretariat estimates, based on national statistics and direct communications from national statistical offices.

Note: Annual averages are calculated on the basis of monthly data, except for employment which are quarterly.

[a] Industry = mining + manufacturing + utilities.

[b] Average gross wages in industry except in Bosnia and Herzegovina: net wages in industry; in Estonia and all the CIS economies: gross wages in total economy; in Yugoslavia and The former Yugoslav Republic of Macedonia: net wages in total economy.

[c] Nominal wages deflated by producer price index.

[d] Gross industrial output deflated by industrial employment.

[e] Nominal wages deflated by productivity.

[f] Real product wages deflated by productivity.

[g] Data for 2001 are for the first three quarters, except Armenia, Azerbaijan, Belarus, Hungary, Republic of Moldova, the Russian Federation and Ukraine, where the data are for the full year.

[h] Data for the first half.

[i] Data for the first three quarters.

in Hungary in 2001. This was due to a large rise in real wages, relatively stable employment in the first half of the year, a substantial increase in agricultural and tourism incomes in the second half, and a rapid growth of household credit. Nevertheless, the strong forint and weaker food prices after mid-year brought the overall inflation rate to just below 7 per cent, some 3 percentage points less than in 2000. In the Czech Republic, inflation remained sticky in 2001, albeit at a relatively moderate rate of some 4 per cent, mainly due to robust household demand which was supported by a combination of expansionary fiscal and income policies and a relatively loose monetary policy. Strong household demand put pressure on consumer prices, particularly of paid services, also in Croatia, Romania, Slovakia and Estonia. In Russia and in most of the other CIS economies the strengthening of household demand in response to increases in real disposable incomes was one of the major sources of inflationary pressure in 2001.

(ii) Producer prices and labour costs in industry in 2001

The annual average rates of *industrial producer price inflation* in general fell sharply in 2001 (table 3.3.2). Furthermore, the disinflation was particularly steep in the second half of the year and the rates of inflation in December 2001 compared with December 2000 were close to zero or even negative in several countries. These 12-month year-end rates remained in double digits only in Romania, Yugoslavia, Belarus and Russia, although the rate was falling during the year even in these economies.

Inflation measured by the PPI was much lower than that measured by the CPI in most of the transition economies in 2001, except Bosnia and Herzegovina, Romania, Slovenia, Belarus, Kyrgyzstan, the Republic of Moldova and probably in Uzbekistan (chart 3.3.2). In fact, smaller increases in the PPI than in the CPI has been the dominant pattern in most of the east European and Baltic countries since the output recovery picked up in the mid-1990s. This trend was interrupted only in 2000 due to the large increase of imported energy and other industrial raw material prices. In 2001, mainly reflecting the favourable effect of significantly weaker world commodity prices on the cost of consumer goods, service prices were again the major source of the increase in the CPI. This generally slower growth of producer prices compared with consumer prices in the fast growing transition economies may be explained to a large extent by two factors. First, real wages tend to increase not only in sectors with rapid productivity growth (mainly cost-efficient, export-oriented manufacturing branches) but also in relatively less productive sectors (i.e. most of the non-tradeable service branches, which are much less exposed to external and domestic competition), the so-called "Balassa-Samuelson" effect.[276] Second, and probably more important throughout the 1990s, the continuing adjustment in relative prices was very much affected by increases in administered or controlled prices, which usually concern rents, utilities, public transport and communications included in the services component of the CPI. In fact, some of the envisaged adjustments to regulated prices were held back in 2000 in order to avoid increasing the headline inflation rates any further.

Wage inflation in industry in 2001 remained strong or even accelerated, rising faster than producer prices except in Bulgaria, Latvia and Lithuania where *real product wages* fell by some 2 per cent (table 3.3.2). Among the east European and Baltic economies, the largest increases in real product wages in 2001 were in Bosnia and Herzegovina, Hungary, Romania, Yugoslavia and Estonia. They also rose relatively fast in Poland (4.4 per cent and up from less than 3 per cent in 2000), but this was mainly due to the sharp drop in producer price inflation as the growth in wages was relatively moderate.

On the other hand, *measured industrial labour productivity* continued to grow, although less vigorously than in 2000, as a result of the slowdown in output growth in most of them. Labour productivity fell only in The former Yugoslav Republic of Macedonia (nearly 5 per cent) and probably remained unchanged in Yugoslavia. In Hungary and Slovenia the increases fell sharply to some 2 per cent in 2001 from some 20 and 7 per cent, respectively, in 2000. This dramatic slowdown in productivity growth in Hungary mainly reflects the sharp weakening of industrial output growth in the second half of the year combined with continued growth of employment in the sector. The slowdown in labour productivity growth was also considerable in Russia, with the rate falling from 9.5 per cent in 2000 to 3.4 per cent in 2001. The best labour productivity performers in 2001, with rates of increases of some 10 per cent or more, were Bosnia and Herzegovina, Romania, Latvia, Lithuania, Kazakhstan, the Republic of Moldova, Tajikistan and Ukraine. But in most of these economies the increases are narrowly based and reflect a spectacular surge of output in just a few industries.[277]

As a result of increased wage pressures and weaker productivity growth, particularly in the second half of the year, industrial *unit labour costs* picked up strongly. There were a few exceptions, where either wage pressures were relatively weak (Bulgaria) or labour productivity growth remained favourable (Czech Republic and Poland) or even sharply accelerated (Bosnia and Herzegovina, Croatia and Kazakhstan). In Latvia and Lithuania both depressed wages and large gains in labour productivity continued to reduce industrial unit labour costs significantly. Nevertheless, the rate of change in *real unit labour costs* (which basically measures the change of labour's share in value added) in eastern Europe and the Baltic countries continued to fall in 2001, albeit, at a generally slower rate than in 2000. The largest falls were in Bosnia and Herzegovina, Bulgaria, Latvia and particularly in Lithuania where the two-year cumulative decline was nearly 40 per cent. Among the east European and Baltic economies real unit labour costs in 2001 rose only in Hungary, The former Yugoslav Republic of Macedonia, Yugoslavia and, to a much lesser extent, in Estonia. In contrast, in the CIS real unit labour costs, which had been falling since the mid-1990s (except in 1998 in the wake of the rouble crisis), accelerated sharply in the course of 2001, a reflection of generally large real wage increases and significantly lower rates of industrial productivity growth. The main exceptions to this unfavourable pattern were Kazakhstan, Kyrgyzstan, the Republic of Moldova and Tajikistan where output and industrial productivity grew strongly in a few industries.

Given the weakening of industrial material input costs, producers' unit operating profits escaped the full impact of the increase in unit labour costs and probably even increased in certain industries in some countries, particularly during the first half of the year in Bulgaria, Croatia, the Czech Republic, Romania, Latvia, Lithuania and Kazakhstan.

---

[276] For evidence of the presence of the "Balassa-Samuelson" effect in the transition economies since 1990, see "Economic transformation and real exchange rates in the 2000s: the Balassa-Samuelson connection", UNECE, *Economic Survey of Europe, 2000 No 1*, chap. 6, pp. 227-239.

[277] Sect. 3.2 above.

CHART 3.3.2

Consumer and industrial producer prices in eastern Europe, the Baltic states and the CIS, 1998-2001
(Year-on-year, monthly percentage change)

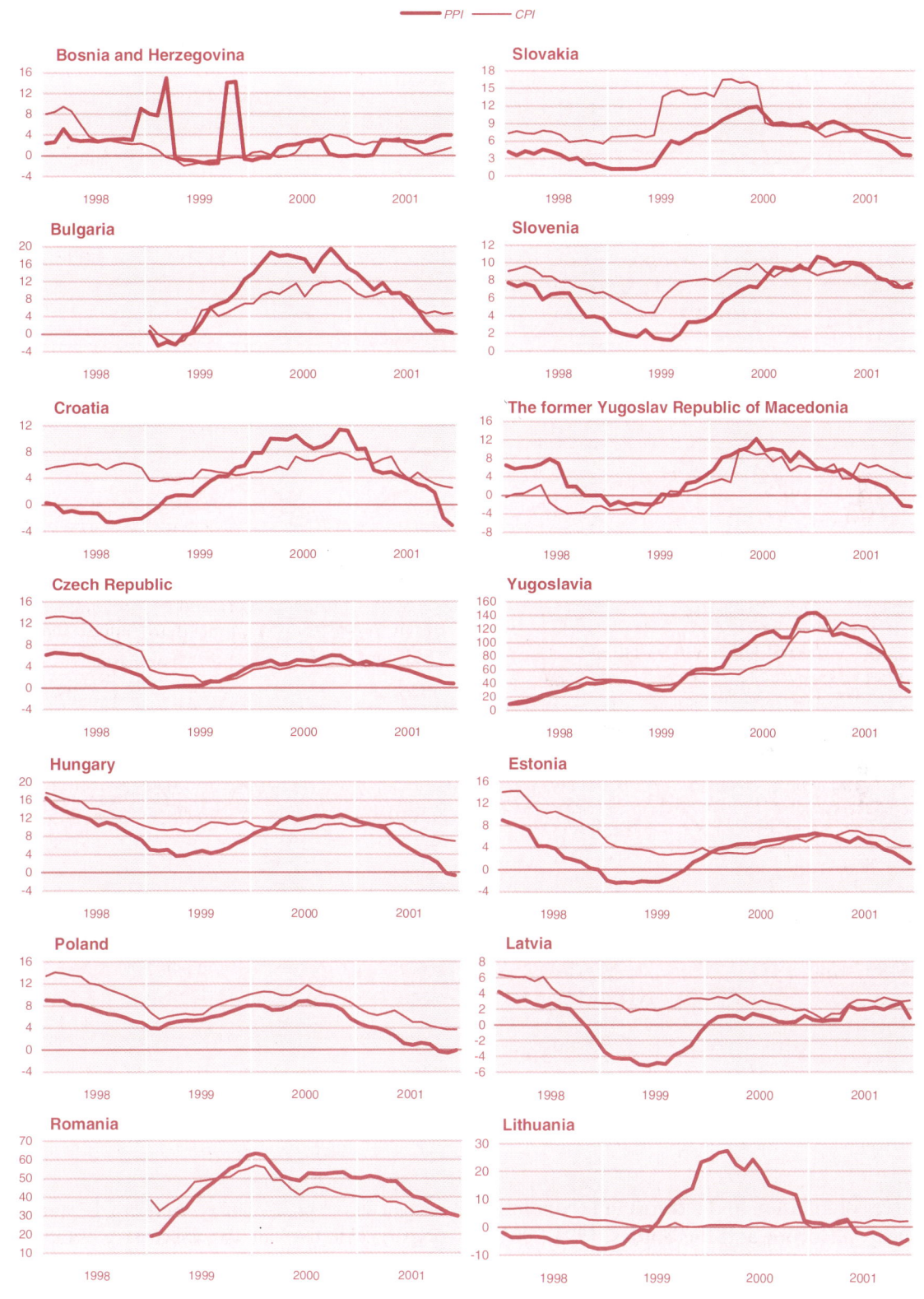

(For source see end of chart.)

CHART 3.3.2 (concluded)

Consumer and industrial producer prices in eastern Europe, the Baltic states and the CIS, 1998-2001
(Year-on-year, monthly percentage change)

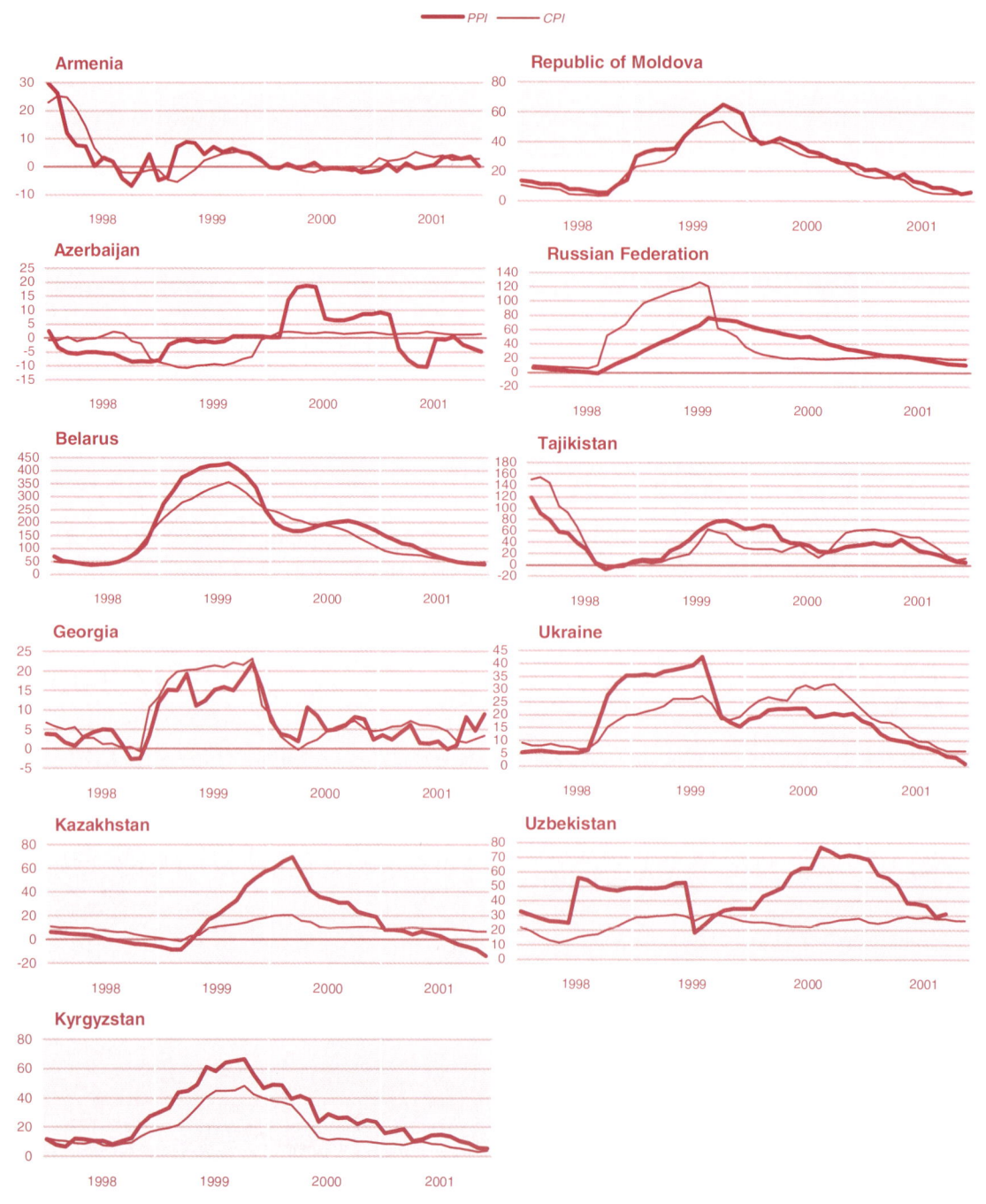

Source: National statistics and UNECE secretariat estimates.

(iii) Sources of inflation in the transition economies of eastern Europe and the Baltics, 1991-2001

Liberalization of prices and the simultaneous large devaluation of domestic currencies at the start of reforms added significantly to the inflationary pressures which were already building up in the final days of the central planning system, formerly characterized by widespread shortages. This "corrective" inflation of the early 1990s was more severe than foreseen in the stabilization programmes. The restoration of price stability, therefore, became one of the primary objectives of the reformist governments of the transition economies. In fact, most of the east European authorities were successful in reducing inflation swiftly from near hyperinflation rates to relatively manageable levels of some 20 to 40 per cent per annum, mainly through the use of the standard measures of financial restraint, that is, tight monetary and incomes policies. However, at these levels of inflation it was still difficult to achieve sustainable growth and the needed adjustment of relative prices to equilibrium levels. Moreover, the process was hampered by the ongoing gradual liberalization of controlled and administered prices (particularly of energy), periodic and ad hoc currency devaluations, as well as reforms requiring one-off policy innovations (such as changes in VAT rates, customs tariffs etc).

This initial disinflation was often achieved at a large cost in terms of output growth, income distribution and overall economic recovery. Apart from the increased risk of social unrest, these efforts to reach quickly price stability at the early stages of enterprise restructuring had negative effects on labour productivity: output collapsed while labour was unable to adjust at the same speed. This dampened, at least partially, the favourable combined effects of the decline in wage inflation and the improvement in the terms of trade on the costs of production. Labour productivity started in general to improve only in the mid-1990s in parallel to the recovery in output and an intensification of the process of enterprise restructuring. This recovery coincided with larger increases in wages and a recovery in unit profits. Thus the nature of inflation towards the mid-1990s was different from that in the initial stages of transition. Inertial factors, various implicit or explicit backward indexation schemes for wages and pensions,[278] wage-drift and inflationary expectations, were now among the major determinants of these moderate but persistent rates of inflation. Thus, the standard instruments of restrictive policies at the early stages of the transition were not very effective in reducing inflation further, mainly because of their negative effects on the fragile economic recovery, and particularly on productivity.[279]

Reduced pressure from import prices (an outcome of both weaker foreign prices and stronger domestic currencies) was the major factor behind the overall rate of domestic price disinflation in most of the east European and Baltic economies between the initial phase of transition until mid-1999. With soaring prices for crude oil and other commodities, the dampening effect of import prices reversed significantly in the closing months of 1999 and throughout 2000. The rate of imported inflation slowed down again only after the early months of 2001 when the decline in world crude oil prices combined with the slowdown in economic activity in the developed market economies, which intensified competition in trade in manufactures.

Although monetary policy remained restrictive in most of the east European and Baltic economies during the second half of the 1990s and early 2000s its effect on inflation was hampered by ineffectual mechanisms and institutions and inadequate support from fiscal policy. Nevertheless it was still helpful in supporting exchange rates. The overall rate of domestic inflation therefore tended to follow closely the trend of import price inflation, a relationship which grew closer over time as the import share of domestic demand increased. As a result, foreign prices remained the principal influence on the development of domestic inflation during most of the transition period, with the exception of the initial phase,[280] when the wages grew much more slowly than price inflation.

This note seeks to identify some of the sources of domestic inflation in 11 east European and Baltic transition economies, over the decade 1991-2001 with an emphasis on the most recent period (1999-2001).

(a) Methodology

One way to analyse the sources of inflation and quantify the major determinants of price changes in an economy in a given period is to disaggregate the implicit GDP and domestic demand deflators into their major components. This is purely an accounting exercise, which does not identify causal relationships or the lagged response

---

[278] These backward indexation schemes (implicit or explicit linkage of wages to actual inflation) were a dominant direct cost-push and indirect demand-pull factor in generating inflation, and offsetting, at least partially, the imported (dis)inflation. Thus, these two counterbalancing effects contributed to the "moderate but persistent" inflation in many of the transition economies. For an empirical analysis of this phenomenon, see T. Pujol and M. Griffiths "Moderate inflation in Poland: a real story", in C. Cottarelli and G. Szapary (eds.), *Moderate Inflation: The Experience of Transition Economies*, IMF and National Bank of Hungary (Washington, D.C.,), 1998; R. Frensch, "Persistent inflation during transition: the Czech and Slovak case", Osteuropa Institute (Münich), 1997, mimeo; B.-Y. Kim, *Determinants of Inflation in Poland: A Structural Co-integration Approach*, BOFIT Discussion Papers, No. 16 (Helsinki), 2001; A. Welfe, "Modelling inflation in Poland", *Economic Modelling*, Vol. 17, Issue 3, August 2000, pp. 375-385.

[279] For an extensive discussion of the characteristics of "moderate" inflation see R. Dornbush and S. Fisher, "Moderate inflation", *The World Bank Economic Review*, No. 7 (Washington, D.C.), 1993, pp 1-44. Based on their case studies of eight countries the authors conclude that "... seigniorage plays no more than a modest role in the persistence of moderate inflations and that such inflations can be reduced only at a substantial short-term cost to growth ... there is unfortunately little encouragement in these case studies for the view that an exchange rate commitment, or incomes policy, allows a country to move at low cost from moderate to low inflation. Governments have successfully reduced moderate inflations through a combination of tight fiscal policy, incomes policy and generally some exchange rate commitment, and taking advantage of favourable supply shocks to ratchet the inflation rate down".

[280] Recently there have been several empirical studies of the role of foreign prices (external transmission effect) in determining inflation in the transition countries. For example see V. Izák, *External Factors in Czech Disinflation (Dynamic Analysis)* Czech National Bank (Prague), 2001; J. Brada and A. Kutan, *The End of Moderate Inflation in Three Transition Economies?*, William Davidson Working Paper, No. 433 (Ann Arbor, MI), 2002. Here the authors update their work done in 1999 under the same title for the Czech Republic, Hungary and Poland. They conclude that "the sharp decline in inflation observed at the end of the 1990s in the three countries should be seen as an exogenously caused event rather than the outcome of monetary policy. ... Thus, inflation remains a threat to these countries and the successes of the late 1990s may not be an accurate indicator of how well the monetary authorities in these countries can control inflation under new international circumstances". But the authors also note that falling import prices also lowered inflationary expectations, which are one of the major obstacles to faster and sustainable rates of disinflation mainly through their adverse effects on excessive wage growth. See also B.-Y. Kim, loc. cit, where the author distinguishes between the three sources of inflation in Poland: wage inflation, monetary inflation and imported inflation, and concludes that interest rates are not important in determining inflation in Poland except through their effect on the exchange rate. On the other hand the labour market and external sectors dominated the determination of Polish inflation during 1990-1999".

> **BOX 3.3.1**
>
> **Contributions to changes in the GDP and domestic demand deflators**
>
> The decomposition of the GDP and domestic demand deflators into their components is based on the following identities of GDP by expenditure (final use) accounts:
>
> $$GDP + M = D + X$$
>
> therefore: $\quad GDP = D + X - M$
>
> and: $\quad D = GDP - (X - M)$
>
> where:
> GDP = gross domestic product
> D = total domestic demand, i.e. the sum of final consumption expenditure and gross capital formation
> X = exports of goods and services
> M = imports of goods and services
>
> If t, c and k subscripts designate time, current prices and constant prices, respectively, and if a dot above the variables denotes growth rates per annum (in per cent), then the contribution of each component to the overall change in the GDP deflator (1) and the domestic demand deflator (2) can be formulated as follows:
>
> $$\left[\frac{\dot{GDP_c}}{GDP_k}\right]_t = \left[\frac{D_c}{GDP_c}\right]_{t-1} * \left[\frac{\dot{D_c}}{D_K}\right]_t \left[\frac{M_c}{GDP_c}\right]_{t-1} * \left[\frac{\dot{M_c}}{M_k}\right]_t + \left[\frac{X_c}{GDP_c}\right]_{t-1} * \left[\frac{\dot{X_c}}{X_k}\right]_t + R_1$$
>
> $$\left[\frac{\dot{D_c}}{D_k}\right]_t = \left[\frac{GDP_c}{D_c}\right]_{t-1} * \left[\frac{\dot{GDP_c}}{GDP_k}\right]_t \left[\frac{X_c}{D_c}\right]_{t-1} * \left[\frac{\dot{X_c}}{X_k}\right]_t + \left[\frac{M_c}{D_c}\right]_{t-1} * \left[\frac{\dot{M_c}}{M_k}\right]_t + R_2$$
>
> R indicates the effect of the interaction of price and compositional (weight) changes. This term in mature market economies is usually very small and frequently ignored. However, it is substantially larger in the transition economies where compositional changes are important; hence, they are shown in table 3.3.3 separately as another component of overall price changes.

of one component to the change in the other. Nevertheless it may help to indicate the transmission belt of price developments from the national accounts (i.e. economy-wide) perspective.

The change in the implicit GDP deflator (the ratio of nominal to real GDP) not only provides a broad measure of domestic cost pressures but is also the broadest measure of inflation which includes not only private (household) consumption[281] but also government consumption, investment goods and exports of goods and services. However, as the GDP deflator measures domestically generated inflationary pressures from all sectors, by definition it also incorporates the price effect of exported goods and services, which should be separated from inflation in domestic markets.[282] On the other hand, not all of the inflation in domestic markets is domestically generated: depending on the degree of openness of the economy, the change in import prices also contributes to the domestic inflation. The question is, therefore, to what extent is the overall rate of domestic inflation actually home-made and to what extent it is imported? The decomposition of the change in the domestic demand deflator into the contributions of the GDP deflator (i.e. domestic costs) on the one hand and the terms of trade effect on the other makes it possible to analyse domestic versus imported inflationary pressures. It is important to note that in this note the import content of exports and the change in the mix of both exports and imports are not taken into consideration.[283] The method of decomposition used for the calculation of contributions is given in box 3.3.1.

---

[281] The implicit private consumption deflator is already a much broader and more internationally comparable inflation indicator than the consumer price index which is limited to a fixed basket of goods which may differ significantly among countries due, *inter alia,* to income levels, habits, geographical location, etc. Furthermore, the implicit private consumption deflator embodies, on a continuous basis, the substitution effect due to relative price changes.

[282] The price of an exported good is not a domestic price, hence does not affect the domestic price level: the same good may have a different price when sold at home rather than abroad due to various reasons, including subsidies, tax breaks, etc.

[283] For changes in the composition of east European and Baltic trade by stages of production in 1996-2000, see sect. 3.5(iii) below, where the new comparative advantages acquired after the mid-1990s in capital- and technology-intensive sectors changed the product composition not only of exports but also of imports. These changes in the composition of trade will have certainly had an effect on the terms of trade.

Table 3.3.3 shows the annual percentage changes in the two deflators (GDP and domestic demand) and the contributions of their major components. The contribution of the price change in domestic demand excluding imports measures the change in the GDP deflator due to the change in the prices of domestic production destined for the domestic markets for consumption and investment. The price change in GDP excluding exports represents that part of domestic inflation due to the changes in domestic factor prices. The contribution of this component to the change in domestic demand deflator is therefore usually called the "domestically generated" or "home-made" part of the overall rate of domestic inflation. The column of import prices in the table shows the contribution of imported inflation to the overall domestic inflation. Import prices are measured in national currencies, which therefore include the combined effect of changes in suppliers' (or foreign) prices and changes in the exchange rates vis-à-vis the trading partners (i.e. nominal effective exchange rates).

(b) Disinflation over the period 1991-1998

The magnitude of the initial price adjustment depended mainly upon the degree of macroeconomic balance and the extent of inherited market distortions at the start of the reforms, such as the degree of hidden, or suppressed, inflation or the level of dependence of the pre-reform production structure on CMEA trade. One of the major adjustments after the initial price liberalization took the form of soaring price inflation. However, after this first stage (during 1989-1990 in eastern Europe and 1991-1992 in the Baltic states) when consumer prices often rose at double-digit monthly rates, the inflation rate whether measured by the GDP deflator or the domestic deflator, fell sharply albeit at different rates among the various countries according to their economic policies and the reforms adopted by governments (chart 3.3.3).

By the mid-1990s, inflation rates were down to single-digit annual rates in Croatia, the Czech Republic and Slovakia. Nevertheless, except in Bulgaria and Romania where soft macroeconomic policies and delayed structural reforms led to excess demand and strong inflationary expectations, all the other countries were also successful in reducing the rate to low double-digit levels in spite of the ongoing price liberalization and other reforms which had adverse effects on prices in the short run. This disinflation was largely due to supply-side factors (basically, falling real wages and squeezed profits) reflecting strong stabilization-cum-liberalization programmes. These domestic supply-side factors were accompanied by some improvement in the terms of trade in several economies particularly in the Czech Republic, Poland, Slovenia and Lithuania (chart 3.3.4). In addition, the recovery in the terms of trade was compounded by the growing shares of imports in domestic demand with the main exceptions of Poland and Slovenia (chart 3.3.5).[284]

Reflecting these developments, the major factor behind domestic price increases in most countries in the mid-1990s were changes in domestic costs (chart 3.3.6). Their contribution ranged between some 80 per cent in Poland, Slovenia and Lithuania to around two thirds in Croatia, the Czech Republic, Hungary and Romania. Only in the other two Baltic states and Slovakia was the contribution of imported inflation bigger than home-made inflation. This reflects, in the main, the very large import share in Estonia (more than 70 per cent), the persistently unfavourable terms of trade in Latvia, and the significantly subdued labour cost pressures in Slovakia. This large contribution of domestic costs can be also seen in chart 3.3.3 which shows that the GDP deflator rose faster than the domestic demand deflator in the majority of countries during the first half of the 1990s.

During 1995-1996 and particularly in 1997 there was a widespread improvement in labour productivity in the majority of east European and Baltic economies reflecting the accelerated output recovery and deep restructuring at the microlevel. This favourable productivity performance offset to a large extent the rise in wage pressures, which were building up after 1993. Industrial real unit labour costs in 1997 stabilized or even declined in many of them. Not all of these reductions in labour and material input cost pressures, however, were transmitted to output prices. Rather, they were reflected in an internal restructuring of total costs with profit margins, having been squeezed for several years, starting to recover.[285]

In 1997 world market commodity prices fell. Furthermore, in most of the east European transition economies, except for those that were hit by economic crisis in late 1996 and early 1997 (Albania, Bulgaria and Romania), the exchange rates remained stable. Import price pressures, both at the retail and the producer levels, therefore weakened and disinflation gathered speed from mid-1997 particularly in those countries where the import share of domestic demand had risen sharply since the beginning of the reforms (reaching some 60 per cent or more for example, in the Czech Republic, Slovakia and the Baltic states).

Disinflation continued in 1998 and was remarkably rapid, and in most countries it overshot expectations. Domestic demand was dampened, albeit to varying degrees, by deliberate or de facto tightening of monetary policy and in some cases by tight fiscal policy as well. However the main factor behind the rapid disinflation in 1998 was the accelerating decline in import prices. Reflecting world demand and supply conditions, international commodity prices in dollar terms fell by more than one quarter and, with intense competitive pressures on world markets, the prices of manufactured goods also fell. In addition, many of the currencies of the east European

---

[284] For a discussion of initial conditions and their role in the Bulgarian case, see UNECE, *Economic Survey of Europe in 1996-1997*, pp. 75-84.

[285] On the growing role of international trade in the east European and Baltic economies in 1991-2000, see sect. 3.2(iii) above.

TABLE 3.3.3

Contributions to changes in the GDP and domestic demand deflators in selected east European and Baltic economies, 1998-2001 [a]

(Percentage, percentage points)

	Changes in GDP deflator	Of which due to changes in:			Changes in domestic demand deflator	Of which due to changes in:		
		Domestic demand deflator excluding imports	Export prices	Composition [b]		GDP deflator excluding exports	Import prices	Composition [b]
Bulgaria								
1998	22.2	15.2	5.9	1.1	20.1	17.2	4.1	-1.2
1999	3.1	1.9	1.1	–	3.0	1.9	1.1	–
2000	5.6	-3.2	8.5	0.4	6.9	-2.6	9.8	-0.3
2001QI	8.6	3.5	5.2	-0.1	7.8	3.1	4.6	0.1
2001QII	8.4	4.5	3.9	–	9.1	4.2	4.9	–
2001QIII	7.4	5.1	2.1	0.1	7.6	5.1	2.7	-0.1
Croatia								
1998	8.4	7.3	1.3	-0.3	6.9	6.1	0.6	0.2
1999	3.9	1.8	2.2	-0.1	4.3	1.6	2.6	0.1
2000	6.6	1.4	5.1	0.1	5.7	1.4	4.4	-0.1
2001QI	5.4	2.7	2.7	–	5.4	2.4	2.9	–
2001QII	4.7	2.6	2.0	–	5.0	2.4	2.6	–
2001QIII	3.8	2.1	1.6	0.1	4.0	2.1	2.0	-0.1
Czech Republic								
1998	10.2	6.9	3.0	0.3	6.4	6.8	-0.1	-0.3
1999	3.6	4.0	-0.6	0.2	3.9	4.1	-0.1	-0.2
2000	0.9	-2.0	2.5	0.4	2.4	-1.5	4.3	-0.4
2001QI	4.0	3.1	0.1	0.8	3.2	3.8	0.2	-0.8
2001QII	4.6	4.0	-0.1	0.7	3.3	4.6	-0.6	-0.7
2001QIII	5.1	4.6	-0.2	0.7	3.3	5.2	-1.2	-0.7
Hungary								
1998	12.6	6.7	5.7	0.2	12.0	6.9	5.3	-0.2
1999	8.4	6.2	2.3	-0.1	9.0	6.0	3.0	0.1
2000	9.1	4.4	5.0	-0.3	10.8	3.9	6.6	0.3
2001QI	10.4	5.8	5.0	-0.4	10.7	5.2	5.2	0.4
2001QII	10.4	7.0	3.8	-0.3	10.6	6.6	3.9	0.3
2001QIII	9.2	7.2	2.3	-0.3	9.4	6.7	2.4	0.3
Poland								
1998	11.8	8.4	3.4	0.1	11.1	8.1	3.1	-0.1
1999	6.8	5.1	1.6	–	7.1	4.9	2.2	–
2000	7.1	7.0	0.4	-0.3	8.9	6.3	2.4	0.3
2001QI	..	..	..	..	..	..	..	..
2001QII	..	..	..	..	..	..	..	..
2001QIII	..	..	..	..	..	..	..	..
Romania								
1998	54.2	46.5	5.9	1.8	48.4	45.1	5.0	-1.7
1999	48.7	33.6	14.6	0.4	47.3	31.5	16.2	-0.4
2000	45.3	33.4	11.6	0.4	43.7	32.4	11.7	-0.4
2001QI	46.5	30.6	16.0	-0.1	48.3	30.1	18.1	0.1
2001QII	44.6	31.3	13.3	0.0	46.0	29.9	16.0	–
2001QIII	36.9	24.9	12.1	-0.1	37.8	24.0	13.8	0.1
Slovakia								
1998	5.1	3.0	1.7	0.4	1.0	3.1	-1.8	-0.4
1999	6.6	3.9	3.4	-0.7	8.1	2.8	4.6	0.6
2000	6.5	-0.6	7.5	-0.5	7.3	-1.0	7.8	0.4
2001QI	6.8	1.6	5.5	-0.3	10.7	1.3	9.1	0.3
2001QII	5.8	0.5	5.5	-0.1	9.2	0.3	8.8	0.1
2001QIII	5.2	0.8	4.5	-1.0	8.2	0.7	7.4	0.1
Slovenia								
1998	7.8	5.7	1.9	0.2	6.3	5.8	0.7	-0.2
1999	6.6	5.0	1.2	0.4	5.7	5.5	0.8	-0.4
2000	5.7	0.4	5.5	-0.2	8.5	0.2	8.0	0.2
2001QI	9.6	3.3	6.6	-0.3	9.9	2.8	6.8	0.3
2001QII	9.5	3.5	6.2	-0.2	10.1	3.2	6.7	0.2
2001QIII	9.8	4.6	5.5	-0.2	9.8	4.2	5.4	0.2

(For source and notes see end of table.)

TABLE 3.3.3 (concluded)

Contributions to changes in the GDP and domestic demand deflators in selected transition economies, 1998-2001 [a]

(Percentage, percentage points)

	Changes in GDP deflator	Of which due to changes in:			Changes in domestic demand deflator	Of which due to changes in:		
		Domestic demand deflator excluding imports	Export prices	Composition [b]		GDP deflator excluding exports	Import prices	Composition [b]
Estonia								
1998	9.3	6.2	2.9	0.3	7.3	5.8	1.7	-0.2
1999	4.5	4.3	1.0	-0.7	4.2	3.2	0.3	0.7
2000	4.7	-0.4	4.5	0.7	4.3	0.2	4.7	-0.6
2001QI	5.6	-2.3	7.5	0.5	5.6	-1.8	7.8	-0.4
2001QII	6.4	-3.2	9.2	0.4	4.6	-2.7	7.7	-0.3
2001QIII	6.5	-2.8	9.1	0.2	3.8	-2.6	6.5	-0.2
Latvia								
1998	5.5	2.2	2.7	0.6	2.2	2.6	0.2	-0.6
1999	7.4	7.9	-0.5	–	4.4	7.0	-2.5	–
2000	4.3	3.8	1.3	-0.8	6.6	2.7	3.2	0.7
2001QI	1.3	0.5	0.8	0.1	2.0	0.5	1.6	-0.1
2001QII	1.5	0.4	1.2	-0.1	2.4	0.3	2.0	0.1
2001QIII	1.6	0.6	1.0	–	2.2	0.6	1.7	–
Lithuania								
1998	6.7	7.8	-2.0	0.8	4.3	7.8	-2.8	-0.7
1999	3.2	4.4	-0.2	-1.0	2.2	3.1	-1.7	0.9
2000	2.1	–	2.7	-0.5	2.1	-0.5	2.1	0.5
2001QI	1.4	-0.3	0.7	1.0	-1.2	0.7	-1.0	-1.0
2001QII	1.2	-0.5	0.9	0.9	-1.0	0.3	-0.5	-0.8
2001QIII	0.5	-0.1	–	0.7	-1.1	0.5	-1.0	-0.6

Source: National accounts.

[a] Quarterly data for 2001 are cumulative.

[b] Calculated as the residual of the change in the GDP and domestic demand deflators minus the contribution of the change in their respective components' prices. It shows the change in the GDP and domestic demand deflators due to the change in the structure of GDP and domestic demand, respectively.

and Baltic economies remained relatively stable in nominal terms and even appreciated against the major currencies (except for a few months following the rouble crisis in August). The currency effect amplified the positive terms of trade effect on the material costs of production and imported finished goods. As a result, in all 11 countries import prices ceased to be a major inflationary factor in 1998 (table 3.3.3). Except in Croatia, domestic inflation, as measured by the domestic demand deflator, decelerated throughout eastern Europe and the Baltic states. In Croatia, however, even though the domestic demand deflator accelerated it still rose by less than the GDP deflator, a reflection of growing domestic cost pressures (measured by the GDP deflator excluding export prices). Also in the Czech Republic, although the change in the domestic demand deflator slowed down in 1998 due to the absence of import price pressures (import prices actually fell and pulled down the domestic rate of inflation slightly), the GDP deflator accelerated rather rapidly, suggesting intensified cost pressures mainly arising from the output recession and its adverse effect on productivity growth.[286] This downturn reflected the painful macroeconomic adjustment after the currency crisis in May 1997.[287]

(c) Change in the nature of inflation in 1999-2001

The favourable terms of trade effect was reversed in mid-1999, with world market prices for crude oil tripling between the beginning of 1999 and September 2000. This sharp rise in dollar prices was compounded by a significant appreciation of the dollar. There was a widespread increase in import price pressure in 2000 (table 3.3.3), except in Romania where the share of imports in domestic demand, together with Poland, is the lowest in the region, at some 35 per cent. In Poland import price pressure remained at about the same level as in 1998 (2.4 percentage points compared with 2.2 points in 1999) and its share in total domestic inflation fell from nearly one third to about one quarter. In the other countries, the sharp increase in the contribution of import prices to the rate of domestic inflation was partly offset by significantly moderated or even reduced domestic cost pressures (Bulgaria, the Czech Republic, Slovakia and Lithuania).

---

[286] The output gap and its dampening effects on wage demands and hence inflation are not considered here as this accounting approach does not include lagged effects, as mentioned above.

[287] On the currency crisis and the macroeconomic policy correction in the Czech Republic, see UNECE, *Economic Survey of Europe, 1998 No. 1*, pp. 75-82.

CHART 3.3.3

Change in the GDP and domestic demand deflators in selected east European and Baltic economies, 1991-2001 [a]
(Percentages)

Source: National accounts.

[a] First three quarters.

CHART 3.3.4

Change in the export and import deflators in selected east European and Baltic economies, 1991-2001 [a]

(Percentages)

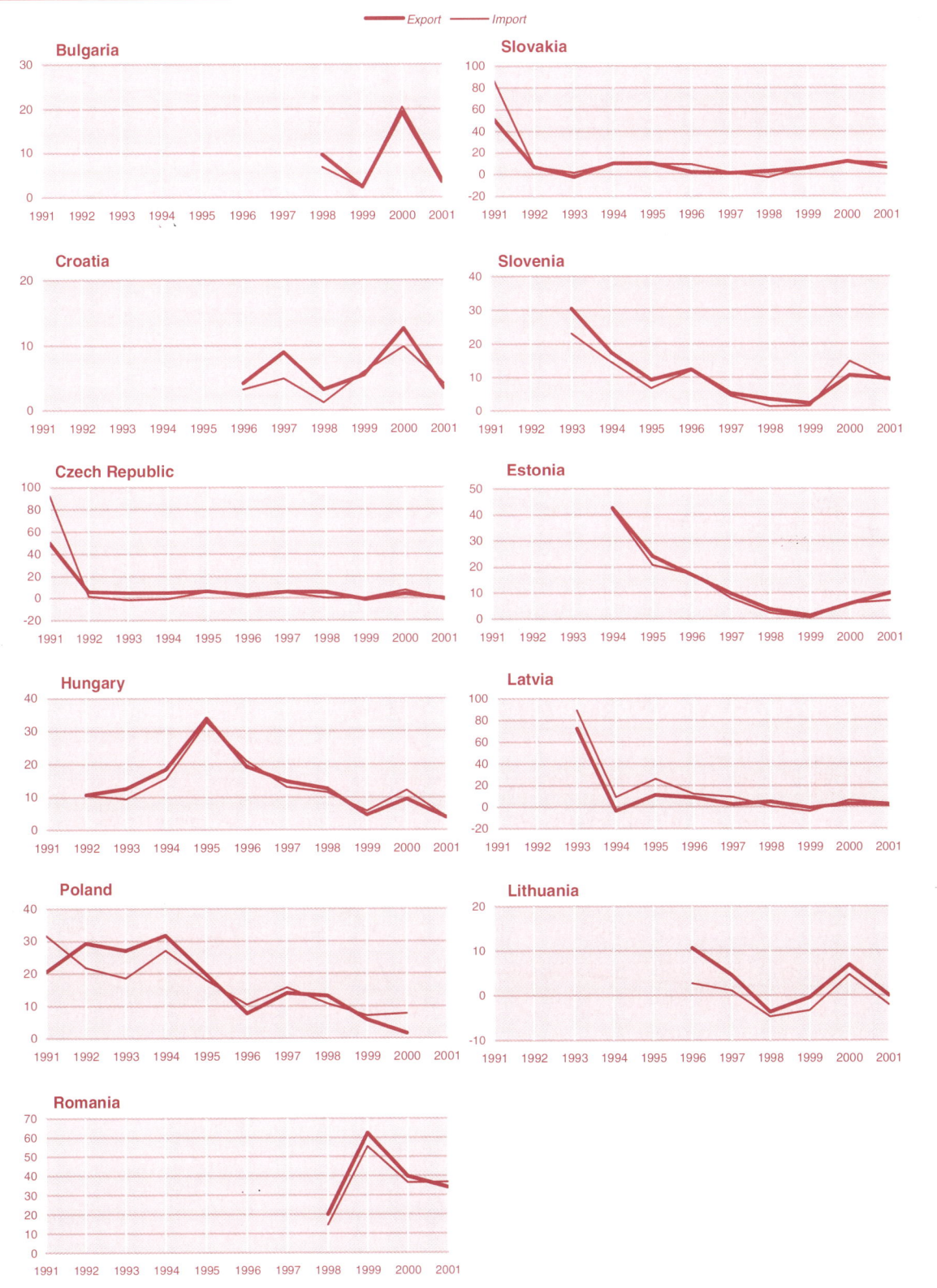

Source: National accounts.

[a] First three quarters.

CHART 3.3.5

Share of imports in domestic demand [a] in selected east European and Baltic economies, 1991-2001 [b]
(Percentages)

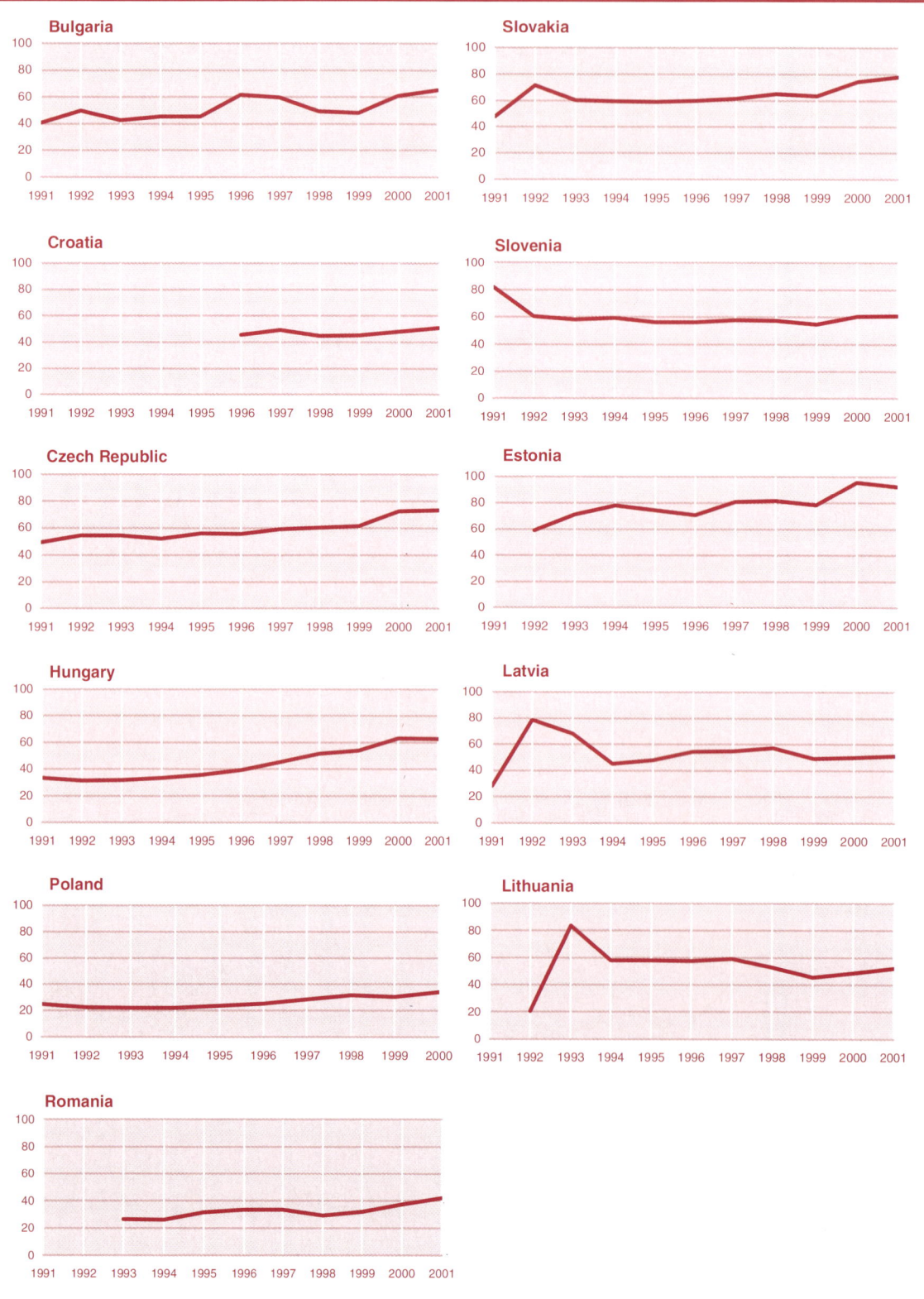

Source: National accounts.

[a] In current prices.

[b] First three quarters.

The Transition Economies 111

CHART 3.3.6

The contributions of domestic costs and import prices to the total domestic inflation rate in selected east European and Baltic economies, 1991-2001 [a]

(Percentages and percentage points)

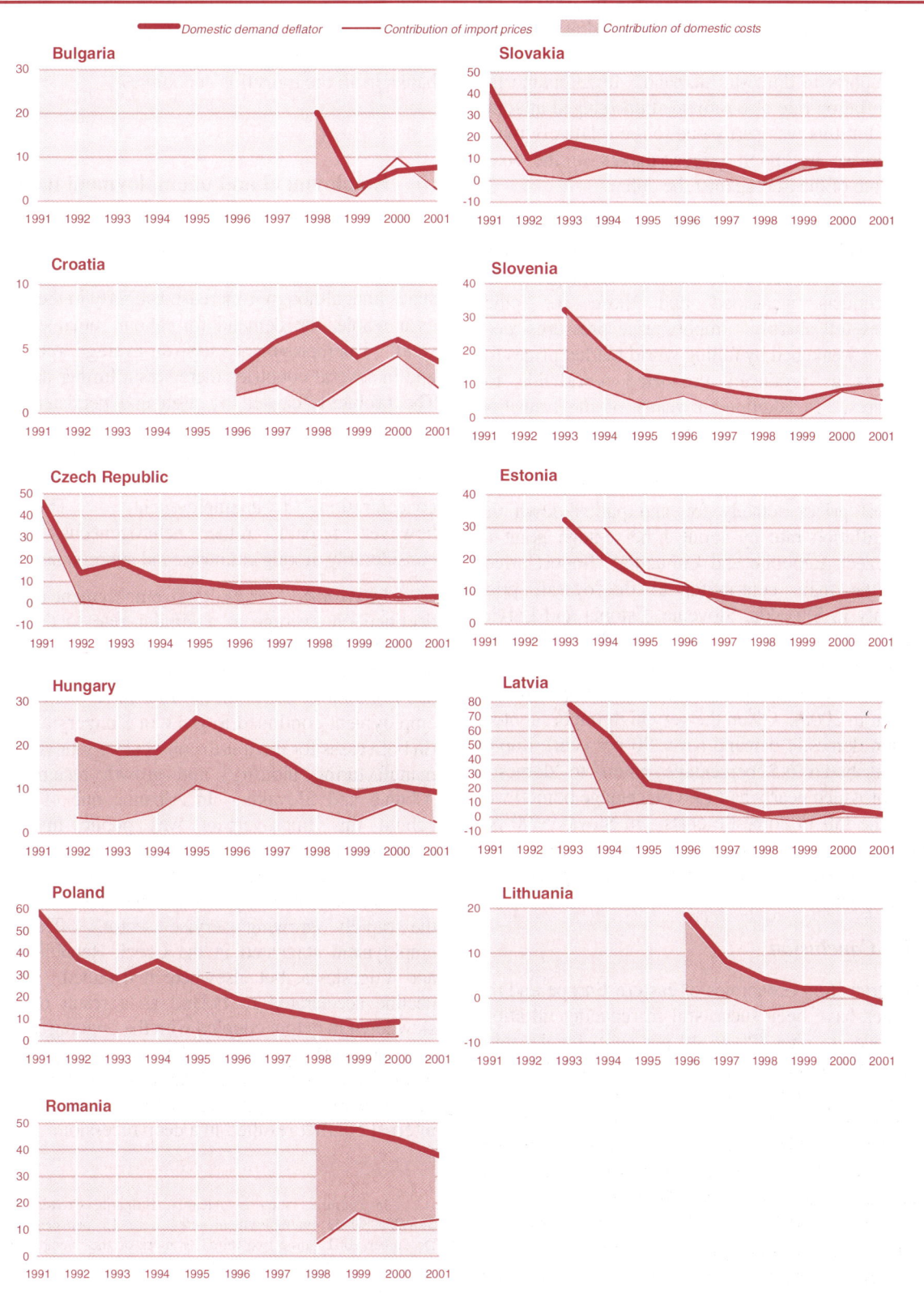

Source: National accounts.

[a] First three quarters.

In general, labour productivity gains were the major offset to the shock of higher import costs in 2000. As a result, despite the strong external shock of higher import prices, inflation measured by the change in the domestic demand deflator continued to decelerate in 2000 in the Czech Republic and Slovakia, and remained at about the same low rate of 1999 in Estonia, in spite of the very high share of imports in domestic demand. In Lithuania the domestic inflation rate also remained unchanged at some 2 per cent, but this was partly due to the relatively much slower increase in import prices, thanks to the fixed exchange rate of the litas against the dollar.

Over the first three quarters of 2001, imported inflation subsided significantly. In the face of the strong downturn in global economic activity, world prices for energy products weakened and those for other commodities fell sharply. Import price pressures were again further weakened by falling world market prices for manufactures as producers competed worldwide for market shares. In addition, in many of the transition economies there was real exchange rate appreciation, particularly against the euro, which further reduced the domestic prices of imports. Over the three quarters of 2001 import prices actually fell and pulled down the domestic inflation rate by about 1 percentage point in both the Czech Republic and Lithuania. Import prices remained the main contributor to the overall price changes only in Slovakia, Slovenia, Estonia and Latvia. In Estonia all of the domestic inflation (3.8 per cent) was due to the acceleration of import prices (they rose by 7 per cent in the first three quarters of 2000, compared with 6.1 per cent in 2000). Given the very high import content of Estonian domestic demand (over 90 per cent), import prices contributed 6.5 percentage points to domestic inflation, of which nearly half (2.6 percentage points) was offset by the fall in domestic costs. In Slovakia 90 per cent of the overall domestic inflation rate (8.2 per cent) was due to the contribution of higher import prices (7.4 percentage points).

(d) Conclusion

The transition economies of eastern Europe and the Baltic states have been successful in reducing inflation, first to moderate rates (20 to 40 per cent) by the mid-1990s and then to single digits by the end of 2001. The above analysis of the origins of this disinflation, based on a national accounting approach, suggests that a significant part of this disinflation was due to the decline in imported inflation, which in turn was only partly due to exchange rate developments. The rest, in fact a large part of it, originated in the easing of foreign suppliers' prices due to the intensified global competition for export markets. Given this significant role of external factors in the disinflation and recognizing that "push" factors due to the continuation of the liberalization of domestic prices and the "catching-up" process are still important, bringing their rates of inflation close to those in the EMU countries still remains a major challenge for most of the transition economies. Given that the growth of labour productivity is the major source of sustainable reductions in home-made inflation, the elimination of the obstacles to faster output growth (*inter alia*, high real interest rates, slow privatization and shallow microlevel restructuring, distortions in the labour and capital markets) has to be given priority both by the governments and the central banks in all the transition economies.

3.4 Labour markets

(i) Employment and unemployment in 2001

The labour markets in the transition economies remained generally tense in 2001. Employment continued to decline in eastern Europe and the Baltic states and, although it increased slightly in the CIS region as a whole, the demand for labour remained generally weak. Unemployment rates, on average, remained high and in several countries there was a further deterioration. The problems caused by high and persistent levels of unemployment were exacerbated in some countries by the continued fall in the number of people eligible for unemployment benefits[288] and by the regional concentration of unemployment.[289] The situation, however, remains rather heterogeneous and differs considerably among countries and subregions.

In *eastern Europe,* total employment (table 3.4.1) declined on average at a similar rate to that in 2000, although the average is strongly affected by the developments in Poland where there was a marked deterioration in 2001. In the first three quarters of 2001, employment continued to grow in Hungary and Slovenia (in both cases for the fourth consecutive year and mainly in manufacturing industry) and growth resumed also in Albania and Slovakia. In Albania, one of the factors behind this improvement was public investment in infrastructure projects and a programme of job creation through the financing of large-scale public works.[290] In Slovakia more than 80 per cent of new jobs were created in the rapidly growing service sector. The level of employment stagnated in the Czech Republic, Romania and Yugoslavia, but elsewhere it declined. The rate of decline accelerated in Poland as a result of the sharp slowdown in output coupled with the ongoing restructuring of unprofitable industries. The fall in employment was particularly pronounced in The former Yugoslav Republic of Macedonia, where escalating ethnic tension and internal conflict resulted in a deep recession.

---

[288] In countries such as Albania, Bulgaria, Croatia, Poland and Slovakia where unemployment rates were 15 per cent or more in December 2001, some two thirds or more of the unemployed were no longer eligible for unemployment benefit.

[289] For a more detailed discussion of regional disparities of unemployment, see UNECE, *Economic Survey of Europe, 2001 No. 2,* p. 27.

[290] Among other measures to fight unemployment, the Albanian Ministry of Labour and Social Affairs allocated $3.5 million to the Labour Encouragement Programme. *Reuters Business Briefing,* 19 September 2001.

TABLE 3.4.1

Total employment and registered unemployment in eastern Europe, the Baltic states and the CIS, 1999-2001
(Percentage change over the same period of the preceding year, per cent of labour force, end-of-period)

	Total employment [a]						Unemployment					
	1999	2000		2001			1999	2000	2001			
	Annual	Annual	QIV	QI	QII	QIII	Dec.	Dec.	Mar.	Jun.	Sep.	Dec.
Eastern Europe	-1.9	-1.5	-0.7	-0.6	-1.2	-0.9	14.6	15.2	15.6	14.9	14.8	15.6*
Albania	-1.8	0.3	–	1.2	1.1	0.8	18.2	16.9	16.0	15.1	14.8	15.0*
Bosnia and Herzegovina [b]	3.1	0.7	0.9	-0.2	-1.3	-1.6	39.0	39.4	39.5	39.3	39.9	40.0*
Bulgaria	-2.1	-4.7	-2.7	-3.4	-4.2	-2.2	16.0	17.9	18.4	17.1	16.5	17.3
Croatia	-0.4	-1.0	-0.7	-1.1	-0.7	-0.2	20.8	22.6	22.9	21.5	22.0	23.1
Czech Republic	-2.5	-2.0	-1.6	-0.1	-0.2	0.1	9.4	8.8	8.7	8.1	8.5	8.9
Hungary	3.1	1.0	1.4	1.4	0.6	0.3	9.6	8.9	9.3	8.4	8.0	8.0
Poland	-2.7	-2.3	-3.2	-3.0	-3.1	-3.6	13.1	15.1	16.1	15.9	16.3	17.4
Romania [c]	-0.6	-0.1	1.0	0.6	-1.1	-0.1	11.5	10.5	10.3	8.7	7.8	8.6
Slovakia [c]	-3.0	-1.4	0.2	0.3	1.7	1.4	19.2	17.9	19.2	17.8	17.4	18.6
Slovenia	1.8	1.3	0.1	1.0	1.2	1.6	13.0	12.0	11.8	11.1	11.3	11.8
The former Yugoslav Republic of Macedonia	1.8	-1.3	-1.9	-2.9	-3.8	-4.5	43.8	44.9	..	..	..	..
Yugoslavia [d]	-8.2	-2.7	-1.3	-1.2	0.1	–	27.4	26.6	27.7	27.1	27.8	27.9
Baltic states	-1.2	-2.0	-2.8	-2.8	-2.3	-1.8	9.1	10.0	10.6	9.8	9.7	10.1
Estonia [e]	-4.1	-0.9	-0.3	-0.1	1.5	0.7	6.7	7.3	8.4	7.5	7.3	7.2
Latvia	-0.5	–	-0.4	-0.2	–	0.2	9.1	7.8	8.1	7.8	7.6	7.7
Lithuania	-0.5	-3.7	-5.2	-5.5	-5.3	-4.0	10.0	12.6	13.2	12.1	12.0	12.9
CIS [f]	-0.5	..	0.4	1.2	0.6	0.2	8.3	..	..	..	..	..
Armenia	-2.9	-1.6	-2.4	-0.7	-0.4	0.2	11.5	10.9	10.7	10.4	10.0	9.8
Azerbaijan	–	–	0.1	0.4	0.3	0.2	1.2	1.2	1.2	1.3	1.3	1.3
Belarus	0.6	–	-0.2	-0.1	-0.1	-0.1	2.0	2.1	2.4	2.2	2.3	2.3
Georgia	-8.9	..	..	..	..	..	5.6	..	..	..	..	..
Kazakhstan	-0.4	1.6	1.6	3.6	2.9	2.7	3.9	3.7	3.8	3.3	2.9	2.8
Kyrgyzstan	3.5	0.2	-0.5	0.3	0.3	0.3	3.0	3.1	3.2	3.2	3.2	3.1
Republic of Moldova	-9.0	1.3	-0.3	-0.3	-1.6	-0.8	2.1	1.8	2.2	2.0	1.8	1.7
Russian Federation [g]	0.2	0.6	1.4	2.2	1.2	0.5	12.2	9.8	9.4	8.6	8.6	9.0
Tajikistan	-3.3	0.5	2.1	1.5	0.5	0.7	3.1	3.0	2.4	2.5	2.5	2.6
Turkmenistan	3.8	..	..	..	..	..	..	..	..	..	..	..
Ukraine	-2.3	-2.5	-2.7	-1.8	-1.4	-1.3	4.3	4.2	4.2	3.8	3.6	3.7
Uzbekistan	1.0	1.1	..	..	..	..	0.5	0.6	..	..	..	0.4
Memorandum items:												
CETE-5	-2.0	-1.7	-1.4	-0.9	-0.9	-1.1	12.5	13.4	13.8	13.6	13.9	14.6
SETE-7	-1.8	-1.2	-0.1	-0.4	-1.3	-0.5	17.1	17.8	17.3	16.8	16.8	17.0*
Former GDR	..	..	..	..	..	..	17.7	17.2	18.6	16.8	16.9	17.6

Source: National statistics; CIS Statistical Committee; direct communications from national statistical offices to UNECE secretariat.

[a] Annual average unless otherwise stated. Changes in employment based on quarterly statistics are not always fully comparable with annual data due to differences in coverage.

[b] Figures cover only the Bosnian-Croat Federation. Data for Republika Srpska are not available.

[c] Labour force survey data.

[d] Data exclude Kosovo and Metohia.

[e] Unemployment: until October 2000 – job seekers, thereafter – registered unemployed as percentage of the labour force.

[f] Regional quarterly aggregates of employment exclude Georgia, Turkmenistan and Uzbekistan.

[g] Unemployment figures are based on monthly Russian Goskomstat estimates according to the ILO definition, i.e. including all persons not having employment but actively seeking work.

Unemployment generally worsened in 2001. The average unemployment rate for the region as a whole was 15.6 per cent in December 2001, more than a year earlier and the highest since the start of economic transformation (table 3.4.1). In the 12 months to December 2001, unemployment declined in Albania, Hungary and Slovenia and there were some falls in unemployment rates in Bulgaria and Romania, although these occurred against a background of stagnating or declining employment which suggests departures from the labour force (the "discouraged" unemployed). The rate of unemployment was relatively stable in the Czech Republic, at about 8.5 per cent in the second half of 2001, but in the rest of the region it continued to rise. The worst deterioration was in Poland, where sluggish growth and continued lay-offs, combined with the increased pressure of newcomers to the labour market, led to a sharp rise in joblessness.[291] In December 2001, the

---

[291] For more details on the economic situation in Poland, see sect. 3.1(iii).

unemployment rate reached a record 17.4 per cent of the labour force, 2.3 percentage points higher than a year earlier and the largest increase among all the transition economies. Unemployment is likely to continue to rise in the coming months as economic growth is expected to remain weak;[292] national analysts believe that by the end of 2002, the unemployment rate in Poland could reach 19 to 20 per cent of the labour force although the government assumed a rise to 18.6 per cent in its 2002 budget.[293]

As was the case in 2000, a strong economic performance in the *Baltic states* did not lead to a radical improvement in the labour markets, although the situation differed considerably among the three countries (table 3.4.1). In Estonia, where there was a strong recovery, employment bottomed out at the end of 2000, and in 2001 it started to grow for the first time since 1997. The improvement was mainly in the rapidly growing manufacturing industry where employment increased for the second year in a row. Despite the strong output growth, the demand for labour remained basically flat for the third consecutive year in Latvia, a reflection of the specific composition of growth.[294] In Lithuania, despite fairly strong output growth, employment fell by nearly 5 per cent in the first three quarters of 2001, although the rate of decline decelerated somewhat in the second half of the year. The decline affected all main sectors of the economy including services, but most of the job losses (nearly 70 per cent of the total) were in agriculture. On the one hand, the divergent trends in output and employment reflect the relatively narrow base of economic growth.[295] On the other hand, the privatization of utilities and other public sector activities in 2001, and the introduction of new bankruptcy and labour laws in March 2001 might have led to widespread job losses.[296]

In the 12 months to December 2001, the average unemployment rate in the Baltic countries was just above 10 per cent,[297] slightly higher than a year earlier (table 3.4.1), although again the situation differed among the countries. In Latvia, the downward trend in unemployment observable since 1999, decelerated in 2001, and the unemployment rate was actually unchanged in the 12 months to December 2001. Also in Estonia the rate was broadly unchanged from a year earlier. In Lithuania, despite the fall in employment, the unemployment rate was falling after reaching a post-independence high of 13.2 per cent in March 2001. It began to rise again in October but at nearly 13 per cent in December 2001 it was only slightly higher than a year earlier, which suggests departures from the labour force and declining activity rates.

The available quarterly data suggest that in the *CIS countries* there was some improvement in the labour markets in 2001. Employment increased slightly in the region as a whole (table 3.4.1) in line with the widespread and strong economic recovery. In the first three quarters of 2001, there was a relatively high rate of employment growth in Kazakhstan, Russia and Tajikistan. In Russia it was accompanied by continued shifts from large- and medium-sized enterprises to other sectors of the economy and in Kazakhstan, it was due to stronger labour demand in construction and particularly in services. In the other countries of the region for which data are available, employment was broadly flat, but it continued to decline in the Republic of Moldova and Ukraine.

Despite these generally positive developments, the official unemployment figures indicate few changes in the 12 months to December 2001. It should be noted, however, that the statistics of registered unemployment in the CIS countries continue to be unreliable as a large proportion of the jobless, although willing to work, do not register for various reasons.[298] According to estimates of the CIS Statistical Committee, the total number of unemployed in the region (i.e. all those who are out of work and searching for a job, not just those who are registered) in 2001, was some 11 million persons in the region as a whole, down from 12.5 million in 2000.[299] The average unemployment rate based on these figures was 9 per cent (compared with 9.5 per cent in 2000). Russia was the main source of this improvement. Goskomstat's estimates of unemployment based on data from the labour force survey indicate a marked fall in unemployment (by more than 600,000 persons in the 12 months to December 2001), while the unemployment rate declined to 9 per cent in December 2001, nearly 1 percentage point less than a year earlier. The situation also improved in Kazakhstan and in Ukraine.[300] In the latter, according to the labour

---

[292] In January 2002, the unemployment rate reached 18 per cent, compared with 15.7 per cent in January 2001.

[293] *Polish News Bulletin*, 22 February 2002, as quoted by *Reuters Business Briefing*.

[294] Although, there was some acceleration of manufacturing output, the growth of the Latvian economy was largely due to the transport sector, and especially from the transit shipment of Russian energy products. For more details, see sect. 3.2.

[295] The nearly 17 per cent increase in industrial output in 2001 was to a large extent based on oil refining, whereas in the relatively large agricultural sector output fell by 8 per cent (sect. 3.2).

[296] EIU, *Country Report, Lithuania* (London), April 2002.

[297] It should be emphasized that the official figures for registered unemployment in the Baltic states tend to underestimate the actual levels of unemployment. In the fourth quarter of 2001, unemployment rates according to the labour force surveys were 11.9 per cent in Estonia, 12.8 per cent in Latvia (November) and 17.5 per cent in Lithuania (November), whereas the registered rates were 7.2, 7.6 and 12.5 per cent, respectively.

[298] Due to the low level of unemployment benefit (in November 2001 it varied in individual countries between some 6 and 30 per cent of average nominal wages), the generally short period during which it is paid and the small chance of finding a job through registration, a large proportion of jobless simply do not register. The proportion is estimated in different countries at between 50 to 80 per cent of total unemployment, so it is clear that registered unemployment is unlikely to reflect the true level of unemployment.

[299] CIS Statistical Committee, *Statistical Bulletin "Statistika SNG"*, No. 2, January 2002, p. 58.

[300] In 2001, Kazakhstan, following Russia and Ukraine, started to conduct a regular quarterly labour force survey. The unemployment rate derived from data in this survey declined during 2001 from 12.7 per cent

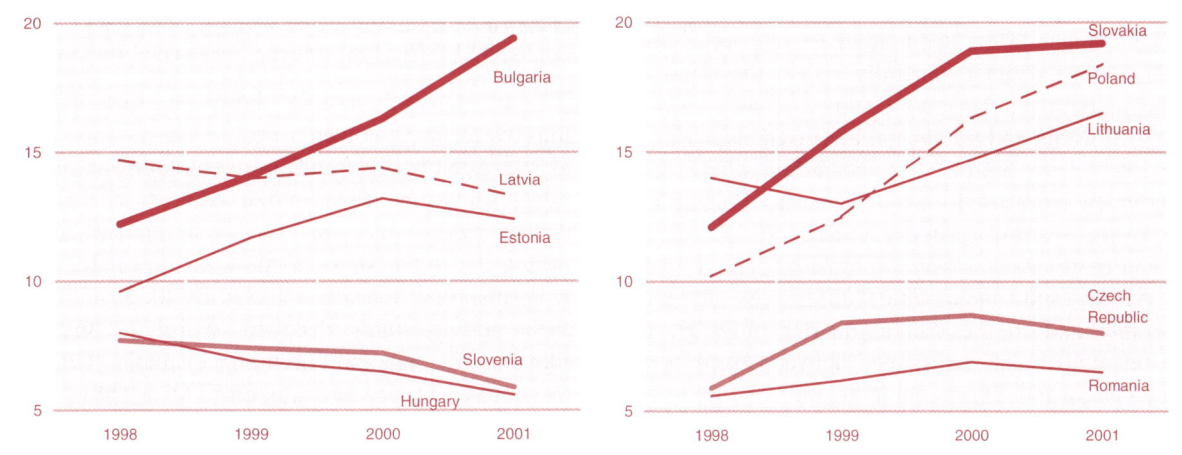

CHART 3.4.1

Unemployment rates in selected central and east European economies, 1998-2001
(Per cent of labour force)

Source: National labour force surveys and direct communications from national statistical offices to UNECE secretariat.
Note: Data refer to second quarter of each year.

force survey, the unemployment rate declined steadily during the year to stand at 10.3 per cent in September 2001, 1.4 percentage points lower than a year earlier.[301]

(ii) Changes in the structure of unemployment, 1998-2001

After its peak at the end of 1993 or early 1994, unemployment started to decline in most central and eastern European countries[302] as a result of a strong economic recovery and, in some of them, active labour market policies.[303] However, since the second half of 1998, due to a combination of factors,[304] the situation has reversed and in many countries of the region there has been a surge in unemployment (chart 3.4.1). Three different patterns in the evolution of unemployment emerged in this period. In the first group of countries (Bulgaria, Poland, Slovakia and, to a lesser extent, Lithuania) unemployment increased sharply from 1998 to 2000 and continued to do so in 2001. In the second quarter of 2001, unemployment rates in these countries approached or exceeded 19 per cent of the labour force, presenting policy makers with a major challenge. In contrast, in Hungary and Slovenia, unemployment declined somewhat between 1998 and 2001, mainly as a result of net job creation;[305] there was a slight downward trend in Latvia as well. In the remaining three countries (the Czech Republic, Romania and Estonia) the increase over 1998-2000 was less pronounced and only in 2001 did the first signs of improvement appear.

The analysis of unemployment at a more disaggregated level suggests that unemployment rates for the main groups of the unemployed (male, female and youth) generally followed the trend of total unemployment. At the same time a number of specific features also emerged in this period. Based on harmonized data[306] this note reviews the main changes which have occurred in the different groups of the unemployed between mid-1998 (i.e. just before the Russian crisis) and mid-2001. In particular, it looks at such issues as gender differentials, the worsening of youth unemployment, and the growing incidence of long-term joblessness. Concerning changes in the position of women in the labour market, this section looks at whether women were disproportionably affected in those countries where there was a sharp increase in unemployment, and whether

---

in the first quarter to 9.2 per cent in the third quarter. *Sotsial'no-economicheskoe razvitie Respubliki Kazakhstan, 1/2002* (Alma Aty), February 2002, p. 118.

[301] Direct communications to the UNECE secretariat from the national statistical office.

[302] The analysis in this section refers to the 10 central and eastern European countries seeking EU membership.

[303] The largest reductions were in Hungary and particularly in Poland (in the latter, the unemployment rate declined between 1993 and 1998 by about 6 percentage points to 10.6 per cent).

[304] Obviously, the importance of different factors responsible for the resurgence of unemployment varies significantly between countries, but in all of them the Russian economic crisis, delayed enterprise restructuring and intersectoral adjustments resulted in a situation when unemployment re-emerged as a serious political and socio-economic problem. UNECE, *Economic Survey of Europe, 2001 No. 1*, pp. 133-136; *Oxford Analytica East Europe Daily Brief*, 11 October 2001.

[305] The continuous growth of employment over 1998-2000, resulted in a cumulative 6 per cent increase in Hungary and more than 3 per cent in Slovenia.

[306] The data used in this note are taken from national labour force surveys conducted according to the ILO's methodology and are internally consistent and comparable across countries. The data relate to the second quarter of each of the respective years 1998-2001. The main sources are the national statistics officially published and/or obtained by the UNECE secretariat through direct communications with the national statistical offices.

they benefited from the improvement in countries with declining unemployment.

(a) Unemployment by gender

The changes in the unemployment rate for women between 1998 and 2001 generally reflect the trend for total unemployment: the rates declined slightly only in Hungary, Slovenia and Latvia and increased elsewhere in the region (chart 3.4.2). In countries where total unemployment increased most (Bulgaria, Poland and Slovakia) there were particularly large (6 percentage points and more) increases in the female unemployment rate. As a result, intercountry differences increased considerably and in the second quarter of 2001, the female unemployment rate ranged between some 5 per cent in Hungary and 20 per cent in Poland.

One salient feature, however, is that in the overwhelming majority of countries, including those with large increases in unemployment, the relative position of male and female unemployment rates during 1998-2001 remained unchanged. In the second quarter of 2001, female unemployment rates were higher than those for men in only 4 out of 10 countries. At the same time, the relative position of women against men[307] improved in most countries (chart 3.4.3). The main exception was Estonia, where the position of women has not only deteriorated considerably but the relationship between male and female unemployment rates was reversed between 2000 and 2001.[308] The largest relative differences (in the range of 20 to 40 per cent) remained, despite some improvement, in the Czech Republic and Poland. At the same time, in Hungary, Latvia and Lithuania, female unemployment rates were some 20 per cent less than those of men. Between 1998 and 2001, the share of women in total unemployment tended to fall in most countries; in 2001, it was higher than that for men only in the Czech Republic, while in Poland and Estonia it was similar to that of men (chart 3.4.5). The lowest shares of female unemployment – less than 40 per cent – were in Hungary and Lithuania.

In most countries, activity rates for both males and females tended to fall between 1997 and 2000. Female activity rates, however, have not declined more than those of men (table 3.4.2). The position of women remained unchanged in Bulgaria and Slovenia, where male and female activity rates declined by approximately the same proportion. In 5 of the 10 countries, female activity rates fell less than those of males and in another 3, the female activity rate increased. In turn, the activity rates of men either increased by a considerably smaller proportion than those of women (Hungary), remained unchanged (Slovakia) or even declined (Lithuania).

It is worth noting, however, that the relative improvement of women's position in the labour market, as suggested by the trends in unemployment and activity rates, should be seen in the context of the previous developments. As was shown in an earlier study,[309] during the initial stage of the transition process women were disproportionately affected by the loss of jobs and their activity rates declined sharply. In Hungary, Slovenia, Estonia and Latvia, for example, the decline in the female activity rate was close to or above one fifth. As a result, by 2000, while women accounted for some 50 to 52 per cent of the population of working age, they had on average much lower activity rates than men[310] and represented a larger part of the inactive population.[311] The data for unemployment and activity rates, however, do not provide a full picture of women's position in the labour market. Thus, for example, a more detailed analysis is needed of the factors behind women's gains in the labour market: whether unemployment and activity rates improved across the board, or whether they were largely due to the increase of female employment in low paid jobs in the public and private service sectors.[312] Overall, the data suggest that in 1998-2001 there was at least a marginal improvement in the relative position of women in the labour markets in some transition economies.

---

[307] As measured by the ratio of the unemployment rate of women to that of men.

[308] Over the entire 1998-2001 period, the female unemployment rate in Estonia was below (by 1 to 2 percentage points) that of men. In the second quarter of 2001 this pattern was suddenly reversed; but in the third quarter the previous relationship was restored. This fluctuation in the ratio in 2001 may reflect some increase in unemployment of young (15-24 aged) women, but it may also be due to the standard error of estimate of the gender specific unemployment rate, which is around 2 percentage points. Direct communications from the National Statistical Office of Estonia to UNECE secretariat.

[309] In a study of the effects of transition on the labour market from a gender perspective over 1985-1997, it was shown that not only did a large number of women lose their jobs, but many of them withdrew from the labour force altogether. In most cases the declining share of women among the unemployed was the result of pushing women out of the labour force rather than into unemployment. Consequently, the female unemployment rate was not significantly higher than that of men in general, and in some economies it was even lower. UNECE, *Economic Survey of Europe, 1999 No. 1*, pp. 135-142.

[310] In 2000, standardized female activity rates (for the working age population of 15-64) varied in most cases between 60 per cent and 65 per cent (with the lowest rate in Hungary (53 per cent) and the highest in Lithuania (68 per cent)). Male activity rates were mostly within a range of 70 to 76 per cent (with the lowest rate in Bulgaria (66 per cent) and the highest in the Czech Republic (79 per cent)). Eurostat, *Employment and Labour Market in Central European Countries, 1/2001* (Brussels), 2001, pp. 48-57.

[311] In Hungary and Poland, for example, where women accounted for some 51 per cent of the working age population in 2000, their share in the inactive population exceeded 60 per cent. Hungarian Central Statistical Office, *Statistical Yearbook of Hungary, 2000* (Budapest), 2001, p. 81; Polish Central Statistical Office (GUS), *Statistical Yearbook of Poland 2000* (Warsaw), 2001, p. 131.

[312] In Hungary, for example, female employment in *Education* and *Other community and personal service activities* (branches M and O in the ISIC classification) grew steadily during 1998-2001. The share of women in these branches increased from some 76 and 48 per cent in 1997, to 78 and 52 per cent in 2000, respectively. Hungarian Central Statistical Office, op. cit., pp. 83, 85.

The Transition Economies 117

CHART 3.4.2

Dynamics of male and female unemployment rates in selected central and east European economies, 1998-2001
(Per cent of labour force)

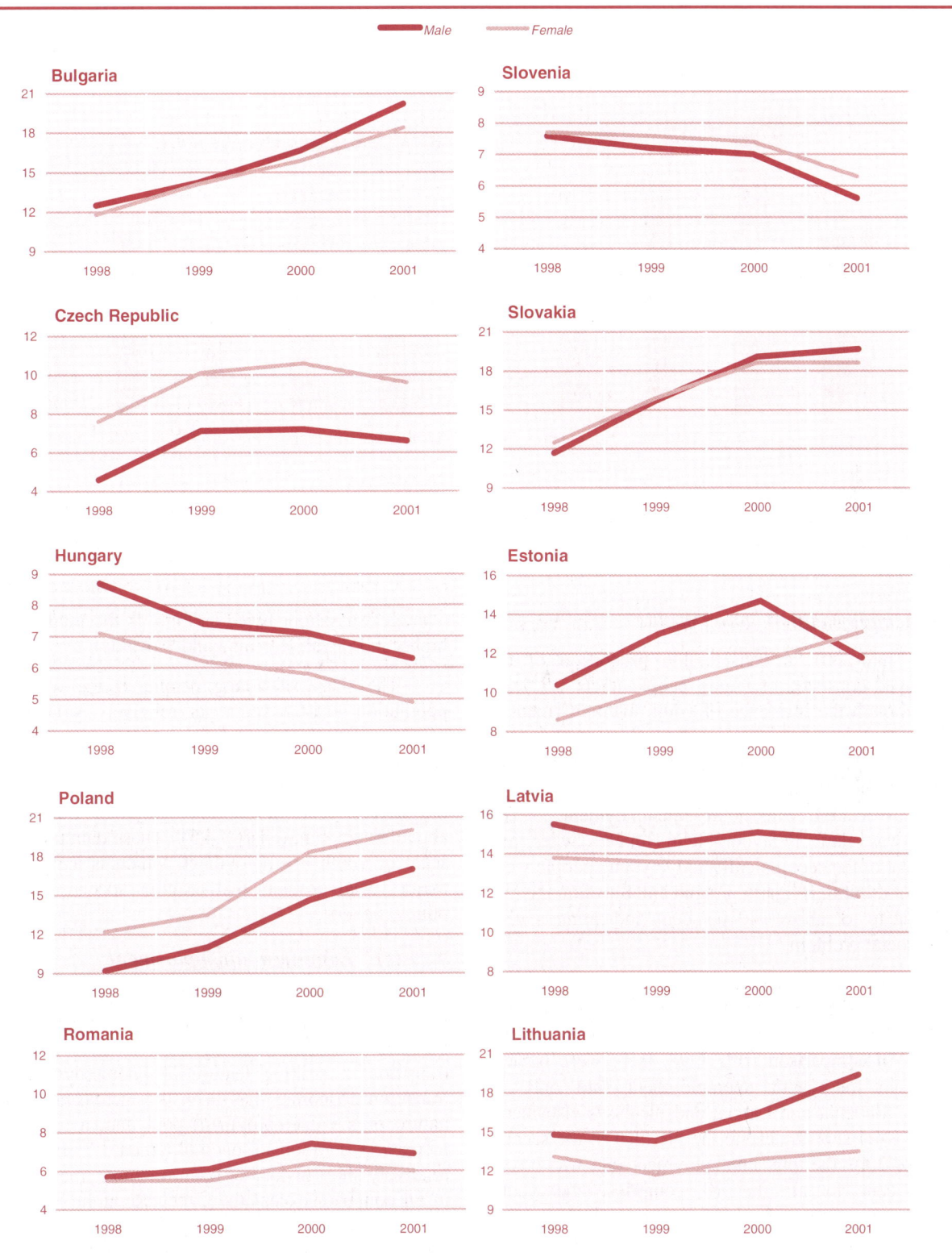

Source: National labour force surveys and direct communications from national statistical offices to UNECE secretariat.
Note: Data refer to second quarter of each year.

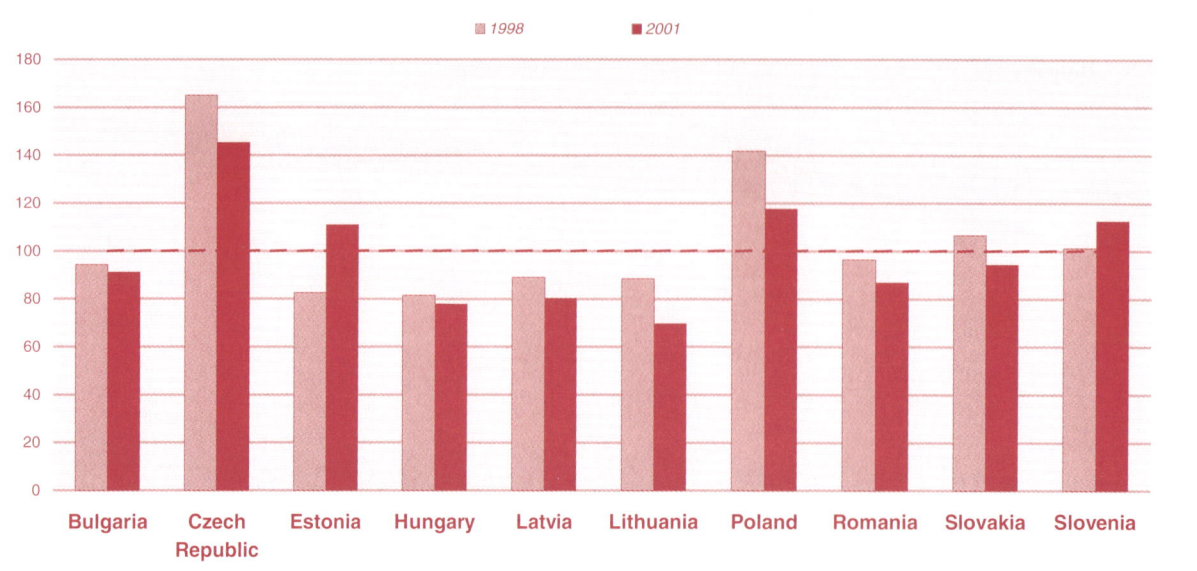

CHART 3.4.3

Ratio of female to male unemployment rates, in selected central and east European economies, 1998QII and 2001QII

(Percentage, male unemployment rate=100)

Source: UNECE secretariat estimates, based on national labour force surveys.

Note: A ratio of 100 means that female and male unemployment rates are equal. A ratio above (below) the reference line indicates that the female rate is higher (lower) than the male.

(b) Unemployment of young people

By mid-2001, a considerable proportion of the unemployed consisted of young people under 25 years old. Although the incidence of youth unemployment is high throughout the region, there are substantial differences between countries (chart 3.4.4). As the chart clearly indicates, the incidence of youth unemployment is closely associated with total unemployment. The situation was relatively favourable in Hungary, where just 10 per cent of the economically active young people were out of work, but in Bulgaria, Poland and Slovakia the rate was close to, or above, 40 per cent, indicating a very serious social problem.

Although youth unemployment in general follows the direction of total unemployment, in most countries the rise in youth unemployment was disproportionately large.[313] In comparison with 1998, there were modest declines in the youth unemployment rate only in Hungary, Slovenia and Latvia, and increases elsewhere. As chart 3.4.4 shows, despite the fact that in the second quarter of 2001 the total unemployment rate was below 20 per cent in all the 10 countries, the youth unemployment rate in 6 of them was well above this level, and in 4 countries it exceeded 30 per cent. In most countries, the youth unemployment rate has increased not only in absolute but in relative terms as well, and on average was about twice as high as the total rate (even higher in Poland, Romania and Slovakia).

The share of young people in the working age population (15-64 years) is relatively similar among countries: in 2000 it varied between some 21 per cent (Bulgaria, Slovenia and Latvia) and 24 per cent (Romania).[314] At the same time, in 5 out of 10 countries the share of youth in total unemployment was close to or above 30 per cent (chart 3.4.5). These data underline the scale of youth unemployment, which is a pressing and major social problem that requires special attention by policy makers.

(c) Long-term unemployment

Long-term unemployment, like youth unemployment, is also a serious problem in all the central and east European countries. From the very first years of the transition, a striking feature of unemployment in the transition economies has been the increasingly stagnant nature of the unemployment pool and, as a result, the increasing duration of unemployment.[315] In spite of a relatively short history of open unemployment, by 1994, in all countries except the Czech Republic some 40 per cent or more of the unemployed had been unemployed for more than one year.[316] Long-term unemployment is

---

[313] In Bulgaria, Poland and Slovakia, for example, the total unemployment rate increased by 7 to 8 percentage points between the second quarters of 1998 and 2001, whereas the increase in the youth unemployment rate was 10, 15 and 20 percentage points respectively.

[314] Eurostat, op. cit., p. 36.

[315] T. Boeri, "Labour market flows in central and eastern Europe", OECD, *Unemployment in Transition Countries: Transient or Persistent?* (Paris), 1994.

[316] UNECE, *Economic Survey of Europe in 1994-1995*, p. 113.

TABLE 3.4.2

Male and female activity rates in selected central and east European economies, 1985, 1997-2000 [a]

(Percentage)

	1985	1997	1998	1999	2000	Changes between 1997 and 2000
Bulgaria						
Male	..	56.8	55.6	54.5	54.0	-2.8
Female	..	46.9	45.6	44.2	44.0	-2.9
Czech Republic						
Male	75.1	71.1	70.8	70.6	69.8	-1.3
Female	59.3	51.8	52.0	52.1	51.6	-0.2
Estonia						
Male	82.6 [b]	73.1	71.9	70.8	71.0	-2.1
Female	71.7 [b]	58.0	57.8	57.0	57.6	-0.4
Hungary						
Male	73.9	60.4	60.0	61.4	61.9	1.5
Female	61.3	42.8	44.1	45.4	45.8	3.0
Latvia						
Male	69.4	68.9	68.6	68.1	65.2	-3.7
Female	68.4	52.6	51.2	50.2	49.9	-2.7
Lithuania						
Male	70.1	70.3	69.6	69.2	67.1	-3.2
Female	65.1	53.9	54.9	55.7	54.8	0.9
Poland						
Male	69.5	65.9	65.4	..	64.1	-1.8
Female	54.9	50.3	50.0	..	49.7	-0.6
Romania						
Male	..	72.5	71.4	70.9	70.6	-1.9
Female	..	57.7	56.3	56.4	56.4	-1.3
Slovakia						
Male	..	68.6	68.9	68.7	68.6	–
Female	..	51.8	51.5	52.0	52.6	0.8
Slovenia						
Male	82.3	66.2	66.3	64.7	64.5	-1.7
Female	65.2	53.2	52.9	51.5	51.7	-1.5

Source: UNECE, *Economic Survey of Europe, 1999 No. 1*, for 1985; CESTAT *Statistical Bulletin*, various issues, for 1997-2000; direct communications to UNECE secretariat from national statistical offices.

Note: Working age population used to calculate activity rates is: 15 and over except for Hungary and Estonia: 15-74.

[a] Labour force ÷ working age population.

[b] 1989: 15-69 years.

---

term unemployed was extremely high (close to 65 per cent) in Bulgaria and Slovenia. Unlike youth unemployment, there seems to be no systematic relation between the incidence of long-term unemployment and the total unemployment rate (chart 3.4.5). Among the other characteristics of long-term unemployment is a rather worrying tendency of an increasing number of long-term unemployed who have been out of work for two years or more.[317] The structure of long-term unemployment tends to be biased towards workers with low skills and education.

The high and increasing incidence of long-term unemployment is not only very painful for individuals but is also a potential source of social instability in the central and east European countries. Moreover, the experience of western Europe demonstrates that the reduction of unemployment during recoveries is much more difficult in countries with a high incidence of long-term unemployment. This suggests that the present high levels of unemployment in eastern Europe are likely to remain a serious problem even if growth remains strong in the next few years.

(iii) The changing patterns of manufacturing employment, 1993-2000

One of the major consequences of the reform process and the accompanying structural adjustments in the transition economies was the sharp decline in output and employment during the early 1990s. In the more advanced transition economies economic growth resumed in 1993-1994, but so far it has not resulted in any notable rise in total employment. In particular, while manufacturing industry has been one of the dynamic sectors in most central European economies, the response of manufacturing employment has almost everywhere been quite modest and has not yet led to sizeable net job creation in this sector. Most of the studies of the development of employment in the transition economies have been limited to the main sectors of economic activity;[318] however, it may well be that the pattern of labour demand has been quite different at lower levels of aggregation of manufacturing activity.

---

very persistent in these countries: even during 1994-1998 – a period of relative improvement in the unemployment situation – there were no positive shifts in this area, and in some countries it even deteriorated further.

Between 1998 and 2001, labour markets in most of the 10 countries were fairly sluggish, with relatively small outflows from unemployment to jobs, as compared with inflows into unemployment. The incidence of long-term unemployment increased in all countries except Hungary, where the share of long-term unemployed diminished substantially, and Estonia, where it remained basically unchanged over the period. In the second quarter of 2001, in the majority of countries, the share of long-term unemployed exceeded 50 per cent of total unemployment (chart 3.4.5). The exceptions were Hungary, Poland and Estonia, where the share varied between some 40 to 45 per cent. The percentage of long-

---

[317] In Bulgaria, for example, according to the results of the labour force survey, 30 per cent of the total number of unemployed had been unemployed for one to three years, and nearly 34 per cent for three years and more. National Statistical Institute, *Employment and Unemployment, 2/2001* (Sofia), 2001, p. 72

[318] Until recently the majority of papers have limited their analysis to the three broad sectors of the economy: agriculture, industry and services. See, for example, A. Nesporova, *Employment and Labour Market Polices in Transition Economies* (ILO, Geneva, 1999); Eurostat, *Central European Countries' Employment and Labour Market Review*, No. 1 (Luxembourg), July 1999; WIIW Structural Report, *Structural Developments in Central and Eastern Europe* (Vienna), 2000. Some papers that have examined the evolution of manufacturing employment at more disaggregated levels have done so in a broader context of analysis of output, productivity, wages and unit labour cost. See, for example, M. Landesmann, "Structural change in the transition economies, 1989-1999", UNECE, *Economic Survey of Europe, 2000 No. 2/3*, pp. 95-117; P. Havlik, "Sectoral patterns of catching up in candidate countries' manufacturing industry", paper presented to the IIASA Workshop (Stockholm), 3-5 May 2001, pp. 9-11.

TABLE 3.4.7

Relative country employment specialization by branch, 1993 and 2000

(Per cent)

Branches	Czech Republic 1993	Czech Republic 2000	Hungary 1993	Hungary 2000	Poland 1993	Poland 2000	Slovakia 1993	Slovakia 2000
Panel A: Within group specialization [a]								
Food products, beverages and tobacco	64.1	69.6	133.8	100.4	110.9	115.3	70.5	76.6
Textiles and textile products	78.3	79.5	102.8	111.5	110.7	105.0	83.8	99.3
Leather and leather products	93.3	78.4	118.8	131.0	91.4	88.3	142.1	181.2
Wood and wood products	69.5	69.0	68.6	68.4	122.4	126.5	87.1	62.7
Paper and publishing	85.4	81.8	122.5	96.2	96.0	108.3	127.6	100.4
Coke and refined petroleum products	98.7	46.0	211.6	210.7	69.4	83.0	120.4	151.0
Chemicals and chemical products	75.8	89.9	138.9	112.2	95.3	97.8	130.8	119.4
Rubber and plastic products	88.9	99.0	94.4	95.6	106.0	104.7	99.5	78.2
Other non-metallic mineral products	106.7	111.9	78.5	67.8	102.1	104.3	103.2	100.0
Basic metals and metal products	127.4	133.0	84.8	79.7	93.9	90.8	91.6	112.4
Machinery and equipment n.e.c.	142.2	132.3	71.8	80.4	83.6	87.6	139.3	135.7
Electrical and optical equipment	106.2	116.9	124.3	182.6	89.4	67.5	112.6	117.7
Transport equipment	128.8	127.5	62.5	78.1	99.9	96.4	85.3	91.6
Furniture/recycling	107.8	92.1	67.5	61.2	107.3	120.6	84.9	53.7
Panel B: Specialization [b] **vis-à-vis EU-15**								
Food products, beverages and tobacco	86.1	95.0	179.8	137.1	149.0	157.3	94.6	104.5
Textiles and textile products	129.9	137.2	170.6	192.6	183.7	181.3	139.0	171.5
Leather and leather products	182.8	115.6	232.7	193.1	179.0	130.2	278.4	267.1
Wood and wood products	115.7	126.9	114.1	125.8	203.6	232.7	144.9	115.3
Paper and publishing	43.5	48.9	62.4	57.5	48.9	64.7	65.0	60.0
Coke and refined petroleum products	193.7	53.8	415.2	246.7	136.1	97.2	236.3	176.8
Chemicals and chemical products	47.9	52.9	87.7	66.0	60.2	57.6	82.6	70.3
Rubber and plastic products	53.5	81.7	56.8	78.9	63.9	86.5	60.0	64.6
Other non-metallic mineral products	136.5	148.6	100.5	90.0	130.6	138.5	132.0	132.8
Basic metals and metal products	128.7	126.2	85.7	75.7	94.9	86.2	92.6	106.7
Machinery and equipment n.e.c.	152.2	111.8	76.8	68.0	89.5	74.1	149.0	114.7
Electrical and optical equipment	64.8	98.0	75.9	153.1	54.5	56.6	68.7	98.7
Transport equipment	85.7	81.1	41.6	49.7	66.5	61.3	56.7	58.3
Furniture/recycling	142.3	123.2	89.2	81.9	141.8	161.2	112.1	71.8

Source: UNECE secretariat estimates, based on national statistics.

[a] Specialization is defined as the ratio of the share of each branch in total national employment to the weighted mean for four countries.

[b] Specialization is defined as the ratio of the share of each branch in total national employment to the EU average.

TABLE 3.4.8

Main features of employment specialization in selected central European economies, 2000

Specialization [a]	Underrepresentation [b]
Czech Republic	
Metal products [c,d]	Food products
Transport equipment [c]	Wood products
Machinery and equipment [c]	Petroleum products
	Leather products
Hungary	
Leather products	Wood products
Petroleum products [c]	Non-metallic minerals
Electrical/optical equipment [c,d]	Furniture/recycling
Poland	
Wood products [c]	Electrical/optical equipment
Furniture/recycling	
Slovakia	
Leather products [c,d]	Food products
Petroleum products [c,d]	Wood products
Electrical/optical equipment [c]	Rubber/plastics
	Furniture/recycling

Source: Based on UNECE secretariat estimates.

[a] Specialization ratio of more than 120.

[b] Specialization ratio of less than 80.

[c] Same specialization as in 1993.

[d] Notably increased level of specialization.

industry (DD). Between 1993 and 2000 the degree of specialization fell only in the leather industry (DC) and particularly in petroleum products (DF). In contrast, the already relatively high degree of concentration of employment in wood products increased further, making it the branch with the highest specialization ratio of more than 180.

Among the underrepresented branches in central Europe (as compared with the EU) in 2000 are chemicals (DG) and transport equipment (DM). Between 1993 and 2000, the east European shares moved closer to the west European pattern in the electrical/optical industry (DL) and rubber/plastics (DH) (in both cases the ratio increasing from some 60 to 85), and the difference narrowed somewhat in the paper/publishing industry (DE). Manufacturing industries are often classified into three groups according to the level of their technology: namely, low-tech, medium- to high-tech and resource- (and scale-) intensive group.[329] In terms of this categorization, the central European countries in 2000 were relatively

---

[329] M. Landesmann, loc. cit., pp. 100-107.

TABLE 3.4.2

Male and female activity rates in selected central and east European economies, 1985, 1997-2000 [a]

(Percentage)

	1985	1997	1998	1999	2000	Changes between 1997 and 2000
Bulgaria						
Male	..	56.8	55.6	54.5	54.0	-2.8
Female	..	46.9	45.6	44.2	44.0	-2.9
Czech Republic						
Male	75.1	71.1	70.8	70.6	69.8	-1.3
Female	59.3	51.8	52.0	52.1	51.6	-0.2
Estonia						
Male	82.6[b]	73.1	71.9	70.8	71.0	-2.1
Female	71.7[b]	58.0	57.8	57.0	57.6	-0.4
Hungary						
Male	73.9	60.4	60.0	61.4	61.9	1.5
Female	61.3	42.8	44.1	45.4	45.8	3.0
Latvia						
Male	69.4	68.9	68.6	68.1	65.2	-3.7
Female	68.4	52.6	51.2	50.2	49.9	-2.7
Lithuania						
Male	70.1	70.3	69.6	69.2	67.1	-3.2
Female	65.1	53.9	54.9	55.7	54.8	0.9
Poland						
Male	69.5	65.9	65.4	..	64.1	-1.8
Female	54.9	50.3	50.0	..	49.7	-0.6
Romania						
Male	..	72.5	71.4	70.9	70.6	-1.9
Female	..	57.7	56.3	56.4	56.4	-1.3
Slovakia						
Male	..	68.6	68.9	68.7	68.6	–
Female	..	51.8	51.5	52.0	52.6	0.8
Slovenia						
Male	82.3	66.2	66.3	64.7	64.5	-1.7
Female	65.2	53.2	52.9	51.5	51.7	-1.5

Source: UNECE, *Economic Survey of Europe, 1999 No. 1,* for 1985; CESTAT *Statistical Bulletin,* various issues, for 1997-2000; direct communications to UNECE secretariat from national statistical offices.

Note: Working age population used to calculate activity rates is: 15 and over except for Hungary and Estonia: 15-74.

[a] Labour force ÷ working age population.

[b] 1989: 15-69 years.

very persistent in these countries: even during 1994-1998 – a period of relative improvement in the unemployment situation – there were no positive shifts in this area, and in some countries it even deteriorated further.

Between 1998 and 2001, labour markets in most of the 10 countries were fairly sluggish, with relatively small outflows from unemployment to jobs, as compared with inflows into unemployment. The incidence of long-term unemployment increased in all countries except Hungary, where the share of long-term unemployed diminished substantially, and Estonia, where it remained basically unchanged over the period. In the second quarter of 2001, in the majority of countries, the share of long-term unemployed exceeded 50 per cent of total unemployment (chart 3.4.5). The exceptions were Hungary, Poland and Estonia, where the share varied between some 40 to 45 per cent. The percentage of long-term unemployed was extremely high (close to 65 per cent) in Bulgaria and Slovenia. Unlike youth unemployment, there seems to be no systematic relation between the incidence of long-term unemployment and the total unemployment rate (chart 3.4.5). Among the other characteristics of long-term unemployment is a rather worrying tendency of an increasing number of long-term unemployed who have been out of work for two years or more.[317] The structure of long-term unemployment tends to be biased towards workers with low skills and education.

The high and increasing incidence of long-term unemployment is not only very painful for individuals but is also a potential source of social instability in the central and east European countries. Moreover, the experience of western Europe demonstrates that the reduction of unemployment during recoveries is much more difficult in countries with a high incidence of long-term unemployment. This suggests that the present high levels of unemployment in eastern Europe are likely to remain a serious problem even if growth remains strong in the next few years.

(iii) The changing patterns of manufacturing employment, 1993-2000

One of the major consequences of the reform process and the accompanying structural adjustments in the transition economies was the sharp decline in output and employment during the early 1990s. In the more advanced transition economies economic growth resumed in 1993-1994, but so far it has not resulted in any notable rise in total employment. In particular, while manufacturing industry has been one of the dynamic sectors in most central European economies, the response of manufacturing employment has almost everywhere been quite modest and has not yet led to sizeable net job creation in this sector. Most of the studies of the development of employment in the transition economies have been limited to the main sectors of economic activity;[318] however, it may well be that the pattern of labour demand has been quite different at lower levels of aggregation of manufacturing activity.

---

[317] In Bulgaria, for example, according to the results of the labour force survey, 30 per cent of the total number of unemployed had been unemployed for one to three years, and nearly 34 per cent for three years and more. National Statistical Institute, *Employment and Unemployment, 2/2001* (Sofia), 2001, p. 72

[318] Until recently the majority of papers have limited their analysis to the three broad sectors of the economy: agriculture, industry and services. See, for example, A. Nesporova, *Employment and Labour Market Polices in Transition Economies* (ILO, Geneva, 1999); Eurostat, *Central European Countries' Employment and Labour Market Review,* No. 1 (Luxembourg), July 1999; WIIW Structural Report, *Structural Developments in Central and Eastern Europe* (Vienna), 2000. Some papers that have examined the evolution of manufacturing employment at more disaggregated levels have done so in a broader context of analysis of output, productivity, wages and unit labour cost. See, for example, M. Landesmann, "Structural change in the transition economies, 1989-1999", UNECE, *Economic Survey of Europe, 2000 No. 2/3,* pp. 95-117; P. Havlik, "Sectoral patterns of catching up in candidate countries' manufacturing industry", paper presented to the IIASA Workshop (Stockholm), 3-5 May 2001, pp. 9-11.

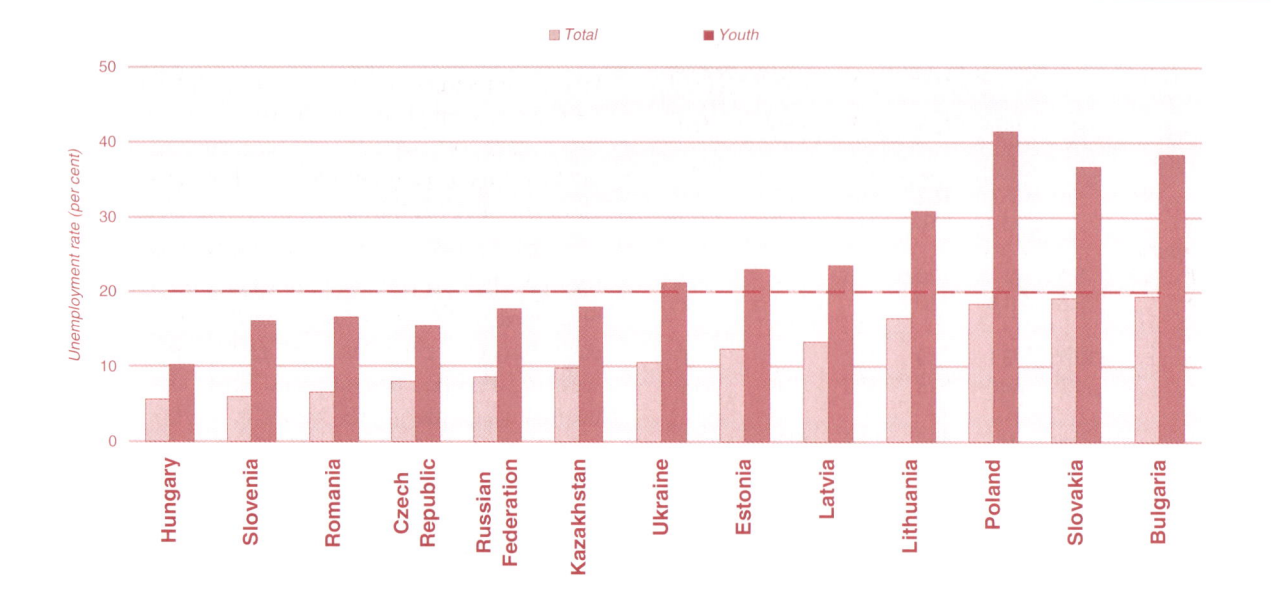

CHART 3.4.4

Total and youth unemployment rates in selected central and east European economies, 2001QII
(Per cent of labour force)

Source: National labour force surveys and direct communications from national statistical offices to UNECE secretariat.
Note: In addition to the 10 central and east European countries, Kazakhstan, Russia and Ukraine, for which data are available for 2001, have been included.

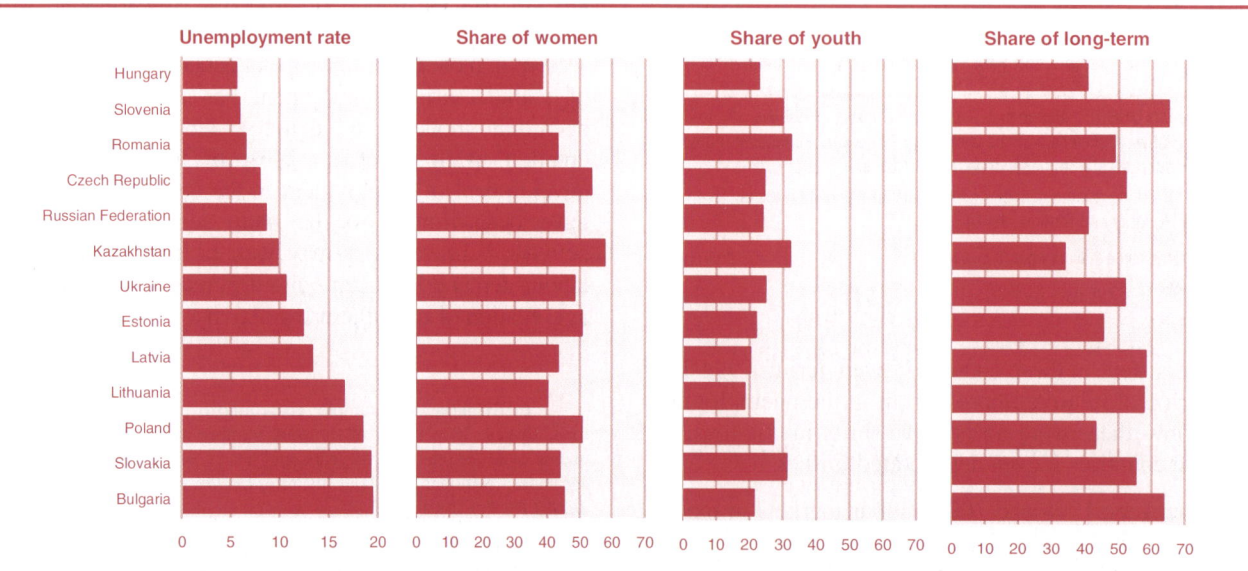

CHART 3.4.5

Total unemployment rate and share of women, youth and long-term unemployed in total unemployment in selected central and east European economies, 2001QII
(Percentages)

Source: National labour force surveys and direct communications from national statistical offices to UNECE secretariat.
Note: In addition to the 10 central and east European countries, Kazakhstan, Russia and Ukraine, for which data are available for 2001, have been included.

This section analyses some of the changes in manufacturing employment at the level of individual manufacturing industries during the past seven years in four central European countries (the Czech Republic, Hungary, Poland and Slovakia), for which consistent, disaggregated data are available from 1993. Among the issues addressed are: changes in the absolute numbers employed in the sector as a whole and in separate industries, identification of the branches where most new jobs were created and of those that lost the most, structural changes in manufacturing employment in a cross-country perspective, and changes in national

TABLE 3.4.3

Cumulative changes in manufacturing employment by branches in selected central European economies, 1993 and 2000

(Percentage change, thousands)

NACE codes	Branches	Czech Republic	Hungary	Poland	Slovakia
D	(DA-DN) Manufacturing	-13.0	0.8	-10.4	-19.5
DA	Food products, beverages and tobacco	-0.8	-20.7	-2.3	-8.2
DB	Textiles and textile products	-23.9	-5.8	-26.8	-17.7
DC	Leather and leather products	-51.3	-26.0	-42.4	-31.6
DD	Wood and wood products	6.7	24.2	14.4	-28.4
DE	Pulp, paper and paper products; publishing and printing	2.5	-2.6	24.3	-22.1
DF	Coke, refined petroleum products and nuclear fuel	-75.0	-38.1	-33.8	-37.7
DG	Chemicals, chemical products and man-made fibres	-9.1	-28.2	-19.0	-35.2
DH	Rubber and plastic products	48.4	56.3	35.5	-3.1
DI	Other non-metallic mineral products	-9.0	-13.2	-8.8	-22.1
DJ	Basic metals and fabricated metal products	-9.2	-5.4	-13.5	-1.3
DK	Machinery and equipment n.e.c.	-37.6	-12.9	-27.6	-39.5
DL	Electrical and optical equipment	27.6	97.0	-10.0	12.0
DM	Transport equipment	-17.0	21.4	-16.7	-16.6
DN	Manufacturing n.e.c.	-14.5	5.0	15.6	-41.5
Memorandum items: (in thousands)					
Cumulative job loss in shrinking sectors		-203.0	-86.0	-415.0	-97.0
Cumulative net job creation in expanding sectors		45.0	92.0	104.0	5.0
Difference		-158.0	6.0	-311.0	-92.0

Source: UNECE secretariat estimates, based on national statistics.

Note: Data coverage: Czech Republic – enterprises with 25 and more employees in 1993 and 20 or more employees in 2000; Hungary – enterprises with 20 or more employees in 1993 and 5 or more employees in 2000; Poland – the whole economy, end-of-year data; Slovakia – enterprises with 25 and more employees.

employment specialization.[319] The main body of data used here are the statistics of employment in 14 branches of manufacturing industry[320] (the 2-digit (alphabetic codes) subsections of the NACE classification – table 3.4.3).

(a) The dynamics of output and employment in the manufacturing industry

Comparing changes in total manufacturing output and employment suggests that between 1993 and 2000 the response of employment to the relatively robust growth of output was very weak, and in some countries employment continued to decline despite the growth of output in this period (chart 3.4.6). As discussed below, the difference between the changes of output and employment was even more marked at the level of individual branches. A possible explanation for this weak labour response in terms of net job creation in the leading reformers of central Europe is a rapid growth of labour productivity due to restructuring and technological innovation embodied in the growth process.[321] It may also partly reflect the absorption of the "overhang" of excess employment, which was a general feature of the transition economies at the start of economic transformation.[322] Consequently, the share of manufacturing in total employment was generally unchanged between 1993 and 2000 and even declined slightly in the Czech Republic and Poland (chart 3.4.7).

During 1993-2000, there were considerable differences in the dynamics of manufacturing employment at the level of individual manufacturing industries. Despite substantial cross-country variations, the available data indicate that two branches – rubber/plastics and electrical/optical equipment had the highest rates of job creation.[323] Employment in the former grew in all four countries except Slovakia (where it declined slightly) with cumulative increases between 1993-2000 of 40 per cent and more (table 3.4.3). Employment also rose in branches such as wood products and paper/publishing in individual countries. However, the situation was not so homogeneous as in the first two branches and varied widely from country to country. At the other end of the scale, the largest declines were in leather products, petroleum products, and machinery and equipment where employment in most cases declined by one third or more.

The absolute changes in manufacturing employment (chart 3.4.8 and table 3.4.4) indicate that for the four countries as a whole employment grew in only 5 of the 14 branches resulting in the creation of 190,000 net new jobs between 1993 and 2000. The largest gains, of nearly 80,000 new jobs (more than 40 per cent of the total) were located in electric/optical machinery followed by the rubber/plastics industry with 58,000 jobs created. The

---

[319] This note follows, both in terms of its subject and methodology, and draws upon the analysis of an earlier UNECE study of 10 west European countries in the 1960s. UNECE, *Structure and Change in European Industry* (United Nations publication, Sales No. E.77.II.E.3), 1977.

[320] The main data sources are the national statistics officially published or obtained by the UNECE secretariat through direct communications with the national statistical offices.

[321] The expansion of manufacturing output was especially notable in some of the more sophisticated and skill-intensive engineering industries. The rapid growth of the leading branches of manufacturing was largely due to the inflow of FDI, which, on the one hand, provided a basis for raising economic growth and created a transmission channel for new technology and know-how and, on the other hand, gradually reshaped the structure of production and exports. For a more detailed discussion of FDI and changes in the pattern of exports in the east European transition economies since 1989 see UNECE, *Economic Survey of Europe, 2000 No. 1*, pp. 125-130 and 145-153.

[322] For a more detailed analysis see UNECE, *Economic Survey of Europe, 2000 No. 1*, pp. 102-103. See also T. Boeri, M. Burda and J. Köllö, *Mediating the Transition: Labour Markets in Central and Eastern Europe* (CEPR and Institute for East-West Studies, London, 1998), p. 42.

[323] For brevity, abbreviated branch names are used hereafter in the text. For the full names of the branches see table 3.4.3.

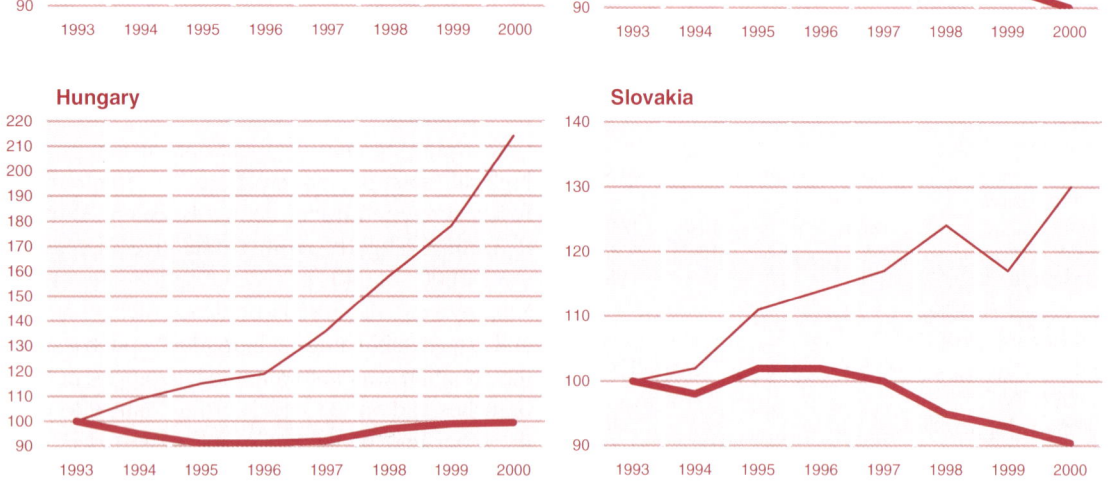

CHART 3.4.6

Output and employment in manufacturing in selected central European economies, 1993-2000
(1993=100)

Source: National statistics and UNECE Common Database.

largest increases (36 per cent and more) were in rubber and plastics. In the other nine branches, however, employment fell by some 744,000, with the largest declines of nearly 205,000 in machinery and equipment, followed by a loss of more than 170,000 jobs in the textiles/clothing industry (31 and 23 per cent, respectively).

(b) The changing composition of manufacturing employment: similarity and divergence

As a result of these changes, within a relatively short period of seven years, major shifts occurred in the distribution of manufacturing employment (table 3.4.5). The two largest changes were in electrical/optical machinery, where its share in total employment increased by as much as 9 percentage points (Hungary), and machinery and equipment, which declined by 5 percentage points (the Czech Republic). Despite the pronounced shifts in the employment structure, the four largest branches (with shares in total employment of some 10 per cent and more) – food products, textiles/clothing, metal products, and machinery and equipment – still retained their leading position in 2000. In the Czech Republic, Hungary and Slovakia, this group of industries was joined by electrical/optical equipment, which increased its share of total employment in all three countries by more than 3 percentage points, with the largest increase in Hungary. In general, the industries with the most significant changes in the four countries were: rubber/plastics and electric/optical equipment (increasing shares); and textile/clothing, leather products and machinery and equipment (decreasing shares).

To place these changes in manufacturing employment in perspective they can be compared with similar changes in the EU. In contrast to the transition economies, the average employment structure in the EU-15 changed relatively little between 1993 and 2000. The only substantial change in the EU was in textiles/clothing, whose share in total employment declined by more than 1 percentage point; in the other branches changes were mostly in the range of 0.2-0.5 percentage points. This comparison underlines the deep and radical structural changes which occurred in the four central European countries in the space of a relatively short period.

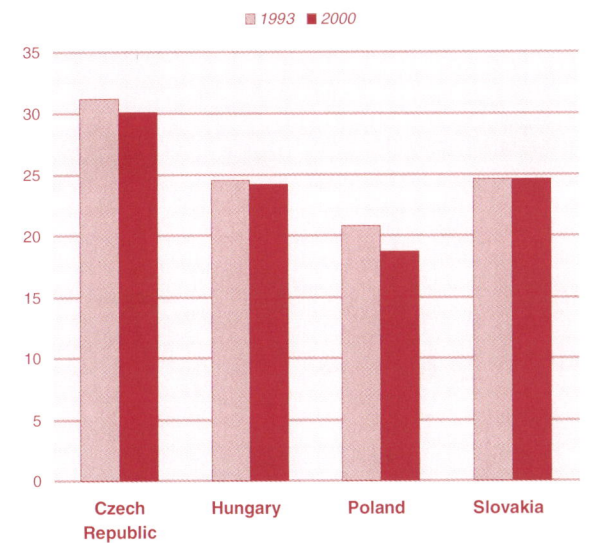

CHART 3.4.7

Share of manufacturing in total employment in selected central European economies, 1993 and 2000
(Per cent)

Source: UNECE secretariat estimates, based on national statistics.

The magnitude of changes in the structure of employment over time and the extent to which they differ among individual countries can be summarized with a simple and widely used index of structural change (C) or similarity (S).[324] The index of structural change (table 3.4.5, last line) indicates that Hungary has undergone the deepest change in manufacturing employment structure between 1993 and 2000 (C = 12.1). In the Czech Republic, Slovakia and particularly Poland, the changes were smaller (with C equal to 9.2, 8.5 and 7.5, respectively). Nevertheless, the index for the EU-15 (3.0), clearly indicates that structural change in all the central European countries during 1993-2000 was relatively large.

In 1993, there was a generally high degree of similarity between the employment structures of all four countries. Their bilateral similarity indices were mostly within the range of 85 to 91 (table 3.4.6). The countries with the closest similarity were the Czech Republic and Slovakia (S = 90.6), while the least similar were the Czech Republic and Hungary (S = 77.4). The countries' bilateral similarity vis-à-vis the EU-15, varied in 1993 between 80 (Hungary and Poland) and 84 (Slovakia). The degree of cross-country similarity remained generally high in 2000, but the structural differences between the countries have also narrowed in general (table 3.4.6).[325] Poland is the exception to this pattern of structural developments. In 2000, its cross-country similarity indices were less than in 1993 implying an increasingly different structure from the other three countries. As in 1993, the Czech Republic and Slovakia were the most similar in the structure of their manufacturing employment. Although the difference between the Czech Republic and Hungary was still the largest among the four countries, it had narrowed appreciably over 1993-2000.

As regards developments over time, there was some convergence between some of the four countries and the EU average over 1993-2000. By 2000, the country with the greatest similarity to the EU-15 was the Czech Republic (the similarity coefficient increasing from 82 to nearly 89). Slovakia also became more similar to the EU structure. In Hungary and Poland, however, the differences between them and the EU-15 remained broadly unchanged and even widened slightly (in the former this mostly reflected the sharp increase in the share of electrical/optical equipment). According to recent studies,[326] the industrial structure of the central European countries (both in terms of output and employment) is positioned somewhere between the less advanced EU-South (defined as an average of Greece, Portugal and Spain) and the more advanced EU-North (an average of France, Germany and the United Kingdom).

(c) The patterns of national employment specialization

The large changes in the structure of employment in all four countries imply considerable changes in the specialization of the employed workforce.[327] The values of the relative specialization coefficients for the branches within each country indicate that between 1993 and 2000 the original specialization in many cases became more pronounced while in others there was a change in the pattern of specialization (table 3.4.7, panel A).

In 2000, there were five branches in which there were very high levels of national specialization: basic metals (the Czech Republic), machinery and equipment (Slovakia), leather products (Slovakia), electrical/optical machinery (Hungary) and petroleum products (Hungary), with ratios of 133, 136, 181, 183 and 211, respectively. In all these cases the countries were already specialized in these industries in 1993 and the ratios were either broadly unchanged or increased considerably between 1993 and 2000. At the other end of the scale, industries such as electrical/optical machinery

---

[324] See the note to table 3.4.5.

[325] In table 3.4.6, the italicized indices are those that declined between 1993 and 2000, implying increasing intercountry differences.

[326] See, for example, P. Havlik, op. cit., pp. 9-11.

[327] The specialization of a country in a specific branch is measured by the so-called relative specialization coefficient, which is calculated as the ratio of branch *i* in country *A* to the weighted average share of this branch in the group of countries. When the ratio is substantially above 100 (in this case, more than 20 points), the country is considered as "specialized" in this branch; when it is substantially below 100, the branch is "underrepresented" in the country.

CHART 3.4.8
Job creation and job destruction in manufacturing industries in selected central European economies, 1993-2000
(Thousand persons)

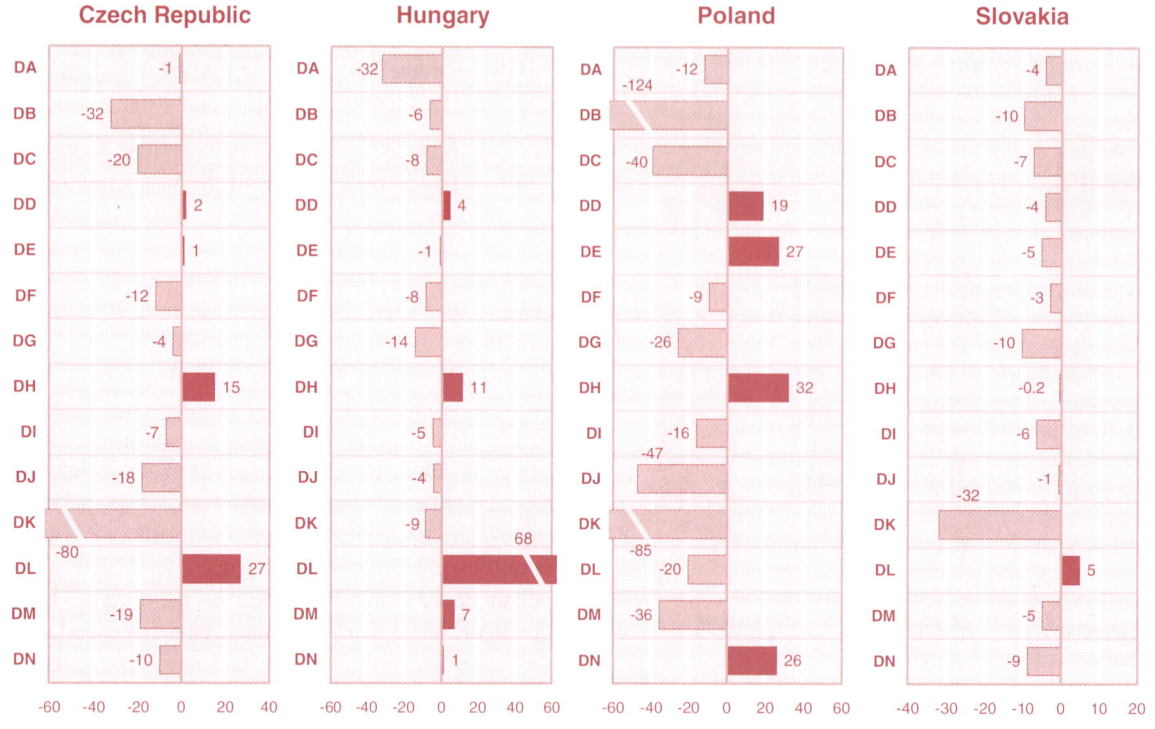

Source: UNECE secretariat estimates, based on national statistics.

Note: The bars denote the absolute number of jobs created or destroyed (net increase or net decrease in employment) by NACE branches. For full branch names, see table 3.4.3.

TABLE 3.4.4
Changes in manufacturing employment by branches, 1993 and 2000
(The four countries combined, thousands, percentages)

	1993	2000	Differences	
	Thousands		Thousands	Percent-ages
Panel A: Branches releasing manpower				
Machinery and equipment n.e.c.	666	461	-205	-30.8
Textiles and textile products	761	589	-172	-22.7
Leather and leather products	186	111	-75	-40.2
Basic metals and metal products	681	610	-70	-10.3
Chemicals and chemical products	258	204	-54	-20.9
Transport equipment	387	334	-52	-13.5
Food, beverages and tobacco	839	790	-49	-5.8
Other non-metallic minerals	325	291	-34	-10.5
Coke and refined petroleum products	72	40	-32	-44.6
Total above	4 174	3 430	-744	-17.8
Panel B: Branches absorbing manpower				
Electrical and optical equipment	410	490	80	19.4
Rubber and plastic products	155	213	58	37.4
Paper and publishing	208	230	22	10.4
Wood and wood products	192	213	21	10.9
Furniture/recycling	285	294	9	3.1
Total above	1 250	1 439	189	15.1
Total	5 424	4 869	-555	-10.2

Source: UNECE secretariat estimates, based on national statistics.

Note: Branches are ranked by the absolute number of decline or increase.

(Poland), wood products (Slovakia), furniture/recycling (Slovakia) and petroleum products (the Czech Republic) were clearly underrepresented with ratios of 68, 63, 54 and 46, respectively (in all cases these were lower than in 1993). At the same time, there were no countries with relative employment specialization (i.e. with ratios above 120) in branches such as: food products, textile/clothing, paper/publishing, chemicals, rubber/plastic and non-metallic minerals.

The changes in relative employment shares varied widely among the countries (main features of specialization in 2000 are summarized in table 3.4.8). The Czech Republic and Slovakia in 2000 were specialized in the same three branches as in 1993 and in the latter the ratios increased markedly. At the same time, Slovakia's previously high specialization in 1993 in paper/publishing and chemicals industry diminished considerably. Poland, in addition to its previous specialization in wood products, acquired a new specialization in furniture and the recycling industry. The largest changes occurred in Hungary: on the one hand, there were considerable increases in the specialization ratios for electrical/optical machinery and for leather products, but on the other hand, there were major reductions in the ratios for food products, paper/publishing and the chemicals industry.

TABLE 3.4.5

Breakdown of employment in manufacturing in selected central European economies and the EU, 1993 and 2000
(Percentage, total manufacturing=100)

	Czech Republic		Hungary		Poland		Slovakia		Memorandum items:			
									CE-4		EU-15	
	1993	2000	1993	2000	1993	2000	1993	2000	1993	2000	1993	2000
Food, beverages and tobacco	9.9	11.3	20.7	16.3	17.2	18.7	10.9	12.4	15.5	16.2	11.5	11.9
Textiles and textile products	11.0	9.6	14.4	13.5	15.5	12.7	11.7	12.0	14.0	12.1	8.5	7.0
Leather and leather products	3.2	1.8	4.1	3.0	3.1	2.0	4.9	4.1	3.4	2.3	1.7	1.5
Wood and wood products	2.5	3.0	2.4	3.0	4.3	5.5	3.1	2.7	3.5	4.4	2.1	2.4
Paper and publishing	3.3	3.9	4.7	4.5	3.7	5.1	4.9	4.7	3.8	4.7	7.5	7.9
Coke and refined petroleum products	1.3	0.4	2.8	1.7	0.9	0.7	1.6	1.2	1.3	0.8	0.7	0.7
Chemicals and chemical products	3.6	3.8	6.6	4.7	4.5	4.1	6.2	5.0	4.8	4.2	7.5	7.1
Rubber and plastic products	2.5	4.3	2.7	4.2	3.0	4.6	2.8	3.4	2.9	4.4	4.7	5.3
Other non-metallic mineral products	6.4	6.7	4.7	4.1	6.1	6.2	6.2	6.0	6.0	6.0	4.7	4.5
Basic metals and metal products	16.0	16.7	10.6	10.0	11.8	11.4	11.5	14.1	12.5	12.5	12.4	13.2
Machinery and equipment n.e.c.	17.5	12.5	8.8	7.6	10.3	8.3	17.1	12.8	12.3	9.5	11.5	11.2
Electrical and optical equipment	8.0	11.8	9.4	18.4	6.8	6.8	8.5	11.8	7.6	10.1	12.4	12.0
Transport equipment	9.2	8.8	4.5	5.4	7.1	6.6	6.1	6.3	7.1	6.9	10.7	10.8
Furniture/recycling	5.7	5.6	3.5	3.7	5.6	7.3	4.5	3.2	5.2	6.0	4.0	4.5
Manufacturing total	100.0	100.0	100.0	100.0	100.0	100.0	100.0	100.0	100.0	100.0	100.0	100.0
Index of structural change [a]	..	9.2	..	12.1	..	7.5	..	8.5	..	7.3	..	3.0

Source: UNECE secretariat estimates, based on national statistics; Eurostat, New Cronos Database (enterprises with 20 and more employees).

[a] The index of structural change C, which measures changes between two periods is defined by $C = \Sigma(q_{i1} - q_{i2})$ where the summation is carried out only over changes of the same sign (for all $q_{i1} > q_{i2}$, or vice versa) and where q_i is the percentage share of branch i, in periods 1 and 2 respectively, and $\Sigma q_i = 100$. The similarity of two percentage distributions is a mirror image of C and is defined by $S = 100 - C$. A similarity index of 100, denotes identical structures.

TABLE 3.4.6

Cross-country similarity indices of manufacturing employment in selected central European economies and the EU, 1993 and 2000 [a]

	Czech Republic	Hungary	Poland	Slovakia	EU-15
1993					
Czech Republic	100	77.4	84.5	90.6	82.1
Hungary		100	87.9	85.0	80.1
Poland			100	86.0	80.4
Slovakia				100	84.1
EU-15					100
2000					
Czech Republic	100	80.3	*82.9*	90.7	88.5
Hungary		100	*85.0*	86.2	*79.7*
Poland			100	*84.1*	*79.4*
Slovakia				100	87.0
EU-15					100

Source: UNECE secretariat estimates, based on national statistics.

[a] Based on the 14-branch NACE distribution of manufacturing employment. For definition of the similarity index, see footnote to table 3.4.5. Italicized indices for 2000 are those which are lower than in 1993.

In order to compare the employment structures in each of the four central European countries vis-à-vis the EU-15, the branch coefficients were calculated in relation to the EU-15 average (table 3.4.7, panel B). The values of these ratios indicate that some manufacturing branches such as: paper/publishing, chemicals, rubber/plastics and transport equipment, remained considerably underrepresented in 2000 as compared to the west European pattern. On the other hand, the central European economies were relatively specialized in textiles/clothing, leather products and petroleum products. With the main exception of the Czech Republic the share of these branches in total manufacturing employment in 2000 was twice as high as in the EU.

The changes in the specialization ratios over time differ considerably between countries, but specialization in two branches – rubber/plastics and electrical/optical equipment – increased considerably between 1993 and 2000 in all four countries contributing to the convergence of the two European patterns. At the same time, while the level of specialization in leather products and petroleum products remained very high, the ratios declined sharply, thereby also contributing to the process of convergence of employment structures.

Chart 3.4.9 shows the branch structure of employment in the four east European countries combined relative to the EU-15 average in 1993 and 2000. The data indicate that in 2000 there were five branches (against two in 1993) where the shares of total employment were similar[328] to the western European pattern, namely, petroleum products (DF), rubber/plastics (DH), metal products (DJ), machinery and equipment (DK) and the electrical/optical industry (DL). The group of industries where there were relatively high degrees of employment concentration in 2000 in the east European countries (with ratios of 150 or more) include textiles/clothing (DB), leather products (DC) and, particularly, the wood products

---

[328] That is, where the specialization ratios did not deviate from the average by more than ± 20 points.

TABLE 3.4.7
Relative country employment specialization by branch, 1993 and 2000
(Per cent)

Branches	Czech Republic 1993	Czech Republic 2000	Hungary 1993	Hungary 2000	Poland 1993	Poland 2000	Slovakia 1993	Slovakia 2000
Panel A: Within group specialization [a]								
Food products, beverages and tobacco	64.1	69.6	133.8	100.4	110.9	115.3	70.5	76.6
Textiles and textile products	78.3	79.5	102.8	111.5	110.7	105.0	83.8	99.3
Leather and leather products	93.3	78.4	118.8	131.0	91.4	88.3	142.1	181.2
Wood and wood products	69.5	69.0	68.6	68.4	122.4	126.5	87.1	62.7
Paper and publishing	85.4	81.8	122.5	96.2	96.0	108.3	127.6	100.4
Coke and refined petroleum products	98.7	46.0	211.6	210.7	69.4	83.0	120.4	151.0
Chemicals and chemical products	75.8	89.9	138.9	112.2	95.3	97.8	130.8	119.4
Rubber and plastic products	88.9	99.0	94.4	95.6	106.0	104.7	99.5	78.2
Other non-metallic mineral products	106.7	111.9	78.5	67.8	102.1	104.3	103.2	100.0
Basic metals and metal products	127.4	133.0	84.8	79.7	93.9	90.8	91.6	112.4
Machinery and equipment n.e.c.	142.2	132.3	71.8	80.4	83.6	87.6	139.3	135.7
Electrical and optical equipment	106.2	116.9	124.3	182.6	89.4	67.5	112.6	117.7
Transport equipment	128.8	127.5	62.5	78.1	99.9	96.4	85.3	91.6
Furniture/recycling	107.8	92.1	67.5	61.2	107.3	120.6	84.9	53.7
Panel B: Specialization [b] **vis-à-vis EU-15**								
Food products, beverages and tobacco	86.1	95.0	179.8	137.1	149.0	157.3	94.6	104.5
Textiles and textile products	129.9	137.2	170.6	192.6	183.7	181.3	139.0	171.5
Leather and leather products	182.8	115.6	232.7	193.1	179.0	130.2	278.4	267.1
Wood and wood products	115.7	126.9	114.1	125.8	203.6	232.7	144.9	115.3
Paper and publishing	43.5	48.9	62.4	57.5	48.9	64.7	65.0	60.0
Coke and refined petroleum products	193.7	53.8	415.2	246.7	136.1	97.2	236.3	176.8
Chemicals and chemical products	47.9	52.9	87.7	66.0	60.2	57.6	82.6	70.3
Rubber and plastic products	53.5	81.7	56.8	78.9	63.9	86.5	60.0	64.6
Other non-metallic mineral products	136.5	148.6	100.5	90.0	130.6	138.5	132.0	132.8
Basic metals and metal products	128.7	126.2	85.7	75.7	94.9	86.2	92.6	106.7
Machinery and equipment n.e.c.	152.2	111.8	76.8	68.0	89.5	74.1	149.0	114.7
Electrical and optical equipment	64.8	98.0	75.9	153.1	54.5	56.6	68.7	98.7
Transport equipment	85.7	81.1	41.6	49.7	66.5	61.3	56.7	58.3
Furniture/recycling	142.3	123.2	89.2	81.9	141.8	161.2	112.1	71.8

Source: UNECE secretariat estimates, based on national statistics.

[a] Specialization is defined as the ratio of the share of each branch in total national employment to the weighted mean for four countries.

[b] Specialization is defined as the ratio of the share of each branch in total national employment to the EU average.

TABLE 3.4.8
Main features of employment specialization in selected central European economies, 2000

Specialization [a]	Underrepresentation [b]
Czech Republic	
Metal products [c,d]	Food products
Transport equipment [c]	Wood products
Machinery and equipment [c]	Petroleum products
	Leather products
Hungary	
Leather products	Wood products
Petroleum products [c]	Non-metallic minerals
Electrical/optical equipment [c,d]	Furniture/recycling
Poland	
Wood products [c]	Electrical/optical equipment
Furniture/recycling	
Slovakia	
Leather products [c,d]	Food products
Petroleum products [c,d]	Wood products
Electrical/optical equipment [c]	Rubber/plastics
	Furniture/recycling

Source: Based on UNECE secretariat estimates.

[a] Specialization ratio of more than 120.

[b] Specialization ratio of less than 80.

[c] Same specialization as in 1993.

[d] Notably increased level of specialization.

industry (DD). Between 1993 and 2000 the degree of specialization fell only in the leather industry (DC) and particularly in petroleum products (DF). In contrast, the already relatively high degree of concentration of employment in wood products increased further, making it the branch with the highest specialization ratio of more than 180.

Among the underrepresented branches in central Europe (as compared with the EU) in 2000 are chemicals (DG) and transport equipment (DM). Between 1993 and 2000, the east European shares moved closer to the west European pattern in the electrical/optical industry (DL) and rubber/plastics (DH) (in both cases the ratio increasing from some 60 to 85), and the difference narrowed somewhat in the paper/publishing industry (DE). Manufacturing industries are often classified into three groups according to the level of their technology: namely, low-tech, medium- to high-tech and resource- (and scale-) intensive group.[329] In terms of this categorization, the central European countries in 2000 were relatively

---

[329] M. Landesmann, loc. cit., pp. 100-107.

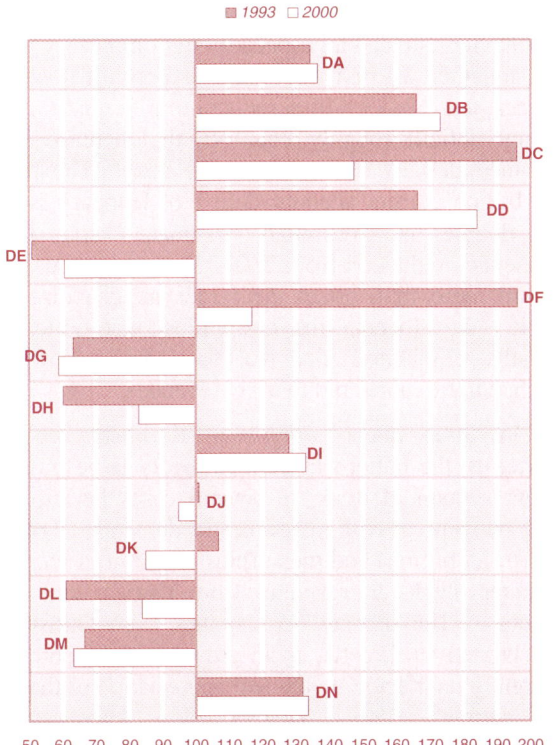

CHART 3.4.9

Relative specialization in manufacturing employment, the four countries combined vis-à-vis the EU average, 1993 and 2000 [a]
(Per cent)

Source: UNECE secretariat estimates, based on national statistics.
Note: For full branch names, see table 3.4.3.

[a] Specialization is defined as the ratio of the share of each branch in the total employment of the four countries to the EU average.

specialized in low-tech branches and relatively underrepresented in the medium- to high-tech sectors of manufacturing industry.

3.5 Foreign trade and payments

(i) Current account developments

The combined current account of the transition countries has been heavily influenced by recent swings in the international price of oil. In 2000, the increase in fuel prices boosted the current account surplus to over $27 billion (table 3.5.1), but it was more than halved in 2001, largely because of the subsequent decline in prices. This last change reflects, above all, the contraction of the Russian current account surplus and, to a lesser extent, a deterioration in the balances of most other countries. Most countries managed to finance their deficits without difficulty, thanks in particular to FDI, although many CIS and some countries of south-east Europe were subject to external financing constraints of varying degrees of severity.

In general the transition economies have faced an unsupportive international economic environment in the past year, but conditions have varied considerably within the region and these, in turn, have influenced the external performance and output of individual countries. Eastern Europe and the Baltic states faced weakening export markets in the west where import demand actually fell in the second half of 2001. In contrast, buoyant output and import growth in Russia, although slowing as the year progressed, provided support for the exports of many CIS and east European countries. The exports of several south-east European and Asian CIS countries have been adversely affected by the crisis in Turkey.

The combined current account deficit of *eastern Europe* and the *Baltic countries* ($20 billion) changed little in 2001.[330] The growth of merchandise exports (section 3.5(ii)) slowed in response to weakening global economic activity, but the expansion of service receipts quickened and, as a result, their growth rate remained roughly the same (10 to 11 per cent) as in 2000. There was a small increase in the growth of imports of goods and services in 2001 (to 9 per cent), which was also due to the acceleration in services. In most countries more buoyant domestic demand contributed to the growth of imports and larger current account deficits. However, in several cases weakening domestic demand led to the opposite outcome. In particular, in Poland there was a large reduction in the current account deficit that had a major impact on the balance of the whole region.

The combined merchandise trade deficit of eastern Europe and the Baltic states increased by almost $3 billion in 2001 (the deficit appears to have stabilized since the late 1990s). Many countries benefited from improvements in export competitiveness[331] and lower import prices of fuels and raw materials (although exporters of raw materials and metal products were disadvantaged), but the influence of these factors were offset by the slowdown in foreign demand. Increased payments abroad associated with FDI have tended to enlarge the investment income deficit. As the stock of FDI and resultant profits have grown, repatriated and reinvested profits have increased. At the same time, net interest payments have diminished due to lower interest rates and, in some cases, reduced levels of net debt. However, there was a large increase in net earnings from services (about $2 billion), and this helped to restore the area's traditional large surplus on this item. This surplus had fallen from a peak of $8 billion in 1997 due to the rapid growth of purchases of business services (in central Europe) and subsequently the losses in tourism and transport revenues associated with the Kosovo Conflict in 1999. However, both of these developments have been reversed in the past two years. The growth of transfers, mostly private remittances, to around $8 billion in 2001 was due entirely to south-eastern Europe. It receives the bulk of this relatively stable source of financing thanks to the diaspora working abroad.

---

[330] This section is based on balance of payments data. The analysis of merchandise trade in sect. 3.5(ii) is based on customs statistics.

[331] Competitiveness is measured by changes in real exchange rates, based on unit labour cost. Real exchange rates based on consumer and producer price indices appreciated in most countries. See sect. 3.1.

TABLE 3.5.1

Current account balances of the ECE transition economies, 2000-2001

(Million dollars, per cent)

	Million dollars			Per cent of GDP	
	2000	Jan.-Sep. 2001	2001	2000	2001
Eastern Europe	-18 885	-12 201	-18 800*	-5.0	-4.5
Albania	-156	-229	-400*	-4.1	-9.6
Bosnia and Herzegovina	-909	-600*	-900*	-21.6	-18.7
Bulgaria	-702	-467	-878	-5.9	-6.7
Croatia	-433	-213	-600*	-2.3	-2.9
Czech Republic	-2 273	-1 485	-2 500*	-4.5	-4.5
Hungary [a]	-1 328	-637	-1 105	-2.9	-2.1
Poland	-9 946	-5 324	-7 081	-6.3	-4.0
Romania	-1 363	-1 378	-2 349	-3.7	-6.1
Slovakia	-713	-1 131	-1 820*	-3.7	-9.0
Slovenia	-612	35	-67	-3.4	-0.4
The former Yugoslav Republic of Macedonia	-113	-232	-400*	-3.2	-11.7
Yugoslavia	-339	-540*	-700*	-4.2	-6.7
Baltic states	-1 483	-894	-1 715	-6.3	-6.9
Estonia	-315	-207	-300*	-6.3	-5.5
Latvia	-493	-390	-765	-6.9	-10.1
Lithuania	-675	-297	-650*	-6.0	-5.4
CIS [b]	47 530	27 765	32 267	13.5	7.8
Armenia	-278	-152*	-220*	-14.5	-10.4
Azerbaijan	-168	18	150*	-3.2	2.6
Belarus	-296	5	-400*	-2.9	-3.3
Georgia	-262	-192	-180*	-8.7	-5.7
Kazakhstan	743	-1 039	-1 800*	4.1	-8.0
Kyrgyzstan	-77	19	-10*	-5.6	-0.7
Republic of Moldova	-126	-91	-130*	-9.8	-8.8
Russian Federation	46 291	28 557	34 157	17.8	11.0
Tajikistan	-62*	-168*	-200*	-6.3	-19.0
Turkmenistan	..	..	..	..	..
Ukraine	1 481	1 237	1 300*	4.7	3.5
Uzbekistan	184	-50	-100*	1.4	-0.9
Total above [b]	27 162	14 670	11 752	3.6	1.4
Memorandum items:					
CETE-5	-14 871	-8 542	-12 573*	-5.1	-3.9
SETE-7	-4 014	-3 659	-6 227*	-4.6	-6.5
Asian CIS [b]	181	-1 944	-2 660*	0.4	-5.0
Three European CIS [c]	1 059	1 151	770*	2.5	1.5

Source: UNECE secretariat, based on national balance of payments statistics; IMF, *Staff Country Reports* (Washington, D.C.), for Bosnia and Herzegovina, Yugoslavia, Tajikistan and Uzbekistan [www.imf.org]; TACIS, *Azerbaijan Economic Trends* (Baku) for Azerbaijan [www.economic-trends.org].

Note: Estimates for 2001 are generally based on three quarters of balance of payments data, changes in merchandise trade in the fourth quarter and trends in the other current account items. A similar approach was used to make estimates for countries where the availability of data differed: Armenia (January-June 2001) and Slovakia (January-November). For Bosnia and Herzegovina and Yugoslavia, the estimates reflect changes in merchandise trade and IMF projections for the other current account items. The current account of Tajikistan in 2000 is an IMF estimate. The UNECE secretariat estimate for 2001 is based on changes in merchandise trade in 2001 and IMF projections in other current account items. No balance of payments data have been available for Turkmenistan since 1998.

[a] Excludes reinvested profits (a net outflow).
[b] Totals include UNECE secretariat estimates for Turkmenistan.
[c] Belarus, Republic of Moldova and Ukraine.

In *central Europe* a faster rate of growth of domestic demand (section 3.2) contributed to the sharp increase in the trade and current account deficits in Slovakia, to some 9 per cent of GDP. In the Czech Republic a marginal increase in the current account deficit[332] was due to larger payments of net investment income abroad, although domestic demand also increased rapidly. The easing of demand contributed to a reduction in the deficits of Slovenia and, in particular, Poland, the latter having pursued an especially tight monetary policy. This, together with an increase of exports of goods and services, led to a $3 billion reduction in Poland's current account deficit, roughly equal to the increase in the combined deficits elsewhere in the area. The revised $1.1 billion current account deficit of Hungary is more than double the preliminary estimate,[333] but it still represents a reduction from 2000. Record revenues from tourism more than offset the larger merchandise trade deficit. In Slovenia the current account deficit was virtually eliminated due to the rapid growth of merchandise exports and tourism revenues.

In 2001 most *south-east European* countries reverted to the pattern of comparatively large trade and current account deficits, in several cases because of a more buoyant growth in domestic demand. Transfers rose (to $5 billion) and the surplus on services increased, the latter thanks to revenues from tourism and "other services". The increase in revenues from tourism was concentrated in Croatia, their growth to a record level apparently unaffected by the real appreciation of the kuna. The current account balance also deteriorated in Yugoslavia despite further increases in transfers, estimated at well over $1 billion – several years ago no remittances at all were reported. Elsewhere there were relatively large falls in transfers, and in The former Yugoslav Republic of Macedonia official and, especially, private transfers, fully account for the worsening of the current account deficit. In Albania and Bosnia and Herzegovina, official transfers had been a major source of funding in the mid-1990s, but their importance has fallen markedly. However, private remittances have held up in Albania and to a lesser degree in Bosnia and Herzegovina.

---

[332] In March the Czech National Bank published balance of payments data for the full year 2001 that reflected substantial revisions for January-September and also for full year 2000. The main changes are an upward revision in FDI-related profit outflows that raise the current account deficit in 2000 from $2,273 million to $2,844 million. A $2,654 million deficit is reported for 2001, above the $2,500 million estimated by the secretariat (on the basis of the pre-revision, January-September data) and presented in table 3.5.1). There are corresponding changes in the financial account in the form of larger FDI inflows (sect. 3.5(iv)(b)).

[333] These data revisions, which also affect 2000, mainly involve increases in merchandise imports (to correct misreported company data) and transfers. The National Bank of Hungary has announced that the balance of payments based on a new methodology (incorporating customs rather than cash flow data) will be published as of 2003. Presumably the new series will also include reinvested profits; these are currently excluded from both the current account, where as an outflow they would raise the deficit, and from the capital account, where they would raise the figure for net inflows of FDI. The magnitude of such an adjustment is uncertain, but estimates made by the IMF suggest that the new methodology would raise the current account deficit in 1997 by over 1 per cent of GDP. IMF Staff Country Report No. 99/27, *Hungary: Selected Issues* (Washington, D.C.), April 1999. Also see UNECE, *Economic Survey of Europe, 2001 No. 1,* table 5.4.3. Higher estimates have been made by some researchers.

In the *Baltic states*, the current account balances of Estonia and Lithuania changed little, increased outflows of investment income offsetting larger revenues from services (mainly tourism). However, a surge in domestic demand in Latvia contributed to the deterioration of the trade and current account deficits. Although revenues from transit facilities continued to rise, their future level is uncertain given that the transport of some Russian goods is to be shifted to new Russian port facilities.

Russia's current account surplus fell by $12 billion to $34 billion in 2001, mainly because of the fall in merchandise export revenues and the increase in imports. Also the deficit on services rose sharply again due to increased expenditures on tourism. However, net investment outflows declined, chiefly because of the rise in interest earned on the growing stock of official reserves. In the other European CIS countries, current account balances also deteriorated as the growth of exports of goods and services (some 7 per cent) was outpaced by the rise in imports. Nevertheless, Ukraine maintained a sizeable surplus, thanks to increased transfers and lower net interest payments (due to falling net debt). In the Republic of Moldova, exports of goods and services continued to expand although by substantially less (14 per cent) than in 2000. Imports expanded slightly more, leaving the comparatively large current account deficit broadly unchanged. (The Asian CIS are discussed in section 3.5(iv)(c) below.)

(ii) International trade

(a) Eastern Europe and the Baltic states

Although economic growth was decelerating in the final months of 2001, the current dollar value of foreign trade of the east European and Baltic countries maintained its momentum expanding at 9 to 11 per cent (year-on-year), only slightly below its rate in 2000.[334] The 8 and 11 per cent growth in the volume of the region's imports and exports, respectively, however, was by nearly half their rates in 2000.[335] The terms of trade improved in a number of countries, as import prices in dollars fell as a result of lower prices for oil and industrial raw materials, while the dollar prices of exports declined less or remained unchanged mainly reflecting the evolution of unit values for manufactured goods.[336] The improvement in the terms of trade helped to slow down but not reverse the deterioration of the region's aggregate merchandise trade deficit, which increased by $0.7 billion to $44.3 billion in 2001.

All the east European and Baltic countries except for The former Yugoslav Republic of Macedonia reported positive growth in their foreign trade. The 11 per cent rise in the aggregate dollar value of the region's exports in 2001, however, conceals a wide range of responses to the global downturn in the various subregions and individual countries (tables 3.5.2-3.5.4, chart 3.5.1). In central Europe, a more diversified commodity mix (chart 3.5.2) and the greater flexibility of the export sector, thanks in part to a somewhat greater capacity to absorb shocks with the help of multinational corporation (MNC) networks, helped to contain the adverse effects of weaker western demand. The Czech Republic and Slovenia actually improved their export performance in 2001, with stronger export growth to both west and east European (including Baltic) markets.[337] Export growth in dollar value in Hungary and Poland, although somewhat slower than in 2000, was still at 9 and 14 per cent, relatively high rates given not only the unfavourable state of external demand but also the appreciation of their currencies in real effective terms (chart 3.1.1).[338] In south-east Europe and the Baltic states, however, aggregate rates of export growth were more than halved compared with those in 2000. For the Baltic states, this reflected the poor performance of Estonia in the second half of the year, since export growth in Latvia and Lithuania remained rather steady throughout 2001 (chart 3.5.3).[339] In south-east Europe, the improvement in the performance of a number of export sectors in Croatia, particularly in trade with the EU, was quite striking. In contrast, the sharp deceleration of export growth from Bulgaria and Romania by the end of 2001, apart from their weak trade with the CIS and some south-east European countries, can probably be explained by the

---

[334] The continued weakness of the euro – in which most of the region's trade is denominated – relative to the dollar depressed export and import values expressed in dollars, but much less than in 2000: in euro terms, east European and Baltic trade rose by 12 to 14 per cent in 2001, following 29 to 31 per cent increases in 2000.

[335] According to UNECE secretariat estimates, based on preliminary data from several east European and Baltic countries (table 3.5.3).

[336] World market prices in dollars for energy resources declined by 12 per cent (year-on-year) in 2001; world commodity prices excluding energy shrank by 9 per cent at the same time (UNECE Common Database). For more details on world market price developments in 2001 see chap. 2.1 and charts 2.1.4-2.1.5 above. Export unit values in dollars for manufactured goods declined in the first nine months of 2001, year-on-year, by 2.4 per cent in western Europe. United Nations Statistics Division, *Monthly Bulletin of Statistics On-line,* (query results for trade-manufactured goods exports, unit value indices in dollars, 1990=100). According to national statistics, dollar export unit values for manufactured goods also declined in Hungary and Poland but increased in the Czech Republic.

[337] Strengthened competitiveness helped: unit labour costs declined in the Czech Republic for the second consecutive year in 2001 (sect. 3.3(ii)), while the real depreciation of the tolar in Slovenia helped to maintain export price competitiveness in the west. In addition, Slovenian companies succeeded in offsetting shortfalls in export earnings of certain sectors in EU markets by considerably expanding exports to Russia, Yugoslavia and Croatia – its major partner among the transition economies.

[338] Since the worldwide downturn in the ICT sector began (chap. 2.1(i)), Hungarian exports of electronics, communication network products and software, which accounted roughly for a third of total exports in 2000, were the most affected: aggregate exports of these products (SITC 75 to 77) declined by 2.5 per cent in dollar value in 2001. In fact, for the first time in several years, Hungarian exports from firms in the traditional domestic sectors grew faster than those from its industrial customs-free zones.

[339] Lithuanian exports increased by more than 20 per cent in both dollar and volume terms, as a result of which exports finally regained their pre-Russian crisis levels. However, the increase was rather narrowly based: excluding exports of mineral products, growth to the EU was only some 5 per cent in dollar value.

TABLE 3.5.2

Trade performance and external balances of the ECE transition economies, 2000-2001

(Rates of change and shares, per cent)

	Merchandise exports (growth rates)		Merchandise imports (growth rates)		Trade balance (per cent of GDP)	
	2000	2001[a]	2000	2001[a]	2000	2001[a]
Eastern Europe	12.8	10.7	11.0	8.6	-10.4	-9.5
Albania	-25.9	12.6	18.5	22.2	-21.6	-24.4
Bosnia and Herzegovina	30.2	18.5	-5.8	2.2	-38.3	-32.1
Bulgaria	20.4	5.7	18.0	11.1	-14.0	-16.2
Croatia	3.0	5.1	1.1	14.7	-18.2	-21.2
Czech Republic	10.5	15.1	14.4	13.7	-6.1	-5.6
Hungary	12.3	8.6	14.5	5.0	-8.6	-6.2
Poland	15.5	14.0	6.6	2.7	-11.0	-7.3
Romania	21.9	9.8	25.6	19.1	-7.3	-10.8
Slovakia	15.9	6.6	12.4	16.0	-3.9	-9.8
Slovenia	2.2	6.0	0.3	0.3	-7.6	-4.8
The former Yugoslav Republic of Macedonia	11.2	-12.4	17.6	-19.0	-21.4	-15.6
Yugoslavia	15.4	10.5	12.7	30.4	-24.6	-27.9
Baltic states	24.4	11.8	15.0	9.1	-17.2	-16.8
Estonia	33.2	4.1	23.7	0.8	-21.5	-18.1
Latvia	8.2	7.3	8.2	10.0	-18.5	-19.9
Lithuania	26.8	20.3	12.9	15.1	-14.6	-14.2
CIS[b]	37.9	4.0	14.1	16.0	20.8	16.2
Armenia	29.7	13.9	9.0	-2.6	-30.5	-26.6
Azerbaijan	87.8	2.4	13.1	10.2	10.9	15.5
Belarus	24.1	2.5	28.5	-10.0	-12.0	-1.6
Georgia	38.5	-3.2	8.1	18.0	-10.6	-13.7
Kazakhstan	63.2	2.7	37.0	34.9	22.3	10.8
Kyrgyzstan	11.2	-7.6	-7.6	-17.1	-3.6	2.7
Republic of Moldova	1.6	24.5	32.3	13.4	-23.7	-22.3
Russian Federation	39.5	2.7	8.9	22.5	26.6	20.8
Tajikistan	13.9	-14.3	1.8	11.6	11.1	-7.3
Turkmenistan	110.6	9.8	20.4	31.6	16.4	3.6
Ukraine	25.8	14.6	17.8	13.1	2.0	2.2
Uzbekistan	0.9	4.8	-5.0	7.5	2.8	0.2
Total above	24.6	7.9	12.0	11.3	3.9	2.8
Memorandum items:						
CETE-5	12.2	11.4	10.2	6.3	-9.1	-7.2
SETE-7	15.9	7.2	14.0	14.7	-14.9	-17.5

Source: UNECE secretariat calculations, based on national statistics and direct communications from national statistical offices.

Note: Foreign trade growth is measured at current dollar values. Trade balances are related to GDP at current prices, converted from national currencies at current dollar exchange rates. GDP values, in some cases, are estimated from reported real growth rates and consumer price indices. On regional aggregates, see the note to table 1.3.1.

[a] Full year 2001 provisional results for eastern Europe and the Baltic states; January-September for CIS and "Total above".

[b] Including intra-CIS trade.

As western import demand faltered in 2001 (chart 3.5.1), the only external support for the transition economies came from the robust import demand in the Russian, CIS and south-east European markets.[341] Although exports to these markets were the fastest growing segment of total exports for Croatia, Poland, Slovenia and the Baltic states (table 3.5.2), maintaining a steady rate of growth of sales to the west was crucial for maintaining the overall level of exports. Despite the stalling of demand in western Europe, several countries in the region succeeded in achieving the latter, thanks to the niches on specific western markets (often as a result of their integration into the production networks of multinational companies), and also in some cases to their wage-cost competitiveness. However, those countries with a high concentration of exports, by market and/or by commodity (for some details, see section 3.5(iii)), were very vulnerable to downturn in particular markets; the experience of Slovakia and Estonia, and to a lesser extent Hungary, are cases in point.

Since the beginning of 2001, Estonia's exports have been especially hard hit by the slowdown in the demand for ICT products, which accounted for more than one fifth of the total in 2000; the situation continued to worsen through the year resulting in a fall of some 6 per cent in annual exports to the EU and a sluggish rate of growth in total export value.[342] Slovak exports were badly affected by the slowdown in new car sales on western markets, the Volkswagen subsidiary being the country's major exporter in 2001, exports of transport vehicles declined by 4 per cent in dollar value, pulling down the overall growth of total exports and in particular those to the EU.[343]

The region's import still continued to rise strongly in 2001, although the aggregate dollar value of east European and Baltic imports increased less than exports, composition of their exports and developments in world market prices. For Bulgaria, the economic crisis in Turkey and the marked deterioration in exports to neighbouring Yugoslavia also played a role.[340]

---

[340] Figures for Albanian exports growth are rather uneven from year to year and dependant on the sources used. According to the Albanian Statistical Institute (INSTAT), Albanian exports grew some 13 per cent in 2001, following a 26 per cent decline in 2000.

[341] Real import demand in south-east Europe soared by some 15 to 18 per cent in 2001, while Russia's imports from non-CIS countries rose by nearly 50 per cent in volume in the first nine months of 2001, a trend that seemed to continue in the last quarter. Non-CIS imports into Kazakhstan and Ukraine also thrived (see next section). Boosted by demand in Russia and other CIS countries, exports from some east European and Baltic countries surged: in 2001, Croatia, Slovenia and Lithuania recorded rises of 45 to 50 per cent and Latvia of 20 per cent, in their export dollar values to the CIS region. In contrast, exports from Romania declined, while those from Bulgaria and Hungary more or less stagnated; in Hungary this was partly due to an unresolved dispute over the former USSR debt.

[342] Estonian exports grew by some 4 per cent in dollar value and shrank by 19 per cent in volume in 2001. Behind this was a 45 per cent decline in the second half of 2001 in the dollar value of exports of machinery and electrical equipment. The withdrawal of Elcoteq's mobile phone production from Estonia seems to be a major cause for this decline (for details, see UNECE, *Economic Survey of Europe, 2001 No. 2*, p. 22, box 1.2.1).

[343] In fact, Slovak exports started to decline year-on-year in the final months of 2001. Shrinking prices for semi-manufactures, which account for a sizeable share in Slovakia's overall exports and in particular in exports to its CEFTA partners (see section 3.5(iii)) contributed to this downturn. Exports to the Czech Republic even declined slightly in the second part of 2001.

TABLE 3.5.3

Foreign trade of the ECE transition economies by direction, 1999-2001

(Value in billion dollars, growth rates in per cent) [a]

Country or country group [b]	Exports Value 2000	Exports Growth rates 1999	Exports Growth rates 2000	Exports Growth rates 2001 [c]	Imports Value 2000	Imports Growth rates 1999	Imports Growth rates 2000	Imports Growth rates 2001 [c]
Eastern Europe, to and from:								
World	133.0	-0.6	12.8	11.7	172.5	-2.1	11.0	9.5
ECE transition economies	27.2	-17.0	13.5	12.4	39.2	-0.8	27.6	8.7
CIS	4.9	-41.2	14.0	12.1	18.4	-3.4	51.2	8.0
Baltic states	1.2	1.3	15.3	26.3	0.5	31.9	50.8	-3.1
Eastern Europe	21.0	-9.2	13.3	11.7	20.4	0.6	11.5	11.7
Developed market economies	97.6	4.8	12.3	12.0	116	-3.7	6.1	8.5
European Union	89.0	4.6	11.7	12.4	100.5	-3.6	4.9	9.1
Developing economies	8.2	1.8	17.0	4.4	17.3	8.5	12.1	17.8
Baltic states, to and from:								
World	8.9	-11.6	24.0	13.6	12.9	-13.6	15.0	10.2
ECE transition economies	2.5	-34.7	18.5	22.9	4.4	-10.5	23.2	11.2
CIS	0.9	-50.9	0.9	37.0	2.7	-13.9	36.7	7.2
Baltic states	1.2	-12.3	27.8	12.5	0.8	-6.3	7.8	16.4
Eastern Europe	0.4	14.6	44.9	24.2	0.9	-3.8	4.3	19.9
Developed market economies	6.0	5.6	25.2	8.0	7.7	-15.1	8.9	2.7
European Union	5.5	5.4	25.8	7.9	6.7	-15.4	9.9	3.5
Developing economies	0.3	8.3	50.3	48.2	0.8	-13.6	37.2	87.4
Russian Federation, to and from:								
World	103.0	2.2	41.3	2.7	33.9	-30.5	12.0	22.5
Intra-CIS	13.8	-21.8	28.8	7.4	11.6	-26.3	39.6	3.0
Non-CIS economies	89.2	7.9	43.5	2.0	22.3	-32.0	1.5	32.8
ECE transition economies	17.9	6.3	61.1	0.0	2.4	-45.3	15.5	24.0
Baltic states	4.9	26.0	73.7	-15.5	0.3	-56.2	12.2	30.9
Eastern Europe	12.9	0.9	56.8	5.7	2.1	-43.0	16.1	22.9
Developed market economies	49.6	4.2	37.6	0.8	15.4	-31.9	2.9	29.6
European Union	36.9	7.1	48.4	5.9	11.1	-28.9	-0.4	32.8
Developing economies	21.7	19.4	44.5	6.7	4.4	-24.5	-9.2	47.9
Other CIS economies, to and from:								
World	41.0	-4.7	34.4	7.2	36.9	-17.1	19.0	10.1
Intra-CIS	15.1	-21.2	38.0	8.3	20.4	-18.3	30.8	9.5
Non-CIS economies	25.9	7.9	32.4	6.7	16.5	-15.9	7.1	10.9
ECE transition economies, to and from:								
World	285.8	-0.7	25.2	7.9	256.2	-9.8	12.4	11.3

Source: National statistics and direct communications from national statistical offices to UNECE secretariat; for the Russian Federation, State Customs Committee data; for other CIS economies, CIS Interstate Statistical Committee.

Note: There were changes in the methodology of foreign trade reporting in several east European and Baltic economies in 1999-2000. In 1999, Poland changed its customs declaration system, substantially increasing its coverage. In 2000, Estonia switched its basic trade statistics to a "special trade" reporting system; this change is reflected in the Baltic states aggregate for January-September 2001 above. For details of changes prior to 1999, see UNECE, *Economic Bulletin for Europe*, Vol. 49, 1997 and Vol. 50, 1998.

[a] Growth rates are calculated on values expressed in current dollars.

[b] For country groups see table 1.3.1.

[c] January-September over same period of 2000.

---

mainly because of weak Hungarian and Polish import growth. In fact, Polish imports, after surging in January, fell slightly below the previous year's value in February and remained rather sluggish through most of the year, with only a slight upsurge in September and October (chart 3.5.3). This reflected the effects of tight monetary policy on Polish domestic demand (see sections 3.1 and 3.2 above), although the strengthening of the zloty in real effective terms (chart 3.1.1) probably helped to support imports. As a result, Poland's merchandise trade deficit shrank to $14.2 billion in 2001, from $17.3 billion in 2000. Hungarian imports were decelerating from the second quarter of 2001 and were flat from September, reflecting both a progressive slowdown in real investment and a markedly weaker demand for imported inputs as industrial output declined under the influence of shrinking export orders. As a result of this slowdown, there was a substantial reduction in the Hungarian merchandise trade deficit.

There was also an improvement in Slovenia's balance: imports remained more or less flat throughout the year as domestic demand weakened and the tolar depreciated slightly in real effective terms. In contrast, the virtual stagnation of Estonia's imports was the result

TABLE 3.5.4

Changes in the volume of foreign trade in selected transition economies, 1998-2001

(Per cent)

	Exports			2001[a]				Imports			2001[a]			
	1998	1999	2000	Jan.-Mar.	Jan.-Jun.	Jan.-Sep.	Jan.-Dec.	1998	1999	2000	Jan.-Mar.	Jan.-Jun.	Jan.-Sep.	Jan.-Dec.
Croatia	11.6	-2.7	-1.3	0.9	4.4	4.7	7.2	-4.4	-4.2	3.7	16.8	22.4	19.1	15.6
CETE-4	11.8	8.4	21.0	17.3	14.7	12.8	..	15.3	7.5	15.2	11.7	9.1	7.4	..
Czech Republic[b]	12.8	9.4	18.7	18.3	15.9	14.3	12.7	8.7	5.1	19.9	21.7	17.2	15.1	13.4
Hungary	22.5	15.9	21.7	16.3	14.2	11.6	8.2	24.9	14.3	20.8	15.6	13.4	8.1	4.1
Transition economies[c]	4.7	-9.3	22.5	17.8	20.4	18.9	14.4	12.1	6.0	17.1	16.2	6.6	3.3	-1.5
European Union	24.1	20.6	21.3	17.8	14.6	12.2	8.3	23.8	14.6	14.4	12.3	11.1	7.4	3.6
Poland	2.3	2.0	25.3	19.2	16.1	13.9	..	14.3	4.4	10.8	5.2	2.9	3.4	..
Transition economies[c]	-5.0	-9.3	25.1	19.0	19.1	19.0	..	12.6	7.8	16.7	7.9	3.8	2.9	..
European Union	8.5	5.4	26.8	20.5	16.9	13.6	..	16.2	4.1	10.5	3.7	2.4	2.8	..
Slovenia[d]	8.1	3.7	11.4	10.0	7.8	7.2	..	11.3	9.2	4.1	-1.4	-0.3	-0.3	..
Baltic states	8.7	-7.1	25.9	15.8	12.6	6.4	..	13.4	-9.3	15.9	19.1	15.7	12.6	..
Estonia[e]	16.7	0.9	41.1	20.9	1.1	-13.0	-19.2	13.1	-9.0	34.9	27.1	18.7	11.2	3.3
Latvia	10.2	-2.1	13.6	9.0	10.9	9.5	8.2	21.3	-3.2	5.1	12.7	10.1	11.1	11.9
Lithuania	1.3	-16.3	19.2	14.8	23.0	21.0	..	9.0	-13.0	7.4	16.6	16.6	14.5	..
Total above	11.0	6.8	20.6	16.5	14.0	11.7	..	14.1	5.4	14.7	12.6	10.4	8.5	..
Russian Federation	-0.3	9.4	10.2	0.2	1.4	2.0	..	-11.0	-15.6	29.2	27.2	36.7	34.4	..
Non-CIS	-0.6	11.3	9.9	1.7	1.9	1.4	..	-8.4	-19.4	28.6	35.3	47.0	49.9	..
CIS	0.8	1.5	12.5	-8.0	-0.1	7.3	..	-18.4	-4.8	30.6	10.4	16.3	8.0	..

Source: UNECE secretariat calculations, based on national foreign trade statistics.

Note: The dollar values of exports and imports in 2000 are used as weights in calculating aggregate growth rates.

[a] Over the same period of 2000.

[b] Volume growth rates for the Czech Republic are calculated on the basis of import and export prices and current value indices reported in the Czech Statistical Office, *Statistical Yearbook 2001* (Prague) and from the Statistical Office website [www.czso.cz], Import and Export Price Indices in the Czech Republic, table 7201-01, December 2001. The differences in volume growth rates as compared with the previous issue of this table are due to the changes in the Czech external statistics methodology, effective from 2000 (for trade values) and January 2001 for the change in the weighting of the price indices.

[c] Transition economies refer to central and eastern European countries, including four European CIS member countries.

[d] Volume growth rates for Slovenia are derived from reported export and import unit value indices and changes in the respective trade values. *Statistical Yearbook of the Republic of Slovenia 2000* and *Monthly Statistical Review of the Republic of Slovenia*, Vol. L, Nos. 6, 9, 12.

[e] Volume growth rates for Estonia calculated on the export and import values according to the special trade reporting system.

---

of a large fall in purchases of inputs for mobile phone production, the major export product in 2000, together with falling prices for industrial raw materials. The impact of these two factors outweighed the marked rise in imports of machinery and consumer goods, a result of booming domestic demand. The effect on Estonia's trade balance, however, was rather small due to the meagre growth in exports.

Among the south-east European countries, there was some improvement in the trade deficits of Bosnia and Herzegovina, where imports remained subdued, and in The former Yugoslav Republic of Macedonia, where imports declined by one fifth because of the deterioration in economic activity.[344] In the rest of south-east Europe, as well as in Latvia and Slovakia, import growth not only accelerated in 2001, but also exceeded that of exports, resulting in a considerable deterioration in their trade deficits.[345] This surge in imports was due to the recovery of domestic demand, particularly of investment, in Bulgaria, Croatia, Slovakia and Latvia, while in the other countries food and beverages and consumer goods were the most rapidly growing products. In some cases this growth in imports was also helped by the further liberalization of trade.[346]

With respect to geographical origin, the fastest growing imports were those from within eastern Europe – predominantly foods, cars and their spare parts, and consumer goods – followed by purchases from the EU markets, especially of investment goods. The value of

---

[344] The steep decline in The former Yugoslav Republic of Macedonia's imports can also be partly explained by the high import values in the base year when there had been an influx of goods destined to refugees and to support selected deliveries to Yugoslavia.

[345] According to preliminary data, in 2001 merchandise trade deficits in relation to GDP rose to 21 per cent in Croatia, 16 per cent in Bulgaria, 11 per cent in Romania and 10 per cent in Slovakia, exceeding the 2000 levels by 2 to 6 percentage points.

[346] In Slovakia, for example, an import surcharge was eliminated as from 1 January 2001, while in Romania, duty free imports of industrial goods from the EU came into force and import protection for some agricultural products in trade with CEFTA partners was relaxed. For more details on the measures taken in 2001 to liberalize trade, see UNECE, *Economic Survey for Europe, 2001 No. 1*, p. 148.

The Transition Economies

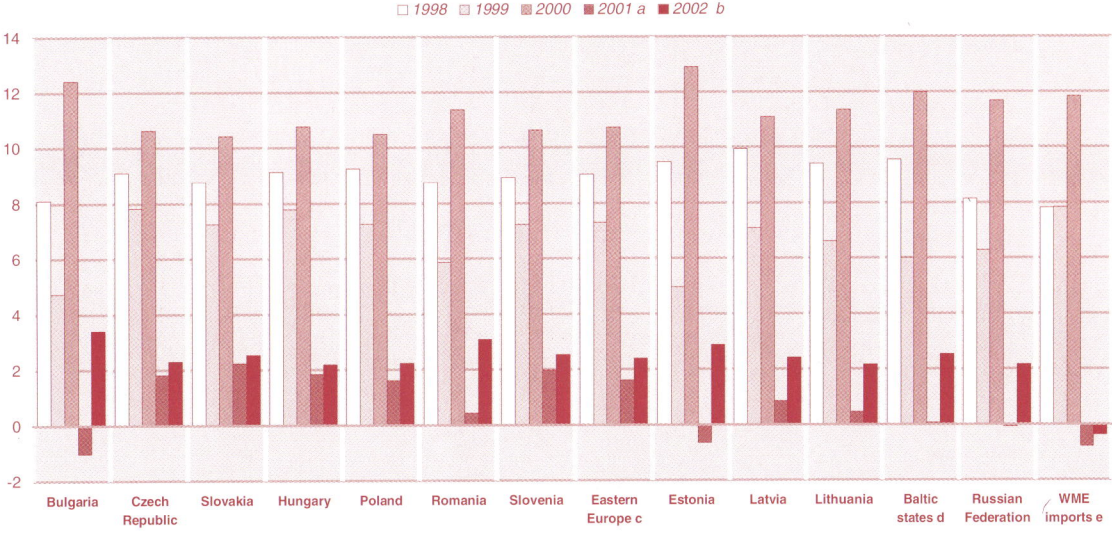

CHART 3.5.1

Specific western demand for selected transition economies' exports, 1998-2002
(Annual percentage change in volume terms)

Source: UNECE secretariat calculations: aggregation of the import volume growth rates of individual western countries weighted by their share in the exports of each eastern country. The western import data refer to goods and services on a national accounts basis.

a Preliminary.
b Forecast.
c Seven countries shown.
d Three countries shown.
e Western market economies (WME) include western Europe, North America, Turkey and Japan.

imports from the CIS slowed noticeably under the impact of lower prices for oil and industrial raw materials.

Although the weakening of west European import demand had little immediate impact on the export revenues of the east European and Baltic region in 2001, the continuing restrained demand in the west in the early months of 2002 will most probably hamper the export growth of these countries, not only in the first half of the year, but also throughout 2002. Even if the anticipated upturn in western markets materializes in the second half of 2002, its initial impact on the east European and Baltic countries is likely to be a rise in imports to replenish stocks of intermediate goods, with export growth picking up only with a lag.

Moreover, the global slowdown in the west appears to be accompanied by some increase of import protection, with new cases of anti-dumping actions being targeted at some transition economies,[347] as well as new restrictions on steel imports by the United States authorities and by the EU.[348] Protectionist sentiment also seems to be on the rise among the transition economies themselves, particularly in the agri-food sector, where opposition to further liberalization within CEFTA shows signs of increasing among new members.

(b) CIS

In the first nine months of 2001, the dollar value of total merchandise exports from the CIS countries increased by 4 per cent, year-on-year (table 3.5.2). This modest increase in export revenues represented a

---

[347] In 2002, the European Commission has initiated anti-dumping measures against eight urea-exporting countries: Belarus, Bulgaria, Croatia, Estonia, Libya, Lithuania, Romania and Ukraine. The levies imposed on urea exports to the EU vary from €9.01 per tonne for Croatia to €21.43 per tonne for Bulgaria. *Dow Jones Reuters Business Interactive*, 6 February 2002.

[348] On 20 March 2002, the United States imposed import tariffs of up to 30 per cent on 21 steel products from countries other than Canada, Israel, Jordan and Mexico. The European Union, in order to protect its market from steel imports diverted after the United States action, adopted temporary safeguard measures on steel: on 27 March 2002 it imposed quotas on imports of 15 steel products (the quotas are based on the average imports between 1999 and 2001 augmented by 10 per cent) with imports beyond these quotas facing tariffs from 14.9 per cent to 26 per cent. For details see European Commission "EU adopts temporary measures to guard against floods of steel imports resulting from US protectionism", *Press Release* (Brussels), 27 March 2002. The EU's switch from the country-by-country quotas considered initially (partly favouring east European accession countries over exports from Asia and South America) to quotas by specific product using a "first come first served" approach might have more serious implications for steel producers in Bulgaria, the Czech Republic, Poland and Slovakia, countries that are among the 10 major steel exporters to the EU. D. McQuaid and M. Newman, "EU steel tariffs seen hitting eastern Europe hardest", *Dow Jones International News*, 27 March 2002.

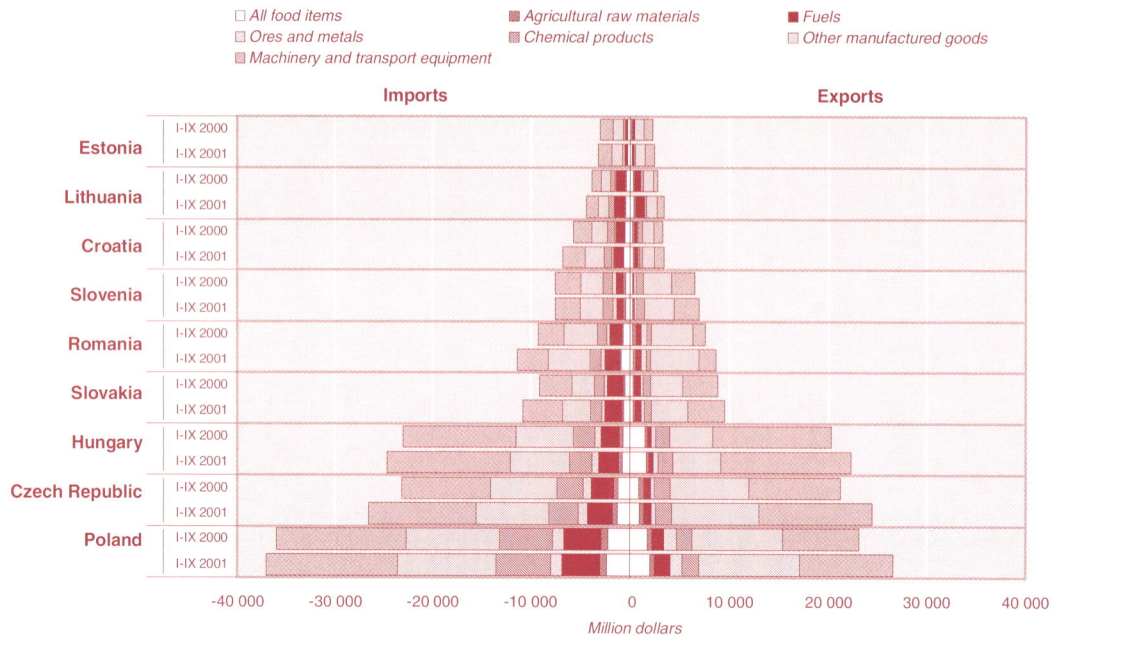

CHART 3.5.2

Exports and imports by commodity groups, January-September 2000 and January-September 2001

Source: National statistics.

Note: Commodity groups based on divisions and groups of SITC Rev.3 as follows: all food items (SITC 0 + 1 + 22 + 4); agricultural raw materials (SITC 2 - 22 - 27 - 28); fuels (SITC 3); ores and metals (SITC 27 + 28 + 68); chemical products (SITC 5); other manufactured goods (SITC 6 + 8 less 68); machinery and transport equipment (SITC 7).

consolidation of gains after a significant turnround in 2000 following the period of depressed commodity prices in 1999. On an individual country basis, exports increased in most countries ranging from 2 per cent in Azerbaijan to 25 per cent in the Republic of Moldova. Only in Georgia, Kyrgyzstan and Tajikistan did they decline (between 3 to 14 per cent).

Recently, there has been no evidence of major changes in the structure of CIS trade as sales of primary commodities have continued to dominate the region's exports. As for prospects for a more diversified industrial base, imports of investment goods remain overwhelmingly focused on enhancing the capacity to produce primary commodities and facilitating their sale by expanding the transportation infrastructure.

As a result, most CIS countries have continued to depend on foreign sales of primary commodities and, in turn, their export revenues are closely related to global commodity prices. In the first nine months of 2001, commodity prices fell across-the-board and affected the export revenues of individual CIS countries in varying degrees.[349] In the countries dependent on shipments of crude oil and refined products, such as Azerbaijan, Kazakhstan and Russia, increased export volumes of oil and oil products helped somewhat to offset the fall in prices. In contrast, lower volumes of metal exports moved in tandem with lower prices reducing the value of regional exports. Overall, almost all the commodity exporters in the CIS registered moderate (2-5 per cent) increases in total exports.

In the foreign exchange markets, in the first three quarters of 2001, the value of the domestic currencies of Armenia, Azerbaijan, Belarus and Georgia declined by 2 to 6 per cent in real terms against the dollar (year-on-year).[350] These depreciations, however, did little to boost exports. Azerbaijan's exports of crude oil are dollar-denominated while Belarus' exports go predominantly to Russia on the basis of roubles or barter. Exports of manufactures and food products from Armenia and Georgia benefited to a limited extent, but for Georgia any benefits were probably offset by the downturn in Turkey – Georgia's main export market – and the depreciation of the Turkish lira. In the other CIS countries such as Kazakhstan, Kyrgyzstan, the Republic of Moldova, Russia and Ukraine, domestic currencies appreciated between 2 and 14 per cent in real terms against the dollar reinforcing the demand for imports in all of them except in highly-indebted Kyrgyzstan.

---

[349] Crude oil prices fell 7 per cent, year-on-year, while export prices for Russia's natural gas decreased by about 16 per cent. Metal prices also fell: nickel was down by a third, copper by 10 per cent and aluminium and gold prices by 5 per cent each. The price of cotton – a major export earner in central Asia – declined by a fifth. For the year 2001 as a whole, commodity prices continued their downward trend. Crude oil prices declined by 14 per cent, aluminium, copper and nickel by 7, 13 and 31 per cent, respectively. Gold fetched 3 per cent less while the price of cotton was 29 per cent lower.

[350] For 2001 as a whole, real exchange rates against the dollar depreciated by about 5 to 7 per cent in Armenia, Kyrgyzstan and Ukraine; and by about 10 per cent in Azerbaijan and Kazakhstan. The domestic currencies appreciated in real terms against the dollar in the Republic of Moldova (7 per cent) and Russia (2 per cent).

CHART 3.5.3

Monthly dollar exports and imports in selected transition economies, 2000-2001
(Year-on-year indices)

Source: UNECE secretariat calculations, based on national statistics.
Note: Monthly indices against same month of previous year are based on a three-month moving average of monthly values in current dollars.

In contrast to the relatively flat export performance, the value of CIS imports increased by 16 per cent in aggregate, largely reflecting increased economic activity. Improved business and consumer confidence led to greater investment by firms and increased consumption by individuals (this was particularly evident in Russia where the volume of imports increased by a third). In value, total imports increased in most CIS countries, except Armenia, Belarus and Kyrgyzstan, where they fell by 3, 10 and 17 per cent, respectively. Imports of investment goods were flat in Belarus, but they declined quite sharply in Armenia (by one third) and in Kyrgyzstan (by one half). Nonetheless, investment spending grew strongly across the CIS region except in Kyrgyzstan and the Republic of Moldova.[351] In particular, in the large and rapidly growing economies of Kazakhstan and Ukraine investment outlays continued to increase at the highest rates in the region. Kazakhstan's imports of machinery and equipment reached about $1.4 billion in January-September 2001 (up by almost 50 per cent, year-on-year) while in Ukraine they rose by 20 per cent. The increased demand for imported consumer goods was driven by the continuing strength of retail trade in virtually all the CIS countries (except Tajikistan). Finally, the demand for foreign investment and consumer goods due to the increased economic activity throughout the CIS was reinforced – as mentioned above – by the real exchange rate appreciation of domestic currencies in the three largest economies of Kazakhstan, Russia and Ukraine.

The region's aggregate merchandise trade surplus fell by about $4 billion to $49 billion mainly because of falling surpluses in Kazakhstan and Russia, the two largest crude oil and metal exporting countries in the CIS.

Trade with non-CIS countries

In the first three quarters of 2001, the value of exports from the CIS countries to the rest of the world increased by 3 per cent (table 3.5.5). Only in Georgia and Kazakhstan did they decline.[352] In the other CIS countries exports to non-CIS markets increased from between 1 per cent in Kyrgyzstan to 15 per cent in the Republic of Moldova. In the latter country the increase was centred on its two most important export products, namely, food and agricultural products and textiles (re-exported under outward processing arrangements). There were also large increases in exports from Tajikistan and Ukraine (13 and 15 per cent, respectively). While the volume of Tajikistan's cotton fibre exports was flat, the volume of aluminium exports to non-CIS markets increased by one half reflecting increased production as well as the redirection of aluminium sales away from CIS markets. In Ukraine, where strong external demand for steel and chemicals – which account for half of total exports – had driven the country's non-CIS export growth in 2000, the value of these exports fell in the first three quarters of 2001. This was due to lower prices and market access constraints mostly related to anti-dumping disputes.[353] The declines, however, were more than offset by across the board increases in exports, in particular, of machinery and equipment and agricultural products.

In other CIS countries exports to the rest of the world increased more moderately. Russian exports to non-CIS markets rose by 2 per cent in value, while the overall volume increased only 1 per cent in the first three quarters of 2001, despite the continued weakening of export prices throughout the year (falling by over 6 per cent in the third quarter of 2001, year-on-year).[354] Russia's non-CIS exports of machinery and equipment (including vehicles) grew by about a third, with exports of arms and military equipment likely to have accounted for most of this growth. In Azerbaijan, Kyrgyzstan, Turkmenistan, and probably Uzbekistan the increased export volumes mainly related to primary commodity production increases.

The total value of CIS imports from non-CIS countries rose by 23 per cent, largely due to the largest – and fast growing – CIS economies. Imports actually fell in Armenia, Belarus and Kyrgyzstan. Generally, the CIS countries buy foodstuffs and machinery and equipment from the non-CIS. Russian imports from non-CIS countries rose 50 per cent in volume, while the dollar value of imports machinery and equipment increased by a third. Large increases in imports of machinery and equipment were also evident in Kazakhstan and Ukraine, both rapidly growing economies. In Azerbaijan, imports of machinery and equipment were $273 million, virtually unchanged from their level in January-September 2000.[355]

Intra-CIS trade

In the first nine months of 2001, the dollar value of intraregional trade grew by 7 per cent (year-on-year). As a result of the growth of the Russian economy – which accounts for about 40 per cent of intraregional trade – exports of the other CIS members to Russia increased by

---

[351] In Kyrgyzstan both capital expenditures and imports of capital goods fell sharply, while in the Republic of Moldova imports of machinery and equipment rose by 20 per cent to $90 million in the first three quarters of 2001, although investment spending fell by a fifth during the same period.

[352] In Georgia, sales of scrap metal were behind a fall of 5 per cent, while in Kazakhstan a 3 per cent decline reflected the fall in the prices of oil and metal products.

[353] An important factor was the imposition of anti-dumping duties on Ukrainian steel products by the United States at the beginning of 2001, resulting in a sharp drop of exports (from $330 million to $82 million) in the first three quarters of 2001. In addition, CIS steel exporters such as Kazakhstan, Russia and Ukraine will be negatively affected by recently introduced United States tariffs on steel products. Russia will be particularly hard hit: its government estimates that the tariffs will affect exports valued at $400 million a year.

[354] The volume of Russian crude oil and oil product exports to non-CIS markets, however, rose by 8 per cent in the January-September 2001 period, tracking closely the increase in crude oil extraction. Similarly, a small decline in natural gas sales to non-CIS markets reflected a commensurate decline in natural gas extraction, but rising domestic metal production in Russia went into inventories or domestic sales as the volume of exports of base metals declined across the board.

[355] The investment-GDP ratio in Azerbaijan is expected to double to about 50 per cent in 2003 as a result of planned expenditures for oil and gas exploration, production and pipeline construction.

TABLE 3.5.5

CIS countries' trade with CIS and non-CIS countries, 1999-2001
(Value in million dollars, growth rates in per cent)

	Export growth 2000	Export growth 2001[a]	Import growth 2000	Import growth 2001[a]	Trade balances 1999	Trade balances 2000	Trade balances 2001[a]
Armenia							
Non-CIS	29.5	8.3	13.9	-7.3	-449	-484	-293
CIS	30.1	32.1	-7.3	16.6	-131	-100	-79
Azerbaijan							
Non-CIS	110.3	3.1	12.1	6.8	8	713	772
CIS	11.4	-3.4	15.4	17.4	-114	-140	-157
Belarus							
Non-CIS	28.0	1.5	7.0	-15.7	-98	376	642
CIS	21.6	3.1	40.4	-7.4	-667	-1 619	-783
Georgia							
Non-CIS	50.9	-4.9	12.3	11.1	-246	-226	-200
CIS	23.4	-0.8	1.0	30.5	-118	-95	-115
Kazakhstan							
Non-CIS	64.7	-2.9	9.9	41.3	2 011	4 456	2 339
CIS	59.2	17.8	72.5	29.7	-106	-381	-499
Kyrgyzstan							
Non-CIS	9.8	1.4	-24.9	-19.9	-70	42	69
CIS	13.1	-19.0	15.2	-14.5	-76	-91	-39
Republic of Moldova							
Non-CIS	-7.2	15.0	50.0	6.9	-134	-321	-238
CIS	8.8	31.3	7.3	26.4	12	16	14
Tajikistan							
Non-CIS	9.8	12.6	-22.3	37.1	225	295	201
CIS	18.7	-39.7	8.7	6.3	-200	-186	-251
Turkmenistan							
Non-CIS	71.4	8.6	10.0	18.7	-300	100	-200
CIS	165.3	10.9	36.0	59.5	-10	620	350
Ukraine							
Non-CIS	21.0	14.5	15.9	14.6	3 227	4 159	3 546
CIS	38.3	14.9	19.2	12.1	-3 491	-3 542	-2 913
Uzbekistan							
Non-CIS	-7.1	7.9	-20.0	9.8	–	290	60
CIS	20.0	-0.6	40.0	3.7	200	90	-45
Total above							
Non-CIS	32.4	6.7	7.2	10.9	4 174	9 400	6 698
CIS	37.3	8.3	30.8	9.5	-4 701	-5 429	-4 517
Russian Federation							
Non-CIS	43.6	2.0	1.5	32.8	40 244	66 995	44 787
CIS	29.0	7.4	39.1	3.0	2 364	2 200	2 196
CIS total							
Non-CIS	40.9	3.1	3.9	23.3	44 418	76 394	51 485
CIS	33.2	7.8	33.7	7.1	-2 337	-3 228	-2 320

Source: CIS Statistical Committee (Moscow); UNECE estimates for Turkmenistan and Uzbekistan.

[a] January-September.

8 per cent in volume and by 3 per cent in value. These increases were also stimulated by the appreciation of the Russian currency in real terms (see chart 3.1.2).

The high, albeit decelerating, rate of economic growth in Russia most benefited some of the smallest CIS economies such as Armenia and the Republic of Moldova.[356] Ukraine's exports to the CIS increased in value by 15 per cent largely due to its booming exports to Russia and despite a third quarter decline of almost one third due to the imposition of value added tax on Ukrainian products sold to Russia.[357] The rapid growth of the Kazakh economy continued to require rising imports of natural gas, pipes and steel products as well as more oil products and machinery and equipment. Kazakhstan imported about $2.5 billion worth of goods from other CIS countries – almost one third more than a year before – due to domestic supply and infrastructure constraints.[358]

While buoyant economic activity in Russia and the strong rouble benefited the other regional trade partners in the first two quarters of 2001, virtually all CIS countries shipped fewer goods to Russia in the third quarter. In that quarter, the overall volume of Russian imports from the CIS fell by 15 per cent on average despite a 15 per cent fall in import prices. This outcome was partly related to Russia's imposition of value added tax on all non-hydrocarbon exports to and imports from CIS countries (except Belarus) on the basis of country of destination (i.e. exports are zero-rated while imports are taxable).[359]

The volume and value of Russia's exports to CIS countries in the first nine months of 2001 increased by 7 per cent. Russia shipped more crude oil and refined products but less natural gas. Natural gas is the most important intra-CIS traded commodity and its sales were quite volatile in 2001 – decreasing in the first half by over 40 per cent, but recovering strongly in the third quarter as a result of a doubling of sales to Ukraine following the resolution of debt problems between the two countries.[360] As a result, Ukrainian imports from Russia in the third quarter of 2001 rose 54 per cent. Russia's machinery and equipment sales to CIS countries increased by 15 per cent in dollars, reflecting a more favourable tax treatment of exports and high rates of economic growth throughout the CIS region.

(iii) Trade specialization by stage of production in eastern Europe and the Baltic states, 1996-2000

(a) Strengthening position in world trade

The growth of the east European and Baltic countries' foreign trade, broken only by a slump in the wake of the Russian crisis in 1998, continues the trend observed since the beginning of transition: the region's

---

[356] Both countries increased their CIS exports by a third on the strength of much higher food and beverage exports to Russia.

[357] While Ukrainian exports to Russia fell by 27 per cent in the third quarter, the country's second quarter exports rose 55 per cent as exporters anticipated the tax change.

[358] For example, the country – abundant in oil and gas – imports annually $100 million of natural gas from Uzbekistan because of a lack of pipelines connecting producers with customers. It is estimated that producers burn about 14 billion cubic meters of excess gas each year (extracted as a by-product of crude oil production) or five times the country's imports.

[359] As other CIS countries have done the same, this move levels the playing field for Russian exporters and import-competing domestic producers whose exports were taxed twice, while imports from the CIS area were not taxed at all. In addition to changing its tax regime, Russia has also continued to reform its customs system with a view to facilitating the country's entry into the WTO and reducing corruption.

[360] In October 2001, the two countries signed an agreement on restructuring of the $1.4 billion natural gas debt. Russia withdrew demand for debt-equity swaps and agreed on a 12-year repayment schedule. As a result, Neftegaz Ukrainy will issue eurobonds for that amount maturing between 2004 and 2013.

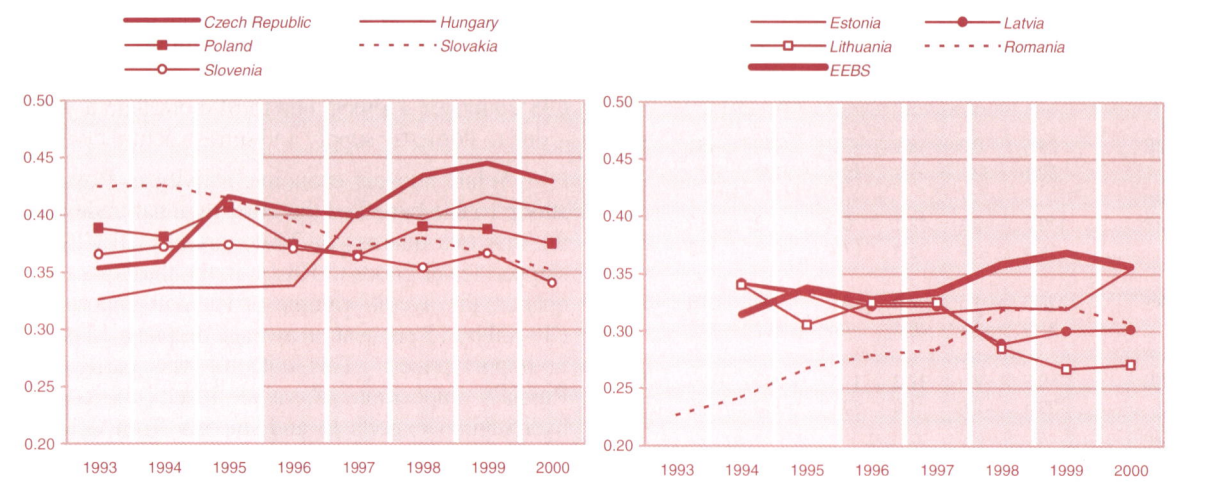

CHART 3.5.4

Geographical concentration of east European and Baltic countries' exports, 1993-2000

Source: United Nations COMTRADE Database.

Note: UNECE secretariat calculations. The indicator is the Hirschmann concentration index, which varies from 0 to 1 (maximum concentration), that is,

$$H_k^{geo} = \sqrt{\sum_{k'=1}^{244} \left(\frac{x_{k'}}{X}\right)^2}$$

where k = reporting country, k' = partner country and x = exports.

exports and imports have been rising in line with overall world trade, but at a somewhat faster pace. Although the east European and Baltic economies play only a marginal role in world trade, their share has increased noticeably in the past five years. With aggregate export and import values amounting to some $142 billion and $186 billion in 2000, eastern Europe (including the Baltic states) accounted for 2.25 and 3.05 per cent of world exports and imports, respectively, a gain of about a quarter of a percentage point for both since 1996.[361] Trade with western Europe, and in particular with the EU, has been even more vigorous: in 2000, 15 east European and Baltic countries accounted for 9.95 per cent of total extra-EU imports and for 13.3 per cent of extra-EU exports, up, respectively, by 1 and 2 percentage points since 1996.[362] This strengthening position of the east European and Baltic countries in world trade, and in particular on west European markets, reflects the ongoing integration of their industries into international production networks, often driven by MNCs.

Although the high initial rate of expan5sion of trade with the European Union (at an average annual growth rate of 20 per cent in dollar values in 1993-1996) has now subsided somewhat, the deepening of trade ties with the EU has continued, with an average annual growth rate of 10 per cent in 1996-2000. In 2000, EU partners accounted for two thirds of east European and Baltic exports and for 59 per cent of imports. The share of trade within the east European region, which had shrunk considerably in the wake of the CMEA collapse, was re-established by the end of 1996, but since then has levelled out, with growth averaging 3.3 per cent annually. Trade with the CIS partners also picked up strongly in 1996-1997, to account for 11 to 12 per cent of the region's total trade, but it was badly affected by the Russian crisis in autumn 1998, which led to a fall in east European and Baltic exports to these markets of 22 and 44 per cent in 1998 and 1999, respectively. Imports from the CIS suffered less, however, due to the large share of primary goods and partly to escalating oil prices in 1999-2000. Penetration of markets in other parts of the world, however, has so far remained rather marginal (for longer-term trends in the geographical distribution of trade see appendix table B.15). Furthermore, the growth of east European and Baltic trade in 1996-2000 was accompanied by a rising degree of partner concentration, which reflected not only the dominance of the European Union as a partner group, but also of a few individual countries within the EU. The highest degree of geographical concentration in exports during this period was reached in 1999 when, in the wake of the Russian crisis, east European and Baltic exporters partly withdrew from CIS markets, but were often unable to quickly divert their goods to other markets (chart 3.5.4).

---

[361] The increase is greater for 1992-1996: from 1.6 per cent of world exports and 1.8 per cent of world imports in 1992 to 2 and 2.7 per cent, respectively, in 1996. UNECE secretariat computations, based on data published in UNCTAD, *Handbook of Trade and Development Statistics 1995* and *Handbook of Statistics 2001* (United Nations publications, Sales Nos. E/F.97.II.D.7 and E.01.II.D.24, respectively).

[362] UNECE secretariat computations, based on data from Eurostat, CD-ROM Theme 6: External Trade, *Intra- and Extra-EU Trade*, Monthly Data and Supplement 2, 2001.

(b) Changes in commodity composition until 1996

A larger share of world trade and the above-mentioned changes in the geographical distribution of trade can be expected to be accompanied by changes in the commodity pattern of trade. However, as shown in earlier UNECE studies, until 1995/1996 changes in the composition of exports and imports of the east European and Baltic countries were rather limited,[363] although quite a few new export products had appeared by the end of that period.[364] By early 1996, the commodity concentration of exports, in general, had diminished somewhat, although exports of industrial processed supplies and of semi-durable consumer goods (largely clothing) were dominant: jointly they accounted for nearly half of total exports from the east European and Baltic countries (charts 3.5.5 and 3.5.6). At the same time, the export mix changed more noticeably only in Romania and Slovakia and in two of the Baltic states (chart 3.5.7).[365] Moreover, until 1996 most of these countries were specialized in a limited number of products that originated mainly in resource-intensive and labour-intensive sectors,[366] although there were important differences among countries in their trade patterns and particularly in the share of the CIS (table 3.5.7). It should be stressed, however, that these relatively moderate shifts are not incompatible with significant changes in the relative importance of individual commodities.

---

[363] UNECE, *Economic Survey of Europe in 1994-1995*, chap. 3.5(ii), *Economic Bulletin for Europe*, Vol. 49 (1997), chap. 2.1(vi), and *Economic Survey of Europe, 1998 No. 1* and *1998 No. 3*, chaps. 3.6(c) and 3.2(iv), respectively. Similar conclusions were drawn also in L. Halpern, "Comparative advantage and the likely trade pattern of the CEECs", in R. Faini and R. Portes (eds.), *EU Trade with Eastern Europe: Adjustment and Opportunities* (London, Center for Economic Policy Research, 1995); B. Hoekman and S. Djankov, "Determinants of the export structure of countries in central and eastern Europe", *The World Bank Economic Review*, Vol. 11, No. 3 (Washington, D.C.), 1997; M. Freudenberg and F. Lemoine, *Central and Eastern European Countries in the International Division of Labour in Europe*, CEPII Working Paper, No. 1999-05 (Paris) April 1999; and in several other studies covering that period.

[364] The observations above are based not only on the 19 BEC categories, but also on data at the 2-digit or 3-digit levels of the SITC (SITC Rev.2 and Rev.3) in the charts and in previous studies mentioned above. Although they remain specialized in a limited number of products, the worldwide exports of east European and Baltic countries are somewhat more diversified if the *number* of exported products is taken into account. In 1996, for example, the Czech Republic reported worldwide exports of some 2,086 manufactured products among the 2,583 on the SITC Rev.3 5-digit list, while in 1993 the number had been 2,056 products. The number was less impressive in other countries, varying from about 1,550 in Hungary and Slovenia to as low as 280 in Latvia, but the change from 1993 was in many cases much longer.

[365] There were similar trends in Bulgaria and Croatia. UNECE, *Economic Bulletin for Europe*, Vol. 49 (1997), chap. 2.1(vi).

[366] In a previous study of the transition economies' export mix with respect to factor intensity (UNECE, *Economic Survey of Europe, 1998 No. 3*) it was shown that during the 1993-1996 period, labour-intensive and resource-intensive products (according to the Krause classification) were more important than the technology-intensive group in the region's exports, with the latter accounting for 15 to 33 per cent of total exports to western Europe.

CHART 3.5.5

Commodity structure of east European and Baltic countries' exports to the world, 1996-2000

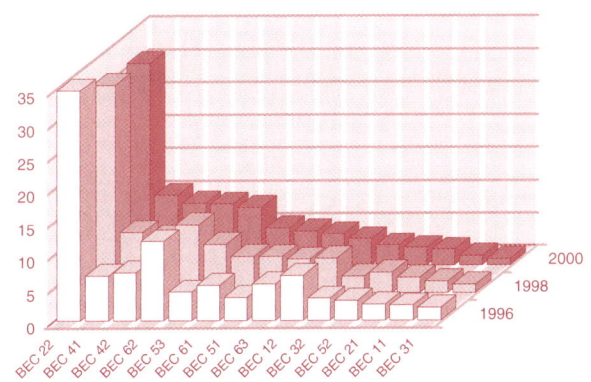

Source: United Nations COMTRADE Database, concordance from SITC to BEC by UNECE Statistics Division. United Nations Statistical Papers, *Classification by Broad Economic Categories Defined in Terms of SITC Rev.3* (United Nations publication, Sales No. E.89.XVII.4).

Note: Commodity classification by broad economic categories (BEC) as follows :

BEC 11: food and beverages, primary (BEC 111: mainly for industry; BEC 112: mainly for household consumption);

BEC 12: food and beverages, processed (BEC 121: mainly for industry; BEC 122: mainly for household consumption);

BEC 21: industrial supplies n.e.s., primary;

BEC 22: industrial supplies n.e.s., processed;

BEC 31: fuels and lubricants, primary;

BEC 32: fuels and lubricants, processed (BEC 321: motor spirits; BEC 322: other);

BEC 41: capital goods (except transport equipment);

BEC 42: capital goods, parts and accessories;

BEC 51: transport equipment and parts, passenger motor cars

BEC 52: transport equipment and parts, other (BEC 521: industrial; BEC 522: non-industrial);

BEC 53: transport equipment and parts, parts and accessories

BEC 61: consumer goods n.e.s., durable

BEC 62: consumer goods n.e.s., semi-durable

BEC 63: consumer goods n.e.s., non-durable

BEC 7: Goods n.e.s.

Thus:

Primary commodities are defined as: BEC 111, 112, 21, 31;

Intermediate goods are defined as: BEC 121, 22, 321, 322, 42, 53;

Capital goods are defined as: BEC 41, 51, 521, 522;

Consumer goods are defined as: BEC 122, 61, 62, 63.

(c) Evolving specialization patterns in 1996-2000

During the period 1996-2000, the specialization patterns of the east European and Baltic countries have become more pronounced, in particular in those countries with a strong foreign capital presence, where the intrafirm trade flows of resident MNCs and outsourcing

CHART 3.5.6

Commodity concentration of east European and Baltic countries' exports, 1993-2000

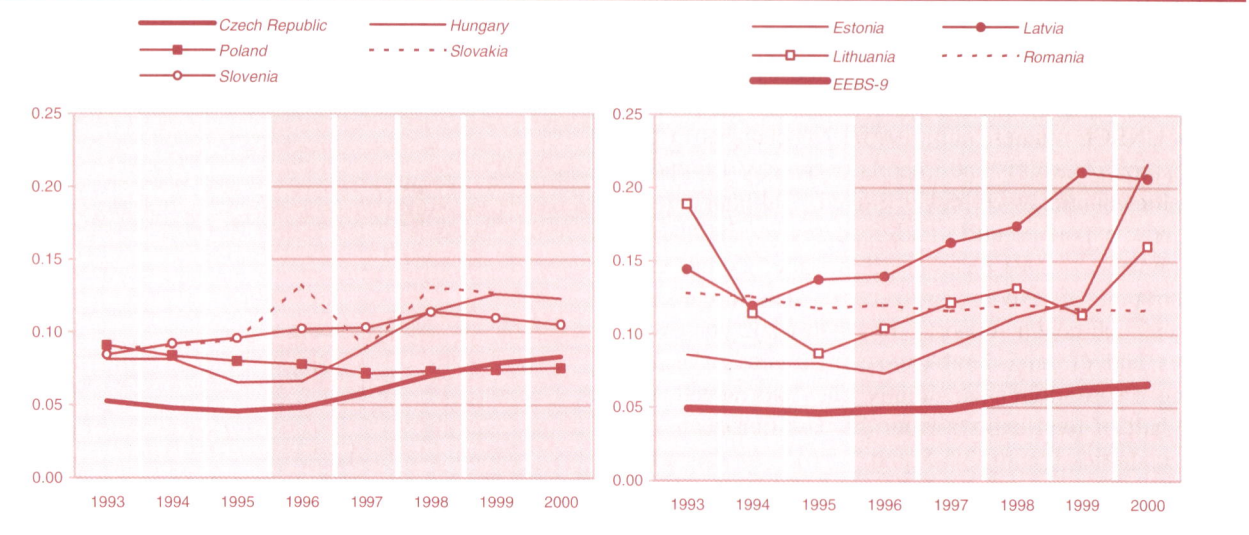

Source: United Nations COMTRADE Database (UNCTAD CD-ROM).

Note: SITC Rev.2 3-digit level, 239 commodities. Concentration is measured by the Hirschmann index normalized to produce values ranking from 0 to 1 (maximum concentration), according to the following formula:

$$H_k^c = \frac{\left(\sqrt{\sum_{i=1}^{239}\left(\frac{x_i}{X}\right)^2}\right) - \sqrt{1/239}}{1 - \sqrt{1/239}}$$

where k = reporting country, i = commodity and x = exports.

by western (and sometimes eastern) companies often have a dominant influence. The fragmentation of industrial processes, which has a major impact on international trade,[367] is not a new phenomenon in the east European and Baltic countries. In 1990-1997 these countries participated to a considerable extent (and with varying degrees of success) in EU outward processing trade in clothing and footwear, but its rising intensity and the larger number of industries involved give it a more critical role in the evolving specialization patterns in the transition economies.

East European and Baltic countries were strongly engaged in outward processing trade (OPT) with the EU during the initial stages of transition. However, this mainly involved unskilled labour and little foreign capital; hence it was rather volatile and highly sensitive to changes in relative labour costs, and moved on to south-east Europe and CIS after labour costs increased in eastern Europe and the Baltics.[368] Integration into the international production networks of the engineering and automotive industries, which, being more capital-intensive, tends to be locationally more stable, became more important in the mid-1990s when, as a result of privatization and greenfield investments, MNCs and other foreign firms acquired many engineering companies in the region and improved the entire supply chain in sectors such as automobiles and their components, precision engineering, tool-making and electronics, thus enhancing their productive and export capacity.

In fact, this shift from east European and Baltic participation in the EU's OPT in clothing, to engagement in the assembly of cars and electronic equipment and/or production of parts and accessories, is well exemplified in chart 3.5.5 (BEC 62, 41, 42, 51, 53). The significant change in the composition of exports since 1996 in many of these countries (chart 3.5.7), which was generally

---

[367] There is an increasing interest in the economic literature in this phenomenon particularly since Feenstra, among others, drew attention to the fact that currently the largest share of world trade is in intermediate goods, and that final consumer goods sold in one country are often the assembly of components processed in many locations. R. Feenstra, "Integration of trade and disintegration of production in the global economy", *Journal of Economic Perspectives*, No. 12, 1998. See also D. Hummels et al., "Vertical specialization and the changing nature of world trade", Federal Reserve Bank of New York, *Economic Policy Review*, No. 4, 1998 and A. Venables, "Fragmentation and multinational production", *European Economic Review*, Vol. 43, No. 4-6, April 1999.

[368] See "Outward processing trade between the EU and the associated countries of eastern Europe: the case of textiles and clothing", in UNECE, *Economic Bulletin for Europe*, Vol. 47 (1995), chap. 5, and the chapters on foreign trade in the 1996-1998 issues of this *Survey*. For econometric evidence that labour cost, along with geographical and cultural proximity, were the most important reasons for the original choice of processing partners, with neither pre-existing comparative advantages of any transition economy nor FDI playing an important role, see S. Baldone et al., "Patterns and determinants of international fragmentation of production: evidence from outward processing trade between the EU and central eastern European countries", *Weltwirtschaftliches Archiv*, Vol. 137, No. 1 (Kiel), March 2001.

The Transition Economies

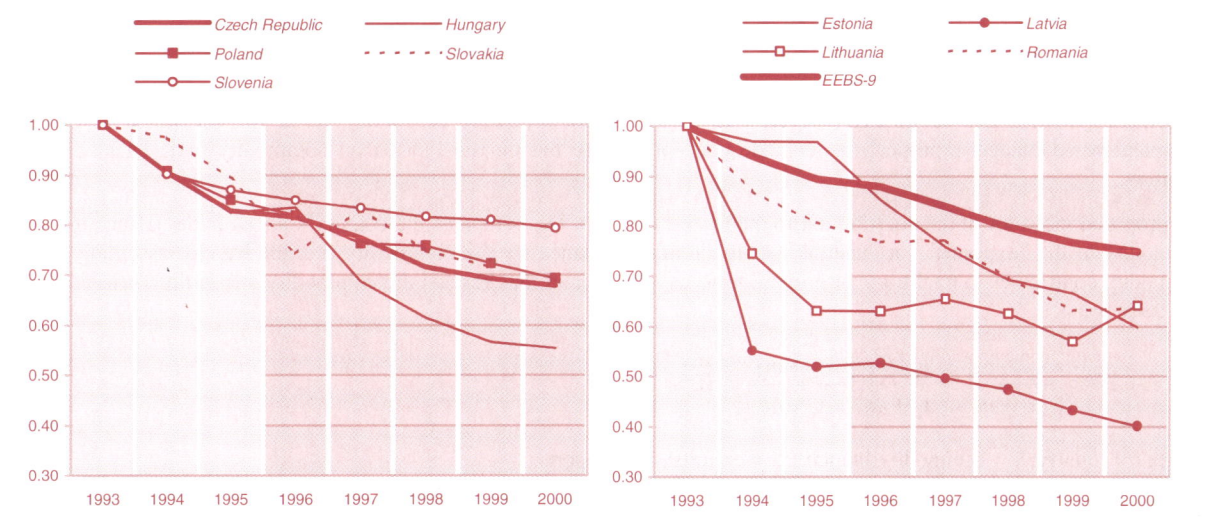

CHART 3.5.7

Changes in the structures of exports [a] **from eastern Europe and the Baltic states, 1993-2000**

Source: United Nations COMTRADE Database (UNCTAD CD-ROM).

Note: SITC Rev.2 3-digit level, 239 commodities.

[a] Structural change is measured with the Finger-Kreinin similarity index according to the following formula: $XS(a,b) = \mathring{a} min.(X_{ia}, X_{ib}).100$, where $XS(a,b)$ is the similarity index between two export vectors, a and b, X_{ia} is the share of commodity i (i = 239 products defined at the 3-digit level of the SITC) in the total exports of a given country in 1993 and X_{ib} is the same share in a subsequent year. It is assumed that the lower index value (higher dissimilarity) points to more serious structural change in the country under consideration. In the present exercise, the lower the value of the index between two years the more dissimilar are the two vectors and this is interpreted as a measure of structural change. The values of the index range between one (identical structures) and zero (completely dissimilar structures).

accompanied by a less important rise in the degree of commodity concentration (chart 3.5.6, except Estonia and Hungary), probably stems for the most part from this sectoral shift in the integration into the international production networks.

These developments in the concentration and commodity mix of exports point to important changes in the specialization patterns of east European and Baltic countries over the last five-year period (1996-2000).[369] To identify these changes, the indicators of revealed comparative advantage for total trade and in respect of the three major partner groups (EU, CEFTA-7 and CIS), as well as measures of vertical and horizontal two-way trade, including intra-industry trade, of the east European and Baltic countries[370] are employed below.

A first step in determining the patterns of specialization is to identify the stages of the production process in which these countries are specialized. For this purpose, product data at the 5-digit level of the SITC Rev.3[371] – at which level most calculations for this study were made – were reclassified into United Nations broad economic categories (BEC),[372] which classifies products according to their end use. The broad classifications adopted are: *primary goods*; *intermediate goods*; *capital goods* (including transport equipment); and *consumer goods*. Goods not elsewhere classified (BEC 7) are not generally taken into account, although in a few countries in some years the share of such goods in the total was not negligible.

Revealed comparative advantage (RCA) is measured by the indicator of contribution to the trade balance (CTB) of the individual country and in trade with its major partner groups proposed by Freudenberg and Lemoine[373] (box 3.5.1).

---

[369] The Freudenberg and Lemoine study (op. cit.), along with papers on the topic of structural change in east European trade by C. Aturupane, S. Djankov, B. Hoekman, B. Kaminski and M. Landesmann are the most comprehensive empirical works to date, and the analysis there will draw on some of their findings and methodological approaches. However, most of these studies concentrate almost exclusively on trade between the EU and the east European countries (only a few include the Baltic states), use data reported by EU countries and cover only the period until 1996/1997.

[370] The analysis below is based on detailed foreign trade data reported to the United Nations COMTRADE Database by nine east European and Baltic countries (EEBS-9: the Czech Republic, Hungary, Poland, Romania, Slovakia, Slovenia, Estonia, Latvia and Lithuania). Data on Bulgaria's foreign trade are available in the necessary detail only for 1996 and 1997; hence Bulgaria is not included in this group of countries. The rationale for each indicator will be discussed below or referred to a source in the notes to tables and charts. The determinants of the change in specialization patterns are not the topic of this note.

[371] The lowest available disaggregation level of the SITC is 5-digits. The United Nations COMTRADE Database reports 3,329 products at five-digits, plus 288 products at 4-digits, which are not further disaggregated. For the chosen sample of countries (reporters), 5-digit products cover from 70 to 80 per cent of their total trade flows in each year, while with the inclusion of the non-disaggregated 4-digit products all trade is covered.

[372] The results reported in this note reflect the state of the United Nations COMTRADE Database as of 31 January 2002 (CD-ROM). Statistical support in extracting the data was provided by the UNCTAD secretariat.

[373] M. Freudenberg and F. Lemoine, op. cit., p. 80.

> **BOX 3.5.1**
>
> **Indicator of revealed comparative advantage**
>
> There are several revealed comparative advantage (RCA) indicators used by trade economists based on the shares of exports, imports, or net trade of individual commodities/branches for a given country with reference to the world/partner group trade.[1] Freudenberg and Lemoine, however, argued that as imports have become an important factor in the explanation of export performance in transition economies, and with intra-industry trade already accounting for a sizeable share of total trade, "comparative advantage is properly a net trade concept", hence the measure based on commodity trade balances is more revealing.
>
> The proposed indicator of the contribution to the trade balance *(CTB)* compares the country's actual trade balance for a given commodity to the "expected" (or "neutral" or "theoretical") balance for the commodity (i.e. the balance assuming that each commodity contributes to total trade in proportion to its weight) and thus reveals the commodity's specific contribution to the total trade balance:
>
> $$CTB = \left[\frac{x_{ik} - m_{ik}}{x_i + m_i} - \frac{x_i - m_i}{x_i + m_i} * \frac{x_{ik} + m_{ik}}{x_i + m_i} \right] * 1000$$
>
> where: i = country; k = commodity(branch); x = exports; m = imports.
>
> A positive contribution is interpreted as a "revealed comparative advantage" (RCA) for trade in that commodity, and a negative one as a "revealed comparative disadvantage" (RCD). By definition, the sum of *CTBs* over all commodities is zero. The indicator is additive, thus the values of commodities can be aggregated to the necessary level.
>
> Interpreting changes in the *CTB* indicator over time a more intense specialization is indicated when RCA or RCD becomes more pronounced, and reduced specialization is present when RCA or RCD declines.
>
> ---
>
> [1] B. Balassa, "Trade liberalization and 'revealed' comparative advantage", *The Manchester School of Economic and Political Studies,* Vol. 33, No. 2, 1965, and "Comparative advantage in manufactured goods: a reappraisal", *Review of Economics and Statistics,* Vol. 68, Issue 2, 1987; K. Abd-El-Rahman, "Firms' competitive and national comparative advantages as joint determinants of trade composition", *Weltwirtschaftliches Archiv,* No. 1 (Kiel), 1991; F. Ng and A. Yeats, *Production Sharing in East Asia: Who Does What for Whom and Why,* World Bank Policy Research Working Paper, No. 2197 (Washington, D.C.), October 1999; M. Freudenberg and F. Lemoine, *Central and Eastern European Countries in the International Division of Labour in Europe,* CEPII Working Paper, No. 1999-05 (Paris), April 1999; B. Algieri, S. Ankurinniemi and L. Zampieri, "Inter-industry specialization versus intra-industry trade: a regional approach", *Economia Internazionale* (Genova), August 2001.

As shown in table 3.5.6, intermediate goods comprise the largest part of trade for each country in the region, accounting on average for 52 per cent of total trade and for nearly 55 per cent of trade with the EU. The rising share of these goods in the region's total trade between 1996 and 2000 suggests an intensified involvement of several of these countries in the international fragmentation of production (and specifically in transport equipment, machinery and electric engineering).[374] The more or less parallel increase in the share of intermediate goods in total imports and of capital goods in total exports (particularly in the Czech Republic, Hungary, Slovakia and Estonia) suggests the same process, since the most labour-intensive part of the production process in these industries is often final assembly; thus, the initial incentive is for MNCs to relocate this part of production to the east European and Baltic countries (which combine a cheap but skilled labour force with geographical proximity to western Europe).[375] It is also worth noting that in Latvia and Lithuania, two countries that have only lately attracted sizeable inflows of FDI, the share of imports of intermediate goods in total trade was the lowest in the region, while on the export side these goods have picked up rather strongly and together with primary goods accounted for 63 to 70 per cent of the total in 2000. Primary goods exports still hold a large share of the total trade in Estonia as well.

The shift in favour of capital goods on the export side has mainly taken place at the expense of consumer goods (mainly semi-durables). The share of the latter has declined in all countries except Romania. However, the contribution to the trade balance (table 3.5.7) shows clearly that the region's main comparative advantage in 2000 remained in consumer goods: in all countries there was a positive contribution of these goods to the total

---

[374] The share of parts and accessories of capital goods and transport equipment (BEC 42 and 53), for example, rose from 12 to 18 per cent of total trade (exports and imports) between 1996 and 2000, and from 14 to 21 per cent, in trade with the EU.

[375] For instance, in Hungary, the share of BECs 42 and 53 in total imports rose from 11 per cent in 1996 to 33 per cent in 2000, while the share in exports of capital goods (including transport equipment) increased from 8 per cent to 25 per cent, respectively. A slightly increased share of exports of durable consumer goods (BEC 61) may also be partly attributed to international production fragmentation (TV sets, household electric appliances, furniture, etc.).

TABLE 3.5.6

Composition of east European and Baltic trade by stages of production, 1996-2000
(Shares in per cent, changes of structure in percentage points)

Partner group	Exports								
	Czech Republic	Estonia	Hungary	Latvia	Lithuania	Poland	Romania	Slovakia[a]	Slovenia
World									
Structure in 2000									
Primary commodities	3.7	11.3	3.7	13.1	7.9	6.0	6.3	3.3	1.7
Intermediate goods	59.5	45.3	51.4	57.0	55.2	49.3	46.5	55.3	51.5
Capital goods	21.9	24.0	25.2	5.4	9.0	15.9	8.0	23.8	19.6
of which: motor vehicles	8.5	1.8	5.2	0.1	2.7	4.6	0.4	13.9	8.6
Consumer goods	14.9	19.4	19.8	24.4	28.0	28.8	39.2	17.6	27.2
Changes 1996-2000									
Primary commodities	-2.3	1.1	-4.3	5.3	0.3	-3.0	0.3	-0.2	0.1
Intermediate goods	0.7	-4.9	3.2	8.5	2.5	5.1	-3.3	-7.3	3.5
Capital goods	5.4	15.3	17.1	-4.0	-1.3	1.8	1.1	9.2	0.1
of which: motor vehicles	4.2	-1.1	4.7	-0.1	-2.0	1.4	-0.3	10.2	-0.9
Consumer goods	-3.7	-11.5	-16.1	-9.9	-1.5	-4.0	1.9	-1.7	-3.7
European Union									
Structure in 2000									
Primary commodities	3.3	10.6	2.9	18.1	9.2	6.0	4.5	2.5	1.6
Intermediate goods	61.2	41.1	54.0	55.4	46.5	50.7	32.4	47.9	54.4
Capital goods	22.1	29.7	24.7	2.7	6.6	15.2	8.2	32.1	22.5
of which: motor vehicles	9.5	0.6	6.4	–	0.7	5.8	–	21.3	11.7
Consumer goods	13.4	18.6	18.4	23.8	37.7	28.1	54.9	17.4	21.4
Changes 1996-2000									
Primary commodities	-3.4	-3.5	-5.4	4.4	-1.3	-2.7	2.2	-0.7	0.1
Intermediate goods	-0.9	-8.5	4.3	–	-7.1	5.8	-3.5	-13.4	7.7
Capital goods	7.2	23.7	17.6	-1.3	2.7	1.2	3.2	15.3	-1.9
of which: motor vehicles	5.0	-1.0	6.4	–	0.6	0.9	-0.3	14.5	-3.0
Consumer goods	-2.9	-11.6	-16.5	-3.1	5.7	-4.3	-1.9	-1.1	-5.9
CEFTA-7									
Structure in 2000									
Primary commodities	6.8	5.2	8.5	6.1	16.8	8.1	8.4	4.5	0.4
Intermediate goods	60.7	55.7	56.8	77.2	68.4	54.1	70.7	70.2	49.0
Capital goods	14.3	22.0	8.3	6.7	4.9	10.1	4.7	7.3	13.2
of which: motor vehicles	5.6	0.1	0.7	0.1	0.2	5.0	1.5	0.8	7.9
Consumer goods	18.2	17.1	26.3	10.0	10.0	27.7	16.1	18.0	37.4
Changes 1996-2000									
Primary commodities	-1.5	1.1	-1.3	-0.5	-5.6	-11.5	-8.7	0.4	-0.4
Intermediate goods	1.5	-12.6	4.8	11.7	4.4	1.4	12.3	2.6	0.2
Capital goods	-0.4	19.5	1.3	-4.4	0.6	2.4	-9.5	-2.3	5.9
of which: motor vehicles	1.8	–	0.3	–	-0.2	4.7	-5.5	0.6	7.9
Consumer goods	0.4	-7.9	-4.8	-6.8	0.5	7.6	5.9	-0.7	-5.7

(For source and notes see end of table.)

trade balance. Within consumer goods, revealed comparative advantage is located mainly in the subcategories of durable and semi-durable consumer goods,[376] while processed foods and beverages contributed positively to the trade balance of 2000 only in Hungary, Poland and Lithuania.

In intermediate goods, the east European and Baltic countries reveal an overall comparative disadvantage.

However, three subcategories of intermediate goods possess a comparative advantage: it is very slight in the case of processed foods and beverages for industrial use and of processed fuels and lubricants (both in traditional resource-intensive sectors), while more pronounced – although quite recently acquired – in the case of parts and accessories for transport equipment.

In 2000, four countries in the sample – the Czech Republic, Hungary, Slovakia and Estonia – are shown to have a comparative advantage in the export of capital goods. Capital goods (except transport equipment) contributed positively to Estonia's trade balance for the first time in 2000. In the Czech Republic, Slovakia and Hungary, their comparative advantage in capital goods had already emerged by 1997/1998, although in the first two countries it was entirely due to the trade in passenger

---

[376] Based on the CTB indicator, in 1996-2000, specialization in durable consumer goods increased markedly in the Czech Republic, Hungary, Poland and Estonia, whereas their specialization in semi-durables became less intensive. In Romania, and to a lesser extent in Lithuania and Latvia, CTB of semi-durable consumer goods increased considerably, due partly to the later involvement in the EU's OPT, while in Slovenia and Slovakia changes were very moderate.

TABLE 3.5.6 (concluded)

Composition of east European and Baltic trade by stages of production, 1996-2000
(Shares in per cent, changes of structure in percentage points)

Partner group	Imports								
	Czech Republic	Estonia	Hungary	Latvia	Lithuania	Poland	Romania	Slovakia [a]	Slovenia
World									
Structure in 2000									
Primary commodities	10.4	7.5	5.1	9.1	26.7	13.4	15.4	15.9	6.7
Intermediate goods	54.5	53.6	62.9	46.1	38.8	49.8	53.8	50.4	56.0
Capital goods	20.7	21.9	19.1	20.5	17.4	22.5	16.7	18.5	21.3
of which: motor vehicles	2.0	4.1	2.4	3.2	4.1	3.5	0.9	3.3	4.8
Consumer goods	14.5	17.0	12.9	24.3	17.2	14.3	14.1	15.2	16.0
Changes 1996-2000									
Primary commodities	-1.8	-0.1	-9.3	-0.8	6.0	-1.6	-7.5	-2.0	-1.0
Intermediate goods	5.0	3.4	11.4	-8.4	-3.8	–	4.9	7.6	2.5
Capital goods	-1.2	5.5	-0.4	5.0	-1.3	1.9	1.0	-5.4	0.1
of which: motor vehicles	-1.0	1.0	-0.4	1.5	-1.9	1.0	0.2	-4.5	-1.9
Consumer goods	-2.1	-8.8	-1.6	4.2	-0.9	-0.3	1.7	-0.2	-1.7
European Union									
Structure in 2000									
Primary commodities	2.0	2.9	1.8	4.9	7.8	3.2	1.8	2.2	3.3
Intermediate goods	61.9	53.4	65.3	42.9	46.7	57.4	62.6	62.3	57.4
Capital goods	22.6	24.9	20.5	28.8	27.1	25.3	19.8	22.1	23.2
of which: motor vehicles	2.5	4.2	2.8	4.2	7.5	4.9	0.9	2.3	5.2
Consumer goods	13.5	18.8	12.4	23.4	18.4	14.0	15.8	13.3	16.1
Changes 1996-2000									
Primary commodities	-1.2	0.1	-0.5	-0.3	3.3	-2.9	-0.3	-1.1	-0.1
Intermediate goods	6.7	1.4	8.3	-3.6	1.9	1.9	3.5	15.6	2.6
Capital goods	-2.3	6.8	-3.6	4.8	-2.8	0.6	-3.4	-13.3	0.3
of which: motor vehicles	-0.9	2.1	-0.7	1.8	-2.2	1.3	0.4	-6.7	-2.4
Consumer goods	-3.2	-8.3	-4.2	-0.9	-2.4	0.4	0.2	-1.1	-2.8
CEFTA-7									
Structure in 2000									
Primary commodities	5.3	4.2	3.6	4.6	3.3	5.4	7.2	9.5	8.9
Intermediate goods	67.4	45.0	64.0	38.1	42.2	63.5	57.4	50.0	63.7
Capital goods	5.4	9.0	11.9	13.9	13.4	10.5	8.7	15.5	11.3
of which: motor vehicles	0.9	1.5	5.4	2.6	2.6	2.1	1.1	7.4	6.8
Consumer goods	21.9	41.8	20.4	43.4	41.1	20.5	26.7	24.9	16.1
Changes 1996-2000									
Primary commodities	-1.9	1.0	-8.1	-1.9	-0.4	-2.7	–	-0.9	-1.7
Intermediate goods	4.3	9.1	-5.6	-0.2	-0.9	1.1	-5.5	-2.6	-1.6
Capital goods	-3.5	0.1	5.0	2.7	-1.9	2.6	-1.9	0.1	2.0
of which: motor vehicles	0.8	-0.6	4.4	1.7	0.5	1.9	0.7	1.8	3.3
Consumer goods	1.2	-10.1	8.7	-0.6	3.1	-1.0	7.3	3.4	1.3

Source: United Nations COMTRADE Database, concordance from SITC to BEC by UNECE Statistics Division. United Nations Statistical Papers, *Classification by Broad Economic Categories Defined in Terms of SITC Rev. 3* (United Nations publication, Sales No. E.89.XVII.4).

Notes: For the description of BEC categories, see chart 3.5.5. BEC 7 – goods not elsewhere specified – are excluded.

[a] 1999 instead of 2000.

motorcars while in Hungary, capital goods (except transport equipment) contributed as well. In fact, in 2000, passenger cars were making a positive contribution to the balance in total trade, and in trade with major partner groups, in all the east European countries (except Romania) but not in the Baltic states. Overall, however, most of the east European and Baltic countries, and the region as a whole, remained at a comparative disadvantage in the export of capital goods.

Although the pattern of comparative advantage of the east European and Baltic countries as a group did not change radically in this five-year period, the shift from structural deficit to structural surplus in several cases testify to the success of some of these countries in acquiring new comparative advantages. What is perhaps more important, these comparative advantages are often located in capital- and technology-intensive sectors. Also a shift from RCA to RCD in intermediate goods for the region as a whole is quite indicative of the underlying change in the pattern of comparative advantage.

There was also a general trend towards a reduction in the intensity of specialization as measured by the relative sectoral contributions to trade balances from 1996 to 2000. Relative sectoral surpluses and deficits

TABLE 3.5.7

Revealed comparative advantage by stage of production, 1996 and 2000

(Measured as deviation between actual and neutral trade balance)

	1996				2000			
	Primary commodities	Intermediate goods	Capital goods	Consumer goods	Primary commodities	Intermediate goods	Capital goods	Consumer goods
Trade with the world								
Czech Republic	-30	46	-26	10	-33	25	6	2
Estonia	12	–	-37	24	19	-41	11	12
Hungary	-32	-17	-56	105	-7	-58	31	34
Latvia	-10	-28	-29	67	18	51	-70	–
Lithuania	-64	50	-42	56	-91	80	-41	52
Poland	-29	-27	-31	87	-35	-2	-31	69
Romania	-82	5	-43	121	-44	-36	-43	124
Slovakia [a]	-71	98	-46	19	-63	25	26	12
Slovenia	-31	-27	-8	66	-25	-22	-8	56
EEBS-9	-39	9	-34	64	-30	-8	-7	44
Trade with EU-15								
Czech Republic	17	34	-49	-2	6	-4	-2	–
Estonia	51	-11	-54	14	39	-61	24	-1
Hungary	30	-36	-84	91	5	-56	21	30
Latvia	39	40	-92	12	65	61	-127	2
Lithuania	27	41	-120	52	7	-1	-100	94
Poland	13	-51	-51	90	14	-33	-50	69
Romania	1	-114	-89	203	13	-150	-58	195
Slovakia [a]	-1	73	-92	20	1	-72	50	20
Slovenia	-9	-40	7	42	-8	-15	-3	26
EEBS-9	13	-21	-57	65	9	-36	-19	46
Trade with CEFTA-7								
Czech Republic	6	-19	28	-14	7	-33	44	-18
Estonia	4	127	-25	-105	4	40	49	-93
Hungary	-10	-88	–	97	24	-36	-18	29
Latvia	–	98	–	-98	4	119	-22	-102
Lithuania	71	80	-42	-109	60	118	-38	-140
Poland	55	-47	-1	-7	13	-46	-2	35
Romania	47	-21	17	-42	6	65	-20	-51
Slovakia [a]	-31	75	-29	-15	-25	100	-41	-35
Slovenia	-48	-81	-10	139	-41	-72	9	104
EEBS-9	1	-4	3	-1	10	-19	10	-1
Trade with CIS countries								
Czech Republic	-355	109	57	190	-250	88	94	69
Estonia	-91	-128	44	175	-13	-77	39	52
Hungary	-301	-12	64	248	-195	-31	43	183
Latvia	-104	-181	67	218	-90	-39	32	97
Lithuania	-240	26	67	147	-247	109	64	73
Poland	-309	-10	35	285	-294	73	41	179
Romania	-202	63	8	131	-220	180	3	37
Slovakia [a]	-315	95	98	122	-183	98	25	60
Slovenia	-156	-57	26	187	-182	-123	95	211
EEBS-9	-299	26	53	219	-241	63	50	127

Source: United Nations COMTRADE Database; UNECE secretariat calculations.

[a] 1999 instead of 2000.

narrowed from 1996 to 2000 in primary and consumer goods, partly indicating that intersectoral complementarities had become less important. The despecialization of the region in capital goods, however, was a net result of recently acquired comparative advantages in that sector by the four countries mentioned earlier, an increasing disadvantage in Latvia and Lithuania (possibly related to FDI) and little or no change in the others.

Individual countries sometimes deviated sharply from the group average. Between 1996 and 2000, Hungary and Estonia, as already noted, succeeded in building revealed comparative advantages in capital goods (other than transport equipment), which was not the case in most of the east European and Baltic economies. Latvia and Lithuania were unique in achieving a major switch in their revealed comparative advantages from primary to intermediate goods.[377]

---

[377] Latvia's shift from a negative to a positive contribution of both primary and intermediate goods to the overall trade balance and the balance with the EU, seems to be exclusively based on the forestry and wood sectors; the near loss of the country's advantage in consumer goods

Slovenia, is also a special case: it is the only country in the sample that has consistently shown a RCD in primary commodities in its trade with the EU, while a comparative advantage in capital goods trade with the EU revealed in 1996-1998 was lost in 1999 but acquired in trade with CEFTA.

With respect to major partner groups, the contrast between trade with the EU and with the CIS remained, particularly as far as primary commodities and capital goods are concerned. In intraregional trade (CEFTA-7) the pattern of specialization changed less clearly: there was a shift from disadvantage to advantage in trade in intermediate goods, while in the consumer goods category only Hungary, Poland and Slovenia succeeded in preserving a positive structural balance. In trade with the EU – the major changes in RCA were the increased negative contribution of intermediate goods and the improved structural balance in capital goods.

All in all, sectors with comparative advantage (among the 18 BEC subcategories) in 2000 accounted for over 40 per cent of aggregate exports in the sample of countries (for 70 to 80 per cent in Estonia, Latvia and Slovakia (in 1999), countries that have the highest degree of commodity concentration (chart 3.5.6)). The sectoral contributions to the trade balance were also noticeably larger in Lithuania and Romania,[378] suggesting that their trade patterns are much more influenced by intersectoral complementarities than in the other countries.

(d) Intra-industry trade developments

The above observations indicate that more intense trade with the west European economies appears to have produced some qualitative changes in the exports and imports of the east European and Baltic economies that have also spilled over into trade with other major partners. Measuring these changes directly is rather difficult, but the fact that an increasing share of trade among these countries and between them and their west European partners is two-way trade (table 3.5.8),[379] with a large component of intra-industry trade, provides some support for this conjecture.[380] As can be seen from chart 3.5.8, the share of intra-industry trade in manufactures trade with the EU peaked in 1995-1996 in most countries and has since fluctuated within the range of 55 to 65 per cent, remaining below 50 per cent only in Latvia and Lithuania.[381] A recent fall in Slovakia's intra-industry trade with the EU coincided with a considerable increase in the share of passenger car exports (in 2000, they accounted for 21 per cent of exports to the EU).[382] In 1996-2000, trade in intermediate goods, and particularly in parts and accessories for transport equipment and other capital goods (BEC 53 and 42) contributed most to the growth of intra-industry exchanges with EU partners across the region, which supports the earlier observation about eastern Europe's increasing involvement in the international fragmentation of production.[383]

In trade with CEFTA-7 partners, the rise of intra-industry trade has been very pronounced in the past five years and has attained similar levels to those in trade with the EU in 1999. The intra-industry trade of the Baltic states with CEFTA partners, however, in this admittedly short period, was much more volatile probably because of the rather low levels of trade and the very limited number of commodities traded both ways.[384]

---

seem to emanate from trade with its east European counterparts. In Lithuania, a much larger negative contribution to the trade balance in the primary commodity sector but strengthened comparative advantage in the intermediate goods category were driven by one major sector, oil refining (partly under the pressure of rising oil prices in 2000).

[378] In 2000, the median of absolute numbers of CTB for four BEC categories was 66 in Lithuania and 44 in Romania, while in all other countries in the sample the median was below 33.

[379] In this note, following Freudenberg and Lemoine, trade is considered to be "two-way" trade, when the value of the minor flow represents at least 10 per cent of the major flow:

$$Min(X_{kk'it}, M_{kk'it}) / Max(X_{kk'it}, M_{kk'it}) > 0.10$$

where X and M stand for values of exports and imports; k for the declaring country; k' for the partner country; i for product; and t for year.

[380] There is, however, an increased questioning of the intra-industry trade theories developed in the past 20-30 years. In fact, the criticism more often concerns not the idea itself but the empirical approaches taken by trade economists to analyze or measure it. See, for instance, D. Davis and D. Weinstein, *What Role of Empirics in International Trade?*, NBER Working Paper, No. 8543 (Cambridge, MA), October 2001. For the most common caveats and ways of avoiding the methodological traps in empirical research on intra-industry trade, see L. Fontagné and M. Freudenberg, *Intra-industry trade: Methodological Issues Reconsidered*, CEPII Working Paper, No. 1997-01 (Paris), January 1997.

[381] UNECE secretariat computations on SITC Rev.3, 5-digit data as reported by the east European and Baltic countries to the United Nations COMTRADE Database. The share concerns only two-way trade flows and is the familiar Grubel-Lloyd index on intra-industry trade, which varies from zero to 100 (see note to chart 3.5.8). It is worth noting however that IIT computed on total trade data reported by east European and Baltic countries is markedly lower than that based on the western mirror data (see for instance C. Aturupane, S. Djankov and B. Hoekman, "Horizontal and vertical intra-industry trade between eastern Europe and the European Union", *Weltwirtschaftliches Archiv*, Vol. 135, No. 1 (Kiel), March 1999.

[382] For example, for bilateral trade between Slovakia and Germany, intra-industry trade in this sector (BEC 51) fell from 54 per cent in 1997 to 20 per cent in 1999, while for parts and accessories of transport equipment (BEC 53) it remained within the range of 60 to 70 per cent in 1998-1999.

[383] In Czech trade with Austria and Germany, the IIT coefficients for the BEC 53 category exceeded 70 per cent in each year between 1996 and 2000, and in Polish and Slovak trade, with the same partners in the same category, the coefficients reached similar levels in 1998. In Estonia's trade with Finland and Sweden, IIT for the BEC 42 category has been near 70 per cent since 1998. According to some researchers, a 70 per cent benchmark in intra-industry exchange seems to be important for accelerating technology transfers between the partners. D. Hakura and F. Jaumotte, *The Role of Inter- and Intra-industry Trade in Technology Diffusion*, IMF Working Paper WP/99/58 (Washington, D.C.), April 1999.

[384] Particularly volatile was Estonia's IIT: although two-way trade was characteristic for 70 manufactured goods at 5-digits in 1996-1998, the number doubled in 2000; two commodities from the construction and pharmaceutical sectors (53354 – resin cements, caulking compounds, etc. and 54293 – medicaments, n.e.s.) were mainly responsible for the fluctuations in the IIT coefficient.

TABLE 3.5.8

Types of trade among the east European and Baltic countries and with the EU, 1996-2000

(Shares in per cent, variation in percentage points)

	Trade with EU						Trade with CEFTA-7					
	Share in 2000			Variation over 1996/2000			Share in 2000			Variation over 1996/2000		
	One-way trade	Two-way trade		One-way trade	Two-way trade		One-way trade	Two-way trade		One-way trade	Two-way trade	
		Horizontal	Vertical		Horizontal	Vertical		Horizontal	Vertical		Horizontal	Vertical
Czech Republic												
All commodities	34.9	12.2	52.9	-5.4	4.8	0.6	34.9	28.5	36.6	-3.5	9.7	-6.3
Manufactures	16.3	15.8	67.9	-7.8	5.5	2.3	30.3	28.4	41.3	-3.6	15.9	-12.3
Estonia												
All commodities	66.3	4.6	29.1	5.6	-2.2	-3.3	83.5	3.2	13.3	0.4	1.4	-1.9
Manufactures	54.4	9.4	36.2	9.1	0.4	-9.5	77.1	4.2	18.7	-1.7	3.2	-1.5
Hungary												
All commodities	46.0	9.9	44.1	-4.2	3.9	0.3	48.0	17.4	34.6	-13.3	7.7	5.5
Manufactures	34.0	11.4	54.6	-4.4	4.3	0.1	37.4	19.2	43.4	-7.6	5.0	2.6
Latvia												
All commodities	85.4	0.9	13.7	-2.3	0.3	2.0	90.8	2.9	6.4	-1.7	2.5	-0.7
Manufactures	78.0	2.2	19.8	-1.3	1.3	-0.1	88.7	3.3	8.0	-3.5	3.2	0.3
Lithuania												
All commodities	80.8	1.7	17.6	-4.4	0.1	4.2	85.0	2.0	13.0	-2.0	0.2	1.8
Manufactures	72.5	1.9	25.6	-6.0	-0.9	6.9	80.0	1.9	18.1	-5.1	0.7	4.4
Poland												
All commodities	53.6	7.5	38.9	-11.3	3.7	7.5	48.5	18.4	33.1	-8.6	-3.8	12.4
Manufactures	39.0	7.2	53.8	-14.0	2.5	11.4	40.5	20.7	38.8	-9.4	-1.8	11.2
Romania												
All commodities	75.1	2.8	22.1	-8.7	0.1	8.6	75.5	7.0	17.5	-6.0	3.2	2.8
Manufactures	68.8	3.1	28.1	-8.4	-0.5	8.9	72.4	5.8	21.8	-4.5	3.1	1.4
Slovakia [a]												
All commodities	58.0	10.2	31.8	-13.4	7.9	5.5	48.2	13.3	38.6	4.3	-1.6	-2.7
Manufactures	37.0	11.8	51.2	-19.3	-0.4	19.6	38.2	16.3	45.5	2.9	0.6	-3.6
Slovenia												
All commodities	48.1	11.9	39.9	-4.8	3.5	1.4	79.3	5.0	15.7	-2.8	2.5	0.3
Manufactures	26.7	21.9	51.4	-2.8	13.3	-10.4	75.7	5.7	18.6	-2.1	2.0	0.1

Source: UNECE secretariat computations, based on 5- and 4-digit SITC Rev. 3 data reported by countries to United Nations COMTRADE Database.

Note: Manufactures are defined as SITC Sections 5 to 8 less Division 68 (non-ferrous metals). Depending on the degree of overlap, both exports and imports are considered to be a part of either one-way or two-way trade; thus trade in a product *i* is considered to be two-way when the value of the minor flow represents at least 10 per cent of the major flow:

$Min (X_{kk'it}, M_{kk'it})/Max (X_{kk'it}, M_{kk'it}) > 10$ per cent

Two-way traded products are considered to be similar (*horizontally differentiated*) if the export and import unit values differ by less than 15 per cent:

$(1/1.15) <= UV^x_{kk'it}/UV^M_{kk'it} <= 1.15$,

where *UV* = unit value; *k* = declaring country; *k'* = partner country; *i* = product; *t* = year; *x* = exports; *m* = imports. When this is not the case, products are considered to be *vertically differentiated*. A 15 per cent range is generally accepted, and 25 per cent is often used as a robustness check.

[a] 1999 instead of 2000.

Although, as clearly shown in table 3.5.8, two-way trade between eastern Europe (including the Baltic states)[385] and the EU was in general gaining momentum in the past five years, the differences in the quality of exports and imports of goods within the same product category seem to be diminishing rather slowly. In order to illustrate this, two-way trade is separated into horizontal and vertical trade, following a widely accepted assumption that relative prices (unit values of exports and imports, in this case) tend to reflect differences in qualities.[386] Thus vertical trade is defined as two-way trade in a commodity (at a 5-digit level) whose unit value of exports relative to its unit value of imports in a specific year falls outside a range of 15 per cent, while trade in commodities whose relative unit values fall within this range of 15 per cent is defined as horizontal.[387]

---

[385] According to the data, only in Estonia did the share of two-way trade with the EU shrink between 1996 and 2000. This coincided with a marked increase in the concentration of exports in the ICT sector. See the story on Elcoteq in UNECE, *Economic Survey of Europe, 2001 No. 2*, p. 22, box 1.2.2.

[386] D. Greenaway, R. Hine and C. Milner, "Country-specific factors and the pattern of horizontal and vertical intra-industry trade in the UK", *Weltwirtschaftliches Archiv* Vol. 130, No. 1 (Kiel), March 1994 and "Vertical and horizontal intra-industry trade: a cross industry analysis for the United Kingdom", *Economic Journal*, No. 105, November 1995 or C. Aturupane et al., op. cit.

[387] A unit value dispersion of 15 per cent for analysis (and in some cases 25 per cent as a check for robustness) was used also by K. Abd-El-

CHART 3.5.8

Share of intra-industry in total trade in manufactures between EU and east European and Baltic countries and in trade with CEFTA-7, 1993-2000

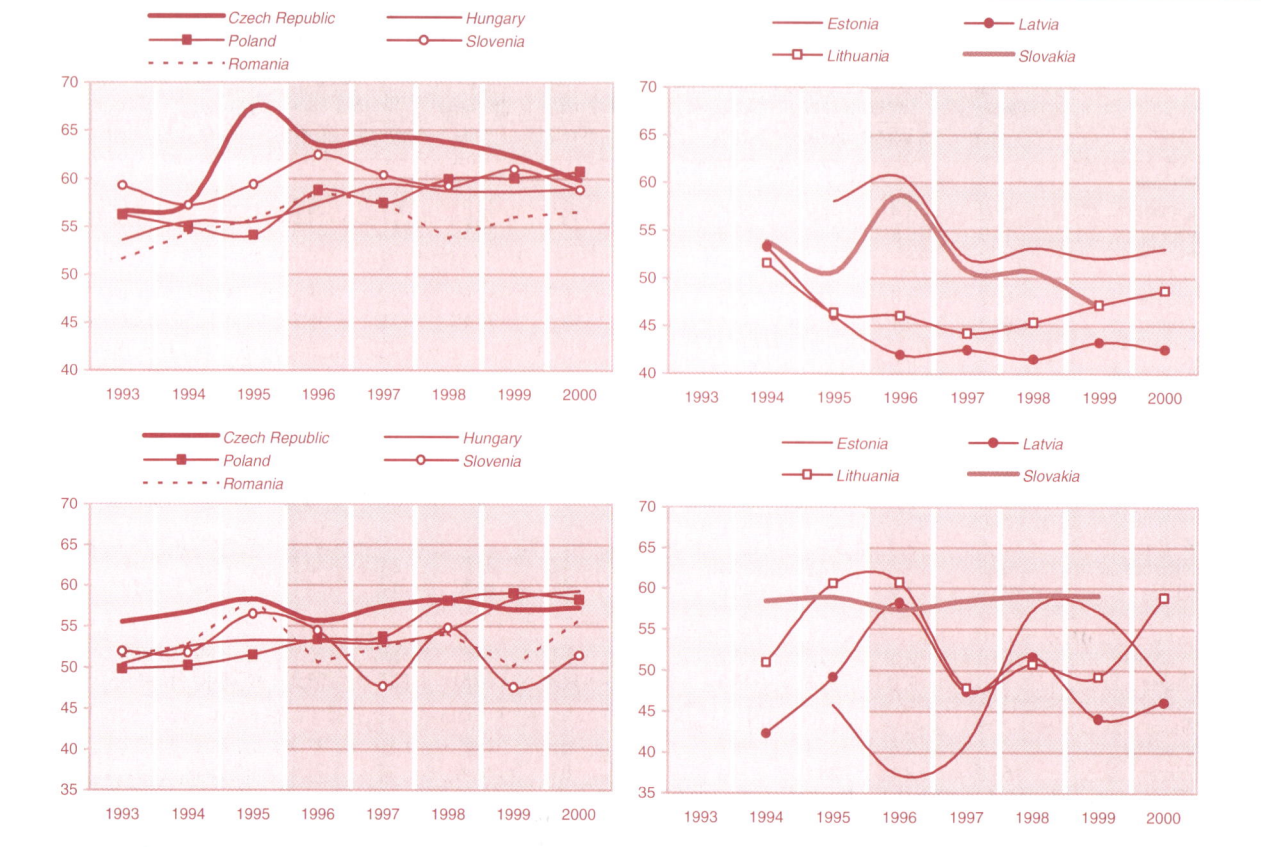

Source: United Nations COMTRADE Database.

Note: UNECE secretariat computations, based on SITC Rev.3 data at 5-digit (adjusted); only two-way trade (see note for table 3.5.8) taken into account. The share is the familiar Grubel-Lloyd index on intra-industry trade (IIT_k) that is,

$$IIT_k = \left\{ 1 - \frac{\sum_i |x_{ik} - m_{ik}|}{\sum_i (x_{ik} + m_{ik})} \right\} * 100$$

where: i refers to the 5-digit SITC product categories (adjusted); k identifies countries; x and m refer to exports and imports. Manufactures are defined as SITC Sections 5 to 8 less Division 68 (non-ferrous metals); at 5-digits (adjusted) around 2,730 commodities are listed, however the number actually engaged in two-way trade varies considerably from country to country (e.g. from some 300 in Latvia to 1,480 in the Czech Republic).

The prevalence of vertically differentiated exchange in manufactured products with the EU in the two-way trade of all the countries shown in table 3.5.8 continues a general trend observable from the outset of the transition,[388] although in some cases horizontal trade has gained some share since 1996 in the Czech Republic, Hungary, Poland and, more considerably, in Slovenia. In contrast, intra-industry exchanges of vertically differentiated products seem to be less important in the trade of several countries with their counterparts in CEFTA-7. In fact, in 2000, one-way trade was still much more important with these partners than with the EU.

Rahman, op. cit., D. Greenaway et al., op. cit., C. Aturupane et al., op. cit., M. Freudenberg and F. Lemoine, op. cit., etc.

[388] C. Aturupane et al., op. cit. and M. Freudenberg and F. Lemoine, op. cit., found similar results for the 1992/1993-1996 period.

The evolution of relative unit values in 1996-2000 in the trade of nine east European and Baltic countries with the EU (table 3.5.9) shows large variations across the countries, although among the countries with higher levels of intra-industry trade they seem to be more stable and on a slightly rising trend. However, as unit values are often affected by changes in the way quantities are measured in the customs statistics, they are notoriously difficult to interpret, and a strong conclusion cannot be drawn from these data.

(e) Conclusions

During 1996-2000, east European and Baltic countries became more closely integrated into international production networks in several sectors mainly related to capital goods (including transport equipment), their parts and accessories, and durable

TABLE 3.5.9

Average ratio and coefficient of variation of export to import unit values in trade with EU, 1996, 1998 and 2000

	1996		1998		2000	
	Mean	Coefficient of variation	Mean	Coefficient of variation	Mean	Coefficient of variation
Czech Republic						
All commodities	1.000	0.862	1.070	0.745	1.069	0.764
Manufactures	0.980	0.843	1.038	0.744	1.040	0.741
Estonia						
All commodities	1.316	0.719	1.653	0.664	1.522	0.774
Manufactures	1.275	0.719	1.685	0.660	1.527	0.754
Hungary						
All commodities	1.181	0.773	1.094	0.871	1.176	0.815
Manufactures	1.166	0.797	1.067	0.864	1.156	0.804
Latvia						
All commodities	0.879	0.627	1.041	0.588	1.196	0.622
Manufactures	0.832	0.675	1.039	0.577	1.116	0.644
Lithuania						
All commodities	1.568	0.609	1.221	0.652	1.204	0.897
Manufactures	1.600	0.605	1.200	0.622	1.222	0.888
Poland						
All commodities	1.034	0.657	1.034	0.672	1.016	0.701
Manufactures	0.972	0.646	0.974	0.693	0.965	0.713
Romania						
All commodities	0.951	0.582	1.152	0.648	1.253	0.628
Manufactures	0.906	0.570	1.137	0.630	1.229	0.615
Slovakia[a]						
All commodities	0.826	0.628	0.997	0.691	1.014	0.848
Manufactures	0.812	0.610	0.981	0.717	0.990	0.956
Slovenia						
All commodities	1.111	0.787	1.066	1.060	1.178	0.775
Manufactures	1.089	0.798	1.044	1.099	1.159	0.755

Source: United Nations COMTRADE Database; UNECE secretariat calculations.

Note: Since unit values may not be dependable for small amounts of trade, they are calculated only for flows exceeding $5,000. While computing an average, extreme ratios of 20 and above between export and import unit values for the same 5-digit level commodity were also excluded.

[a] 1999 instead of 2000.

consumer goods. This development has been shaping new specialization patterns for these countries. Trade statistics clearly confirm the active involvement of these countries in the fragmentation of international production: in the past five years, the share of intermediate goods in their total trade, has risen to above 50 per cent and the share in exports of capital goods and durable consumer goods increased fairly closely in line with the share in imports of intermediate goods. The latter reflected the relocation of final assembly processes from the west to eastern Europe and the Baltics. In consequence, there was a shift from a revealed comparative advantage in intermediate products in 1996 to a comparative disadvantage in this sector in 2000 for the region as a whole, and a shift into a comparative advantage in capital goods in four out of nine countries. Also worth mentioning are recently acquired comparative advantages by east European countries in the subcategory of parts and accessories of transport equipment (within intermediate goods) and in passenger motorcars.

These changes in revealed comparative advantage are mainly located in trade with the EU, as weakened trade with the CIS remains focused on primary commodity imports and consumer goods exports, while trade within the region does not exhibit any strong pattern on its own, but rather reflects developments in trade with the EU.

Interindustry trade with the EU seems to be giving way to intra-industry trade although the process is relatively slow in the case of the Baltics and Romania. Trade in vertically differentiated products, which is most often fragmentation-driven, is considerably stronger than two-way trade in horizontally differentiated products, although in several countries of the region the latter has been gaining ground in the past few years.

(iv) External financing, FDI and debt issues

(a) Total capital flows

During 2001 and early 2002 emerging markets continued to face a challenging and uncertain environment, punctuated by the financial crises in Argentina and Turkey. The difficulties of these two countries and the slowing of the global economy increased risk aversion and in general reduced investors' interest in the securities of emerging market economies. The attacks of 11 September sparked higher risk premia and caused a temporary suspension of new international bond issues. However, toward the end of 2001 concerns eased, helped by increasing confidence about containing any spillovers from the two countries in distress. Despite this unpropitious environment, the financial situation of many transition economies improved and many of them were upgraded by the international rating agencies (section 3.5(iv)(d)). This includes Russia where unsustainable capital inflows and weak liquidity had set the stage for the rouble crisis in 1998. However, the financial situation within the region remains uneven with a sizeable number of countries lacking access to private (or even much official) capital and doubts persist about the capacity of several to service their debts.

Although global financial flows slowed in 2001[389] the majority of *east European* and *Baltic countries* attracted more capital (tables 3.5.10 and 3.5.11), which generally financed their current account deficits and increased official reserves (table 3.5.12). The combined net financial inflows of roughly $27 billion in 2000-2001 fell short of the $30-$31 billion received in the peak years 1998-1999, but the difference largely reflects the disappearance of the large, unsustainable short-term

---

[389] Net external financial flows to the emerging market economies are estimated to have fallen from $167 billion to $135 billion in 2001. Much of the decline, but certainly not all, was due to outflows from Argentina and Turkey. IIF, *Capital Flows to Emerging Market Economies* (Washington, D.C.), 30 January 2002.

TABLE 3.5.10

Net capital flows into the ECE transition economies, 2000-2001
(Million dollars, per cent)

	Capital and financial account flows[a]					Changes in official reserves[b] (million dollars)	
	Million dollars			Capital flows/GDP			
	2000	Jan.-Sep. 2001	2001	2000	2001	2000	2001
Eastern Europe	24 932	17 009	24 848*	6.6	5.9	6 047	6 048
Albania	288	330	510*	7.7	12.3	132	110*
Bosnia and Herzegovina	923	680*	908*	21.9	18.9	14	8*
Bulgaria	1 111	174	1 178	9.3	9.0	409	300
Croatia	1 015	1 237	2 000*	5.3	9.7	582	1 400*
Czech Republic	3 092	2 108	4 100*	6.1	7.3	819	1 600*
Hungary[c]	2 388	1 411	1 008	5.2	1.9	1 060	-97
Poland	10 565	6 110	6 579	6.7	3.7	619	-502
Romania	2 291	2 544	3 833	6.2	9.9	928	1 484
Slovakia	1 537	844	2 020*	8.0	9.9	824	200*
Slovenia	790	605	1 352	4.4	7.2	178	1 285
The former Yugoslav Republic of Macedonia	349	256	440*	9.8	12.9	236	40*
Yugoslavia	585	710*	921*	7.2	8.8	246	221*
Baltic states	1 754	1 050	2 240*	7.5	9.0	271	525
Estonia	437	66	200*	8.7	3.7	122	-100*
Latvia	511	437	1 070	7.1	14.1	18	305
Lithuania	806	547	970*	7.1	8.1	131	320*
CIS[d]	-30 205	-16 805	-21 978*	-8.6	-5.3	17 325	10 289
Armenia	298	162	260*	15.5	12.3	19	40*
Azerbaijan	442	-168	-200*	8.4	-3.5	274	-50*
Belarus	372	-21	410*	3.6	3.4	76	10*
Georgia	242	209	220*	8.0	7.0	-20	40*
Kazakhstan	-602	1 361	2 200*	-3.3	9.8	141	400*
Kyrgyzstan	98	-44	50*	7.1	3.3	21	40*
Republic of Moldova	172	92	140*	13.4	9.5	46	10*
Russian Federation	-30 281	-18 938	-25 938	-11.7	-8.4	16 010	8 219
Tajikistan	91	178	..	9.2	..	29	..
Turkmenistan	..	..	..	..	..	..	..
Ukraine	-1 083	-42	300*	-3.5	0.8	398	1 600*
Uzbekistan	-153	136	200*	-1.1	1.7	31	100*
Total above[d]	-3 519	1 254	5 111*	-0.5	0.6	23 643	16 863
Memorandum items:							
CETE-5	18 371	11 077	15 059*	6.3	4.6	3 500	2 486*
SETE-7	6 561	5 931	9 790*	7.5	10.3	2 547	3 563*
Asian CIS[d]	615	2 104	3 110*	1.3	5.8	795	450*
Three European CIS[e]	-539	29	850*	-1.3	1.7	520	1 620*
Russian Federation[f]	-21 040	-12 751	-17 551	-8.1	-5.7	16 010	8 219

Source: UNECE secretariat, based on national balance of payments statistics; IMF, *Staff Country Reports* (Washington, D.C.), for Bosnia and Herzegovina, Yugoslavia, Tajikistan and Uzbekistan [www.imf.org]; TACIS, *Azerbaijan Economic Trends* (Baku) for Azerbaijan [www.economic-trends.org].

Note: Estimates for 2001 are generally based on three quarters of balance of payments data for 2001 and fourth quarter changes in 2000. July-December 2000 data were used to make estimates for 2001 for Armenia, for which only January-June 2001 data are available. Similarly, December 2000 data were used for Slovakia for which only January-November 2001 are available. IMF projections were used for Bosnia and Herzegovina and Yugoslavia. IMF estimates and projections are used for Tajikistan in 2000 and 2001. No balance of payments data have been available for Turkmenistan since 1998.

[a] Includes errors and omissions; excludes changes in official reserves.
[b] A negative sign indicates a decrease in reserves.
[c] Excludes reinvested profits (net inflow).
[d] Totals include secretariat estimates for Tajikistan and Turkmenistan.
[e] Belarus, Republic of Moldova and Ukraine.
[f] Excluding errors and omissions.

flows into Poland in the late 1990s. Poland was also a major influence on the aggregate net inflow in 2001 when it made a large voluntary debt repayment (see below). FDI flows, which have proved to be a relatively stable source of financing for eastern Europe, declined somewhat in 2001, but they continue to dominate capital inflows (table 3.5.11). Portfolio investment increased sharply, due to new eurobond issues and net flows into the domestic

The Transition Economies

TABLE 3.5.11

Net capital flows by type of capital into eastern Europe, the Baltic states and selected members of the CIS, 1999-2001

(Billion dollars)

	Eastern Europe				Baltic states				Three European CIS[a]			
			Jan.-Sep.				Jan.-Sep.				Jan.-Sep.	
	1999	2000	2001	2001	1999	2000	2000	2001	1999	2000	2000	2001
Capital and financial account	24.4	22.2	14.7	20.0*	2.2	1.6	0.9	1.1	0.1	-0.6	-1.1	0.3
Capital and financial account [b]	28.7	24.9	17.0	24.8*	2.2	1.8	1.0	1.0	-1.1	-0.5	-1.0	–
of which:												
FDI	18.8	19.6	11.8	19.0*	1.0	1.1	0.7	0.9	1.0	0.8	0.6	0.8
Portfolio investment	2.6	2.2	4.8	5.5*	0.8	0.1	0.3	0.2	-0.2	-0.1	-0.1	-0.6
Medium-, long-term funds	4.9	5.1	0.4	-1.8*	0.5	–	-0.1	0.1	1.0	-1.2	-1.7	0.3
Short-term funds	-2.8	-5.6	-2.8	-3.4*	-0.1	0.5	-0.1	-0.2	-1.8	-0.3	-0.1	-0.2
Errors and omissions	4.3	2.7	2.3	4.8*	–	0.1	0.1	-0.1	-1.2	0.1	0.1	-0.3
Memorandum item:												
Short-term investment [c]	4.1	-0.6	4.2	6.9*	0.6	0.6	0.4	–	-3.3	-0.2	–	-1.1

Source: UNECE secretariat estimates, based on national balance of payments statistics.

[a] Belarus, Republic of Moldova and Ukraine.

[b] Including errors and omissions.

[c] Portfolio investment (including international bond issues), short-term funds and errors and omissions.

---

securities markets.[390] There was relatively little new medium- and long-term foreign borrowing (private and official loans), as suggested by the data showing a net repayment of debt. Short-term funds had a differential impact on payments balances within the area. Large outflows were reported by the Czech Republic, Hungary and Poland, but Slovakia, Slovenia and most of the south-east European countries attracted substantial short-term funds. In the latter, these funds, which are subject to sudden reversals, were channelled into official reserves, which suggests that the vulnerability of these countries has not increased.

Within *central Europe*, financial flows into Slovakia, Slovenia and the Czech Republic increased. In the Czech Republic sizeable foreign investments and the prospects for further large privatization-related FDI have led to an appreciation of the koruna. The Czech authorities have created a special foreign currency account with the central bank to "park" privatization revenues and thus avoid their conversion in the currency market. Although foreign investment in Hungary increased, net financial capital inflows declined due to outflows of short-term capital and a net repayment of loans. The latter reflects repayments of state debt, in line with government policy to reduce external indebtedness. In Poland, inflows also fell as a result of smaller inflows of FDI and net repayments of medium- and long-term loans. The discounted buy-back of its $3.3 billion of Paris Club obligations to Brazil (a major outflow) was a debt-reducing operation at the end of 2001, partially funded from the official reserves.

The large increase in capital inflows into *south-east Europe* in 2001 consisted mainly of short-term capital, unrecorded flows[391] (mainly into Romania) and portfolio investment.[392] Net FDI changed little and has remained the major source of financing. On average, FDI covered 55 per cent of the current account deficits in 2001 and led to a record increase in official reserves. This has been a very positive development since these countries have traditionally lacked sufficient liquidity; foreign exchange reserves are now sufficient to cover almost four months of imports of goods and services for the region as a whole. Nonetheless, liquidity still remains inadequate in several countries (table 3.5.12). Yugoslavia has made considerable progress in normalizing its international financial relations. Early in 2001, it rejoined the IMF and World Bank and became a member of the EBRD. An IMF stand-by agreement followed in June 2001 (the same month in which international donors assembled a $1.2 billion financing package). In November, the government reached an agreement with the Paris Club to restructure $4.5 billion of bilateral debt, of which $3 billion is to be written off.[393] The former Yugoslav Republic of Macedonia has been dependent on official assistance (although it received record FDI flows in 2001; see below). In March 2002 international donors made a further commitment to the country's economic reform programme, pledging additional assistance of $515 million.

---

[390] Although stock markets in several east European countries (and Russia) have boomed, any foreign investment in equities appears to have been swamped by debt flows.

[391] Errors and omissions.

[392] Portfolio investment includes eurobond issues by Bulgaria (€750 million), Croatia (€500 million) and Romania (€975 million).

[393] Among other provisions, 51 per cent of the debt (in net present value terms) is to be cancelled after the country has obtained an EFF from the IMF, and a further 15 per cent reduction is possible after successful completion of the programme. Paris Club, *Press Release*, 19 November 2001 [www.clubdeparis.org].

TABLE 3.5.12

Selected external financial indicators for eastern Europe, the Baltic states and the CIS, 1999-2001

(Million dollars, per cent)

	Gross debt (million dollars)		Gross debt/exports (per cent) [a]			Gross debt /GDP (per cent)		Official reserves				Net FDI/ current account [b]	
								Millions		Months of imports [a]			
	2000	2001	1999	2000	2001	2000	2001	2000	2001	2000	2001	2000	2001
Eastern Europe	187 046	194 804	122	113	107	49	46	70 411	75 316	4.3	4.2	104	101
Albania	1 033	1 139[c]	155	126	123	28	27	352	362[d]	2.8	2.4	92	45
Bosnia and Herzegovina	2 584	2 800[c]	199	171	173	61	58	497	653[d]	2.3	2.9	17	18
Bulgaria	10 364	9 894	168	142	126	86	75	3 342	3 390	4.8	4.4	143	73
Croatia	11 002	11 049	118	122	109	58	53	3 524	4 696	4.1	4.8	251	133
Czech Republic	21 386	21 825[e]	66	57	52	42	39	13 019	14 337	3.9	3.8	197	176
Hungary	30 742	33 386	104	93	91	66	65	11 190	10 727	3.9	3.4	84	190
Poland [a]	69 497	71 781[e]	207	204	194	44	40	26 562	25 648	6.4	6.1	82	96
Romania	10 625	11 820	92	85	86	29	31	3 922	5 442	3.2	3.8	77	49
Slovakia	10 804	10 973[e]	84	75	71	56	54	4 022	4 141	3.2	2.8	289	106
Slovenia	6 217	6 717	49	56	57	34	36	3 196	4 330	3.2	4.3	18	505
The former Yugoslav Republic of Macedonia	1 488	1 440	100	89	98	42	42	429	739[d]	2.2	4.3	149	105
Yugoslavia	11 304	11 980[c]	598	435	414	140	114	355	850[c]	1.0	1.9	7	13
Baltic states	12 575	13 611[e]	97	92	90	53	54	3 083	3 587	2.4	2.5	74	73
Estonia	3 007	3 358[e]	70	61	66	60	62	921	820	2.1	1.8	103	137
Latvia	4 711	5 044[e]	124	135	135	66	67	851	1 149	2.5	3.0	81	32
Lithuania	4 857	5 209[e]	104	92	83	43	44	1 312	1 618	2.5	2.7	56	91
CIS	211 577	208 119	176	124	121	60	50	32 203	42 225	3.1	3.6	..	..
Armenia	854	930	179	155	149	45	44	318	321	3.8	3.9	37	32
Azerbaijan	1 170	1 250[c]	73	54	46	22	22	680	889	3.4	4.0	77	..
Belarus	2 299	2 251[c]	37	28	26	22	19	350	330[d]	0.5	0.4	39	25
Georgia [f]	1 610	1 690[c]	225	191	204	53	54	109	160	1.1	1.6	50	67
Kazakhstan [g]	12 525	14 148[e]	171	115	133	68	63	1 594	1 997	1.9	1.9	..	144
Kyrgyzstan	1 704	1 809[c]	306	292	317	125	119	239	264	3.9	5.0	..	200
Republic of Moldova [g]	1 562	1 532[e]	203	194	159	121	104	230	253[d]	2.5	2.5	109	115
Russian Federation	171 800	167 000	210	143	140	66	54	24 264	32 542	3.9	4.6	..	..
Tajikistan	1 205	1 127[c]	171	145	159	122	107	87	114[c]	1.1	1.5	35	10
Turkmenistan [h]	..	..	..	..	..	..	..	..	..	..	..	..	..
Ukraine [f h]	10 348	10 282[e]	72	53	49	33	27	1 353	2 955	0.8	1.7	..	..
Uzbekistan [h]	4 200	3 800*	120	125	115	31	33	1 273	1 400*	4.8	4.9	..	100
Total above	411 197	416 534	145	117	113	55	48	105 697	121 128	3.8	3.9	..	..
Memorandum items:													
CETE-5	138 646	144 682	113	107	101	47	44	57 989	59 184	4.6	4.4	107	124
SETE-7	48 400	50 122	158	137	130	55	53	12 422	16 132	3.4	3.9	90	55
Asian CIS	25 568	27 054	156	116	121	52	51	6 005	6 144	3.2	2.9	..	115
Three European CIS [i]	14 209	14 065	67	50	46	33	27	1 933	3 538	0.8	1.3	..	..
Georgia [g]	1 792	..	..	..	..	..	..	..	..	..	..	..	..

Source: National statistics; IMF, *International Financial Statistics*, March 2002 and *Staff Country Reports*, various issues (Washington, D.C.); press reports; UNECE secretariat estimates. Debt ratios reflect the latest available debt data, available as of mid-March 2001.

[a] Exports of merchandise and services, and income receipts. Total imports of merchandise and services, and income payments. For Poland, exports exclude net receipts from non-classified current account items.

[b] Per cent. In this table (..) indicates that the current account balance was positive and/or net FDI negative.

[c] IMF projections. For Yugoslavia, debt prior to Paris Club write-off, half of which will be implemented in 2002.

[d] October for Bosnia and Herzegovina and Republic of Moldova, November for Albania, The former Yugoslav Republic of Macedonia and Belarus.

[e] September; November for Slovakia.

[f] Gross debt excludes cross-border inter-enterprise arrears.

[g] Gross debt includes cross-border inter-enterprise arrears.

[h] Government and government guaranteed debt only.

[i] Belarus, Republic of Moldova and Ukraine.

The financial situations of the *European CIS* have tended to diverge. The payment positions of Belarus and the Republic of Moldova have remained difficult: both lack access to the private capital markets, and only the Republic of Moldova has been able to qualify for IMF funding (although disbursements were suspended in 2001). In the past two years, the Republic of Moldova has been able to attract more FDI, which has financed the current account deficit, and capital flight appears to have been reduced. However, there is a persistent concern about the sustainability of its debt (section 3.5(iv)(c)). Ukraine's financial situation appears to have eased further in 2001. As noted above, exports (and imports) and private transfers increased and there was a net inflow of capital in 2001 (in contrast to an outflow in 2000), consisting of near-record levels of FDI and new official loans. Despite some capital flight, official reserves increased sharply, although they remain below recommended levels. After restructuring its outstanding bonds in 2000, Ukraine benefited from a Paris Club rescheduling in July 2001,[394] which contributed to a credit upgrading. However, in February 2002, Ukraine and the IMF failed to reach agreement on resuming disbursements from an Extended Fund Facility (EFF). Maintaining a sizeable current account surplus is essential if Ukraine is to meet the scheduled repayments on its external debt.

In *Russia*, the current account surplus continued to fund a huge, although diminishing, net outflow of capital (table 3.5.10).[395] Nearly all the individual capital account items remained in deficit in 2001, including net FDI. Significant capital flight persists. Repayments of $7-$8 billion (about one half of the $13 billion debt servicing bill) were made and $2.8 billion of IMF credits were prepaid, both operations reducing external debt. Despite these outflows, official reserves increased again, to some $33 billion (excluding gold), which is equivalent to 4.6 months of imports of goods and services. International creditors are increasingly confident that Russia will be able to manage its $17 billion debt service obligation in 2003.[396] Concern about its ability to do so had arisen because of the possibility of weaker oil and other commodity prices and their adverse impact on Russian export revenues. It now appears that even if the current account surplus diminishes somewhat, Russia has sufficient resources at its disposal to avoid a rescheduling in 2003.[397] These include expected privatization revenues and new eurobond issues, fiscal reserves (estimated at $3.5 billion at end 2001) and foreign exchange reserves. The firming of world oil prices in the first quarter of 2002 improves the prospects for full debt servicing.

Partial-year financial data, available for only a few Asian CIS countries, suggest some strengthening in their financial positions in 2001. These countries are characterized by a lack of or low credit ratings, little or no access to the capital markets and relatively low FDI (except for some of the energy producers). Kazakhstan is by far the largest importer of capital, mainly in the form of FDI. Accelerated FDI inflows financed a larger current account deficit and increased portfolio outflows, but official reserves increased. Armenia and Georgia have received relatively large amounts of official assistance (relative to their GDP), but in most other countries, data suggest meagre capital inflows and some small increases in reserves. (The Asian CIS are also treated in section 3.5(iv)(c).)

Table 3.5.12 suggests an improvement in the financial indicators of many transition economies. Output has been stronger than expected (section 3.2) and debt burdens have continued to decline, thanks to the growth of GDP and of exports of goods and services.[398] The structure of current account financing (mainly FDI) has remained favourable, and liquidity has improved (including in several countries where it has been below recommended levels). All these developments (among others) point to a reduced vulnerability to external shocks.

(b) Foreign direct investment

FDI in the transition economies has proved resilient in the wake of the Asian and Russian financial crises and the recent global slowdown, increasing to nearly $28 billion in 2001 (table 3.5.13). There had been some concern that foreign investment in these countries would diminish as global economic prospects weakened. While this may have occurred in some cases, MNCs have strategic goals – privatization-related acquisitions, lower cost production facilities,[399] building capacity in EU candidate countries, and so on – that may dominate cyclical considerations. The results in 2001 were affected by delays in key privatizations in some countries (e.g. Bulgaria, Croatia, Poland, Slovakia and Ukraine), which in several cases resulted in smaller FDI inflows. It should be borne in mind that FDI flows are expected to diminish as privatizations draws to a close.

Preliminary data indicate that the Czech Republic received $4.5 billion, about the same flow of FDI as in 2000.[400] Estimates for 2002 vary, but inflows could amount to nearly $6 billion if all scheduled privatizations

---

[394] The rescheduling is expected to reduce debt service during 2001-2002 from an initial amount of $800 million to $285 million. Paris Club, *Press Release*, 13 July 2001 [www.clubdeparis.org].

[395] Financial outflows from Russia were roughly equivalent to the flows into all the other transition economies (table 3.5.10).

[396] The debt servicing bill may actually be smaller, perhaps less than $15 billion, after the prepayment to the IMF.

[397] Separately, Prime Minister M. Kasyanov stated that debt servicing will not be a problem in 2003 when payments are due to rise to $17 billion (down from earlier forecasts of $18-$19 billion). *Financial Times*, 21 February 2002.

[398] In several countries, including moderately-indebted Estonia and Latvia, there was some deterioration in one or more debt indicators. In Latvia, a higher than expected current account deficit was financed partly by foreign borrowing (including a €200 million eurobond issue) that led to a further increase in external debt (FDI inflows were less than had been expected).

[399] Such efficiency seeking investments may even accelerate during an economic downturn when profits come under pressure.

[400] According to revised data released by the Czech National Bank (see footnote in sect. 3.5(i)), the Czech Republic received an FDI inflow of $4,986 million (compared to the $4,595 million originally reported) in 2000. The reported figure for 2001 is $4,916 million instead of the preliminary estimate of $4,500 shown in table 3.5.13. The change appears to reflect higher profits associated with FDI.

TABLE 3.5.13

Foreign direct investment in the ECE transition economies, 2000-2001

(Million dollars, per cent)

	Inflows[a]		Net[a]	Inflows/GDP (per cent)	
	2000	2001	2001	2000	2001
Eastern Europe	20 433	19 856	19 040	5.4	4.7
Albania[b]	143	180*	180*	3.8	4.3
Bosnia and Herzegovina	150	164	164	3.6	3.4
Bulgaria	1 002	651	641	8.4	5.0
Croatia	1 115	900*	800*	5.9	4.4
Czech Republic	4 595	4 500*	4 400*	9.1	8.0
Hungary[c]	1 649	2 443	2 101	3.6	4.7
Poland (cash basis)	8 294	6 929	6 823	5.3	3.9
Romania	1 040	1 137	1 154	2.8	2.9
Slovakia	2 075	2 000*	1 930*	10.8	9.8
Slovenia	176	442	338	1.0	2.4
The former Yugoslav Republic of Macedonia[b]	170	420*	420*	4.7	12.3
Yugoslavia	25	90	90	0.3	0.9
Baltic states	1 173	1 457*	1 245*	5.0	5.8
Estonia	387	600*	410*	7.7	11.0
Latvia	408	257	245	5.7	3.4
Lithuania	379	600*	590*	3.3	5.0
CIS	5 367	7 021	3 877	1.5	1.7
Armenia[b]	104	70*	70*	5.4	3.3
Azerbaijan[b]	129	20*	20*	2.4	0.3
Belarus	116	100*	100*	1.1	0.8
Georgia	131	120*	120*	4.3	3.8
Kazakhstan	1 245	2 600*	2 600*	6.8	11.6
Kyrgyzstan	-2	20*	20*	-0.2	1.3
Republic of Moldova	138	150*	150*	10.7	10.2
Russian Federation	2 714	2 921	-203	1.0	0.9
Tajikistan[b]	22	20*	20*	2.2	1.9
Turkmenistan[b]	100*	100*	100*	2.3	1.7
Ukraine	595	800*	780*	1.9	2.1
Uzbekistan[b]	75	100*	100*	0.6	0.9
Total above	26 973	28 334	24 162	3.6	3.3
Memorandum items:					
CETE-5	16 789	16 314	15 591*	5.7	5.0
SETE-7	3 644	3 542	3 449*	4.2	3.7
Asian CIS	1 804	3 050	3 050*	3.7	5.7
Three European CIS[d]	848	1 050	1 030*	2.0	2.0

Source: UNECE secretariat, based on national balance of payments statistics; IMF, *Staff Country Reports* (Washington, D.C.), for Bosnia and Herzegovina, Yugoslavia, Tajikistan and Uzbekistan [www.imf.org]; TACIS, *Azerbaijan Economic Trends* (Baku) for Azerbaijan [www.economic-trends.org].

Note: IMF estimates and projections are used for Tajikistan in 2000 and 2001. IMF projections are used for Bosnia and Herzegovina and Yugoslavia in 2001. Secretariat estimates for 2001 are generally based on three quarters of balance of payments data for 2001 and fourth quarter changes in 2000. The FDI flow into the Czech Republic in 2001 is a preliminary official estimate. Changes in coverage are available in UNECE, *Economic Survey of Europe, 2001 No. 1*, box 5.3.1.

[a] Million dollars. Inflows into the reporting countries. Net signifies net inflow (inflows – investment abroad).

[b] Inflows are net flows.

[c] Excludes reinvested profits.

[d] Belarus, Republic of Moldova, Ukraine.

are completed. The sharp increase in FDI in Hungary (to $2.4 billion) was due entirely to intercompany loans of $1.7 billion (which are debt creating, unlike the equity component of FDI). It should be noted that FDI data currently reported by Hungary understate actual inflows because reinvested profits are excluded (section 3.5(i)). The sale of a stake in the local telecommunications company resulted in a record $400 million inflow in The former Yugoslav Republic of Macedonia. Slovenia recently changed its FDI policy to offer several investment incentives. As a result, investments of €500 million are expected in 2002-2003 (somewhat more than was received in 2001). FDI inflows have remained weak in the European CIS. They increased somewhat in Russia, but investments abroad by residents were even larger, resulting in net outflows for the second year running. Russia has a huge potential for attracting FDI, especially in the oil and gas sectors, but doubts about the investment climate persist, despite the recent significant reforms.[401] FDI outflows are bound to continue as cash-rich Russian oil companies invest abroad.[402] FDI in Kazakhstan increased sharply in 2001, but the bulk went into the oil and gas industry, disappointing hopes for a better sectoral distribution of foreign investment.

(c) The Asian CIS and special support for the CIS7

Incomplete data for the Asian CIS[403] indicate that their combined current account (table 3.5.1) deteriorated owing to the near stagnation in the growth of exports of goods and services and the continuation of buoyant import growth. The current account balances of two important energy exporters – Kazakhstan and Turkmenistan[404] – worsened partly because of weaker oil prices, although the current account of Azerbaijan moved into surplus. The strong growth of imports into Kazakhstan appears to be associated with FDI-related purchases of oil and gas field equipment and services.

Poverty, debt and the problems of fiscal sustainability, slow progress in economic reform and proximity to an area of military conflict are only some of the issues that have prompted the international community to focus on seven low-income CIS economies – Armenia, Azerbaijan, Georgia, Kyrgyzstan, the

---

[401] A $12 billion multi-year investment in the oil and gas sector on Sakhalin Island has been recently announced. However, conclusion of this major agreement does not necessarily presage a boom in new production sharing agreements (PSAs) because important obstacles to the use of this investment instrument remain in place.

[402] For example, Yukos, the second largest Russian oil company, recently announced that it planned to make foreign investments of up to $4 billion in downstream activities in the next three years. *Financial Times*, 7 February 2002.

[403] Balance of payments data are available for only part of 2001 and only merchandise trade data are available for Turkmenistan. See notes to table 3.5.1.

[404] No balance of payments data have been available for Turkmenistan since 1998. The statement relating to its current account reflects secretariat estimates based on changes in merchandise trade. While these estimates may give a good indication of the direction of change, the levels are likely to be unreliable, and thus they are not shown in table 3.5.1.

Republic of Moldova, Tajikistan and Uzbekistan (CIS7).[405] In these countries there were particularly sharp falls in output after the break-up of the former Soviet Union. About 20 million people, or half their total population, live in extreme poverty. They have little or no access to international credit markets[406] and, with the exception of Uzbekistan, they are eligible for the IMF Poverty Reduction and Growth Arrangement.[407] FDI inflows have been meagre aside from some investment in natural resource extraction projects (especially oil in Azerbaijan). A debt sustainability analysis has been carried out by the IMF and World Bank using two sets of macroeconomic assumptions. The baseline scenarios suggest the following:[408]

- Georgia will require continued concessional lending and grants from abroad, as well as concessional rescheduling of bilateral debts.

- The Kyrgyz Republic will require debt relief in order to have any chance of achieving external sustainability by the end of the decade.

- The Republic of Moldova is likely to need short-term debt relief. The near-term problem is a eurobond maturing in 2002, but the debt service will remain high until 2006.

- Tajikistan faces a particular liquidity problem until late in the decade, and in fiscal terms the burden is very high (partly owing to low revenue levels). A rescheduling of maturities coming due would provide only modest relief, but a reduction of debt in 2004 would help significantly.

- Armenia is the only one of these countries that does not need a restructuring of bilateral or commercial debt.

Under the low-case scenarios, reflecting lower assumptions about the growth of GDP and exports of goods and services, the projected external situation of these countries becomes even more difficult and even Armenia would require a rescheduling of official bilateral debt, or alternatively, greater fiscal adjustment than assumed in the base scenario.

In 2001,[409] compared with the scenario assumptions, the economic performance of these five countries was mixed. In all cases actual GDP growth exceeded the assumptions, in Armenia and Tajikistan by substantial margins. Armenia and the Republic of Moldova achieved a relatively rapid growth of exports of goods and services and some reduction in their current account deficits (in relation to GDP; in both cases the deficits remain large). However, in Georgia, services alone buoyed export receipts, which together with lower merchandise imports and higher remittances contributed to an improved current account balance. In Kyrgyzstan, and especially Tajikistan, the value of exports of goods and services declined, in the latter case contributing to the considerable increase in the current account deficit. These results recall the fact that the achievement of sustained export (and GDP) growth will depend not only on domestic policies but also on the favourable development of factors beyond their control – e.g. the weather, the absence of armed conflict in neighbouring countries and access to international transport links. Several countries were able to increase their exports of food because droughts had abated and harvests improved. However, Kyrgyzstan's exports of hydroelectricity were reduced because of low reservoir levels, while various problems affecting Tajikstan's exports were exacerbated by a ban on transit through Kazahkstan as a result of the conflict in Afghanistan.

Regarding Azerbaijan and Uzbekistan (not covered by the IMF-World Bank paper),[410] a strong growth of Azerbaijani exports of goods and services moved the current account into surplus. The country has one of the most comfortable liquidity positions among the CIS, which is also the case of Uzbekistan. In 2001, the latter's exports of goods and services declined, and the current account balance reverted to deficit.

Although many of these low-income countries have made considerable progress, further policy and institutional reforms are essential if they are to achieve sustained growth. Adequate support from the international community in the form of market access, generous financing, debt relief where appropriate and continuing policy advice will be indispensable. Georgia's near-term financial situation has been eased with the rescheduling in March 2001 of its Paris Club debt coming due in 2001-2002. In March 2002, Kyrgyzstan and the Paris Club also reached a rescheduling agreement that will eliminate most of its debt service obligations to Paris Club creditors for three years.[411] The Republic of Moldova and Tajikistan are likely to request similar assistance during 2002.

---

[405] A seminar dealing with the challenges faced by these countries was hosted by the United Kingdom and organized by the IMF, World Bank, ADB and EBRD in London, 21-22 February 2000. See, *Poverty Reduction, Growth and Debt Sustainability in Low-income Countries*, a paper prepared for the seminar by IMF and the World Bank (Washington, D.C), 4 February 2002.

[406] There was virtually no net bank lending to the Asian CIS in the first half of 2001. BIS, *BIS Quarterly Review of International Banking and Financial Market Developments* (Basle), December 2001. Only Azerbaijan has obtained a (sub-investment grade) international credit rating (sect. 3.5(iv)(d)), but it has not issued any bonds.

[407] The IMF and World Bank are the largest multilateral creditors, and Russia and Turkmenistan are the largest bilateral creditors.

[408] IMF and World Bank, op. cit., table 13. The baseline scenario for 2001-2010 assumes average annual growth rates of GDP and exports of goods and services for Armenia of 5.5 and 9.3 per cent, respectively, for Georgia (3.8 and 7.3 per cent), Republic of Kyrgyzstan (4.4 and 6.2 per cent), Republic of Moldova (4.5 and 11.9 per cent) and Tajikistan (4.6 and 5.9 per cent).

[409] The IMF-World Bank paper is based on data through 2000.

[410] Unlike the other CIS7 low-income countries, Azerbaijan and Uzbekistan are considered less-indebted countries by the World Bank.

[411] The agreement is expected to reduce debt service during 2002-2004 from an initial amount of $101 million to $5.6 million. Paris Club, *Press Release*, 7 March 2002 [www.clubdeparis.org].

(d) Changes in credit ratings and bond issues

The international asset rating agencies[412] provide assessments of the creditworthiness of countries, including of the risk of default on their long-term foreign currency bonds. More broadly, the ratings are seen to reflect countries' economic capacity, the investment climate and their resilience to shocks. Only Hungary was rated prior to the transition, but during the past decade 17 countries have obtained ratings and, in general, perceptions of their creditworthiness have steadily improved. In the wake of the Asian (1997) and Russian (1998) financial crises, investor sentiment turned against all emerging markets. Although most of the rated transition economies were affected only temporarily, the Republic of Moldova, Romania, Russia, Slovakia and Ukraine were eventually downgraded, Russia and Ukraine because of payments defaults (but not on sovereign eurobonds). In the others, beset by lagging reforms and the accumulation of imbalances, the shocks triggered currency devaluation.

At the beginning of 2002, the so-called "first wave" EU accession countries[413] – the Czech Republic, Hungary, Poland, Slovenia and Estonia – have the highest ratings among the transition economies (table 3.5.14). The highest rated, Slovenia, is on a par with Greece and Malta. Of the "second wave" countries, Slovakia, Latvia and Lithuania have also received (somewhat lower) investment grade ratings, but Bulgaria and Romania are perceived as speculative risks (sub-investment grade). Croatia, which has signed a Stabilization and Association Agreement with the EU, also benefits from an investment grade rating. Of the CIS, Azerbaijan, Kazakhstan and Russia are currently rated sub-investment grade and enjoy different degrees of access to the capital markets. In the past two years there has been an extensive upgrading of these transition economies,[414] especially among the sub-investment grade category, and Croatia and Lithuania have moved up a notch to investment grade. The upgradings are particularly noteworthy because they occurred during a period of market turbulence (although it was not as virulent as the bouts during 1997-1998).[415] In several cases they contributed to improved borrowing terms (see below).

---

[412] The three major agencies are FitchRatings, Moody's, and Standard and Poor's. For reasons of presentation, only Fitch ratings are presented in table 3.5.14, but there are considerable similarities between the ratings of the three agencies (and the recent changes). The differences between agency ratings are given in the note to table 3.5.14.

[413] The first and second wave countries were defined in the European Council's Agenda 2000, July 1997.

[414] Table 3.5.14 highlights only upgrades since the beginning of 2000, but Poland (November 1998), Hungary (October 1999) and Slovenia (December 1999) were upgraded in the wake of the rouble crisis.

[415] In its September assessment of sovereign credit risk, the *Institutional Investor* found that the transition economies group was the only one of its regional subgroupings to receive a higher evaluation during the previous six-month period.

TABLE 3.5.14

Fitch credit ratings [a] for the transition economies and changes in 2000-2002

Fitch, S&P/Moody's [b]	Country	Date	Outlook
Investment grade [d]			
A/A2	Slovenia[e]	Dec. 1999	Stable
A-/A3	**Hungary**[e]	Nov. 2000	Stable
	Estonia	Aug. 2001	Stable
BBB+/Baa1	Czech Republic[e]	Nov. 1997	Stable
	Poland[e]	Nov. 1998	Stable
	Estonia	Sep. 2000	
BBB/Baa2	Latvia[e]	Jun. 1998	Positive
BBB-/Baa3	**Croatia**	Jun. 2001	Stable
	Lithuania	May 2001	Stable
	Slovakia[f]	Oct. 2001	Stable
Sub-investment grade [d]			
BB+/Ba1	Slovakia[g]	Dec. 1998	Positive
BB/Ba2	**Kazakhstan**[g]	Jul. 2001	Stable
BB-1a3	**Bulgaria**	Jan. 2002	Stable
	Azerbaijan	Jul. 2000	
B+/B1	**Russia**	Oct. 2001	Stable
	Azerbaijan	Jun. 2000[h]	
B/B2	**Romania**[e]	Nov. 2000	Positive
	Russia	Aug. 2000	
B-/B3	Ukraine	Jun. 2001[h]	Stable
	Russia	May 2000	
Downgrades			
CCC+/Caa1	Republic of Moldova	May 2001	
CCC-/Caa3	Turkmenistan	May 2001	N/A
CC/Ca2	Republic of Moldova	Jun. 2001	Negative

Source: Fitch, Moody's and Standard and Poor's rating services.

Note: Countries in bold print were upgraded during 2000-2001. Moody's rates Russia (Ba3), Slovakia (Baa3) and Ukraine (B2) one notch higher and Lithuania (Ba1) one notch lower than does Fitch. Standard and Poor's rates the Czech Republic (A-), Slovakia (BBB-) and Ukraine (B) one notch higher than Fitch does. Ratings above A (i.e. AAA and AA) are not shown in the table.

[a] Ratings on long-term sovereign bonds.

[b] The first rating is the one used by Fitch and Standard and Poor's, the second is used by Moody's.

[c] Date of ratings change.

[d] An investment grade rating implies a strong (in the case of a single A rating) or adequate (in the case of a BBB rating) ability of a country to service its obligations. In the speculative grades (BB and below) the servicing of obligations is probable at best.

[e] The rating was affirmed in 2001.

[f] Moody's and Standard and Poor's rate Slovakia an investment grade risk.

[g] The rating was affirmed in 2002.

[h] First rating.

The relatively high ratings and recent upgrades of the "first wave" countries reflect a variety of factors, including macroeconomic fundamentals and policies, progress in economic reform, including meeting the requirements of EU accession (*acquis communautaire*), the adequacy of official reserves, political risk, and so on. The justifications for the recent individual upgradings are also grounded in these factors, but they also reflect country-specific situations (conditions for the further upgrades of fast reformers tend to differ from those for

The Transition Economies

slow reformers). For example, in Hungary, the upgrade was associated with the adoption of monetary and exchange rate policies consistent with the requirements of ERM-2. In the case of Bulgaria, it reflected the government's prudent fiscal targets, the ambitious structural reform programme and the strong liquidity position. In Romania, the key has been the improved external financial position, which should enable the country to finance a larger current account deficit and meet its debt servicing obligations.

The Republic of Moldova and Turkmenistan have been downgraded but for different reasons. Since the rouble crisis, there have been doubts about the sustainability of the former country's debt and fiscal positions, but the downgrade was triggered by its failure to make an interest payment on a bond in June 2001.[416] In Turkmenistan a broad range of economic and political factors have been responsible for the lowering of its rating, and despite the country's economic potential.[417]

The upgradings in table 3.5.14 have reinforced a perception of increasing gaps in creditworthiness within the region. The highest rated countries have continued to consolidate their positions, and several sub-investment rated countries have moved up. However, for various reasons the low-income CIS7 (section 3.5(iv)(c)), Belarus, Turkmenistan and several south-east European countries have made little or no progress towards gaining access to the private capital markets.

The upward adjustment of credit ratings is broadly consistent with the improvement in financial indicators (section 3.5(iv)(a)) and the performance of the securities of the transition economies in the international markets (chart 3.5.9).[418] Moreover, terms on the new bond issues of several countries continued to improve despite the unpropitious climate and many were heavily oversubscribed. Chart 3.5.9 indicates that spreads on the foreign currency bonds of the transition economies are often smaller than those of emerging market economies with comparable credit ratings. The impact of the crises in Argentina and Turkey on spreads was subdued and

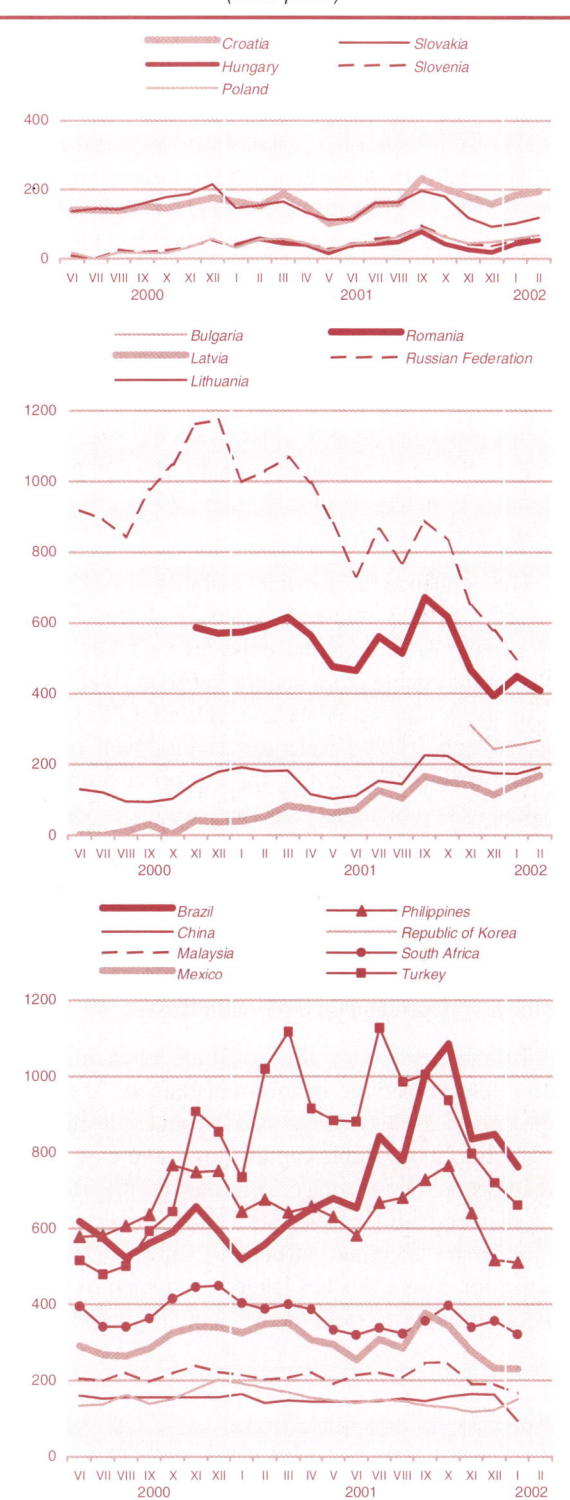

CHART 3.5.9

Yield spreads [a] on the international bonds of selected transition and other emerging market economies, 2000-February 2002

(Basis points) [b]

Source: Financial Times; Datastream.

a The bonds of all the transition economies are denominated in euros, except those of Russia, which, like those of the other emerging market economies, are in dollars. All spreads on bonds are relative to United States treasury bills of the same maturity. It should be noted that spreads on euro-denominated bonds are generally lower than those on dollar-denominated bonds.

b One basis point equals 0.01 per cent.

---

[416] Although the Republic of Moldova eventually made the semi-annual coupon payment, concerns persist as to whether it will be able to manage the full $75 million redemption due in June 2002. FitchResearch, 3 July 2001.

[417] Fitch has attributed the current low credit rating (down from B in 1998) to Turkmenistan's "trade profile [and] political and structural deficiencies that heighten its vulnerability to domestic and external shock ... combined with a poor credit history, the lack of reliable economic information and the reluctance of international financial institutions to provide support". Fitch, *Republic of Turkmenistan Sovereign Report*, 18 May 2001. It may be noted that Turkmenistan is the only transition economy that does not publish balance of payments data (see above).

[418] It has been observed that credit rating agencies have been fairly conservative in the aftermath of the Asian crisis, typically upgrading countries only once the improvement in the economic situation has been validated by data. IMF, *Assessing the Determinants and Prospects for the Pace of Market Access by Countries Emerging from Crises* (Washington, D.C), 6 September 2001.

temporary, without seriously hindering access to the capital markets. Spreads did rise after 11 September (as did margins on the bonds of most countries),[419] but they have since narrowed to pre-attack levels. At the beginning of 2002, there was a tendency for spreads to increase, due partly to the Enron effect and concern about corporate debt, especially that of heavily-indebted telecommunications firms.

Although the emerging market economies as a whole have suffered from heightened risk aversion, creditworthy transition economies have benefited from the increased willingness of investors to differentiate between borrowers. The combination of economic restructuring and macroeconomic stabilization, improved financial indicators, credit rating upgrades and closer association with the EU is likely to make the accession countries less vulnerable to external shocks and financial crises.[420] While these improvements do not guarantee immunity to external shocks, they do seem to have reduced the risks of contagion.

The enhanced financial stability of Russia is noteworthy from a regional viewpoint since it was the sharp growth of its external exposure and low reserves that led to the rouble crisis and the subsequent collapse of the country's import demand (section 3.2(iv)). This increased stability is a welcome development in and of itself and despite the fact that the transition countries are probably less vulnerable than they were in 1998. The absence of serious contagion from Argentina and Turkey suggests that financial institutions and investors have been successful in eliminating some of the potential channels/mechanisms of contagion. Also the east European countries (including the Baltic states) have become less dependent on trade with Russia.

To the extent that the transition economies have become less vulnerable to external financial shocks and are perceived as less risky, their stronger position may translate into more stable capital flows and ease the task of macroeconomic management during the run-up to EMU membership (see chapter 5). However, they will still need to continue structural reforms (including privatization where this has lagged) and keep imbalances in check (in order to meet the Maastricht criteria). Other necessary measures, also frequently cited as conditions for further credit upgrades, include reductions in government indebtedness, improvements in fiscal transparency (section 3.1(ii)), and measures to tackle corruption.

The volume of bond issues by the transition economies amounted to $6.3 billion in 2001, about $1 billion more than in 2000. The majority were sovereign and municipal issues, but there were also a few corporate bonds from energy producers in Kazakhstan, Romania and Russia. Terms on the new issues of several countries continued to improve[421] despite the unpropitious climate and many were heavily oversubscribed. This reflects international investors' search for higher yields, while seeking what are perceived as less risky investments. The modest overall level of issuance by creditworthy countries reflects the availability of alternative finance, especially FDI, but several countries have borrowed to finance government budget deficits. Hungary, a major issuer of bonds in the past, has recently borrowed only to refinance maturing debt. In early 2002, the terms on Poland's €750 million bond (at the same goods terms as in 2001)[422] also reinforces the view that there has been little if any negative spillover from the crisis in Argentina.

---

[419] According to some analysts Poland was seen as a safe haven during the financial crisis in Argentina (this was already the case during the Russian crisis). More generally, east European issues are seen to have benefited from the post 11 September "flight to quality". For example, according to Gian-Carlo Perrasso, senior economist at JP Morgan, "The only credits that have performed well are eastern Europe, where the spread with respect to the United States has actually narrowed", *Financial Times*, 16 October 2001 and 19 February 2002.

[420] It has been pointed out that from the viewpoint of foreign currency bonds, most of the investment-rated transition economies are no longer considered as emerging markets (although this may not be the case from the standpoint of local debt and stock markets). The spreads on their eurobonds, 60-75 basis points over German bunds, is considered too small for investors in emerging markets.

[421] This pertains to Croatia, Kazakhstan and Lithuania.

[422] The Polish bond was issued at 75 basis points over German bunds, the same spread as in early 2001.

PART TWO

POLICIES FOR ADJUSTMENT AND GROWTH

161

CHAPTER 4

TECHNOLOGICAL ACTIVITY IN THE ECE REGION DURING THE 1990s

While the task of estimating the contribution of technical change to economic growth is fraught with difficulty, there is a clear positive relationship for the ECE region as a whole between variables such as the share of high-tech products in total exports and of educational expenditure and attainment levels on the one hand, and levels of GDP per head on the other. There is also a marked tendency in the developed industrial economies for gross expenditure on R&D (GERD) to increase over time as GDP increases. The relationship between GERD and GDP is obscured in the case of the transition economies by the fact that in the late communist period GERD was largely devoted (military purposes apart) to the maintenance of networks of research institutes, which made little or no contribution to economic development. The restructuring of these networks is still going on. As a result, formal R&D provided few inputs into the economic recovery recorded in the CEECs from the mid-1990s, and continues to weigh on the CIS countries as a dead weight loss rather than a source of benefits. R&D policy for the transition countries has to be developed against that specific background. Effective R&D policy may require further cuts in formal, public sector R&D expenditure, and should in any case be focused on developing the private sector's capacity to assimilate new technology, maximizing the technological impact of FDI, and helping to build supply networks as vehicles for technological upgrading.

4.1 Introduction

Technical change and technological learning are essential for long-term growth and economic convergence. While the relationship between aggregate and sectoral trends in technology is a complex one, the case of the United States illustrates the kinds of pattern that may emerge. Having grown through the 1990s at an average rate of almost 4 per cent per year in terms of real GDP per capita, the United States stood at the forefront of the major economies of the ECE region at the turn of the century.[423] It was the information and communications technology (ICT) sector that provided an important stimulus to growth, which helped the United States pull further ahead of most other ECE region economies. Without the technological and organizational capability to develop, or at least to absorb, these new technologies, some countries in the region may fall even further behind.

Technological and organizational capabilities can be defined as the "resources needed to generate and manage technical change, including skills, knowledge and experience, and institutional structures and linkages".[424] Building technological capabilities is a cumulative, path-dependent activity that generates technical change, investment in new capacity and ultimately growth. It is also a complex and diverse activity that involves interaction between users and producers, and between firms and other organizations, engendering different patterns of technological accumulation and innovation[425] depending on the learning structure. The incentive structures and competences of institutions support and sustain the rate and direction of technological learning, and suggest a role for public policy to influence the innovation process.

At the level of the firm, technological capability can be defined in terms of absorptive capacity – the capacity to assimilate technical knowledge from both home and abroad, from inside the firm and outside it. In the global economy foreign direct investment, joint ventures, strategic alliances, technology licensing, subcontracting and embodied technology transfer all play an important

---

[423] In this chapter GDP is converted to international dollars using purchasing power parity rates. An international dollar has the same purchasing power over GDP as the dollar has in the United States. For the purpose of comparing countries across the ECE region and across time, all data on real GDP are taken from World Bank, *World Development Indicators* (Washington, D.C.), January 2002.

[424] M. Bell and K. Pavitt, "Technological accumulation and industrial growth: contrasts between developed and developing countries", *Industrial and Corporate Change*, Vol. 2, No. 2, 1993, pp. 157-210.

[425] That is, R&D which has produced a concrete, commercially applicable result, in terms of new products or processes.

role in this process of assimilation, and therefore in the growth process. These knowledge flows appear as externalities or technological spillovers in neoclassical endogenous growth theory, and as a joint product in the classical theory of production. The absorptive capacity of a firm is defined as its "ability to identify, assimilate and exploit knowledge from the environment".[426] When firms want to apply knowledge acquired from technological spillovers, they must enter into a time-consuming and costly process of investing in their absorptive capacity. Thus the idea of absorptive capacity becomes a connecting device between the potential for catching up (technological opportunities) and its realization (appropriability conditions).[427]

Technological opportunities can arise from changing patterns of demand, changes in the size of markets, the product cycle, or developments in science and technology (S&T). The technological capabilities of the business enterprise lie in its engineering, design, research and marketing resources and assets. The realization of these technological opportunities will depend on the factors that facilitate the creation and diffusion of knowledge, the growth of demand (both global and national), and the strategic behaviour of the firm, including the capability to take advantage of these opportunities. "Being backward in the level of productivity carries a potential for rapid advance",[428] but the realization of this potential depends on whether firms in the economy can develop the requisite technical competence.

This chapter describes the patterns of technological activity in the ECE region and considers the link between R&D activity and industrial innovation, on the one hand, and the level of economic development on the other. After summarizing the patterns of convergence with and divergence from the United States per capita GDP level in the 1990s, the chapter describes the patterns of education, R&D and inventive activity in the region. Emphasis is placed on the rapid decline of R&D in the transition economies in the early 1990s, and its implications for catching up with the United States. Attention then shifts to the innovation surveys carried out in the European Union and some of the transition economies, and some conclusions are drawn about the innovation process during economic transformation. The chapter then assesses the patterns of technology transfer and diffusion, and closes with a brief discussion of S&T policy.

4.2 Convergence and divergence in per capita income levels in the 1990s

Real GDP per capita provides a rough indication of how successful the ECE region has been in managing technological and organizational change.[429] If the exceptional case of Luxembourg, which depends heavily on high value added services, is put to one side, the United States is the leader on this indicator, and therefore the target for closing the income (or technology) gap. Chart 4.2.1 shows real GDP per capita relative to the United States for the period 1993 and 2000.[430] It presents the ECE member states in the order they appeared in 2000, and shows their position relative to the United States in 1993 as solid lines. Lines that appear within each bar indicate that a country is catching up with the United States, whereas lines that appear outside the bar indicate that a country is falling behind. For example, Ireland caught up by 29 percentage points from 1993 to 2000 and is now ranked fifth in terms of real GDP per capita. By contrast, Switzerland fell behind by 12 percentage points and is now ranked sixth.[431]

There were two main reasons for choosing this time period. First of all, this is the period when ICT began to emerge as a new "growth mode", globally and more particularly in the United States.[432] Thus the OECD estimates that the contribution of ICT investment to GDP per capita doubled in the United States and Finland during the second half of the 1990s.[433] Almost all of the west European countries that kept pace with or surpassed the United States growth rate in these seven years had a relatively large share of value added and employment in ICT (manufacturing and services).

---

[426] W. Cohen and D. Levinthal, "Innovation and learning: the two faces of R&D", *Economic Journal*, No. 99, September 1989, pp. 569-596.

[427] This chapter builds mostly on M. Knell, "Patterns of technological activity in CEECs", in M. Landesmann (ed.), *Structural Developments in Central and Eastern Europe* (Vienna, WIIW, 2000).

[428] M. Abramovitz, *Thinking About Growth* (Cambridge, Cambridge University Press, 1989).

[429] Differences in GDP per capita can also be attributed to differences in labour productivity, or GDP per hour worked. Demographic patterns and utilization of the labour force can have an important impact on real GDP. For example, in 1999, Belgium, France, Italy and the Netherlands had higher levels of GDP per hour worked than the United States. OECD, *Science, Technology and Industry Scoreboard: Towards a Knowledge-based Economy* (Paris), 2001.

[430] UNECE, "Catching up and falling behind: economic convergence in Europe", *Economic Survey of Europe, 2000 No. 1*, chap. 5, describes the patterns of convergence and divergence relative to United States per capita GDP from 1950 in more detail.

[431] It is important to add that when net income from abroad is taken into account (gross national income, or GNI), Switzerland moves up in rank to third and Ireland down to twelfth. The main reason is that Ireland pays more than 12 per cent of its GDP as license fees and profits expatriated by TNCs. Switzerland is a net recipient of foreign income. The differences between GDP and GNI are much smaller in other ECE member states.

[432] This so-called "new economy" is the fifth growth mode since the Industrial Revolution began in the late 1700s. C. Freeman and L. Soete, *The Economics of Industrial Innovation,* Third Edition (London, Pinter, 1997).

[433] The contribution of ICT to economic growth is analysed in more detail in OECD, *Science, Technology and Industry Outlook: Drivers of Growth: Information Technology, Innovation and Entrepreneurship* (Paris), 2001.

CHART 4.2.1
Real GDP per capita in current PPPs in the ECE region, 1993 and 2000
(United States=100)

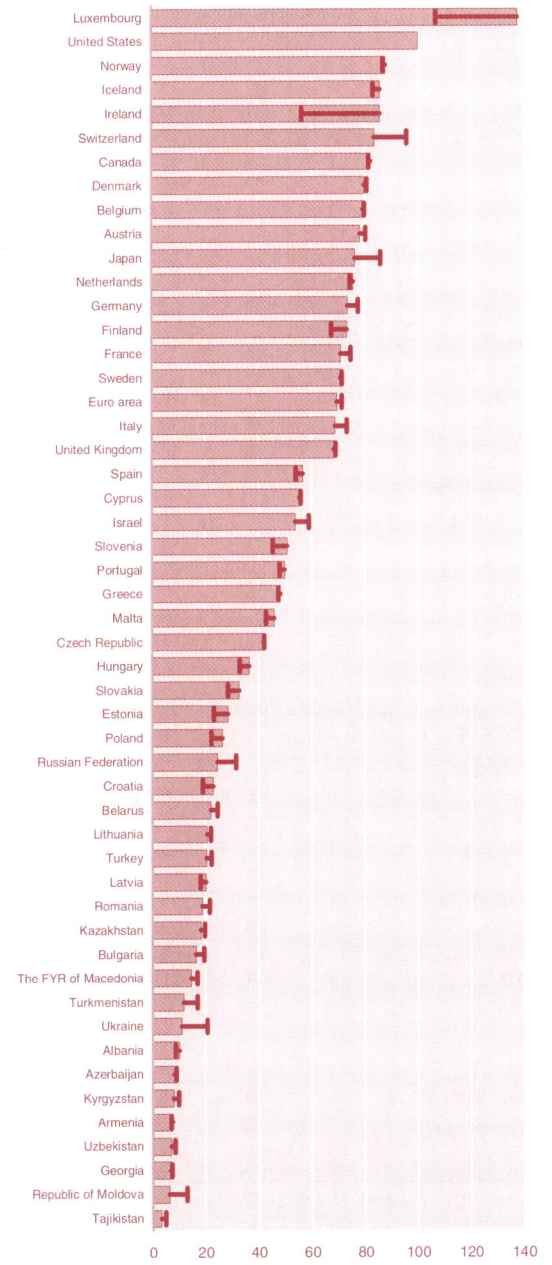

Source: UNECE calculation, based on World Bank, *World Development Indicators* (Washington, D.C.), 2002.

Note: The relative position of countries in 1993 is shown by the solid lines (see text).

CHART 4.2.2
High-technology exports and per capita GDP in the ECE region, 1999

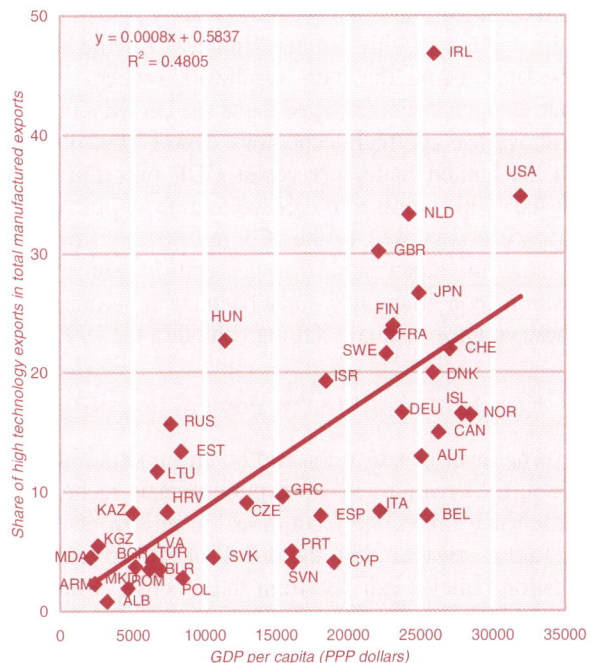

Source: UNECE calculation, based on World Bank, *World Development Indicators* (Washington, D.C.), 2002.

Note: The following country codes are taken from the ISO 3166 Country Code List. Albania: ALB; Armenia: ARM; Austria: AUT; Azerbaijan: AZE; Belarus: BLR; Belgium: BEL; Bosnia and Herzegovina: BIH; Bulgaria: BGR; Croatia: HRV; Cyprus: CYP; Czech Republic: CZE; Denmark: DNK; Estonia: EST; Finland: FIN; France: FRA; Georgia: GEO; Germany: DEU; Greece: GRC; Hungary: HUN; Iceland: ISL; Ireland: IRL; Italy: ITA; Kazakhstan: KAZ; Kyrgyz Republic: KGZ; Latvia: LVA; Lithuania: LTU; Luxembourg: LUX; The former Yugoslav Republic of Macedonia: MKD; Republic of Moldova: MDA; Netherlands: NLD; Norway: NOR; Poland: POL; Portugal: PRT; Romania: ROM; Russian Federation: RUS; Slovak Republic: SVK; Slovenia: SVN; Spain: ESP; Sweden: SWE; Switzerland: CHE; Tajikistan: TJK; Turkey: TUR; Turkmenistan: TKM; Ukraine: UKR; United Kingdom: GBR; Uzbekistan: UZB; Yugoslavia: YUG; Canada: CAN; Israel: ISR; Japan: JPN; Republic of Korea: KOR; United States: USA.

The second reason relates more specifically to the transition economies. By 1993 most of the east European economies had already moved to a new phase of economic transformation (somewhat later in the Baltics). In this phase, enterprises started to become more assertive in their search for new products and markets. The increase in global aggregate demand, partly linked to the rapid growth of ICT, combined with the establishment of new market institutions, stimulated growth and encouraged enterprises to follow a more strategic approach and reorganize their asset structures. The process of industrial restructuring accelerated during this phase as enterprises concentrated on eliminating redundant labour and developing the capability to reorganize and transform themselves into global competitors. Even with the high growth rate in the United States, several countries in eastern Europe narrowed the income gap in the late 1990s.

The share of high-technology exports in total exports illustrates the relationship between technology and economic development. High-technology products, which include aerospace, computers, pharmaceuticals, scientific instruments and electrical machinery, generally have a high R&D intensity. While these sectors vary enormously in terms of spin-off or linkage potential (from computers at the top to aerospace at the bottom), their aggregate share of exports provides some indication of the technological capabilities of a country. Chart 4.2.2 reveals that countries having the highest rate of catch up with the United States tend to be well above the trend line for this indicator, with Estonia, Hungary, Ireland and the Netherlands being the most significant outliers. Some countries, such as France, Russia and the United Kingdom, had above average high-technology exports, but diverged from the United States per capita GDP over the period analysed. This seems to be related to a heavy reliance on aerospace as opposed to the ICT industries. Structural problems in Japan and Russia (and in some other transition economies) may have been an important factor in these countries actually falling behind in the 1990s.

4.3 Education and technological learning

The level of education can be an important driving force in the creation and absorption of new technology. By the same token, the institutional arrangements of the educational system and the level of effectiveness of education policies can constrain the process. Education is an investment in human skills and competences that, over time, become part of the human capital stock, or social capabilities, of a country. The rich human capital stock in the transition countries is a key element in their underlying resource endowment, although it does not, of course, translate directly into capabilities that can be directly deployed in the process of economic transformation. General indicators of the human capital stock include various measures of the educational attainment of the population and workforce. Investment in human capital can be measured as the resources that each country puts into education and, in particular, its spending per student at each educational level.

Despite being crude, narrow and insensitive to differences in the quality of formal education, educational attainment remains the best indicator available of the human capital stock of a country. Table 4.3.1 provides a breakdown of the educational attainment of the workforces of the ECE region in 2000, based on national labour market surveys. About 86 per cent of the workforce in the United States had at least an upper secondary education, one of the highest levels in the ECE region. By contrast, only slightly more than 20 per cent of the workforce in Portugal and Turkey had at least an upper secondary education. Both eastern Europe and the Russian Federation had attainment levels similar to the United States, reflecting the fact that education was an important objective under central planning. Although many of the competences taught under the old system have become obsolete, especially in the social sciences, the general level of education has made it easier for companies, in the process of investment, to establish technological congruence between the human capital stock and the production process, a key condition of effective redeployment. This in turn has helped to attract transnational corporations (TNCs) to take advantage of

TABLE 4.3.1

Educational attainment of the workforce in the ECE region, 2000 [a]

(Percentage share)

	First level and below	Lower secondary	Upper secondary	Applied tertiary	Degree tertiary
European Union					
Austria	x	21	69	2	7
Belgium	11	21	36	21	10
Denmark	x	23	52	8	17
Finland	x	26	43	x	31
France	x	29	46	x	25
Germany	2	16	58	10	14
Greece	31	11	30	13	15
Ireland	14	20	28	20	14
Italy	13	36	40	1	10
Luxembourg	19	12	34	12	20
Netherlands	9	22	45	x	24
Portugal	64	15	12	3	6
Spain	27	27	19	x	27
Sweden	7	14	49	16	14
United Kingdom	12	6	47	9	17
Other western Europe					
Cyprus	24	9	41	10	17
Iceland	3	41	28	13	15
Israel	7	9	35	25	23
Norway	x	15	56	11	17
Switzerland	x	18	58	x	24
Turkey	68	10	14	x	8
North America					
Canada	4	14	31	32	19
United States	5	8	51	8	27
Eastern Europe					
Croatia	7	18	58	7	10
Czech Republic	x	10	78	x	11
Estonia	1	11	46	23	18
Hungary	1	18	65	x	16
Lithuania	4	15	44	21	15
Poland	x	16	69	3	12
Romania	14	21	52	4	9
Slovakia	9	4	41	35	11
Slovenia	3	19	63	7	9
CIS					
Georgia	x	49	7	14	27
Russian Federation	2	10	34	x	54
Memorandum item:					
Japan	x	20	49	12	19

Source: ILO, *Yearbook of Labour Statistics* (Geneva), 2001, except for United States data, which come from OECD, *Education at a Glance* (Paris), 2001.

Note: An x indicates that the data are included in the adjacent right-hand column.

[a] 1997 for Japan; 1998 for Croatia, Denmark, Norway and the Russian Federation; 1999 for Finland, Georgia, Switzerland, Turkey and the United States.

the relatively high skills and low unit labour costs in these countries.

One alarming trend in the transition context is that enrolment in higher education has been declining in most of the CIS countries, whereas it has been rising in Europe and North America. Table 4.3.2 shows gross enrolment ratios for 1989 and 1996.[434] This trend in the CIS

---

[434] The gross enrolment ratio is the total enrolment in a specific level of education, regardless of age, expressed as a percentage of the official school-age population corresponding to the same level of education in a given school year.

TABLE 4.3.2
Gross enrolment ratios in the ECE region, 1989 and 1996
(Per cent)

	Secondary 1989[a]	Secondary 1996[b]	Tertiary 1989[c]	Tertiary 1996[d]		Secondary 1989[a]	Secondary 1996[b]	Tertiary 1989[c]	Tertiary 1996[d]
European Union					**Eastern Europe**				
Austria	103	103	33	48	Albania	77	36	8	12
Belgium	103	146	39	56	Bulgaria	77	77	26	41
Denmark	109	121	34	48	Croatia	85	82	23	28
Finland	114	118	45	74	Czech Republic	93	99	16	24
France	95	111	37	51	Estonia	126	104	26	42
Germany	98	104	34	47	Hungary	75	98	14	24
Greece	93	95	25	47	Latvia	97	84	25	33
Ireland	99	118	28	41	Lithuania	95	86	35	31
Italy	81	95	30	47	Poland	82	98	20	25
Luxembourg	74	88	3	10	Romania	101	78	9	23
Netherlands	119	132	37	47	Slovakia	88	94	16	22
Portugal	60	111	20	39	Slovenia	90	92	25	36
Spain	102	120	35	51	The FYR of Macedonia	..	..	19	20
Sweden	90	140	31	50	Yugoslavia	..	..	18	22
United Kingdom	83	129	27	52					
Other western Europe					**CIS**				
Cyprus	81	97	13	17	Armenia	..	90	24	12
Iceland	98	109	26	38	Azerbaijan	91	77	23	17
Israel	83	88	32	41	Belarus	98	93	48	44
Malta	84	84	11	29	Georgia	95	77	36	42
Norway	101	119	39	62	Kazakhstan	101	87	40	33
Switzerland	99	100	25	33	Kyrgyzstan	102	79	15	12
Turkey	46	58	12	21	Republic of Moldova	83	81	36	27
					Russian Federation	95	87	54	43
North America					Tajikistan	102	78	22	20
Canada	81	97	89	88	Turkmenistan	..	..	12	..
United States	94	97	73	81	Ukraine	95	91	46	42
					Uzbekistan	101	94	17	..
Memorandum item:									
Japan	96	103	29	41					

Source: UNESCO, *World Education Indicators* (Paris), current 2002 data (http://www.uis.unesco.org).

[a] 1990 for Germany, Georgia and Tajikistan; 1992 for Slovakia.

[b] 1993 for the Russian Federation and Ukraine; 1994 for Cyprus and Uzbekistan; 1995 for Albania, Belgium, Canada, Denmark, Hungary, Israel, Japan, Kyrgyzstan, Poland, Portugal and the United States.

[c] 1990 for Germany; 1991 for Yugoslavia; 1992 for Slovakia; 1996 for Luxembourg.

[d] 1994 for Cyprus, Japan and the Russian Federation; 1995 for Belgium, Canada, Denmark, Israel, Kyrgyzstan, Poland, Portugal and the United States.

countries may be due partly to the decline in real GDP per capita over the decade. Lower enrolments in higher education could, nevertheless, indicate potential long-term difficulties, especially since many of the skills acquired before 1990 have become obsolete. They may equally represent, to a degree, the process of structural adjustment within the educational sectors of these countries, as courses with no relevance to modern, knowledge-based activities are cut back or discontinued.[435] The rising trend in enrolment across Europe (including eastern Europe) indicates that the stock of human capital has risen over the decade, and in fact, the educational attainment profile across age groups confirms that younger workers have higher attainment levels than older ones. Even so, there are large differences in educational systems across the region. In some countries, for example, applied subjects appear at the upper secondary level and not at the tertiary level. And there is strong evidence that the social sciences, and especially business education, has improved significantly in eastern Europe, and indeed also in the CIS countries.

Table 4.3.3 shows direct public and private expenditure (in PPPs) by country and educational institutions in relation to the number of full-time equivalent students enrolled in these institutions. As in the previous table, investment in human capital is shown to vary considerably across countries. At the secondary level, spending per student varies by a factor of 6.5, ranging from Switzerland with the highest expenditure level to Poland with the lowest. However, these differences do not appear

---

[435] With annual tuition fees in state universities in Russia now running at over $1,000, such restructuring processes have been strongly reinforced by market disciplines.

TABLE 4.3.3

Expenditure per student in the ECE region by level of education, 1998

(Current PPP dollars)

	Primary	Secondary	Tertiary
European Union			
Austria	6 065	8 163	11 279
Belgium	3 743	5 970	6 508
Denmark	6 713	7 200	9 562
Finland	4 641	5 111	7 327
France	3 752	6 605	7 225
Germany	3 531	6 209	9 481
Greece	2 368	3 287	4 157
Ireland	2 745	3 934	8 522
Italy	5 653	6 458	6 295
Netherlands	3 795	5 304	10 757
Portugal	3 121	4 636	..
Spain	3 267	4 274	5 038
Sweden	5 579	5 648	13 224
United Kingdom	3 329	5 230	9 699
Other Western Europe			
Israel	4 135	5 115	10 765
Norway	5 761	7 343	10 918
Switzerland	6 470	9 348	16 563
North America			
Canada	..	..	14 579
United States	6 043	7 764	19 802
Eastern Europe			
Czech Republic	1 645	3 182	5 584
Hungary	2 028	2 140	5 073
Poland	1 496	1 438	4 262
Memorandum item:			
Japan	5 075	5 980	9 871

Source: OECD, *Education at a Glance* (Paris), 2001.

Note: Public and private institutions based on full-time equivalents.

CHART 4.3.1

Public expenditure per secondary student in relation to GDP per capita in the ECE region, 1995

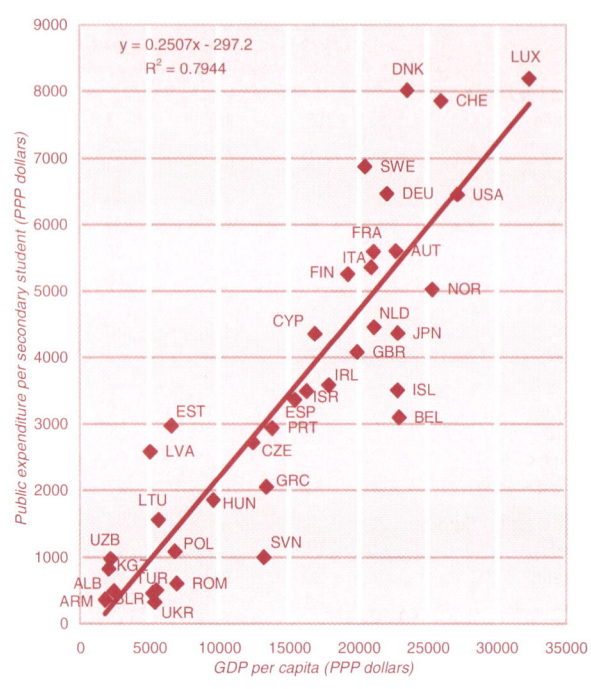

Source: UNECE calculation, based on World Bank, *World Development Indicators* (Washington, D.C.), 2002.

Note: As for chart 4.2.2.

CHART 4.3.2

Public expenditure per tertiary student in relation to GDP per capita in the ECE region, 1995

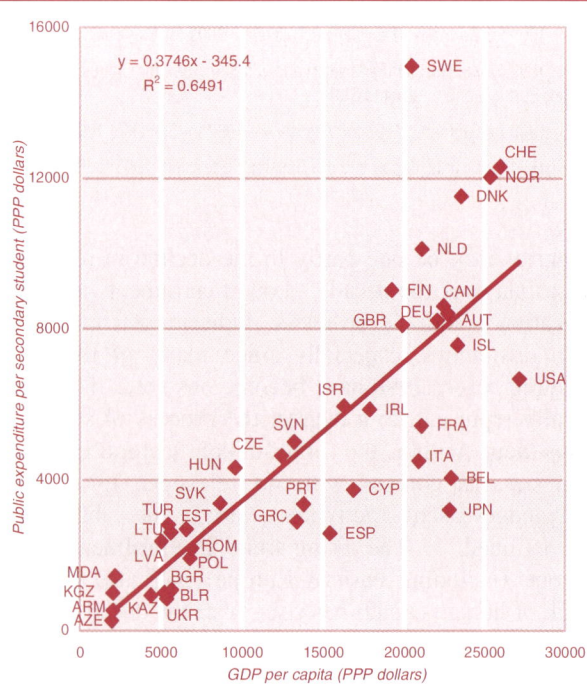

Source: UNECE calculation, based on World Bank, *World Development Indicators* (Washington, D.C.), 2002.

Note: As for chart 4.2.2.

to be very important when compared with levels of scientific literacy or the growth rate of GDP per capita.[436] As charts 4.3.1 and 4.3.2 show, public expenditure per secondary and tertiary student was not only highly correlated with real GDP per capita in 1999, but also positively related. This relationship seems to confirm the proposition that education has a positive impact on economic growth over the long run. It could also simply mean that the relative price of education is higher in the high-income countries or that education is a luxury good (i.e. with positive and significant income elasticity) in the low-income countries. Whether the emphasis is placed on the investment or the consumption aspect of education, however, the key conclusion is that it is important to improve funding, access and the quality of education, if wealth and social well-being are to improve in the ECE region.

---

[436] OECD, *Literacy in the Information Age* (Paris), 2001. A World Bank study by L. Pritchett, *Where Has all the Education Gone?*, World Bank Policy Research Working Paper, No. 1581 (Washington, D.C.), March 1996, suggests that human capital has an insignificant and even negative effect on economic growth.

4.4 R&D activity in the ECE region during the 1990s

Financial and human resources devoted to R&D are two of the most commonly used indicators of technological inputs. Finance for R&D is generally channelled to basic research, applied research or experimental development. Researchers, including scientists and engineers, with at least tertiary level qualifications in an S&T field of study, carry out these activities. Together, they measure the scientific inputs that go into technological activities, but they do not capture innovative activities that go beyond R&D.[437]

Table 4.4.1 shows the evolution of gross expenditure on R&D (GERD) as a percentage of GDP, and the number of researchers per thousand of labour force. These indicators show a clear long-term trend toward greater R&D intensity in western Europe and North America over the past two decades, and a clear downward trend in the transition economies during the 1990s, especially during the first years of transition. Against the background of the rapid decline in aggregate output during this period, the absolute decline in the level of R&D expenditure was much more dramatic than the decline in the GERD/GDP ratio. In eastern Europe, Bulgaria and Slovakia experienced the sharpest decline in R&D intensity. But by the late 1990s, R&D intensity was increasing again in many of the east European countries, as the demand for knowledge-based products increased and the innovation system improved. In Hungary, many of the old research institutes have reappeared as consultancy and research service firms.

In all of the countries of the CIS there were sharp falls in GERD, and in research intensity (despite the large reductions in absolute levels of GDP) through the 1990s. In some, such as Russia and Ukraine, the falls in the number of researchers were much less dramatic, as research institutes, particularly those under the Academy of Sciences, managed to survive through a mixture of financial improvisation and hand-to-mouth restructuring, often involving the redeployment of scientists as instrument-makers, service engineers, etc. In others, such as Armenia and Uzbekistan, GERD and the number of researchers both declined sharply, as entire research institutes were shut down. Over the CIS as a whole, much of the old system remains intact, and in some of the CIS countries the organization of R&D has not been restructured in any meaningful way.

The disintegration of the old industrial ministries after the collapse of central planning left most of the industrial R&D institutes in the former communist countries without finance. This came at a particularly difficult time, when exposure to international competition was highlighting the fact that the bulk of their competences were obsolete.[438] The fact is, however, that in the centrally planned economy formal R&D often had only a tenuous connection with innovation, diffusion and productivity gains. The private sector is the largest source of finance for formal R&D in a market economy, and this helps to integrate it with production and facilitates the transfer of technology at the enterprise level. The rapid decline in R&D expenditure in eastern Europe and the CIS countries might, therefore, be interpreted as reflecting a process of making the S&T system more efficient and responsive to the market. In relation to the former Soviet Union, however, a caveat must be entered here. The rapid demise of the departmental system in the early 1990s meant that valuable concentrations of human capital in the industrial R&D institutes were lost. The institutes of the Academies of Sciences survived much better. As a result, the Russian and Ukrainian scientific systems became even more skewed towards "blue-skies" research than they had been under the Soviet system, and to that extent less rather than more oriented to commercially meaningful innovation.

In the more advanced economies changes in GERD over time tend to be procyclical. As chart 4.4.1 shows, the growth of R&D tends to follow the same trend as GDP growth, but with a lag. In the United States, growth in R&D generally moves more sharply than GDP growth, declining more rapidly than GDP during recession and increasing more rapidly during periods of high growth. This may reflect the tendency of investors, particularly in the technology markets, to avoid risky projects that require large volumes of R&D in times of economic downturn. The large fluctuations in R&D expenditures in Hungary through the 1990s indicate the continuing restructuring of the S&T system during that period (see chart 4.4.2). But the amplitude of these fluctuations diminished as the decade came to a close, indicating that the Hungarian S&T system is behaving increasingly like those of western Europe. Similar trends are observable in the other transition economies.

(i) The structure of R&D

As table 4.4.2 shows, the levels and structure of R&D expenditure differ considerably across the ECE region. There are also considerable differences in terms of who finances these activities and who performs them. Business enterprises finance and perform most R&D: in 1999, they provided more than 60 per cent of the finance

---

[437] The main guide for measuring resources devoted to R&D activities is the "Frascati Manual". OECD, *Standard Practice for Surveys of Research and Experimental Development, Frascati Manual 1993* (Paris) [www.oecd.org]. It is currently undergoing its sixth revision, which should be ready by the end of 2002. The OECD also provides guidelines for the measurement of human resources devoted to S&T activity and the analysis of such data. OECD, *Manual on the Measurement of Human Resources Devoted to S&T, Canberra Manual* (Paris), 1995 [www.oecd.org].

[438] K. Pavitt, "Transforming centrally planned systems of science and technology: the problem of obsolete competencies", in D. Dyker (ed.), *The Technology of Transition* (Budapest, Central European University Press, 1997).

TABLE 4.4.1

R&D intensity in the ECE region, 1981-2000

(Percentage share)

	GERD[a] as a percentage of GDP				Researchers[b] per thousand labour force				GERD[a] as a percentage of GDP				Researchers[b] per thousand labour force		
	1981[c]	1991[d]	1995[e]	2000[f]	1981[g]	1991[h]	1999[i]		1981[c]	1991[d]	1995[e]	2000[f]	1981[g]	1991[h]	1999[i]
European Union	1.69	1.90	1.81	1.86	3.3	4.4	5.5	**Eastern Europe**							
Austria	1.13	1.47	1.56	1.80	2.1	4.2	4.8	Bulgaria	..	1.53	0.62	0.52	..	13.9	3.8
Belgium	..	1.62	1.74	1.98	3.1	4.3	6.5	Croatia	..	1.01	0.96	0.98	..	6.7	4.2
Denmark	1.06	1.64	1.84	2.06	2.5	4.1	6.4	Czech Republic	..	2.02	1.01	1.35	..	4.0	2.6
Finland	1.17	2.04	2.29	3.31	3.7	5.5	9.9	Estonia	..	..	..	0.75	..	7.1	6.3
France	1.93	2.37	2.31	2.15	3.6	5.2	6.1	Hungary	..	1.07	0.73	0.82	..	2.7	3.1
Germany	2.47	2.53	2.26	2.46	4.4	6.1	6.3	Latvia	..	0.59	0.52	0.42	..	2.8	2.5
Greece	0.17	0.36	0.49	0.68	..	1.6	3.3	Lithuania	..	0.43	0.48	0.52	..	3.6	3.4
Ireland	0.68	0.93	1.34	1.39	1.6	3.8	5.1	Poland	..	1.05	0.69	0.70	..	2.7	3.3
Italy	0.88	1.23	1.00	1.03	2.3	3.1	3.3	Romania	..	0.79	0.80	0.41	..	3.5	2.0
Netherlands	1.78	1.97	1.99	2.05	3.4	4.0	5.1	Slovakia	..	2.25	0.98	0.69	..	4.1	3.6
Portugal	0.30	0.51	0.57	0.76	0.7	1.2	3.1	Slovenia	..	1.60	1.69	1.51	..	4.0	4.6
Spain	0.41	0.84	0.81	0.90	1.4	2.6	3.7	The FYR of Macedonia	..	..	..	0.43	..	..	..
Sweden	2.21	2.79	3.46	3.80	4.1	5.9	9.1	Yugoslavia	..	1.15	1.11	1.25	..	4.7	5.3
United Kingdom	2.38	2.08	1.98	1.87	4.7	4.4	5.5								
Other western Europe								**CIS**							
Iceland	0.63	1.16	1.54	2.32	3.1	4.9	10.1	Armenia	..	1.09	0.08	0.26	..	9.1	3.4
Norway	1.18	1.65	1.71	1.70	3.8	6.3	7.8	Azerbaijan	..	0.75	0.31	0.35	..	4.0	2.7
Switzerland	2.18	2.66	2.73	2.73	..	4.5	5.5	Belarus	..	1.43	0.95	0.81	..	10.1	4.2
Turkey	..	0.53	0.38	0.63	..	0.6	0.8	Georgia	..	1.10	0.11	0.19	..	9.2	5.8
								Kazakhstan	..	0.56	0.27	0.17	..	2.9	1.6
North America								Kyrgyzstan	..	0.33	0.26	0.13	..	2.8	1.3
Canada	1.24	1.60	1.74	1.66	3.2	4.7	5.8	Republic of Moldova	..	1.03	0.75	0.58	..	5.1	2.6
United States	2.37	2.72	2.50	2.76	6.2	7.5	8.1	Russian Federation	..	1.89	0.79	1.09	..	11.9	6.8
								Tajikistan	..	0.44	0.11	0.07	..	1.8	1.5
								Turkmenistan	..	0.48	0.26	0.10	..	2.9	..
Memorandum item:								Ukraine	..	1.81	1.34	1.14	..	9.7	4.3
Japan	2.32	3.00	2.98	2.93	6.9	9.2	9.7	Uzbekistan	..	1.16	0.39	0.36	..	3.8	1.4

Source: OECD, *Main Science and Technology Indicators,* Vol. 2001, Issue 2 (Paris), 2001; national sources.

[a] Gross expenditure on R&D.

[b] Full-time equivalent.

[c] 1982 for Portugal.

[d] 1990 for Portugal; 1992 for Croatia, Latvia and Switzerland; 1993 for Lithuania and Slovenia.

[e] 1996 for Switzerland.

[f] 1996 for Switzerland; 1997 for Bulgaria, Ireland and Latvia; 1998 for The former Yugoslav Republic of Macedonia, Turkmenistan and Yugoslavia; 1999 for Belgium, Croatia, Denmark, Estonia, the European Union, Greece, Iceland, Italy, Japan, Lithuania, the Netherlands, Norway, Portugal, Romania, Slovenia, Turkey and Uzbekistan.

[g] 1992 for Portugal; 1993 for Finland.

[h] 1989 for Austria and the Netherlands; 1990 for Portugal; 1992 for Croatia, the Czech Republic and Switzerland; 1993 for Estonia, Lithuania, Romania and Slovenia; 1994 for Latvia, Poland and Slovakia.

[i] 1996 for Switzerland; 1997 for Ireland, Italy and the United States; 1998 for Austria, Belgium, Poland and the United Kingdom.

for R&D and performed almost 70 per cent of these activities in the OECD member states.[439] The government is also an important provider of finance, although enterprises and universities perform most of the activity funded by government. Portugal relies most heavily on government finance (70 per cent), whereas Ireland is at the other extreme in relying mainly on industry (69 per cent).

The countries of eastern Europe tend to rely more on the government for performing and funding R&D. The relatively low percentage of R&D financed by the government in the Czech Republic and Slovakia reflects the abrupt withdrawal of financial support to the majority of research institutes and the attempt to privatize some of the R&D institutes in these countries. Yet the relatively high percentage of R&D being performed by government across the region indicates that remnants of the old S&T system still survive. The Hungarian and Polish governments, for example, have pursued a strategy of gradually restructuring R&D, an approach that has included subsidization of certain research institutes, many of which are under the control of the old academies of sciences. On the other hand, the relatively higher percentage of R&D performed by the universities in these two countries shows that some parts of the S&T system are being substantially restructured.

---

[439] OECD, *Science, Technology and Industry Scoreboard:* ..., op. cit.

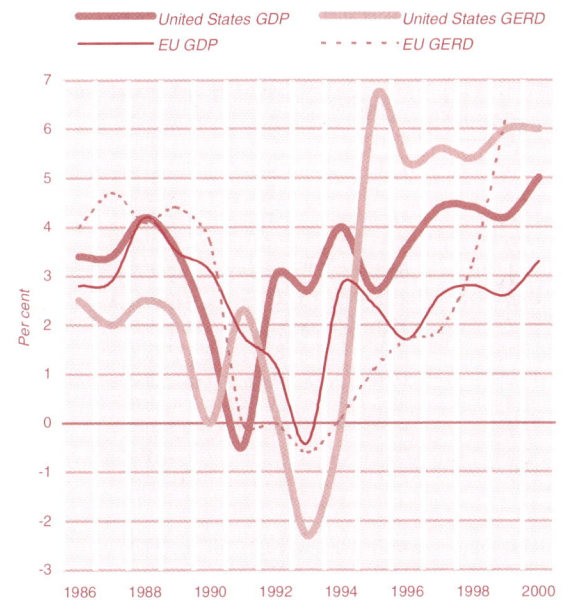

CHART 4.4.1

Real GDP and GERD growth rates in the European Union and the United States, 1986-2000

Source: UNECE calculation, based on Eurostat, New Cronos Database; OECD, *Main Science and Technology Indicators*, Vol. 2001, Issue 2 (Paris), 2001; UNECE Common Database.

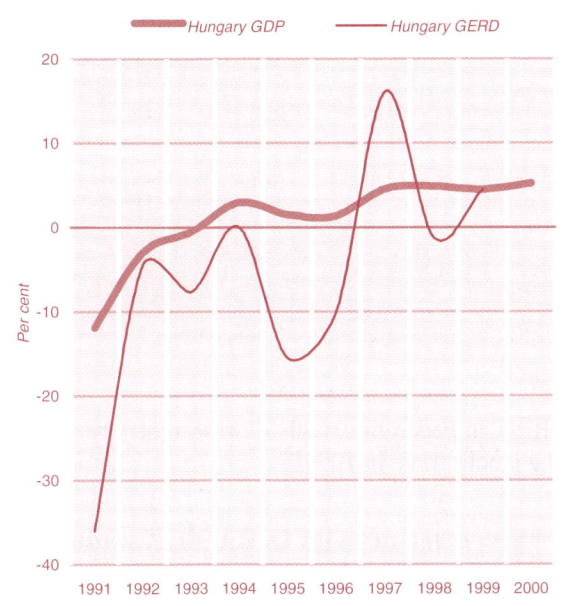

CHART 4.4.2

Real GDP and GERD growth rates in Hungary, 1991-2000

Source: UNECE calculation, based on OECD, *Main Science and Technology Indicators*, Vol. 2001, Issue 2 (Paris), 2001 and UNECE Common Database.

While the Russian Federation also relies heavily on the government for funding R&D expenditure, the business enterprise sector performs comparatively more R&D than most of the countries of eastern Europe. Many of the Russian high-tech enterprises were forced to search for funds abroad in the late 1990s, with the result that external financing increased from 2 per cent of GERD in 1994 to almost 17 per cent in 1999.[440] The high proportion of R&D performed by the business sector in Russia reflects the continued importance of "state orders" for R&D related to defence and other government programmes, rather than any dynamism in firm-level R&D programmes as such. It is also indicative of the fact that the higher education sector is still not playing a significant role in Russia in the performance of R&D. Across the transition region, firms still rely on central sources for R&D funding to an abnormal extent. This again confirms the picture of an S&T system in the transition region, which still retains many features of the old system.

(ii) **Shifting spending priorities or improving efficiency?**

Charts 4.4.3 and 4.4.4 show a clear positive relationship between R&D intensity and researchers per thousand of the population and real per capita GDP. The trend-lines indicate that the level of finance and human resources devoted to R&D activities is associated with a certain level of overall economic development. It may be the case that R&D leads to a higher GDP per capita; it is also plausible that a higher level of economic development allows firms to spend relatively more on R&D. It is likely that there are processes of cumulative (two-way) causality at work here. At the same time, countries with a relatively lower level of GDP per capita generally rely on R&D to help increase absorptive capacity.

In 1999 R&D intensity levels in both eastern Europe and Russia were similar to those of the less developed economies of the European Union (Greece, Portugal and Spain). Chart 4.4.3 also shows the relative positions of selected east European and CIS countries in 1991. The reduction in R&D expenditures in eastern Europe appears as a shift in the expenditure level to one more comparable with countries at a similar level of economic development. This tends to confirm the hypothesis, mentioned above, that R&D expenditures at the time of the collapse of central planning were not optimal, in the sense that there was substantial inefficiency in the use of R&D capacities. The transition towards a market economy has meant a shift in priorities away from research narrowly defined, towards experimental development and creating absorptive capacity, although the trend is blurred by the tendency in some countries for applied research to be cut back more sharply than basic research, as discussed above.

---

[440] OECD, *Main Science and Technology Indicators,* Volume 2001, Issue 2 (Paris), 2001.

TABLE 4.4.2

Gross expenditure on R&D (GERD) by source of financing and performing sector in the ECE region, 1999

(Million dollars, per cent)

	Million current PPP dollars	Per cent financed by:				Per cent performed by:			
		Industry	Government	Other national sources	Abroad	Industry	Government	Higher education	Non-profit
European Union									
Austria	3 646	40.1	39.7	0.3	19.9	63.6	6.4	29.7	0.3[a]
Belgium	5 025	66.2	23.2	3.3	7.3	71.6	3.3	23.9	1.2
Denmark	2 969	58.0	32.6	3.5	5.3	63.4	15.2	20.3	1.2
Finland	3 752	66.9	29.2	0.9	3.0	68.2	11.4	19.7	0.7
France	29 240	54.1	36.9	1.9	7.0	63.2	18.1	17.2	1.5
Germany	47 574	65.0	32.5	0.4	2.1	69.8	13.8	16.5	..
Greece	1 084	24.0	48.7	2.5	24.8	28.5	21.7	49.5	0.3
Ireland[b]	1 084	69.2	22.2	2.0	6.7	73.1	7.0	19.2	0.7
Italy	13 830	44.0	51.3	..	5.1	52.8	22.0	25.2	..
Netherlands	8 395	49.7	35.7	3.4	11.2	56.4	16.5	26.2	0.9
Portugal	1 269	21.3	69.7	3.7	5.3	22.7	27.9	38.6	10.8
Spain	6 375	48.9	40.8	4.7	5.6	52.0	16.9	30.1	1.0
Sweden	7 756	67.8	24.5	4.2	3.5	75.1	3.4	21.4	0.1
United Kingdom	25 463	49.4	27.9	5.1	17.6	67.8	10.7	20.0	1.4
Other western Europe									
Iceland	170	43.4	41.2	1.5	13.9	46.7	30.2	20.9	2.2
Norway	2 140	49.5	42.5	1.6	6.3	56.0	15.4	28.6	..
Switzerland[c]	4 868	67.5	26.9	2.5	3.1	70.7	2.5	24.3	2.5
Turkey	2 636	43.3	47.7	4.2	4.8	38.0	6.7	55.3	..
North America									
Canada	14 727	42.6	32.3	9.3	15.8	57.0	12.1	29.9	1.0
United States	244 699	66.8	28.8	4.5	..	74.7	7.7	13.9	3.6
Eastern Europe									
Czech Republic	1 751	52.6	42.6	0.8	4.0	62.9	24.3	12.3	0.5
Hungary	776	38.5	53.2	0.3	5.6	40.2	32.3	22.3	..
Poland	2 496	38.1	58.5	1.7	1.7	41.3	30.8	27.8	0.1
Romania	573	50.2	46.7	..	..	74.4	18.6	7.0	..
Slovakia	402	49.9	47.9	–	2.3	62.6	27.5	9.9	–
Slovenia	480	56.9	36.8	..	..	55.0	28.5	15.9	..
CIS									
Russian Federation	10 784	31.6	51.1	..	..	69.9	25.2	4.8	..
Memorandum item:									
Japan	95 085	72.2	19.5	7.9	0.4	70.7	9.9	14.8	4.6

Source: OECD, *Main Science and Technology Indicators,* Vol. 2001, Issue 2 (Paris), 2001.

[a] 1998 instead of 1999.

[b] 1997 instead of 1999.

[c] 1996 instead of 1999

Differences in the unit labour costs of researchers explain some of the differences between charts 4.4.3 and 4.4.4. A number of the countries of eastern Europe and the former Soviet Union have similar numbers of scientists and engineers per thousand employees to countries with much higher levels of overall development. The three least developed countries of the European Union (Greece, Portugal and Spain), and three of the more advanced countries of eastern Europe (the Czech Republic, Hungary and Poland) all have about three researchers per thousand of the labour force. Even Austria and Italy are within the range found in much of eastern Europe for that indicator. Estonia is an outlier among the countries undergoing economic transformation, but the Estonian record is surpassed by the Russian Federation, with 7.7 researchers per thousand of the labour force in 1999.

4.5 Inventive activity in the ECE region

Patents are a means of protecting inventions, and as such can be taken as a measure of inventive activity or the output of R&D.[441] Table 4.5.1 presents the average number of resident patent applications per million of the population from 1995 to 1998 in all countries of the ECE region, and the average number of utility patents, i.e. "patents for

---

[441] For a more detailed discussion of patents as an indicator of innovative activity see, OECD, *Using Patent Data as Science and Technology Indicators, Patent Manual* (Paris), 1994 [www.oecd.org].

CHART 4.4.3

Research intensity and GDP per capita in the UNECE region, 1991 and 1999

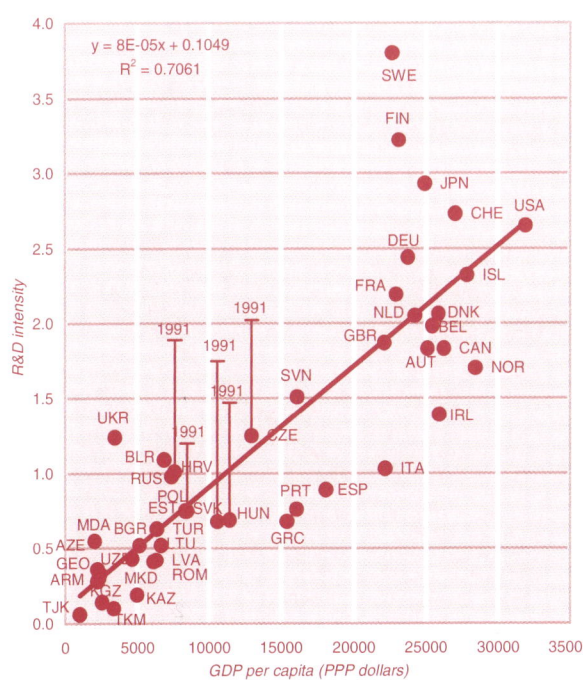

Source: UNECE calculation, based on OECD, *Main Science and Technology Indicators*, Vol. 2001, Issue 2 (Paris), 2001 and UNECE Common Database.

Note: As for chart 4.2.2.

CHART 4.4.4

Researchers per thousand of the labour force and GDP per capita in the ECE region, 1991 and 1999

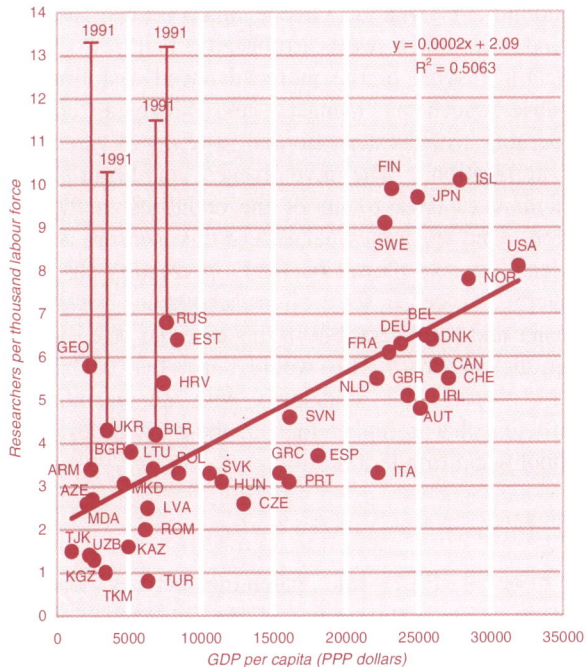

Source: As for chart 4.4.3
Note: As for chart 4.2.2.

inventions", per million of the population granted in the United States from 1998 to 2000. Since the United States has the largest market for technology, the number of patents granted in the United States provides one of the best technological "output" indicators. Because the United States is the home country of American inventors, the number of United States patents per million of the population is biased upward, but the table does show clearly the differences between western Europe and the transition economies. As with other indicators, the more advanced east European economies have similar patterns of patenting activity to the southern EU countries, but Hungary and Slovenia are well above Greece and Portugal.

Table 4.5.1 also shows the growth of patenting in the United States during the 1990s. In all the countries in western Europe there was growth in patenting activity in the 1990s, with Iceland growing the fastest and Switzerland the slowest. For those transition economies for which the time series data are available, there was a tendency for patenting to decline in the 1990s. Romania is the exception, but it started from a very low base. Russia also indicates some growth in recent years, but this represents only a partial recovery from the 50 per cent decline in patenting in the period 1980-1990.[442] The general decline in patenting in the transition region from the late 1980s to the early 1990s mainly reflects a shift towards more imitative activities in technological accumulation after the exposure to western competition. The low number of patents per capita indicates that eastern Europe and the CIS countries are not pushing forward the technological frontier.

Resident patent applications in each home country provide an alternative view of the patenting process, but they do not take into account the number of patents actually granted. The difference between the number of resident patent applications in the United States and the number of patents granted to residents of the United States in table 4.5.1 provides a rough indication of the rate of success of patent applications (about 65 per cent). It is clear that there are many inventions for which researchers file patents in their home countries but not in the United States. Nevertheless, the general pattern of patent applications within each country is similar to the pattern of patents granted in the United States. Chart 4.5.1 shows a clear positive relationship between resident patents per million of the population and real per capita GDP. Thus countries with a higher level of overall development generate more new technology than those at lower levels of development – not unexpectedly, because countries with relatively lower levels of GDP per capita generally acquire or absorb technology that is already patented.

---

[442] United States Patent and Trademark Office, *All Technologies Report* (Washington, D.C.), 19 March 2001 [www.uspto.gov].

TABLE 4.5.1

Patenting activity in the ECE region, 1990-2000
(Per million population, percentage growth)

	Resident patent applications		United States utility patents granted	
	Average 1995-1998	Researchers per application	Average 1996-2000	2000 (1990=100)
European Union				
Austria	329.1	7	55.9	128
Belgium	158.3	16	66.8	222
Denmark	488.9	7	83.0	276
Finland	713.7	6	120.2	203
France	309.3	9	64.0	133
Germany	727.8	4	116.3	134
Greece	23.9	41	1.8	225
Ireland	274.1	7	26.2	228
Italy	70.7	19	27.8	136
Luxembourg	469.6		64.1	235
Netherlands	326.2	7	78.7	129
Portugal	10.4	123	0.9	157
Spain	69.4	19	6.2	208
Sweden	847.4	5	157.6	205
United Kingdom	453.6	6	60.7	131
Other western Europe				
Iceland	102.8	46	41.9	567
Israel	300.6	..	124.6	262
Malta	32.0	..	2.6	..
Norway	341.5	11	50.3	221
Switzerland	680.3	4	176.1	103
Turkey	4.1	69	0.1	200
North America				
Canada	127.8	24	103.9	184
United States	468.0	8	300.8	180
Eastern Europe				
Albania	0.3	..	0.1	..
Bulgaria	41.1	50	0.2	4
Croatia	58.7	47	2.6	..
Czech Republic	60.4	20	1.9	82[a]
Estonia	11.6	316	1.2	..
Hungary	85.4	13	4.1	39
Latvia	78.2	19	0.7	..
Lithuania	31.5	57	0.4	..
Poland	63.6	22	0.4	76
Romania	73.7	18	0.2	400
Slovakia	43.3	43	0.7	..
Slovenia	151.0	15	7.4	..
The former Yugoslav Republic of Macedonia	38.1	..	0.0	..
Yugoslavia	52.3	49	0.2	..
CIS				
Armenia	26.6	70	0.1	..
Azerbaijan	24.7	52	0.1	..
Belarus	73.0	35	0.5	..
Georgia	53.8	73	0.1	..
Kazakhstan	72.3	13	0.1	..
Kyrgyzstan	27.1	26	0.1	..
Republic of Moldova	77.0	22	0.0	..
Russian Federation	114.7	32	1.3	106[b]
Tajikistan	5.2	69	0.0	..
Turkmenistan	11.4	52	0.0	..
Ukraine	90.7	31	0.4	..
Uzbekistan	37.3	16	0.1	..
Memorandum item:				
Japan	2 755.9	2	245.7	160

Source: Eurostat, New Cronos Database.

[a] Includes Czechoslovakia. [b] Includes the USSR.

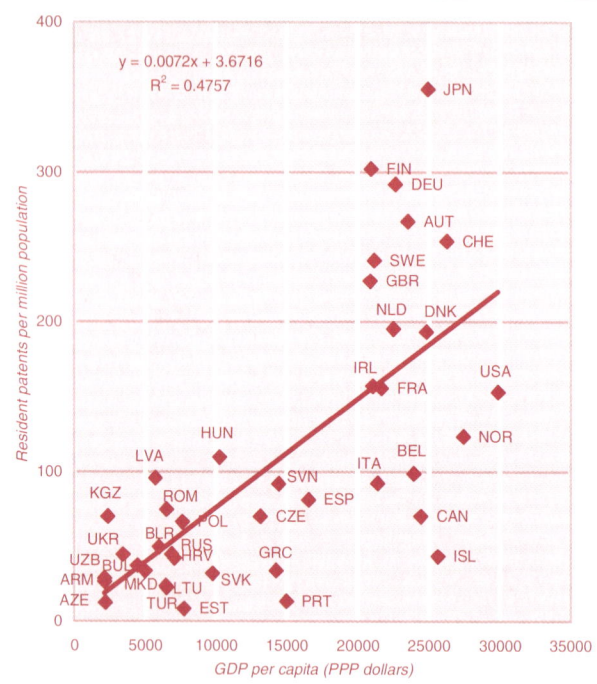

CHART 4.5.1

Resident patents per million of the population and per capita GDP in the ECE region, 1996-1998 average

Source: UNECE calculation, based on World Bank, *World Development Indicators* (Washington, D.C.), 2002 and OECD, *Main Science and Technology Indicators*, Vol. 2001, Issue 2 (Paris), 2001.

Note: As for chart 4.2.2.

The number of researchers per patent is an indicator of the inventiveness of the R&D system. Table 4.5.1, column 2 provides evidence that many of the transition economies, as well as Greece and Portugal, are well behind the other west European countries on this indicator. Some of the more advanced east European countries, such as Hungary and Slovenia, compare favourably with western Europe in terms of researchers per application. The inventiveness of Japan's R&D system is superior to all of the countries in the ECE region. One apparent weakness of this measure is that it neglects the role played by R&D in creating absorptive capacity. License fees often accompany technology transfer and spillovers, but in this case the patent resides with another firm. This would imply that catching up could properly be associated with an inventiveness coefficient that is higher than in those countries on the technological frontier.[443]

4.6 Innovative activity in Europe

Formal R&D and patenting activity may not provide an accurate picture of the innovation process. Firm-specific capabilities are reflected in the ability of

---

[443] It should be noted, however, that Ireland has an inventiveness coefficient that is similar to the United States, although, unlike the United States, it is a net importer of technology.

CHART 4.6.1

Manufacturing innovation and per capita GDP in Europe, 1996

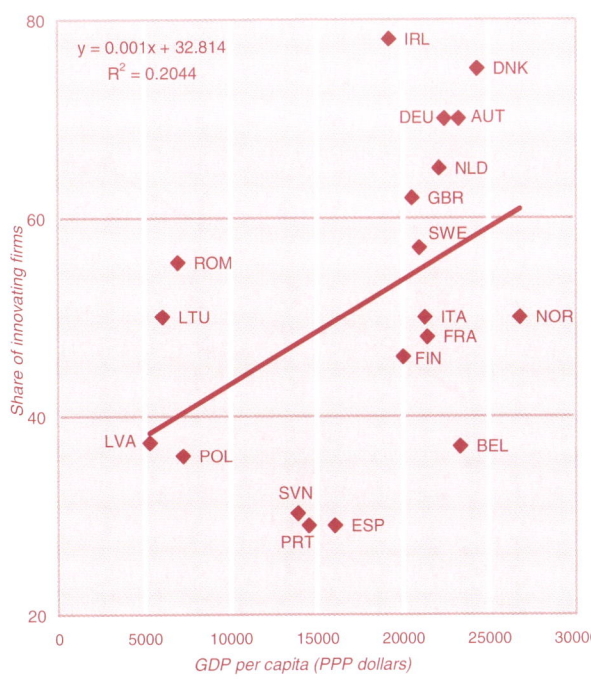

Source: UNECE calculation, based on World Bank, *World Development Indicators* (Washington, D.C.), 2002 and Eurostat, New Cronos Database.

Note: As for chart 4.2.2.

TABLE 4.6.1

Innovative activity in European firms, 1994-1996 [a]

(Per cent of total)

	Manufacturing			Services
	Total innovating firms	Product innovators	Process innovators	Total innovating firms
European Union	54	44	39	40
Austria	70	60	49	55
Belgium	37	31	22	13
Denmark	75	57	51	30
Finland	46	29	25	24
France	48	38	31	31
Germany	70	65	53	46
Ireland	78	66	54	58
Italy	50	37	41	..
Luxembourg	44	32	29	48
Netherlands	65	56	46	36
Portugal	29	15	23	28
Spain	29	24	25	..
Sweden	57	48	38	32
United Kingdom	62	52	37	40
Other western Europe				
Norway	50	35	40	22
Eastern Europe				
Latvia	37	33	31	..
Lithuania	50	46	37	..
Poland	36	33	26	..
Romania	56	51	50	..
Slovenia	32	30	27	..
Slovenia	30	29	28	..

Source: Eurostat, New Cronos Database and *Community Innovation Survey, 1997.*

[a] 1996-1998 for Latvia, Lithuania and Slovenia.

the firm to introduce higher quality products, cost-saving processes, and improved organizational and managerial processes. Such capabilities are often not captured in the traditional measures because they do not require R&D and are covered by other forms of intellectual property rights (IPRs). This may explain why there is no strong correlation between the percentage of innovative firms, as identified by innovation surveys, and real per capita GDP (chart 4.6.1). Innovation surveys collect data on the input to, and output from, innovation at the firm level, and identify the main factors influencing the innovative behaviour of firms. Eurostat carried out two innovation surveys in the 1990s, one in 1994 and another in 1997.[444] Latvia, Lithuania, Poland, Romania, the Russian Federation and Slovenia carried out their own surveys in the late 1990s, using a similar methodology. The aggregate results of the two Eurostat surveys were very similar: about 50 per cent of all manufacturing enterprises in the European Union introduced either a product or process innovation from 1990 to 1992 and about 53 per cent of firms introduced some kind of innovation from 1994 to 1996. About 40 per cent of all European Union firms in the service sector innovated from 1994 to 1996.

Table 4.6.1 summarizes some of the results from the second survey, together with data from the transition economies that carried out a similar survey. Although there were considerable differences across countries, the survey confirmed that product innovation and process innovations tend to occur together. The main reason is that the introduction of a new product almost always entails the introduction of a new method of production.[445] Countries with a large proportion of dynamic high-tech industries also had a stronger tendency to innovate. One interesting observation is that Austria and Ireland had higher than average rates of innovation, although they also had lower than average R&D intensities. This indicates that global markets are an important source of technology for these countries.

East European firms in almost every industry were more likely to introduce new products than new processes. This is generally the case among the medium-

---

[444] The "Oslo Manual" provides the guidelines for collecting and interpreting innovation data. OECD, *Proposed Guidelines for Collecting and Interpreting Technological Innovation Data, Oslo Manual* (Paris), 1997 [www.oecd.org]. The *Community Innovation Survey* of 1993, carried out by Eurostat, was the first attempt to collect comparable firm-based data on the sources, output and impact of innovation, the obstacles to innovation and technological diffusion, and corporate strategy in Europe.

[445] Freeman and Soete, op. cit.

TABLE 4.6.2

Structure of innovative expenditure in manufacturing and services in the ECE region, 1996

(Percentage share of total innovation expenditure)

	R&D	Acquisition of: Embodied technology	Acquisition of: Disembodied technology	Other [a]
Manufacturing				
European Union	53	22	4	21
Austria	47	33	2	17
Belgium	42	35	4	18
Denmark	35	44	6	21
Finland	43	27	2	27
France	65	12	1	22
Germany	63	13	2	21
Ireland	33	44	4	18
Italy	27	45	5	22
Netherlands	46	33	1	21
Portugal	7	68	8	16
Spain	37	32	7	24
Sweden	50	17	11	23
United Kingdom	31	41	7	20
Other western Europe				
Norway	31	39	4	26
Eastern Europe				
Poland	9	53	3	34
Romania	6	88	2	4
Slovenia	41	5	5	50
CIS				
Russian Federation	15	45	2	28
Service sector				
European Union	46	16	15	23
Austria	10	32	27	31
Belgium	23	25	8	45
Denmark	35	19	6	40
Finland	51	21	8	20
France	51	9	24	16
Germany	57	13	13	17
Ireland	66	8	3	23
Netherlands	21	39	8	31
Portugal	5	35	45	15
Sweden	19	9	9	63
United Kingdom	15	25	24	36
Other western Europe				
Norway	27	32	11	30
Eastern Europe				
Poland	9	43	21	28
Slovenia	8	38	30	24

Source: Eurostat, New Cronos Database and *Community Innovation Survey, 1997.*

[a] Includes extramural R&D; industrial design or preparations to introduce new services; training directly linked to technological innovation; and market introduction of innovation.

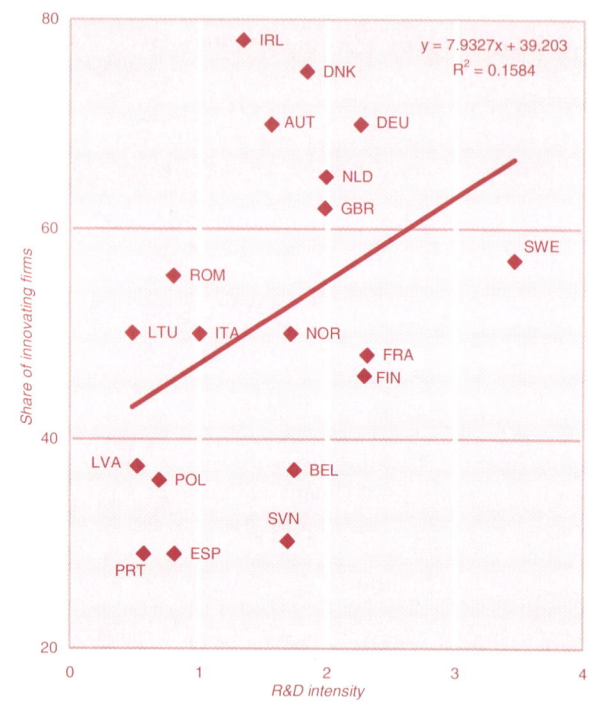

CHART 4.6.2
Manufacturing innovation and R&D intensity in Europe, 1996

Source: UNECE calculation, based on OECD, *Main Science and Technology Indicators*, Vol. 2001, Issue 2 (Paris), 2001 and Eurostat, New Cronos Database.

Note: As for chart 4.2.2.

the European Union average in Romania, while in Poland and Slovenia it was below the average. A second survey made in 1998 indicates some improvement in Slovenia, and in Poland the share of technologically new and improved products in total sales steadily increased between 1996 and 1999.[446] However, Russia appears to have serious problems in introducing new products and processes.[447] This confirms the belief that there has been little reform of the S&T system in Russia.

The innovation surveys also show that expenditure on R&D is not the only innovative activity carried out by firms. Table 4.6.2 shows that R&D only accounts for about 50 per cent of total investment in innovative activity in the European Union. Chart 4.6.2 also suggests that the relationship between R&D spending and innovative activities is not very strong, particularly in the transition economies. Acquisition of embodied technology (i.e. machinery and equipment) is more important in the

to high-tech industries of Europe as a whole, and it indicates that a process of quality upgrading is taking place. In general, larger firms innovate more frequently than small firms – because they can spread the risk more easily. They also tend to introduce more process innovations than product innovations. But the vast majority of firms that innovated said they introduced both sorts of innovation. Innovative activity, as reported in the 1994-1996 innovation survey, appeared slightly above

---

[446] Central Statistical Office of Poland (GUS), *2000 Statistical Yearbook of the Republic of Poland* (Warsaw). In contrast, there was a dramatic decline in innovative activity between the first survey carried out in 1993 and the second in 1997. Almost twice the number of firms innovated during the first three years of the transition process as during 1994 to 1996. The reason may be that a higher share of firms introduced innovations at the outset of the transition process, but it could also reflect differences in the questionnaire.

[447] Data from Russian innovation surveys are not included in the table because of inconsistencies in coverage.

transition economies, and in those economies with high innovative activity and lower R&D intensity. This suggests that the technology acquired is embodied in investment goods. Acquisition of disembodied technology (i.e. licenses and patents) does not appear to be very important for manufacturing industries, but it is for services. All this suggests that in-house R&D is not as important for services as for manufacturing.

4.7 International technology transfer and domestic spillovers in eastern Europe

International technology transfer and local diffusion has been especially important for small countries such as those in eastern Europe attempting to catch up with the technological leader. There are at least five ways that technology can be transferred across countries: (1) foreign direct investment; (2) joint ventures; (3) strategic alliances; (4) technology licensing; and (5) capital goods imports (or embodied technology transfer). Most often, domestic R&D efforts complement technological knowledge obtained from abroad. R&D undertaken by the enterprise is necessary for identifying, assimilating and utilizing existing knowledge from abroad.[448]

Inward foreign direct investment (FDI) frequently includes a transfer of technology. Certain activities by TNCs can help to shape the speed and direction of the economic transformation of eastern Europe by transferring technology directly to affiliates and indirectly to domestic firms through technology spillovers. But this may not always be the case. A sample of enterprises from 10 European Union candidate countries shows that technology often transfers through the parent-subsidiary relationship and trade, but that in practice the expected spillover benefits to purely domestic enterprises rarely materialize.[449] In addition, there is evidence that there are significant crowding-out effects for local firms in competing industries in some countries. For the Czech Republic, Poland, Romania and Slovenia, there is evidence that trade may be a more important channel of technology transfer than FDI. A small innovation survey carried out during 1996 in Hungary suggests that foreign affiliates introduce process innovations more often than local enterprises, but that the local enterprises are more likely to introduce new products.[450]

TABLE 4.7.1

International royalties and licence fees in the ECE region, 2000
(Million dollars, per cent)

	Receipts	Payments	Balance	Payments/ GERD
European Union				
Austria	162	547	-385	16
Belgium	786	894	-108	20
Finland	1 138	547	591	14
France	2 310	2 050	260	7
Germany	2 820	5 450	-2 630	12
Greece	5	203	-198	27
Ireland	504	7 899	-7 395	602
Italy	563	1 198	-635	11
Luxembourg	118	125	-7	..
Netherlands	2 176	2 565	-389	34
Portugal	21	255	-234	32
Spain	403	1 681	-1 278	34
Sweden	1 275	900	375	10
United Kingdom	7 220	6 010	1 210	23
Other western Europe				
Israel	500	349	151	..
Norway	131	391	-260	15
North America				
Canada	1 412	2 879	-1 467	25
United States	38 030	16 100	21 930	6
Eastern Europe				
Bulgaria	4	10	-6	16
Czech Republic	44	82	-38	12
Estonia	2	8	-6	21
Hungary	112	257	-145	69
Latvia	2	12	-10	40
Lithuania	..	12	..	21
Poland	34	554	-520	50
Romania	3	45	-42	30
Slovakia	16	58	-42	44
Slovenia	12	49	-37	18
The former Yugoslav Republic of Macedonia	3	6	-3	40
CIS				
Belarus	1	2	-2	1
Kazakstan	..	11	..	35
Kyrgyzstan	1	1	–	59
Republic of Moldova	1	2	-1	27
Russian Federation	91	31	60	1
Ukraine	1	663	-662	181
Memorandum item:				
Japan	10 230	11 010	-780	8

Source: UNECE calculation, based on IMF, *Balance of Payments Statistics Yearbook* (Washington, D.C.), 2001.

The technology balance of payments measures the international trade in scientific or technological knowledge.[451] Royalties and payment for license fees make up a large part of the technology balance of payments. Table 4.7.1 contains data on royalty payments

---

[448] Cohen and Levinthal, op. cit.

[449] J. Damijan, M. Knell, B. Majcen and M. Rojec, "Is technology transferring to the accession countries? Evidence from panel data for ten transition countries" (University of Ljubljana), 2001, mimeo. For a more general discussion of FDI and growth in the context of eastern Europe, see UNECE, "Economic growth and foreign direct investment in the transition economies", *Economic Survey of Europe, 2001 No. 1*, chap. 5.

[450] P. Tamas, *Egy Távlatosabb Nemzeti Technologiapolitika Aalapvetese a Piaci Korulmenyek Kozott Mukodo Gazdasagi Szervezetek Oldalarol* (The Bases of a Long-Term National Technology Policy from the Standpoint of Economic Organizations Operating in a Market Environment) (Budapest, OMFB, 1997).

[451] The technology balance of payments (TBP) measures all intangible transactions related to international technology transfer. Royalties and license payment fees, which are included in the IMF balance of payments statistics, make up a large part of the TBP. For a more precise definition of TBP, see OECD, *Proposed Standard Method of Compiling and Interpreting Technology Balance of Payments Data, TBP Manual* (Paris), 1989 [www.oecd.org].

CHART 4.7.1

High-technology imports and per capita GDP in the ECE region, 1999

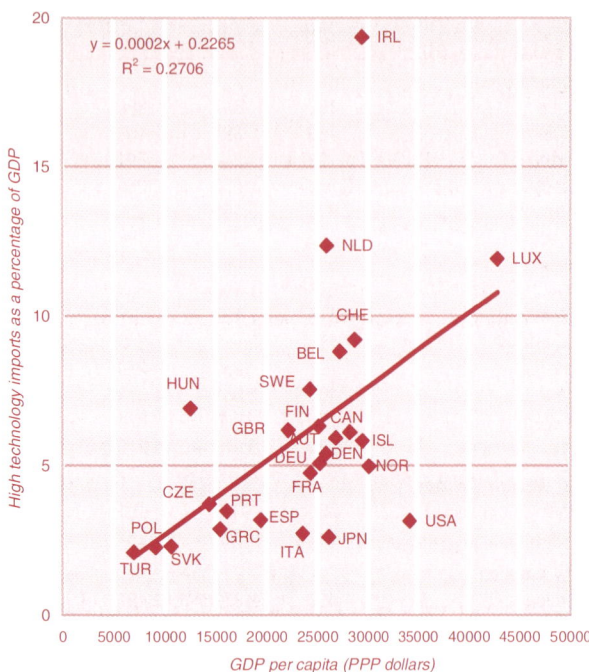

Source: UNECE calculation, based on World Bank, *World Development Indicators* (Washington, D.C.), 2002 and OECD, *Main Science and Technology Indicators*, Vol. 2001, Issue 2 (Paris), 2001.

Note: As for chart 4.2.2.

and licence fees for the ECE region in 2000. A negative balance indicates a net import of intangible assets related to technology. Within the European Union, only Finland, France, Sweden and the United Kingdom are net exporters of technical knowledge. All of the transition economies are net importers except for the Russian Federation, which is strong in the aerospace industries. Even so, the total technology receipts of the Russian Federation are smaller than those of Hungary, and less than 20 per cent of those for Ireland. Ireland is interesting because its payments are the largest of all the member states of the European Union. These payments are more than six times the amount spent in Ireland on R&D, and explain much of the difference between its GDP per capita and GNI per capita. Hungary also has a relatively high ratio of payments abroad to R&D expenditures, confirming that, like Ireland, it depends heavily on imports of technology.

Another important channel of technology transfer is the import of advanced capital goods and turnkey industrial plants. As an indicator of embodied technology transfer, imports of high-technology goods represent the demand for new knowledge, and exports from these industries represent the supply of technical knowledge. Such imports are often accompanied by inward FDI and license and royalty payments abroad. Chart 4.7.1 shows the share of high-tech imports in GDP in relation to real GDP per capita. Although the relationship is weak, it does show a positive relationship between high-technology imports and the level of development. Ireland is the most significant outlier, with the share of high-technology imports exceeding 19 per cent. This is not surprising, given the high penetration of foreign capital into Ireland and the high royalty and license fees paid abroad. Hungary and Netherlands are also well above the trend line, which indicates that these countries are also obtaining significant volumes of technology through trade.

4.8 Is there a role for science and technology policy?

In the market-based economy the social returns to R&D tend to be well above the private returns. This observation supports the idea of an innovation policy to encourage and support research and institution building, and ensure that there are incentives for private industry to take initiatives. The basic sciences, for example, have considerable scope for knowledge spillovers and consumer-surplus effects in the long run, but little possibility for making a profit until a commercial use is found for the idea. This is one reason why subsidies to research in universities and collaborative efforts between universities and private business are so important.

The centrally planned economy had the opposite problem, namely, excessive R&D spending, especially in the military sphere, by the Soviet Union and many of the east European states before 1990.[452] The problem was exacerbated by chronic shortages in the economy, and the belief that *any* expenditure on R&D would increase the growth rate. Since the organization of industry provided little scope for knowledge spillovers, R&D encountered diminishing returns. This, in turn, was one of the main reasons for the decline in productivity growth, which lay at the root of the descent into stagnation of those economies. While there was a conscious attempt by central planners to reduce the static inefficiencies present in the market economy, they ended up by eliminating the dynamic efficiencies that create increasing returns and long-run productivity growth. Falling further and further behind, the S&T systems of eastern Europe and the former Soviet Union were exposed to the global market during the rapid transition from autarky to free trade. This exposure highlighted the legacy of obsolete technical competences and social capabilities left by the collapse of central planning, and set in sharp relief the need to develop specific S&T policies to address the essential problems of the transition.[453]

---

[452] P. Hanson and K. Pavitt, *The Comparative Economics of Research Development and Innovation in East and West: A Survey* (New York, Harwood Academic Publishers, 1987).

[453] K. Pavitt, op. cit.

As has been argued on different grounds throughout this chapter, the relationship between technological activities and economic growth, although an intuitively obvious one, is not clear-cut. R&D can have a direct effect in creating new products and processes and an indirect effect in helping to create the capacity to absorb technology from abroad. Regression analysis suggests that product and process innovation have an important influence on the growth of real per capita GDP. But it is equally the case that economic development can itself be a factor influencing S&T activity. Transition is essentially about restoring productivity growth in a region of the world where productivity had stagnated for a generation. And there is no productivity growth without technical change and technological learning in the broadest sense.

R&D may be particularly important for productivity growth in a transition context for several reasons. First, the sheer ineffectuality of the old communist R&D systems means that there is considerable scope for Gerschenkronian catch up[454] in terms of simply establishing an operationally efficient R&D system. On the other hand, the transition economies had ample room for productivity catch up based on X-efficiency gains without necessarily increasing their R&D intensity. Thus this pattern reflects a *strengthening*, not a weakening of R&D potential in the countries concerned. But the scope for productivity increases through rationalization of R&D is far from exhausted. Public policy, including donor policy, should continue to focus on the fundamental task of building a basic framework for S&T in the transition countries, as a precondition for effective economic catch-up. There might be room within that basic framework for some elements of the institutional framework inherited from the old system. As a general rule, however, the big institutes left over from communism represent a liability rather than an asset in a transition context. The same must be said of the legacy of socialist-era coordinating mechanisms for science and technology, notably the academies of science. An effective R&D system for the transition countries will contain public, private and mixed elements, but the main nexus of R&D activity will shift from institutes to firms.

There is much more to R&D, however, than research and development work in a narrow sense. Particularly in a transition context, a key objective of R&D in the broadest sense is to enhance the ability of firms to assimilate and take advantage of technical knowledge from abroad, and to internalize new process and product innovations efficiently. Such a relationship may explain why there is a high correlation between R&D and industrial innovation in eastern Europe, and suggests that well-directed public policies could increase the rate of industrial innovation by encouraging specific, absorption-oriented aspects of R&D, e.g. through the development of technology transfer "roadshows", fiscal policies designed to underpin R&D-innovation linkages, and the like. Again, R&D is often sector-specific, and formal aggregate indicators tend to underestimate technological activities related directly to production, and located in small firms.[455] For that reason, policies to support supply-network building, as a vehicle for technological upgrading, can yield substantial returns. Last but not least, governments can use national science and technology policies to enhance the benefits of FDI for the local economy. Many transition economies have attracted substantial amounts of FDI,[456] and while its direct impact on economic performance has often been significant, its indirect effects (technological spillovers) generally appear disappointing.[457] However, there is some evidence that local firms that have focused R&D resources on increasing their capacity to absorb technology have been able to benefit from intra-industry FDI spillovers.[458] This suggests the need to develop a coherent policy linking science and technology policies with foreign investment promotion strategies.[459] National policies often aim to attract MNCs that perform R&D and tend to favour local spending on R&D-oriented to leading-edge innovation. However, it might be useful to promote R&D investment that specifically enhances the ability of local enterprises to absorb foreign technologies, particularly if the rates of return on such investments are higher than on leading edge R&D investment.[460] This is consistent with the view, supported by European experience, that small countries (as most transition economies are) should have the capability to use advanced technology and should emphasize promoting its internal dissemination rather than developing entirely new cutting-edge technologies.[461] In practice, this kind of foreign investment policy can be integrated with supply networking policy, as discussed above.

---

[454] A. Gerschenkron, *Economic Backwardness in Historical Perspective* (Cambridge, Harvard University Press, 1962).

[455] P. Patel and K. Pavitt, "Patterns of technological activity: their measurement and interpretation", in P. Stoneman (ed.), *Handbook of the Economics of Innovation and Technical Change* (Oxford, Basil Blackwell, 1995), pp. 14-51.

[456] UNECE, "Economic growth and foreign direct investment in the transition economies", *Economic Survey of Europe, 2001 No. 1*, chap. 5, pp. 185-225.

[457] UNECE, "The environment for FDI spillovers in the transition economies", paper presented at the UNECE/EBRD Expert Meeting on Financing for Development, *Enhancing the Benefits of FDI and Improving the Flow of Corporate Finance in the Transition Economies* (Geneva), 3 December 2001 [www.unece.org/ead/ead_h.htm].

[458] Y. Kinoshita, *R&D and Technology Spillovers through FDI: Innovation and Absorptive Capacity*, CEPR Discussion Paper, No. 2775 (London), May 2001.

[459] UNECE, "The environment for FDI spillovers ...", op. cit.

[460] Y. Kinoshita, op. cit., found that the rate of return on absorption-related R&D in the Czech Republic has been considerably greater than that on innovation-related R&D.

[461] M. Blomstrom, *Host Country Benefits of Foreign Investment*, NBER Working Paper, No. 3615 (Cambridge, MA), February 1991.

TABLE 4.8.1

Government budget appropriations or outlays for R&D (GBAORD) in the ECE region, 2000 [a]

(Million current PPP dollars)

	Total GBAORD	Per cent total GBAORD					
		Defence budget R&D	Economic development	Health and environment	Space	Non-related	University funding
European Union	62 038	14.5	20.2	13.2	5.8	14.4	30.4
European Commission	2 646	–	61.3	21.4	0.7	6.5	–
Austria	1 222	–	11.7	9.0	0.1	14.9	64.2
Belgium	1 561	0.4	29.5	9.0	11.7	22.0	18.7
Denmark	1 031	0.6	22.7	17.2	2.8	17.8	39.0
Finland	1 258	1.3	41.4	16.1	2.1	12.3	26.8
France	13 109	22.6	14.2	8.5	10.7	21.2	17.7
Germany	16 442	8.4	19.8	11.9	4.6	16.8	38.5
Greece	545	0.8	21.6	16.1	0.6	7.2	44.7
Ireland	278	–	52.2	11.5	–	12.6	23.8
Italy	8 162	0.9	13.3	14.6	8.0	10.5	42.6
Netherlands	3 228	3.1	23.6	11.1	2.9	10.9	44.2
Portugal	1 108	1.6	33.1	16.9	0.3	7.6	36.1
Spain	4 127	26.2	26.6	10.3	4.8	5.3	25.7
Sweden	1 639	7.1	17.1	10.0	3.3	11.5	50.9
United Kingdom	9 427	38.0	6.9	22.3	2.3	11.3	18.7
Other western Europe							
Iceland	79	–	34.6	..	..	7.7	29.1
Norway	896	5.0	25.5	19.2	2.3	8.6	39.4
Switzerland	1 337	1.9	3.6	1.6	..	..	58.0
Eastern Europe							
Poland	1 430	..	..	..	..	..	..
Slovak Republic	233	–	28.4	16.9	0.3	29.0	17.8
Slovenia	190	–	23.5	7.5	–	65.7	3.3
Romania	208	2.2	52.8	10.2	1.5	30.1	3.3
CIS							
Russian Federation	5 222	37.1	28.1	9.9	9.4	14.5	–
North America							
Canada	3 937	5.6	31.3	21.5	8.2	6.1	25.6
United States	75 415	50.0	7.4	24.9	11.2	6.6	..
Memorandum item:							
Japan	21 461	4.1	32.0	7.3	5.6	14.0	35.5

Source: OECD, *Main Science and Technology Indicators*, Vol. 2001, Issue 2 (Paris), 2001.

[a] 1998 for Canada and Switzerland; 1999 for the European Commission, the European Union, Ireland, Romania, the Russian Federation, Slovenia and Spain.

It should always be borne in mind that demand is at least as important a determinant of productivity growth as R&D. Increasing demand encourages technical change and technological learning because it induces firms to invest and rationalize production.[462] This helps to explain why there is a high correlation between output growth and productivity growth in the most advanced countries of eastern Europe. But rationalization of production becomes increasingly complex and costly once the "easy pickings" of the early transition period have been exhausted. To sustain the relationship between aggregate demand and productivity growth in the long run, therefore, will require a better-focused and more efficient R&D system than any of the transition countries currently possess.

What does the experience of the ECE region as a whole suggest about public spending on R&D activities as a way of fostering growth and innovation? Table 4.8.1 provides a breakdown of government budget appropriations by socio-economic objective for a number of countries in the region. Two patterns appear in the table: (1) the larger countries, and in particular the United States and the Russian Federation, devote a large share of their R&D budgets to defence and space-related projects; and (2) the small states devote the largest shares to economic development and university funding. (The European Commission, it should be added, supplements the R&D efforts of its individual member states by putting considerable funds into technology-related programmes in economic development, health and environment.) Germany and Japan fall between the two patterns, partly because of restrictions on military production and/or spending.

---

[462] N. Kaldor, *Causes of the Slow Rate of Growth of the United Kingdom* (Cambridge, Cambridge University Press, 1966).

Sectoral allocation apart, the effectiveness of government spending may also depend on the policy instruments used. There are at least three main policy instruments: government and university research, grants and subsidies to the business sector, and fiscal incentives.

Universities play a key role, in terms of provision of laboratory services, development of research instrumentation and the training of post-graduate students in research project management. United States university laboratories have, for instance, played a key role in the development of contemporary biotechnology. In addition to universities, public research is carried out in national laboratories (such as NASA) and pan-national laboratories (such as CERN) as well as in academies of science (eastern Europe). A key objective of this kind of public funding is to generate basic knowledge that can eventually be used by firms in their own research. Universities also pursue this basic objective, but are more independent in forming their research agenda.

Grants and subsidies are another way to increase technological activities. Subsidies may also take indirect forms, such as tax breaks and other fiscal measures, which may be less discriminatory but tend to reward enterprises for past effort instead of encouraging the development of new technology.[463] A key issue here is whether intellectual property rights remain with the government (and become a public good), or accrue to the performer – a particularly important issue for the transition countries. Generally, direct support is only given for specific projects or goals that yield high social returns. The strongest criticism of this policy instrument is that the government may not make the right choice as to what technology should be funded, *a fortiori* when it faces the financial and human limits to the implementation of public policy common among the countries of the transition economies. Still, the strong industry-based policy followed by Japan, and more recently, the more R&D-based policy followed by Hungary and Ireland, have clearly been important in sustaining high rates of economic growth. These arguments point strongly to the need for the ECE member states, and in particular the transition economies, to strengthen their technology and innovation policies. Policy initiatives should not focus only on the appropriate level of public funding, but also on strengthening the S&T system. Direct support for R&D as such is, however, only one element of S&T policy. Indeed, given the legacy of ineffectual, over-sized research institutes, which still burdens most of the transition economies, direct support for R&D should probably be de-emphasised in a transition context. More important are policies to promote cooperation between government, industry and education, strengthen the education and training system, introduce measures to enhance technical change and technological learning in private business, introduce alternative and more flexible rules governing IPRs and encourage multinationals to transfer technology to their foreign affiliates, while at the same time making local partners more able to absorb technology transfer.[464] Moreover, science and R&D policy must stress the market-based determination of output irrespective of where the research activity takes place. While governments should resist the temptation to interfere in the commercial application of innovations, the scope for public sector support and reinforcement of market mechanisms in this area will remain substantial, providing a key focal point for long-term economic policy-making.

---

[463] The impact of these policy instruments are discussed in more detail in OECD, *Science, Technology and Industry Outlook* (Paris), 2000.

[464] European Commission, *Green Paper on Innovation*, COM(95)688 (Brussels), December 1995, also points to the need for Europe to develop a more comprehensive innovation policy that includes an easily accessible venture capital market and closer ties between universities and industry [www.europa.eu.int].

CHAPTER 5

ALTERNATIVE POLICIES FOR APPROACHING EMU ACCESSION BY CENTRAL AND EAST EUROPEAN COUNTRIES

The choice of a strategy for EMU accession is one of the most important policy decisions facing the candidate countries. As regards monetary policy, this involves the choice of exchange rate regime before joining the exchange rate mechanism (ERM-2) and the determination of exchange rate fluctuation margins (wide or narrow band) within the framework of the ERM-2. This is a complex issue and numerous factors such as the volatility of short-term capital flows have to be taken into consideration in formulating an efficient and sustainable policy course. A major dilemma inside the ERM-2 could be the potential conflict between nominal exchange rate stability and meeting the Maastricht inflation criterion on account of the trend appreciation of the real exchange rate related to the Balassa-Samuelson effect. This may entail a trade-off between faster nominal convergence and output growth (i.e. real convergence), at least in the short run. On the other hand, the long-run benefits of being part of the euro zone may outweigh the short-term adjustment costs associated with meeting the Maastricht convergence criteria. Policy makers in each candidate country must carefully assess and weigh the specific costs and benefits of alternative policy strategies that lead to adopting the euro.

5.1 Introduction

The European Union is facing the biggest round of enlargement in its history, at least in terms of the number of countries involved.[465] From the viewpoint of the candidate countries, EU accession and entry to the euro zone are crucial steps towards their ultimate goal of closing the development gap with the west European economies. Catching up will be a long-term process, which is expected to be accelerated by the adoption of EU regulations and standards, by closer economic integration with the more developed parts of Europe and by adopting the common currency, the euro.

Compared with previous enlargements, the forthcoming one differs in a number of respects. First, of the 12 associated countries, 10 belong to the group of central and east European transition economies (CEEC-10). The income gap between them and the EU average is much wider than for previous new entrants or within the history of the Union itself, a fact underlining the importance of harmonizing both real and nominal convergence. Second, the candidate countries have to conduct their pre-accession stabilization policies under the potential pressure of large capital flows in a globalized world. In the period between now and their joining the euro zone they could face adverse shocks originating either from other emerging markets or from capital inflows attracted by their convergence on the EU. The CEEC-10 have already dismantled most of their capital controls, a precondition of EU entry, which leaves them vulnerable to conditions in the global capital markets and to speculative capital flows. In contrast, the present euro members retained capital controls for stabilization purposes, and used them quite actively, until quite late in the run-up to monetary union.

Third, the next enlargement will be of a "new quality" EU, as the economic and monetary union (EMU) entered its Stage III on 1 January 1999 with the introduction of the euro. When the new candidates join the EU, having met the Copenhagen criteria for accession, they will also enter the EMU with the status of "member states with derogation", since at the time of entry they are not expected to be able to meet the Maastricht criteria for accession to the monetary union. However, as the new members are committed to entering the euro area as soon as they fulfil the nominal convergence criteria, they will not have the opt-out option that was available to some of the earlier EU member countries (Denmark and the United Kingdom).

---

[465] "The negotiations do not yet allow the Commission to conclude that the conditions of accession are fulfilled by any of the candidates. ... The Union should therefore be prepared for negotiation by the end of 2002, in view of accession in 2004, with all countries meeting the necessary conditions". European Commission, *Strategy Paper and Report of the European Commission,* on the progress towards accession by each of the candidate countries (Brussels), November 2001, p. 29.

Accession to the euro zone – EMU membership[466] – will be based on the assessment by the EU authorities of the sustainability of convergence, each candidate country being treated equally.

The accession countries will have to decide on a strategy for meeting the demands of their future membership in the single currency area by taking into account a number of fundamental challenges to the conduct of their economic policies and by facing several possible policy dilemmas during the convergence process. Although the macroeconomic framework for the period of nominal convergence needs to be analysed on a case-by-case basis and with an understanding of the underlying mechanisms, there are a number of policy options – as well as constraints – which are common and these are the focus of the discussion in this chapter.

Section 5.2 looks at some of the macroeconomic policy challenges in fast-growing, catching-up economies as well as their implications for EMU accession. The role of exchange rate policy, one of the central issues in the run up to EMU accession, is discussed in section 5.3. Section 5.4 addresses the problem of the timing of accession in the context of the expected costs and benefits of alternative accession scenarios. The main policy conclusions are summarized in section 5.5.

5.2 Catching up and EMU accession

Accession to the EU and entry to the euro zone can be seen as steps to promote a faster and more balanced closing of the development gap. The decision about euro zone membership will be based on a country-by-country examination and although the candidates are definitely committed to adopting the common currency[467] they will be relatively free to decide the date of their application to join the euro zone. Thus, apart from fulfilling the nominal convergence criteria, such a decision implies a careful assessment of the expected benefits and costs of different paths to EMU entry.

The basic question is how the adoption of the euro can influence the real convergence process (in the sense of reducing the gap in per capita incomes) and vice versa? The present euro member states underwent a lengthy, multistage convergence process and most of them achieved a high degree of income convergence prior to the monetary union. Whether that is the best approach for the CEEC-10 depends on the links between EMU accession, on the one hand, and growth, structural transformation and policy reforms in the acceding countries, on the other. The choice of an accession strategy for the candidate countries should also reflect the evolution in the EU's regulatory framework and the changes in the external environment as compared with the past.

(i) Income gap – structural gap

According to annual reports on progress towards EU accession published by the European Commission in November 2001, the CEEC-10 can be divided into two groups: the advanced CEEC-8 candidates – the Czech Republic, Estonia, Hungary, Latvia, Lithuania, Poland, Slovakia and Slovenia – are judged to have functioning market economies although they are not yet fully prepared to cope with market competition in the Union, while Bulgaria and Romania still have to make more progress to satisfy the eligibility criteria of EU membership.

Despite the relatively high rates of growth during the second half of the 1990s the average per capita GDP in the CEEC-10 as a percentage of the EU average, and measured in terms of purchasing power parity (PPP), was 39 per cent in 2000. Extrapolating the trends at the end of the 1990s in conjunction with the Pre-Accession Economic Programmes, the European Commission estimates (table 5.2.1.) an 11 to 34-year catching-up period – 11 years for Hungary, 34 years for Poland – before per capita GDP in the CEEC-10 reaches 75 per cent of the present EU-15 level.[468]

A rough comparison of the CEEC-10 with the catching-up economies already in the euro zone highlights the fact that although most of the candidate countries are significantly poorer in terms of GDP per capita than Greece, Ireland, Portugal and Spain were at the time of their entry into the EU or the euro area,[469] their basic structural indicators (openness, share of industry and agriculture in GDP and employment, etc.) do not differ greatly. An important lesson from EU's past experience is that the income gap was not an obstacle to entry into the EMU by the less developed EU member states that met the nominal convergence criteria (e.g. Greece and Portugal).

Recovery in the transition economies from the shocks of the early 1990s demanded a rapid reorientation of their economies towards western markets, based on a restructuring of their production and trading patterns. By the end of the 1990s, the CEEC-10 had already achieved a relatively high degree of integration with the EU in

---

[466] For the sake of brevity and simplicity, EU and EMU membership/accession is often used in the literature to distinguish between EMU membership with derogation for the monetary union and fully-fledged EMU membership, respectively. These terms will be used in this sense throughout this chapter, although, in view of the above, this is obviously not a precise distinction.

[467] Once in the EU, the new members will be committed to the rules set by the Treaty and to following a path towards adoption of the euro; they will also have to treat their exchange rate policy as a matter of common interest.

[468] Slovenia had already reached 72 per cent of the EU-15 level by 2000.

[469] J. Pelkmans, D. Gros and J. Ferrer, *Long-run Economic Aspects of the European Union's Eastern Enlargement*, Scientific Council for Government Policy, Working Documents 109 (The Hague), September 2000.

TABLE 5.2.1

The speed of catching up by the CEEC-10,[a] 1996-2004

	GDP real compound annual growth rates		GDP per capita at current exchange rates	GDP per capita in PPP dollars				
				Level (as per cent of EU-15)			Years required to reach 75 per cent of the average of:	
	1996-2000	2001-2004	2000	1996	2000	2004	EU-15	EU-27[b]
Bulgaria	-1.3	6.1	7	25	24	31	31	31
Czech Republic	0.9	3.8	24	65	60	68	15	6
Estonia	5.1	5.8	17	33	38	48	19	16
Hungary	4.0	5.3	22	47	52	64	11	7
Latvia	4.7	5.7	15	26	29	36	27	25
Lithuania	3.2	4.7	15	29	29	35	31	30
Poland	5.2	3.5	20	36	39	45	33	33
Romania	-1.6	5.0	8	33	27	33	34	33
Slovakia	4.6	4.5	17	46	48	56	20	16
Slovenia	3.9	1.4	44	66	72	85	1	..

Source: Real convergence in candidate countries – past performance and scenarios in the Pre-accession Economic Programs (PEPs), European Commission, *DG Economic and Financial Affairs: Exchange Rate Strategies for EU Candidate Countries*, ECFIN/521/200 (Brussels), 16 November 2001.

a The projection was based on growth rates of PEPs up to 2004, PEP figures for 2004 thereafter; EU forecasts by the European Commission up to 2003 and average growth 1995-2003 thereafter.

b EU-15 plus 12 associated countries. The projection takes account of the fact that the EU average is changing, of the low initial income levels of the CEEC-10 and their above average growth rates.

terms of foreign trade (table 5.2.2, columns 1-4).[470] The correlation of business cycles between the more advanced CEEC-8 and the EU suggests that closer trade relations, especially growing shares of intra-industry trade, are fostering business cycle convergence and helping to create the preconditions for policy integration and joining the monetary union.[471] These developments indicate that the candidates have moved closer to meeting the optimal currency area criteria with the EU.[472] Experience after the 1998 Russian crisis, during the oil price shock in 2000 and the quick reactions to the recent slowdown in the world economy in 2001 provide further evidence of the more synchronized developments in the CEEC-10 and the EU.

The transformation of financial systems – both public and private – has somewhat lagged behind, but reforms have recently accelerated in many of the candidate countries with significant progress in the privatization and restructuring of their banking sectors. Although the level of financial intermediation is still relatively low (table 5.2.2, column 7) and financial markets are not yet very deep, this mainly reflects the level of development of these economies, which nevertheless is now sufficient to attract increasingly large numbers of foreign investors. In the more advanced candidate countries, the prudential indicators have also improved substantially.[473]

(ii) Equilibrium real appreciation and inflation in catching-up economies

In an "ideal" scenario the catching-up economies would have higher growth rates than the EU average, while at the same time inflation would fall to a rate defined as price stability, the fiscal conditions would be met and interest rates would be falling. However, such a scenario is likely to be upset by the nature of macroeconomic structural changes in catching-up economies. The latter are revealed in the dynamics of the real exchange rate, which connects the real and nominal sides of the convergence process.

In fast-growing, catching-up economies, such as the CEEC-10, the equilibrium real exchange rate, measured on the basis of consumer prices, can be expected to appreciate in the medium term. This is due to the systematically higher rate of increase in productivity of the tradeable sector vis-à-vis both the non-tradeable (services) sector and the country's main trading partners. The faster productivity growth in the tradeable sector leads – via the wage-setting process – to a rise in the relative price of non-tradeables, resulting in higher consumer price inflation and a real appreciation of the exchange rate. This is known as the Balassa-Samuelson effect,[474] which is an equilibrium

---

[470] In addition it is worth mentioning that Hungary and the Czech Republic are sixth and ninth, respectively, among the OECD countries by the share of high and high-medium technology manufactures in GDP. OECD, *Science, Technology and Industry Scoreboard* (Paris), 2001, p. 125.

[471] Chap. 3.2(iv) in this *Survey*.

[472] J. Fidrmuc and F. Schardax, "Pre-ins, ante portas? Euro area enlargement, optimum currency area and nominal convergence", Oesterreichische Nationalbank, *Focus on Transition*, No. 2 (Vienna), 2000, pp. 28-48.

[473] More detailed analysis is presented in N. Wagner and D. Iakova, "Financial sector evolution in the central European economies: challenges in supporting macroeconomic stability and sustainable growth", in R. Feldman and C. Watson (eds.), *The Road to EU Accession* (IMF, Washington, D.C., November 2001).

[474] B. Balassa, "The purchasing power parity doctrine: a reappraisal", *Journal of Political Economy*, Vol. 72, 1962, pp. 584-596; P. Samuelson, "Theoretical notes on trade problems", *Review of Economics and Statistics*, Vol. 46, 1964, pp. 145-164.

TABLE 5.2.2

Structural indicators in the CEEC-10, 1995-2000

	As per cent of GDP in 2000 [a]		Share of exports [a] to EU-15		Share of agriculture, 2000		Bank lending to the private sector (per cent of GDP)	Stock of accumulated FDI [b] (euros)	
	Imports	Exports	1995	2000	In GDP	In total employment	2000	Billions	Per capita
Bulgaria	64.1	58.5	..	51.2	14.5	..	14.6	2.0	239
Czech Republic	75.2	71.4	38.4	68.6	3.9	5.1	48.3	22.7	2 213
Estonia	100.4	95.4	..	76.5	6.3	7.4	26.2	2.8	1 980
Hungary	66.7	62.5	42.1	75.1	4.1	6.5	25.6	18.0	1 790
Latvia	54.5	45.8	..	67.6	4.5	13.5	18.6	2.2	943
Lithuania	51.9	45.5	..	47.9	7.6	19.4	11.6	2.5	683
Poland	38.1	32.0	52.7	69.9	3.3	18.8	26.0	26.0	671
Romania	39.9	34.1	..	63.8	12.6	42.8	7.2	7.1	371
Slovakia	76.0	73.5	40.8	59.1	4.5	6.5	30.7	5.4	1 000
Slovenia	62.7	59.1	64.8	63.8	3.2	9.9	38.0	2.7	1 348

Source: Annual reports on the progress of the candidate countries towards EU accession, published by the European Commission, *Enlargement Monitor*, available at [http://europa.eu.int/comm/enlargement/report2001/], November 2001.

a Imports and exports of goods and services.

b Foreign direct investment comprises privatization income and inter-company loans.

phenomenon and is a fundamental feature of a fast-growing, catching-up economy. Hence the higher rate of consumer price inflation in the catching up economy is not the consequence of macroeconomic disequilibria or lax policy; it just reflects the fact that convergence in productivity across countries is likely to bring about convergence in incomes and price levels. A continuation of the process of catching up will imply higher rates of inflation in the candidate countries to reduce this gap in relative price levels. At present in the CEEC-10 there is a considerable gap between per capita GDP measured at PPPs and at current exchange rates (see table 5.2.1), which suggests that the relative prices of non-traded to traded goods in the candidate countries are much lower than in the EU.

As displayed in chart 5.2.1, there is a clear positive relation between relative productivity growth and relative price increases in both the CEEC-10 and the EU, suggesting that the differences between the rates of price increase in the non-tradeable and tradeable sectors in the candidate countries can be largely explained by the gaps between productivity growth in the two sectors. There are a number of estimates of the Balassa-Samuelson effect in the CEEC-10, ranging from 0.8 to 3 percentage points of annual consumer price inflation,[475] but its size is uncertain for the accession period. A faster rate of catch up may be accompanied by larger productivity gaps between the two sectors and in this case the inflation rate determined by the structural characteristics of the economy might be substantially above the rate in the more developed EU countries.[476]

(iii) Macroeconomic constraints and risks during the catching-up period

Most accession countries are small, open economies with largely liberalized capital markets. Due to their potential for catching up they are attractive to foreign investors but at the same time are sensitive to events in other emerging markets, a category in which they are generally placed by international investors. The management of a sustainable regime of high growth and capital flows, while avoiding volatility in economic performance, constitutes a major challenge for policy makers in the CEEC-10. Sustainability requires careful manoeuvring between the fundamental trade-offs characterizing the accession process: rapid growth and growing imbalances, long-term convergence and short-term stabilization, foreign versus domestic pressures on monetary conditions and price versus exchange rate stability.

---

[475] For a discussion of the Balassa-Samuelson effect in transition economies and panel estimations for the region, see L. Halpern and C. Wyplosz, "Equilibrium exchange rates in transition economies", *IMF Staff Papers*, Vol. 44, No. 4 (Washington, D.C.), 1997, pp. 430-461 and "Economic transformation and real exchange rates in the 2000s: the Balassa-Samuelson connection", UNECE, *Economic Survey of Europe, 2001 No. 1*, chap. 6, pp. 227-239. Estimates for Hungary are contained in M. Kovacs and A. Simon, *The Components of the Real Exchange Rate in Hungary*, National Bank of Hungary Working Paper, 1998/3 (Budapest), March 1998. For Slovenia see P. Rother, "The impact of productivity differentials on inflation and the real exchange rate: an estimation of the Balassa-Samuelson effect in Slovenia", IMF Staff Country Report No. 2000/56, *Republic of Slovenia: Selected Issues* (Washington, D.C.), February 2000, pp 26-38. For comparison with the EMU countries see E. Alberola-Ila and T. Tyrvainen, *Is There Scope for Inflation Differential in EMU? An Empirical Evaluation of the Balassa-Samuelson Model in EMU Countries*, Banco de Espana Working Paper, No. 9823 (Madrid), 1998.

[476] There are, however, some factors that might reduce the extent of the equilibrium appreciation in the accession period. Many of the CEEC-10 have already achieved a considerable degree of productivity growth in the service sectors, due to large inflows of foreign direct investment. Furthermore, the information technology explosion is also likely to bring about a sharp upturn in the productivity of the non-tradeable sector. Labour market conditions, with differing degrees of rigidity among the acceding countries, may also affect the actual size of adjustment.

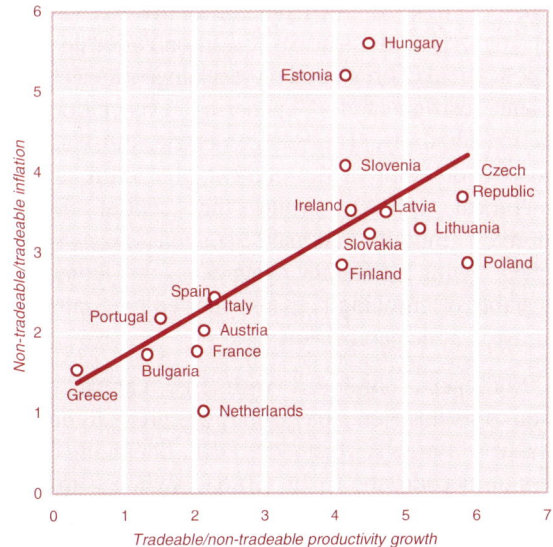

CHART 5.2.1

Productivity growth and inflation differentials in CEEC-8 and the EU, 1987-1998

(Percentage points)

Source: UNECE Common Database; H.-W. Sinn and M. Reutter, *The Minimum Inflation Rate for Euroland*, NBER Working Paper, No. 8085 (Cambridge, MA), January 2001.

Due to the abundance of attractive investment opportunities, investment demand is likely to remain high in the accession countries. At the same time domestic savings may not be sufficient to finance these growing demands for investment; in particular, the corporate sector is likely to see a persistent surge in borrowing. Since domestic savings are further constrained in most of the CEEC-10 by the large claims of public spending arising from the costs of transformation reforms, the process of catching up is likely to be accompanied by increasing current account deficits and, possibly, external debt. As long as there is an exchange rate risk, increasing levels of foreign currency denominated debt may be seen by market participants as a sign of vulnerability. Thus, in the run-up to monetary union, the sustainability of the current account deficits together with the level of foreign indebtedness could become an important constraint, especially if the increase in foreign debt is not matched by increased export capacity.

The debt constraint can be alleviated by attracting non-debt-creating foreign direct investments (FDI). Experience in the second half of the 1990s shows that even large current account deficits can be financed – sometimes over-financed – largely by FDI. At the same time, several transition countries also received substantial short-term capital inflows during the 1990s, in the form of portfolio investment and bank loans. These flows were quite volatile during the emerging market turmoil of the late 1990s. Large capital movements could be a major source of vulnerability during the accession period.

The fragility of the financial sector is another source of vulnerability that might lead to currency crises. Despite the progress in financial reforms and in prudential regulation, the exposure of the financial system to sudden exchange rate changes due to currency and maturity mismatches may still be a major risk during the accession period. In particular, the expansion of short-term liabilities denominated in foreign currency may increase the risks of financial and exchange rate instability.

(iv) **Policy challenges**

Maintaining an equilibrium path during the process of catching up requires a careful balance between the needs of long-term convergence and short-term stabilization. A trend appreciation of the equilibrium real exchange rate implies that, in fast growing economies, the equilibrium real interest rate, consistent with the uncovered interest parity (UIP) condition, could be substantially below that of the advanced countries, especially if the market-determined risk premium is falling.[477] At the same time, domestic conditions in the acceding countries may require the central banks to keep interest rates relatively high to support disinflation and promote household savings. Therefore, the basic monetary policy dilemma is how to reconcile the external and domestic constraints, both in devising a longer-term strategy and in conducting everyday policy, which is key to harmonizing the real and nominal sides of convergence.

In acceding countries with good records of stabilization and credible policies, interest rate policy is largely constrained by the changes in regional and emerging market risk. The required interest rate premium might sometimes be strongly influenced by external factors that are beyond the control of domestic policy makers (e.g. the so-called "sudden stop" contagion from emerging market crises[478] or the "flight to quality" by anxious investors). In this case the interest rate implied by the parity condition might exceed significantly the level appropriate for domestic conditions. The jump in the required risk premium and/or capital outflows might force policy responses – exchange rate depreciation or higher interest rates – that run against the smooth development of fundamentals.

---

[477] According to the UIP condition, $r - r^* = \Delta e/e - (\pi - \pi^*) + \mu$, where π = domestic inflation; r = domestic real interest rate; π^* = foreign inflation; r^* = foreign real interest rate; $\Delta e/e$ = exchange rate change; μ = required risk premium. The term $[\Delta e/e - (\pi - \pi^*)]$ represents the change in the real exchange rate. The required country risk has three elements: the default risk, representing the creditworthiness of a country; the currency risk, which depends partly on economic fundamentals and partly on the "insurance" guaranteed by the exchange rate regime; and the regional risk, determined by external factors.

[478] Due in part to the prospects of EU and EMU membership, the CEEC-10 have been less prone to sudden stops than the Latin American emerging economies. G. Calvo, "The case for hard pegs in the brave new world of global finance", paper presented at the ABCDE Europe Conference (Paris), 26 June 2000.

Minimizing the destabilizing impact of external disturbances in the run-up to accession is one of the major challenges for policy makers in the acceding countries as they will continue to be prone to nominal as well as real asymmetric shocks. Concerning nominal shocks, it can be conjectured that contagion from emerging markets will be the most serious danger until accession takes place, although domestic fiscal policy and the vulnerability of the financial sector should also be considered as potential sources of instability.[479] EU entry itself and the prospect of EMU membership can also be considered as a potential source of future asymmetric shocks as the period until EMU accession is likely to trigger large amounts of financial investments arbitraging the convergence between yields on domestic and EMU assets.

Managing real exchange rate appreciation is a major challenge. Domestic policy makers have to choose between accommodating the real appreciation by an inflation differential or by a nominal appreciation of the exchange rate. On the one hand, aiming at a relatively stable nominal exchange rate may result in a rate of inflation which is higher than that required by the price stability criterion of the Treaty. On the other hand, a nominal appreciation could help to bring about a faster rate of disinflation, but this can be risky in a small, open economy, because it may hurt the tradeable sector and endanger export-led growth. Thus a balanced catching-up scenario coupled with a relatively stable nominal exchange rate might imply a higher rate of inflation in the CEEC-10 than that required by the EMU price stability criterion. This indicates that there may be a conflict between simultaneously meeting both the price and exchange rate stability criteria for EMU membership (see below).

There are two main pillars that pre-accession stabilization policy can rely on: (i) the exchange rate which, depending on the flexibility of the regime, can absorb the impact of capital movements and shocks, and (ii) fiscal policy, which is the key to managing a balanced and steady process of catching up.[480] The management of effective catching up presupposes a balanced fiscal and monetary policy mix,[481] and in this context the characteristics of the exchange rate and monetary regimes are of crucial importance.

5.3 The role of the exchange rate regime in a catch-up process

Since the start of the transition the CEEC-10 have been running a variety of exchange rate and monetary regimes.[482] The revealed preference for exchange-rate-based regimes at the start of transition reflected the fact that policy makers regarded the exchange rate as the most efficient instrument for anchoring expectations in a transition economy. The Baltic countries introduced hard pegs, two of them in the context of a currency board arrangement (CBA), while the central European economies initially adopted pegged exchange rate regimes with different degrees of commitment (adjustable or crawling pegs), supported by controls over the most volatile capital flows.

Soft pegs combined with free capital mobility proved to be unsustainable as they are prone to becoming speculative targets; thus, by the beginning of the accession decade, the CEEC-10 had switched to "corner solutions", running either flexible exchange rate regimes or very hard pegs.[483] The Czech Republic, Poland, Romania, Slovakia and Slovenia have floating or managed floating regimes, while Hungary is operating a wide band regime (±15 per cent) with a central parity pegged to the euro. The Baltic countries continue to maintain CBAs (in Estonia and Lithuania) or a very hard peg (with a zero fluctuation band in Latvia). Bulgaria also introduced a CBA in 1997 after a crisis and a bout of hyperinflation.

The experience with CBAs during the last decade has highlighted the most important conditions for a hard peg to work efficiently: namely, a relatively small economy with a large export sector, flexible labour markets, strong financial institutions and a supportive fiscal policy.[484] In the pre-accession period, this group of applicants face the challenge of maintaining their balanced approach so as to become eligible for fully-fledged membership of the EMU. In contrast, the central European countries with flexible exchange rates face the additional challenge of adjusting their regimes to meet the nominal convergence criteria for EMU entry.

---

[479] The probability of unbridled fiscal spending will be moderated by the fact that most candidates had already experienced, during the 1990s, the risks and effects of inadequate fiscal control. The candidate countries are therefore well aware of the conditions for sustainability and their interest in stability-oriented macropolicy, backed by a balanced fiscal stance, is independent of their desire for EMU accession.

[480] "Balancing fiscal priorities – challenges for the central European countries", in R. Feldman and C. Watson (eds.), op. cit.

[481] The fiscal implications of EMU accession are not discussed specifically in this chapter. On some of the fiscal policy challenges in the candidate countries, see chap.3.2 in this *Survey*.

[482] "Monetary and exchange rate regimes in central European economies on the road to EU accession and monetary union", in R. Feldman and C. Watson (eds.), op. cit.

[483] The elimination of the crawling peg regimes in the CEEC-10 should not be regarded as failure of the intermediate regimes. In fact, the crawling regimes in Hungary and Poland played a leading role in gaining credibility in a certain phase of stabilization. However, the crawling regimes should be seen as a temporary solution and the exit from them in the central European economies was "pre-programmed", although accelerated by speculative flows. From the wider perspective of EU enlargement the experience of the 1990s may be an important lesson for future accession candidates.

[484] A. Gulde, J. Kahkonen and P. Keller, *Pros and Cons of Currency Board Arrangements in the Lead-up to EU Accession and Participation in the Euro Zone*, IMF Policy Discussion Paper PDP/00/1 (Washington, D.C.), January 2000.

An important policy question is whether flexibility can help the candidate countries to minimize the risk of crises in the run-up to monetary union and to meet the requirements for ERM-2. In the light of the usual policy advice – "if shocks are real, then float, if they are nominal, then fix" – the choice of floating regimes in acceding countries is far from obvious, as the asymmetric shocks that they are likely to face in the accession period are mostly nominal and of external origin. Moreover, the introduction of a flexible exchange rate regime strongly affects the monetary policy framework in a candidate country. The floaters all changed from exchange rate targeting to an inflation targeting regime, although their medium-term objective is to join the ERM-2, which implies a return to exchange rate targeting.

(i) Flexible exchange rates as a policy tool

A flexible exchange rate regime increases the degrees of policy freedom. Thus, the possibility of testing for an equilibrium level of the exchange rate before joining the ERM-2 is often used as an argument in favour of such a regime. In flexible regimes, exchange rate appreciation may also support the disinflation policy, while monetary policy, due to its greater independence, can cope better with speculative flows. A floating exchange rate can also support economic stabilization by absorbing the impact of external shocks and resisting contagion from emerging markets. In the event of a crisis, triggered by capital outflows and accompanied by a rise in risk premia, the authorities might choose to let the exchange rate depreciate, thereby minimizing the loss of reserves and avoiding an excessive increase in interest rates.

There are also counter arguments against exchange rate flexibility. Thus in practice, it may be difficult to identify clearly the factors behind a nominal appreciation: whether it reflects an equilibrium adjustment or is a consequence of speculative capital inflows. In addition, as noted earlier, nominal appreciation may curb export-led growth. As to the absorption of nominal shocks, in practice the exchange rate cannot easily absorb the transmission of country risk caused by contagion, as depreciation might create expectations of further depreciation.[485]

The recent experience of the CEEC-10 is quite varied. Since the introduction of floating, the Czech koruna and the Polish zloty have tended to appreciate, while for the Slovak koruna the tendency has been less clear. By contrast, the Slovenian tolar has been steadily depreciating, as was the Hungarian forint before its exit from the narrow band crawling regime in May 2001 (chart 5.3.1).[486]

The emerging market crises of the late 1990s (in Asia and Russia) provided a test of how the advanced candidate countries' exchange rate regimes respond to shocks. In general, exchange rate flexibility increased substantially the currency risk, as reflected in higher interest premia (chart 5.3.2). The risk premia on domestic assets reacted very differently in the various countries: in Hungary (with a quasi-fixed regime) there was a jump in the premium, while in the Czech Republic and Poland it declined. At the same time the Czech koruna and the zloty depreciated significantly, while the depreciation of the forint was constrained by a narrow band. The evidence generally supports the view that flexible regimes make it easier to cope with external shocks but at the cost of higher interest rates.

International experience suggests the following conditions have to be met if a flexible regime is to work efficiently: a high credibility of economic and monetary policies; the exchange rate pass-through coefficient should be low or declining; the private sector should behave with great caution in handling exchange rate risk; and currency mismatches and financial market volatility should not increase. It is also important to find a credible nominal anchor, assuming an incomplete pass-through of the greater exchange rate volatility into domestic prices and inflationary expectations.

(ii) The role of nominal anchors

The candidate countries that introduced flexible exchange rates in the late 1990s opted for inflation targeting (IT) as a nominal anchor. Successful IT presupposes a greater credibility of monetary policy, a clear communication of goals and means to the public, and also adequate knowledge of the monetary transmission mechanism. Disinflation policy under this regime can only be successful if the inflation forecast can convincingly replace the exchange rate as the nominal anchor for expectations and economic decisions. In addition, the weight of direct and indirect exchange rate transmission channels should be considerably reduced as compared with that prevailing under an exchange rate target. To be effective, IT also requires that interest rates have a significant impact on the economy.

In this regard, the candidate countries face more uncertainty than more developed economies. In small, open economies the effects of exchange rate movements may be relatively strong as compared with the impact of interest rates, even if the credibility of the inflation target has increased. In addition, the effect of real interest rates on savings may be ambiguous: in a transitional environment there are a number of distorting factors that affect the propensity of households to save while, as

---

[485] Empirical evidence for Latin America indicates that countries with floating rates had to raise interest rates even higher than those with pegs: that is, they had to pay a price for the lack of exchange rate commitment. R. Hausmann, "Latin America: no firework, no crisis?", paper presented at the conference on *Lessons from the Recent Global Financial Crises*, co-sponsored by the Bank for International Settlements and Federal Reserve Bank of Chicago (Chicago), 30 September-2 October 1999.

[486] In both cases policy played a role in the depreciation.

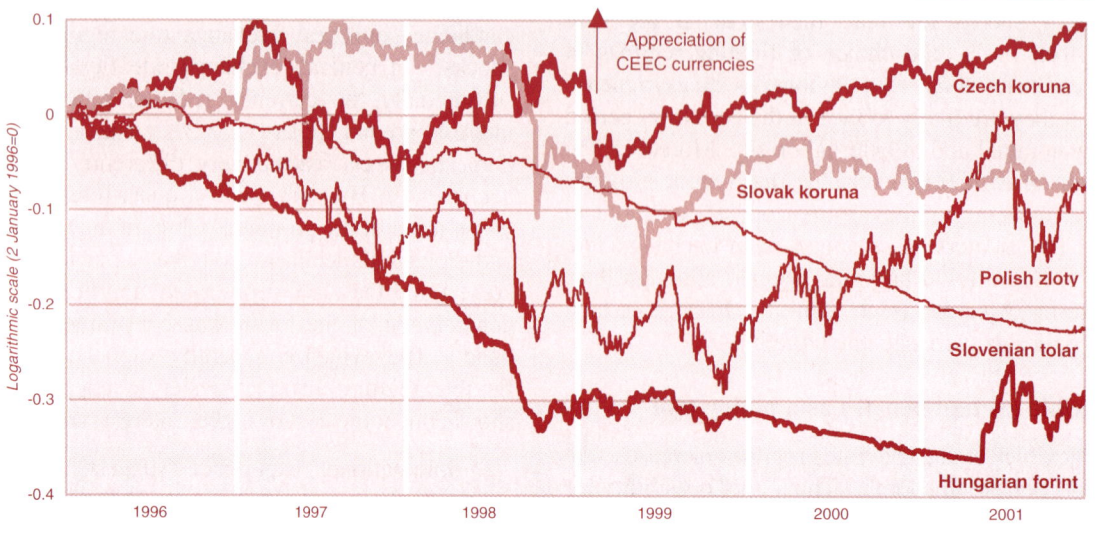

CHART 5.3.1

Nominal exchange rates of CEEC-5 currencies against the deutsche mark, 1996-2001

(Indices in logarithmic scale, 2 January 1996=0)

Source: National Bank of Hungary.

noted earlier, the share of debt financing in corporate investment is expected to increase, partly independently of the level of interest rates. Hence the effectiveness of monetary policy may be limited. On the other hand, under a flexible exchange rate regime, the monetary authorities can intervene in both the money and foreign exchange markets in order to balance domestic and foreign constraints and to avoid large disturbances in nominal values.

(iii) The exchange rate mechanism (ERM-2) as a flexible tool for managing convergence

Before accession to the monetary union the applicants are required to spend at least two years in the exchange rate mechanism (ERM-2). In principle, this regime is not a simple "ante-chamber" of the euro, but a monetary framework which aims at several objectives: (i) facilitating nominal convergence (towards the Maastricht criteria); (ii) allowing a market test for exchange rate stability; (iii) ensuring that a given country enters the euro zone at an appropriate exchange rate; and (iv) preparing the central banks to operate as members of the System of European Central Banks.[487]

The framework of ERM-2 differs from a "simple" flexible exchange rate regime in two important respects. The central parity can be seen as a nominal anchor, representing – in principle – the exchange rate that should be irrevocably fixed. The wide-band version of ERM, in its version after the 1992 crisis, provides a framework for exchange rate management. In addition, safeguarding the limits of the band in the ERM-2 will be a joint commitment by the candidate country and the ECB, which should help to reduce the danger of self-fulfilling exchange rate expectations.

The compatibility of currency board arrangements with the ERM-2 has already been acknowledged by the EU authorities. But for the candidate countries with floating exchange rate regimes to enter the ERM-2, they have to agree with the ECB on a central parity and on a band width, which will vary across countries but is likely to be wide.[488]

The application of an ERM-2 type regime over a long period could raise several problems of sustainability in the candidate countries. The ERM-2 does not tolerate a devaluation of the central parity. Consequently, the applicant countries have an interest in starting the ERM-2 with a somewhat devalued central parity (as did Greece), allowing them to move into the stronger part of the band even in the face of depreciation pressures stemming from unanticipated external shocks.[489] On the other hand, the ERM-2 allows a revaluation of the central parity. However, if the ERM-2 is maintained for a long time and the real appreciation of the equilibrium rate is

---

[487] P. van der Haegen and C. Thimann, "Monetary policy in transition and towards accession", in G. Tumpel-Guggerell, L. Wolfe and P. Mooslechner (eds.), *Completing Transition: The Main Challenges* (Berlin, Springer, 2001), pp. 161-184; J. Kröger and D. Redonnet, "Exchange rate regimes and economic integration", *CESINFO Forum* (Munich), Summer 2001, pp. 6-13.

[488] The case of Hungary may be worth following because, after abandoning the crawling band, it introduced unilaterally, in October 2001, a euro-peg regime with a ±15 per cent width band.

[489] Otherwise, whenever the exchange rate depreciates, especially if it diverges from the central parity, market participants may begin to doubt the sustainability of the ERM-2, and speculation might become self-fulfilling.

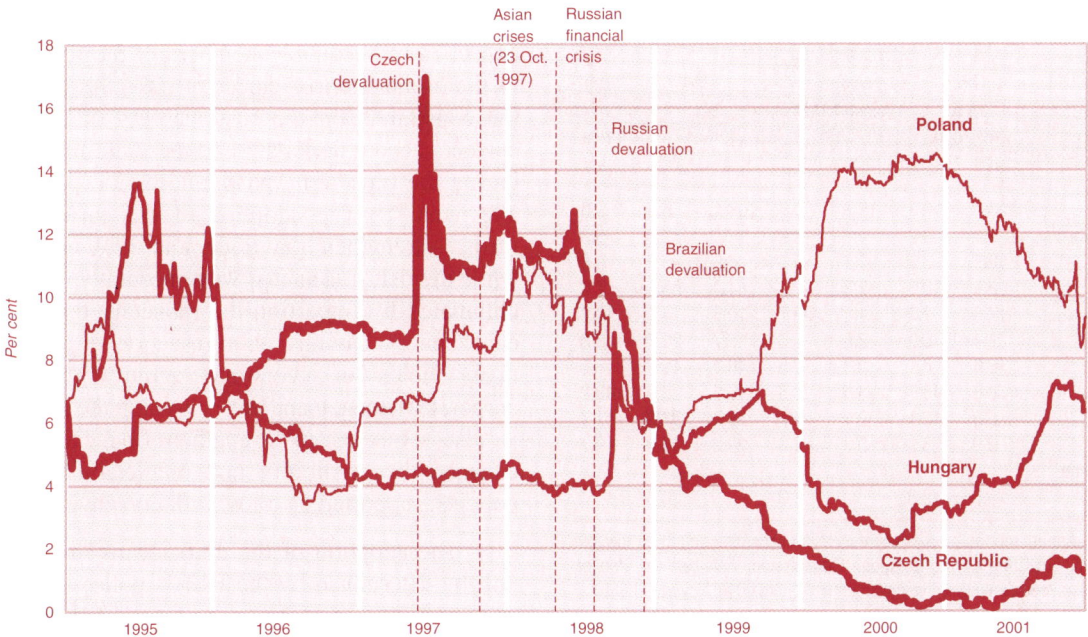

CHART 5.3.2

Interest rate premia on the Czech koruna, Hungarian forint and Polish zloty, 1995-2001
(Per cent)

Source: Computation based on data from national central banks' databases and IMF, *International Financial Statistics* (Washington, D.C.), various issues.

Note: The interest rate premium is calculated on the basis of three-month interbank rates and pre-announced rates of depreciation compared to three-month LIBOR. No adjustment for exchange rate changes is made for the Czech Republic where the currency was either fixed or floating. The Polish zloty was left to float in April 2000 and the crawling depreciation of the Hungarian forint ended in October 2001, so no adjustment for exchange rate changes is made after these dates.

accommodated by nominal exchange rate appreciation, even a wide band might become "narrow", depending on the size of the Balassa-Samuelson effect.[490] Regular realignments, as well as an exchange rate fluctuating in the upper part of the band, could generate self-fulfilling expectations of appreciation, and thus endanger a smooth catching-up process.

An ERM-2 regime where there is systematic upward pressure on the exchange rate towards the upper part of the band (as has been the case in Hungary since the introduction of its ERM-2 compatible regime), or where the central parity is steadily appreciating, is not fulfilling its role as an effective anchor; instead the regime is operating as a soft inflation target. This suggests that the candidate countries with flexible regimes are likely to make their official commitments to the ERM-2 only at the "last moment" (i.e. two years before the examination for EMU accession), when there is a higher probability of sustainability and when the central parity can be chosen more accurately to fulfil its role of anchoring expectations.

5.4 EMU accession scenarios

(i) The policy framework of EMU accession

(a) The prospects for nominal convergence

Although the economic rationale behind the Maastricht criteria was fairly clear for the original members of the euro area, the long catching-up process facing the new candidates raises some new issues. The contradiction between the price and exchange rate stability criteria has been acknowledged by allowing a revaluation of the central parity in the ERM-2, but a reinterpretation of the price stability requirement[491] is so far not considered as a feasible option by the EU authorities. Thus, there remain a number of policy issues that need to be addressed by the candidate countries.

The CEEC-8 have already achieved a substantial degree of nominal convergence with the EU (table 5.4.1) and the prospects for continued disinflation are favourable. The fiscal indicators suggest that in most of

---

[490] In the case of Hungary, assuming around a 2 per cent annual nominal appreciation of the forint, the exchange rate can be expected to breach the wide band before the official commitment to an ERM-2 with the EU is made.

[491] This issue has been raised by W. Buiter and C. Grafe, *Anchor, Float or Abandon Ship: Exchange Rate Regimes for Accession Countries*, CEPR Discussion Paper, No. 3184 (London), January 2002, and G. Szapáry, *Maastricht and the Choice of Exchange Rate Regime in Transition Countries During the Run-up to EMU*, National Bank of Hungary Working Paper, 2000/7 (Budapest), July 2000, among others.

TABLE 5.4.1

Indicators of nominal convergence of CEEC-10 with the Maastricht criteria, 1996-2001

(Percentage share)

	CPI[a] annual average		General government budget[b] balance to GDP		Public debt[c] to GDP		Long-term interest rates
	1996	2000	1996	2000	1997	2000	2001
Bulgaria	123.0	10.3	-15.3	-0.7	107.4	76.9	5.0
Czech Republic	9.1	3.9	-1.7	-4.2	13.0	17.3	5.3
Estonia	19.8	3.9	-1.6	-0.7	6.8	5.3	6.8
Hungary	23.5	10.0	-3.2	-3.1	64.2	55.7	7.0
Latvia	17.6	2.6	-1.3	-2.7	10.6	14.1	10.2
Lithuania	27.4	0.9	-2.8	-3.3	15.7	23.7	6.3
Poland	18.6	8.4	-2.3	-3.5	46.9	40.9	8.8
Romania	38.8	45.7	-3.5	-3.8	16.5	22.9	49.2
Slovakia	5.8	12.1	-2.1	-6.7	29.7	37.3	7.7
Slovenia	9.9	8.9	0.3	-2.3	23.2	25.8	9.7
EMU reference value	..	2.8	..	3.0		60.0	..

Source: Annual reports on the progress of the candidate countries towards EU accession, published by the European Commission, *Enlargement Monitor*, available at [http://europa.eu.int/comm/enlargement/report2001/], November 2001.

[a] These are the first results of the attempt to produce HICP compatible consumer price indices for the CEEC-10.

[b] The general government balances include the central budget, municipalities and extrabudgetary funds, but exclude privatization revenues. The fiscal indicators are provisional in the sense that they do not yet fully comply with EU methodological requirements. The general government deficit/surplus refers, for the first time, to the national accounts concept of consolidated general government net borrowing/net lending of ESA 95. General government debt is defined as consolidated gross debt at end-year nominal value.

[c] Yields on 10-year government bonds at the end of 2001. Shorter maturities taken for: Bulgaria, Estonia, Latvia, Lithuania, Romania and Slovakia.

the CEEC-10 budget deficits are under control while public debt ratios are below the Maastricht ceiling.[492] However, since major restructuring programmes still lie ahead, even in some of the more advanced candidate countries, increases in deficits and public debt levels cannot be ruled out. Thus, there is little room for further increases in public debt in these economies.

Comparing the Maastricht indicators of the candidates with those of the less developed EMU-12 countries suggests that the advanced applicants are somewhat better placed today than the current euro zone members were at a comparable stage before joining the EMU. The task of nominal convergence facing the more advanced candidates does not seem to be any more difficult than that facing Greece in the mid-1990s.[493] At the same time, the probability of shocks to the CEEC-10 during the accession period is considerably higher due to the elimination of capital controls.

(b) Benefits and costs of EMU accession

The main benefits that the acceding countries expect from becoming members of the EMU are related to the elimination of currency risk. As a consequence, the constraints on the sustainability of relatively large current account deficits – and these are major constraints during the transition and the accession period – will be alleviated.[494] The adoption of the euro will probably further strengthen the trade and FDI creation effects, while the risks associated with speculative inflows will be eliminated by importing the monetary conditions of the euro zone. Savings on transaction costs may benefit many fields of economic activity (trade, tourism, corporate management, etc.). Joining the monetary union is also expected to strengthen financial stability and enhance the credibility and predictability of domestic policies, supported by a low inflation environment.

The adoption of the euro means that interest rates will be determined by the ECB's policy decisions based on the average performance of the EMU countries but, given the relatively small size of their economies and financial sectors, the candidate countries are unlikely to have a significant influence on monetary conditions in the euro area. Thus, the "imported" interest rate is likely to be lower than that consistent with the cyclical position of the fast-growing new members. The interest rate premia will also be substantially reduced by the elimination of exchange rate risk. Hence, real interest rates can be expected to fall, resulting in lower interest payments on public debt, on the one hand, and increased investment, largely driven by exports, on the other. As a result, the potential growth rate of the new euro members could increase, allowing them a faster rate of catch up with the more developed countries.[495]

On the cost side, the loss of control over the exchange rate and monetary policy is the most important item. However, as discussed above, there is unlikely to be much scope for policy to influence exchange and

---

[492] Despite their rapid decline in the second half of the 1990s, public debt ratios in Bulgaria and Hungary need to be further reduced in order to bring them into line with their growth potential. S. Gomulka, "Policy challenges within the (enlarged) European Union: how to foster economic convergence?", paper presented at the conference on *Convergence and Divergence in Europe*, Oesterreichische Nationalbank (Vienna), 5-6 November 2001.

[493] J. Pelkmans et al., op. cit.

[494] The experience of the less advanced euro members provides evidence for this. During the convergence period, the annual current account deficit of Portugal was around 3 per cent of GDP, while after adoption of the euro it rose to close to 9 per cent. A. Alesina, O. Blanchard, J. Gali, F. Giavazzi and H. Uhlig, *Monitoring the ECB 3: Defining a Macroeconomic Framework for the Euro Area* (London, CEPR, 2001). However, it should be noted that although the risk of exchange rate crisis is eliminated after EMU accession, large current account deficits remain a hazard for the economy due to the increase of debt.

[495] There are differing estimates of the impact of the currency union on trade and growth in the new member countries. See e.g. F. Frankel and A. Rose, *Estimating the Effect of Currency Unions on Trade and Output*, CEPR Discussion Paper, No. 2631 (London), December 2000; and P. Havlik, *EU Enlargement: Economic Impacts on Austria, the Czech Republic, Hungary, Poland, Slovakia and Slovenia*, The Vienna Institute for International Economic Studies (WIIW), Research Report No. 280 (Vienna), October 2001.

interest rates in the run-up to accession. It should also be noted that, after EU accession, the new members will not be allowed to resort to competitive devaluation as this would threaten the convergence achieved so far. The "loss of monetary independence" will not bring about fundamental changes in policy-making, as monetary conditions were already strongly influenced by external conditions during the 1990s. Consequently, fiscal policy will continue to be the main instrument for preserving stability in the new member countries.

One of the central issues that needs to be addressed in the context of a cost-and-benefit analysis is the possible temporary output loss associated with EU accession. There is now a broad understanding that meeting the nominal convergence (Maastricht) criteria, may imply a cost for the CEEC-10 in the form of a loss of potential output. The reason is that, according to the present rules, euro zone accession may require an inflation rate that is lower than that implied by the Balassa-Samuelson effect.[496] To meet the required inflation rate, the candidates may have to combine an excessively tight fiscal adjustment with some appreciation of the exchange rate.[497] Such a policy is likely to generate in the short run a loss of potential output and, consequently, higher rates of unemployment; thus, in this way, the Maastricht definition of price stability could penalize fast-growing, catching-up economies.[498] Disinflation achieved this way will probably not add much to the credibility of policy in a given country, but investors will be able to enjoy large profits from the arbitrage opportunities created in the short-term securities markets.

At the same time, the output loss is likely to be only temporary. After accession to EMU, the candidate countries should benefit from the growth enhancing factors outlined above and a fast rate of catch up can be expected to resume. Indeed, for the economies that are already part of EMU, there are no binding nominal restraints on a possible catch-up process, provided their macroeconomic fundamentals are sound and stable.[499] In the longer run the benefits of being part of the euro zone are likely to outweigh the short-run costs associated with accession.

(ii) The timing of accession

While in the present debate there is little doubt about the appropriateness of monetary union as an ultimate target for the CEEC-10, there is much more ambiguity as regards the timing of accession. This is probably the most debatable issue within the candidates' strategy for EMU accession. The debate generally focuses on two main alternative scenarios: that of rapid nominal convergence – "the early accession scenario", and that of a more gradual nominal convergence – "the postponed accession scenario".[500]

The decision as to the timing of EMU accession should be made in the wider context of the long-run catch-up process. The optimal timing of entry is likely to be different for each of the candidate countries. Still, there are some general considerations, which are relevant to the discussion regarding the choice of strategy.

(a) Early accession

This scenario supposes immediate entry to the ERM-2 after joining the EU, fulfilling without delay the Maastricht criteria (including that for inflation) and adopting the euro as soon as possible (two years after EU accession at the earliest).

Countries that have already declared a preference for early accession expect the eventual net gains from the adoption of the euro, in terms of trade, growth and stability, to more than offset any short-term output loss. Those among the more advanced candidate countries with high ratios of public debt to GDP and large current account deficits may have serious concerns about the sustainability of an ERM-2 compatible regime in the face of possible large capital inflows and adverse shocks. Thus a preference for early accession reflects a conviction that the main risks during the catching-up process are likely to stem from external factors beyond domestic policy control. Early membership of the monetary union is thus seen as contributing to the stability and development of financial markets.

There are candidate countries – in particular those with CBAs – that have already been pursuing quite conservative fiscal policies during the transition. Since

---

[496] This can be illustrated by the development of the inflation rate in Greece around the time it was being assessed for EMU entry.

[497] Appreciation can be achieved by maintaining short-term interest rates high enough to attract short-term capital inflows.

[498] Under a different interpretation of the price stability criterion this output loss in the candidate countries might be avoided. Thus it has been suggested that for fast-growing, catching-up economies a more appropriate measure of price stability could be the dynamics of traded goods prices rather than that of consumer prices (according to the Maastricht criteria). The rationale is that if inflation is only driven by the Balassa-Samuelson effect (and is hence an equilibrium phenomenon) then it would not show up in traded goods prices but only in non-traded goods prices (and the CPI). Such a redefinition could allow a candidate country to qualify for nominal convergence (and accede to EMU) without any temporary loss of potential output. W. Buiter and C. Grafe, op. cit.

[499] The experience of Ireland is quite instructive in this regard. UNECE, "Ireland: regional economic adjustment in a monetary union", *Economic Survey of Europe, 2001 No. 1*, chap. 2.6, pp. 64-67.

[500] Apart from these two main scenarios, there is a lively (but mostly academic) debate on the possibility of unilateral euroization – i.e. adopting the euro before EU membership – as a policy option which might avoid some of the difficulties anticipated by the candidate countries. However, since the EU authorities have clearly and strongly rejected the possibility of unilateral euroization by the accession candidates – as a violation of the stages foreseen by the Treaty for the adoption of the euro – it is difficult at present to regard unilateral euroization as a realistic policy alternative for EMU accession by the present candidates.

the hard peg remains a unilateral commitment under the ERM-2, these candidates will have to continue to safeguard their credibility and maintain a balanced fiscal stance.[501] More generally, the Maastricht fiscal criteria will probably not be a problem for a large group of accession candidates. In all these countries, meeting the inflation target would not require in principle fundamental changes in their fiscal stance. Setting more ambitious targets for fast nominal convergence – which may be painful in the short run – depends mainly on political will, but they promise substantial long-run advantages in terms of a faster rate of catching up led by the private sector.

However, rapid adoption of the euro also involves serious risks. The viability of this scenario presupposes a high degree of flexibility in the setting of public sector wages and of prices. If budgetary policy were too rigid, a global recession coupled with a sudden collapse of foreign capital inflows could lead to a deep crisis. An important condition for this option, therefore, is that a candidate country should have sufficiently flexible economic structures and labour markets to enable it to cope with external shocks.

(b) Postponed accession

Under this scenario, after joining the EU, the country continues to maintain a flexible exchange rate regime combined with IT or a CBA for some time to come; at an appropriate time it first enters ERM-2 and, subsequently, the euro zone.

In the aftermath of EU accession, such a scenario can potentially accommodate a faster rate of real convergence based on a slower pace of nominal convergence. Countries with a relatively weak capacity to absorb FDI, and/or with low public debt ratios, would be able to benefit from some room for manoeuvre in fiscal policy. They might be able to foster a faster rate of economic growth by increasing public spending and by running somewhat larger budget deficits than required to meet the Maastricht criterion. Given that the capacity to absorb capital will play a crucial role during the catch-up process, an advantage of this scenario is that it would allow governments to increase their infrastructural investment in order to promote development.

In this scenario the "more generous" fiscal policy would be combined with a "less ambitious" rate of disinflation. Thus, candidates that have already reduced inflation close to the "optimal" rate for catching up could shift from a floating exchange rate to an ERM-2 compatible regime and then try to pursue a path of equilibrium real appreciation by allowing a somewhat higher inflation rate than that required for accession to the EMU.

However, this scenario presupposes that a candidate country is able to handle all the risks that might occur during the accession period and prior to adopting the euro. It might require coping with destabilizing capital flows and a convergence play for several years. Policy makers pursuing this scenario would need to be convinced that sound macroeconomic policies could sustain a relatively fast rate of catch up over the long run. In addition it should be noted that despite the relatively low levels of public debt, the tolerable size of government deficits is likely to continue to be constrained by the sustainability of the external position during the accession period, and this might significantly limit the scope for fiscal relaxation.

(iii) The choice of accession strategy

There are no general and clear-cut arguments in favour of an early or postponed accession. The academic literature is split on this issue: some experts see more benefits than risks from an early entry,[502] but others support postponement.[503] Thus the decision on the accession strategy should reflect the specificity of the economic environment as well the political preferences in individual countries.

One of the implicit arguments in favour of early accession is that it gives clear and unambiguous signals to investors and market participants. In a way, given the fact that the CEEC-10 are still struggling with the legacies of their past, EMU accession would be the final stamp verifying their status as mature market economies. On the other hand, there may be sound economic arguments in favour of postponing accession if there are reasonable expectations that this would give a boost to real convergence. Coupled with some degree of monetary independence, such a strategy could help the new EU members to minimize the risks and effects of asymmetric shocks.

The present exchange rate regimes in the CEEC-10 will not be a formal impediment in the choice of either of the two main policy courses.[504] However, stability and sustainability concerns (both in the run-up to and during

---

[501] Fast-growing countries with CBAs, however, are also likely to have a higher inflation rate than that required by the Maastricht criterion, even if they pursue a prudent fiscal policy.

[502] F. Coricelli, "Exchange rate arrangements in transition to EMU: some arguments in favour of early adoption of the euro", in G. Tumpel-Guggerell, L. Wolfe and P. Mooslechner (eds.), *Completing Transition: The Main Challenges* (Berlin, Springer, 2001), pp. 203-215.

[503] D. Begg, B. Eichengreen, L. Halpern, J. von Hagen and C. Wyplosz, "Sustainable regimes of capital movements in accession countries", paper presented at the conference, *How to Pave the Road to the E(M)U? The Monetary Side of the Enlargement Process (and its Fiscal Support)* (Eltville), 26-27 October 2001; W. Buiter and C. Grafe, op. cit.

[504] However, prior to accession, countries with a CBA will need to agree with the EU authorities the exchange rate parity that will be irrevocably fixed.

ERM-2) may be among the important factors that will affect the choice of accession strategy. Under such circumstances early accession may be considered as a way of shortening the period of exposure to excessive risks.

Fiscal policy considerations are likely to have a significant weight in the decision on the timing of accession. If – for different policy reasons outlined above – fiscal policy has limited room for manoeuvre, postponing EMU accession may not make much sense since these constraints will effectively rule out the growth enhancing policy options. In any case, fiscal prudence is a *sine qua non* for EMU accession, whatever the decision on timing.

Most of the advanced candidates (the CEEC-8) have already expressed their preference for early accession to the euro zone. The EU authorities, however, consider adoption of the euro by the future new members as a medium term rather than an immediate objective after EU accession; they stress that adoption of the euro will pose a number of significant challenges to candidates concerning the speed of real and nominal convergence, as well as the need to strengthen their market structures and financial sectors.[505] These considerations reflect the concern of the current EMU member countries that the possible structural and inflation risks created by the new entrants be kept to a minimum.

5.5 Conclusions

The choice of a strategy for EMU accession, including the timing of accession, is one of the most important policy decisions facing the candidate countries. This is a highly complex issue and numerous factors have to be taken into consideration in formulating an efficient and sustainable policy course. The decision should be preceded by a careful analysis of the alternative policy options and a balanced assessment of their pros and cons, as well as of their short- and long-run implications.

The ability of the financial sector in the candidate countries to cope with market forces and risks have been, and will remain, a major consideration in making judgements about the maturity of these economies. The candidate countries still fall into the category of emerging markets, and during the transition a number of them have experienced the destabilizing effects of volatile capital flows. Mobilizing external resources while avoiding a dangerous degree of volatility of capital flows presents a major challenge to policy makers in the candidate countries in the pre-accession phase. Their ability to cope with shocks will be an important factor in deciding about the timing of the application for EMU accession.

The sustainability of the exchange rate regime in the pre-accession phase is another major challenge for the candidate countries. The problem with a flexible exchange rate regime is the continuing risk of speculative and contagion shocks. On the other hand, a fixed exchange rate precludes the possibility of making real exchange rate realignments through nominal realignments of the exchange rate. There are no clear-cut recipes for choosing the most appropriate exchange rate regime but the capacity to maintain its credibility until accession should be among the key issues to be considered.

One of the important policy dilemmas facing the candidate countries is the conflict between the goals of achieving real and nominal convergence if these are to be pursued simultaneously. There may be a trade-off between the rate of nominal convergence (and hence accession to EMU) and the speed of real convergence, at least in the short run. That is, a fast growing candidate country may have to sacrifice some of its growth in the pre-accession phase in order to meet the strict nominal conditions for participating in the EMU.

On the other hand, the benefits of being part of the EMU can be expected to give an extra boost to growth after accession. Hence, the long-run benefits of a fast EMU accession may outweigh the short-term costs associated with accession.

In deciding about the timing of EMU accession, the acceding countries have to weigh the long-term advantages of adopting the euro against a temporary output loss in the short run, whatever the date of EMU entry. The size of the loss will be determined by several factors, including the timing of entry. Before coming to a decision on the EMU accession date, it would be advisable to compare the net present values over the long run of different paths to EMU rather than focusing only on its immediate effects.

Summing up, it is obvious that there are many uncertainties and risks surrounding the choice of a strategy for EMU accession. There are no universal, "one-fits-all" guidelines for the formulation of macroeconomic policies and strategies and this also holds for policies related to EMU accession. In pursuing their goal to be part of the euro zone, policy makers in each candidate country must carefully assess and weigh the specific costs and benefits of alternative policies in their own country before opting for EMU accession.

---

[505] "The ECB and the accession process", speech delivered by the President of the European Central Bank, W. F. Duisenberg, at the Frankfurt European Banking Congress (Frankfurt), 23 November 2001; European Commission, *DG Economic and Financial Affairs: Exchange Rate Strategies for EU Candidate Countries*, ECFIN/521/200 (Brussels), 22 August 2000.

PART THREE

SOCIAL DIMENSIONS OF ECONOMIC DEVELOPMENT

CHAPTER 6

NEW FORMS OF HOUSEHOLD FORMATION IN CENTRAL AND EASTERN EUROPE: ARE THEY RELATED TO NEWLY EMERGING VALUE ORIENTATIONS?

The profound economic and political changes in central and eastern Europe since the transition to democracy and a market economy got underway around 1990, have been accompanied by a sharp decline in fertility, which, in relative terms, has been steeper than in western Europe during any period of comparable length since the postwar baby boom ended in the middle of the 1960s. This has left central and eastern Europe as the region with the lowest fertility in the world.

There were three competing explanations – political, economic and social – of these developments, which were examined in a tentative and speculative manner in the Economic Survey of Europe, 1999 No. 1.[506] These were political instability and the attendant "fear of the future"; the drop in living standards on account of the deep economic crisis; and the spread of new forms of family and reproductive behaviour favouring smaller families, that have prevailed in western Europe since the middle of the 1960s.

A second study,[507] published in the Economic Survey of Europe, 2000 No. 1, suggested that the social and economic crisis of the 1990s was a major driving force behind the fertility decline and an econometric fertility model was developed to articulate the role of the various social and economic factors at work, including changes in incomes, unemployment, the educational attainment of women and the increase in the average age of motherhood.

The present study is complementary to the previous two papers. It acknowledges the role of the economic and social crisis in the fertility decline but also finds evidence that new forms of fertility and family behaviour have taken root in central and eastern Europe, which are associated with changes in social norms and values consistent with the new economic and political order, and many of which, having originated in the west, are branded as "western." The implication is that the economic recovery underway in central and eastern Europe should be expected to only partly reverse the earlier decline in fertility rates.

6.1 Introduction

Starting in the 1960s, there was a drastic transformation in the pattern of household formation and reproduction in north-western Europe. The age at first marriage rose again after falling to an unprecedented low during the 1960s. Premarital and postmarital cohabitation increased, and procreation in such informal unions soon followed. Divorce rates continued to rise in tandem with high separation rates among cohabitants. Also starting in the late 1960s was a pronounced postponement of fertility, which was followed by only a partial catching up at later ages.[508] In the 1970s, total fertility rates (TFRs) in western countries essentially reflect differential postponement; in the 1990s, national TFRs mainly capture differential rates of catching up after age 30.[509]

---

[506] UNECE, "Fertility decline in the transition economies, 1982-1997: political, economic and social factors", *Economic Survey of Europe, 1999 No. 1*, chap. 4, pp. 181-194.

[507] UNECE, "Fertility decline in the transition economies, 1989-1998: economic and social factors revisited", *Economic Survey of Europe, 2000 No. 1*, chap. 6, pp. 189-207.

[508] A detailed analysis of these tempo shifts in successive cohorts is given in T. Frejka and G. Calot, "Cohort reproductive patterns in low-fertility countries", *Population and Development Review*, Vol. 27, No. 1, 2001, pp. 103-132. See also R. Lesthaeghe, "Postponement and recuperation – recent fertility trends and forecasts in six western European countries", paper presented to the IUSSP Seminar on International Perspectives on Low Fertility, National Institute of Population and Social Security Research (Tokyo), 21-23 March 2001.

[509] On the repercussions of shifts in cohort fertility patterns on TFRs, see R. Lesthaeghe and G. Moors, "Recent trends in fertility and household formation in the industrialized west", *Review of Population and Social Policy*, No. 9, 2000, pp. 121-170.

At first it was thought that the economic recession following the 1974 oil crisis was responsible for later marriage and postponement of childbearing,[510] but there were already some suspicions that the roots of the new forms of household formation were to be found in the 1960s, and more particularly in the marked shift in values that occurred during that decade. Demographic changes were linked to (i) the accentuation of individual autonomy in ethical, moral and political spheres; (ii) to the concomitant rejection of all forms of institutional controls and authority; and (iii) to the rise of expressive values connected to the so-called "higher order needs"[511] of self-actualization and the quest for recognition. This connection between the demographic and value transformations became known as "Europe's second demographic transition".[512]

Towards the end of the 1980s, several features of this "second transition" seemed to stop at the Alps and Pyrenees. Italy, Portugal and Spain had started the postponement phase with respect to marriage and fertility, but the other two features, i.e. cohabitation and procreation outside wedlock, had either failed to gain ground (Italy) or were just beginning to spread (Portugal, Spain). Until 1990, earlier patterns of marriage and fertility had also been maintained in central and eastern Europe. As yet there were no clear signs of postponement or of the diffusion of premarital cohabitation. It thus seemed that the "second demographic transition" was a northern and western European phenomenon, which had crossed the oceans (Australia, Canada, New Zealand, United States) but not the old European cultural and political divides.

After 1990 this picture changed completely. In the Iberian Peninsula, the proportions of births outside marriage rose more rapidly, signalling that both cohabitation and procreation within informal unions were spreading. In central and eastern Europe (but not in the CIS countries), the postponement of marriage and childbearing started and progressed to the point of causing a fall in national TFRs to levels below 1.5 children and even 1.3. A new term was coined: "lowest-low fertility".[513] A direct connection was made between marriage and fertility postponement on the one hand and the effects of the difficult economic transition on the other. In particular, these demographic changes were directly linked to rising unemployment, a reduction in activity rates especially for women, the end of life-long employment guarantees, the drop in real household incomes, the decline of state support for families and the enhanced visibility of poverty.[514]

Not everyone in central and eastern Europe, however, was convinced that the economic crisis was the sole explanation for the demographic changes. Mainly younger members of the demography profession in the Czech Republic, Hungary and Russia suspected that "western values" had penetrated their societies. They felt that the younger generations, which were to marry and start childbearing during the 1990s, had different priorities and aspirations compared with those of the older cohorts who had spent much of their lives during the communist era.[515] The outcome was a debate between the "crisis thesis" and the "second demographic transition thesis". As is common with such debates about the essentials of life, the two explanations were pitted against each other, and were viewed as mutually exclusive by their respective proponents.

The present contribution does not subscribe to such an "either/or" proposition. There is nothing mutually exclusive about the operation of both economic and cultural factors. In fact, they may be interwoven and mutually reinforcing.[516] To state it simply and metaphorically, the cart of demographic change can be pulled by two horses simultaneously. At the onset, it may well be that the horse of the economic crisis is doing

---

[510] In an initial article on the second demographic transition produced, Lesthaeghe and van de Kaa still considered that the tempo shifts in fertility and nuptiality were enhanced by the economic recession of the 1975-1985 decade. Hence they envisaged the possibility of a joint operation of economic and cultural factors. R. Lesthaeghe and D. van de Kaa, "Twee demografische transities?", in R. Lesthaeghe and D. van de Kaa (eds.), *Groei of Krimp?*, book volume of "Mens en Maatschappij" (Deventer, Van Loghum-Slaterus, 1986), pp. 9-24.

[511] The term was introduced by the psychologist A. Maslow, *Motivation and Personality* (New York, Harper and Row, 1954). His "lower order needs" mainly pertain to subsistence needs (not luxury goods!), safety and longer-term material security.

[512] The term first appears in the already cited Dutch language journal, but it spread following van de Kaa's subsequent article. D. van de Kaa, "Europe's second demographic transition", *Population Bulletin*, Vol. 42, No. 1, 1987.

[513] H.-P. Kohler, F. Billari and J. Ortega, "Towards a theory of lowest-low fertility", paper presented to the IUSSP General Conference (Salvador, Brazil), 18-24 August 2001.

[514] UNECE, "Fertility decline in the transition economies, 1989-1998: economic and social factors revisited", *Economic Survey of Europe*, 2000 No. 1, pp. 189-207.

[515] S. Zakharov, *Fertility Trends in Russia and the European New Independent States: Crisis or Turning Point?* (ESA./P/WP.140), United Nations Population Division, Expert Group Meeting on Below-Replacement Fertility (New York), 4-6 November 1997, pp. 271-290; S. Zakharov and E. Ivanova, "Fertility decline and recent changes in Russia: on the threshold of the second demographic transition", in J. Davanzo (ed.), *Russia's Demographic Crisis* (Santa Monica, CA, Rand Corporation, 1996), pp. 36-82; E. Fratczak, "Declining fertility in Poland during the transition period 1989-1997", paper presented to the Workshop on Lowest-low Fertility, Max Planck Institute for Demographic Research, (Rostock), 10-11 December 1998; D. Philipov, "Low fertility in central and eastern Europe – culture or economy?", paper presented to the IUSSP seminar on International Perspectives on Low Fertility, National Institute of Population and Social Security Research (Tokyo), 21-23 March 2001; K. Zeman, T. Sobotka and V. Kantorova, "Halfway between socialist greenhouse and postmodern plurality: life course transitions of young Czech women", paper presented to the Euresco Conference on the Second Demographic Transition (Bad Herrenalb), 23-28 June 2001, session 2B; J. Rychtarikova, "The second demographic transition and the transformation of fertility and partnership in the Czech Republic and other eastern European countries", ibid., session 2A; L. Rabusic, "On marriage and family trends in the Czech Republic in the mid-1990s" (in Czech), *Demografie*, Vol. 38, No. 3, 1996, pp. 173-180.

[516] R. Lesthaeghe, "On theory development and applications to the study of family formation", *Population and Development Review*, Vol. 24, No. 1, 1998, pp. 1-14. See also R. Lesthaeghe and J. Surkyn, "Cultural dynamics and economic theories of fertility change", ibid, Vol. 14, No. 1, 1988, pp. 1-45.

much of the pulling, while the other is quietly trotting along. But in the longer run, i.e. when the transitional recession has been fully overcome in central Europe and when there has been a sustained improvement of the economic situation, the second horse may take over. This proposition has important consequences for the future: if it is correct, an improvement of the economic situation would not lead to a restoration of the old demographic pattern of early marriage and fertility schedules, but to patterns of family formation that tend to converge on those of the west. In addition to later marriage and fertility postponement, features of the other "second demographic transition" can also be expected to emerge: premarital and postmarital cohabitation, procreation within cohabitation and possibly longer spells of single living.

The aim of the present chapter is to look for more precise indicators that signal the presence of the second horse in the above metaphor.[517] A new source with a lot of information about types of living arrangements and values is the 1999 round of the European Values Surveys. But this source is not without problems, as will be shown in the next section.

6.2 The European Values Surveys of 1999

Since 1980 the European Values Surveys (EVS) have become a major source of information on changing values and their covariates.[518] There have now been three rounds of the EVS (1981, 1990, 1999) in a fairly large number of countries. Attitude and value measurements cover a broad variety of domains: marriage and family, gender, religion, civil morality and ethics, political preferences, trust in institutions, the propensity to protest, "postmaterialism",[519] social distance and tolerance for minorities, qualities valued in socialization and in work, world orientation, economic ideology (free enterprise versus state intervention), community involvement and organization membership, etc. Most of these topics are covered by multiple questions or items, which improve the validity of their measurement. In the 1999 round, many countries also fine-tuned the household questions, *inter alia*, by inserting a probe for earlier premarital cohabitation. As a consequence, a finer typology for living arrangements could be constructed from this latest round of data.

The major drawback of the EVS has always been its small national sample sizes. The EVS standard practice is that a sample of 1,000 respondents suffices to cover the entire population, i.e. both sexes and all ages from 18 to 80. Only a few countries have larger sample sizes.[520] Such small samples are generally inadequate for crucial topics such as the study of the value orientations of the newly arriving cohorts of young adults, or for addressing any questions pertaining to more narrow age groups or subcategories.

The present study has also been hampered by the small national EVS samples, and as a result it was necessary to pool information for countries. For the present purpose, three pooled groups are formed:

- *WEST-8*: Austria, Belgium, Denmark, France, Germany, Portugal, Spain and Sweden;
- *CENTRAL-7*: Croatia, Czech Republic, Hungary, Lithuania, Poland, Slovakia and Slovenia;
- *EAST-5*: Belarus, Bulgaria, Romania, the Russian Federation and Ukraine.[521]

In order to avoid possible idiosyncrasies in a small sample for a large country overshadowing the overall pattern, the samples have not been weighted according to national population sizes. Hence, in everything that follows for the three groups, all countries are given an equal weight of unity.

As already mentioned, the 1999 EVS permits a more meaningful classification of respondents according to household situation than was possible in the earlier EVS rounds. More specifically, use is made of the following eight categories:

- *Respar*: respondents residing in the parental household without a partner or spouse. Most of them are never married or were never in a union, and never left home either (88 per cent). The rest have returned to the parental household after a different history;
- *Single*: Respondents who are not living with their parents, have never married and are not currently in a partnership either. Some had an earlier relationship, but none have children;

---

[517] Demographic surveys such as the rounds of Family and Fertility Surveys (FFS) commonly measure detailed event histories and socio-economic indicators, but at most only include a short module on value orientations. Furthermore, this module is rarely incorporated in the national surveys. At the concluding conference of the 1990s FFS round, not a single paper covered this subject.

[518] For an overview of the indicators and national results, see L. Halman, *The European Values Study – A Third Wave* (Tilburg, WORC Tilburg University, 2001). UNECE would like to acknowledge the permission given by the EVS Consortium for the use of the 1999 data files. Most of the national data sets are now in the public domain and can be obtained from Halman@kub.nl.

[519] R. Inglehart's term "postmaterialism" has been a constant source of misinterpretation. Inglehart coined the term largely as an expression of Maslow's "higher order needs" in the political sphere (democratic participation, grass-roots democracy, concerns related to environmental quality, freedom of speech, emancipation, new political ideas, etc.). The "materialist" orientation in Inglehart's formulation deals with income security, safeguarding of the social security system, political stability and "law and order". This concept has nothing to do with consumerism or conspicuous consumption of luxury goods. R. Inglehart, *The Silent Revolution: Changing Values and Political Styles among Western Publics* (Princeton, N.J., Princeton University Press, 1977).

[520] Sample sizes of 2000-2500 are used only in Belgium, Germany, Italy and Russia.

[521] The results of the present analysis are also available for smaller groups of countries or for countries with larger sample sizes. At present the analysis is completed for Belgium, the Czech Republic, France, Germany, Denmark+Sweden, Spain+Portugal, Slovakia+Hungary, Poland+Lithuania and Slovenia+Croatia. The outcomes will be put on the following web site: www.vub.ac.be/SOCO/Interface Demography/publications.

TABLE 6.2.1

European Values Surveys, 1999: sample size and relative proportion of household positions in three regional groups of countries [a]

(Number, per cent)

	WEST-8		CENTRAL-7		EAST-5	
	Number	Per cent	Number	Per cent	Number	Per cent
Respar: resident in parental households	783	15.1	984	22.9	602	19.2
Single: living alone	474	9.1	154	3.6	97	3.1
Coh0: cohabiting without children	719	13.9	337	7.8	102	3.3
Coh+: cohabiting with children	385	7.4	198	4.6	127	4.1
Mar0: married without children	278	5.4	145	3.4	154	4.9
Mar+N: married with children, never cohabited	1 548	29.8	2 114	49.2	1 622	51.8
Mar+E: married with children, ever cohabited	740	14.3	198	4.6	188	6.0
FmNu: Formerly married/in union; not in new union	259	5.0	164	3.8	242	7.7
Total	5 186	100.0	4 294	100.0	3 134	100.0

Source: European Values Surveys Consortium, national data sets.

Note: For definitions of household categories and country groups, see section 6.2 of the text.

[a] Respondents aged 18 to 45.

- *Coh0*: currently unmarried but cohabiting respondents without children, irrespective of earlier histories;
- *Coh+*: currently cohabiting respondents with children, again irrespective of earlier histories;
- *Mar0*: currently married respondents with a spouse present but without children;
- *Mar+N*: currently married respondents with a spouse and children, but who never passed through premarital cohabitation (N = *never* cohabited);
- *Mar+E*: currently married with spouse and children, but who passed through premarital cohabitation (E = *ever* cohabited);
- *FmNu*: formerly married or cohabiting respondents who are currently divorced or separated, but not yet in a new union. The majority of these respondents (85 per cent) have children and many women among them form a lone parent household.

The sample sizes for the eight household types in each of the three groups of countries are given in table 6.2.1 (absolute numbers and percentage distribution). These pertain to respondents aged 18 to 45. Despite the pooling of national samples, sample sizes are still small for several household categories in central and eastern Europe. This obviously reflects their smaller prevalence in the population. But, for work relating household positions to value orientations, these sample sizes are adequate.

The question concerning current cohabitation in the 1999 EVS also gives a rough idea of the incidence of premarital cohabitation among younger adults. In table 6.2.2 the 1999 figures are compared with those for the 1990s in the Family and Fertility Surveys (FFS).[522] In table 6.2.2 it can be seen that the EVS sample sizes are only a fraction of those used in the FFS. Furthermore, the FFS survey dates are heterogeneous and spread over 6 years, i.e. between 1991 and 1997. Finally, the procedures for gathering information in the two sources are different. This implies that the data in this table are merely *indicative* of the prevailing trend, and especially that *the 1999 EVS orders of magnitude are definitely subject to confirmation or correction by later and more representative sources.*

With this major caveat in mind, the comparison of FFS and EVS 1999 nevertheless suggests that there has been an increase in the percentages of women aged 20-24 and 25-29 who are currently cohabiting. Unmarried cohabitation has become a significant new household type in most Baltic and central European populations since the early 1990s. This is especially the case for Estonia and Latvia, and for Croatia, the Czech Republic, Hungary and Slovenia. The 1999 EVS suggests more modest rises in Poland and Lithuania. In Slovakia cohabitation among young women is still exceptional. In most of these central European countries, cohabiting women aged 20-29 are still childless, but in Estonia, Latvia and Slovenia there are significant proportions of cohabiting women with children.

In eastern Europe the incidence of cohabitation is still lower than in central Europe. The low figures in the 1997 FFS in Bulgaria are confirmed by the 1999 EVS, but there may have been a rise in Belarus, Romania and Russia, particularly for women aged 25-29. In Ukraine, as in Bulgaria, households of young cohabitants are still rare.

Taken at face value, the latest EVS results show a large increase in cohabitation in four countries (Estonia, Latvia, Hungary and Slovenia) that brings its prevalence (30 to 50 per cent) to western levels. Pending confirmation, this would mean that the two Baltic states increasingly resemble the Scandinavian situation, and that Hungary and Slovenia are moving toward the Austrian example.

---

[522] The FFS results are published in the form of a series of country reports with standardized graphs and tables. UNECE and UNFPA, *Fertility and Family Surveys in Countries of the ECE Region. Standard Country Report*, Economic Studies, No. 10 (various issues).

TABLE 6.2.2

Unmarried cohabitation of women in the transition economies in the 1990s
(Per cent, number)

		Percentage shares						Sample size
		Age group 20-24			Age group 25-29			
		Total	Without children	With children	Total	Without children	With children	
Baltic states								
Estonia	FFS 1991	13	9	4	19	5	14	659
	EVS 1999	42	33	9	22	4	18	99
Latvia	FFS 1995	8	5	3	6	2	4	778
	EVS 1999	40	26	14	37	19	18	73
Lithuania	FFS 1994	3	2	1	1	–	1	990
	EVS 1999	10	5	5	10	4	6	91
Central Europe								
Croatia	EVS 1999	30	30	–	13	13	–	146
Czech Republic	FFS 1997	10	8	2	9	3	6	601
	EVS 1999	24	22	2	17	11	–	146
Hungary	FFS 1992	7	5	2	2	1	1	1 456
	EVS 1999	33	28	5	27	16	11	87
Poland	FFS 1991	–	–	–	–	–	–	1 194
	EVS 1999	16	11	5	3	–	3	85
Slovakia	EVS 1999	6	6	–	3	3	–	125
Slovenia	FFS 1994	15	6	9	14	4	10	875
	EVS 1999	37	29	8	31	15	16	109
Eastern Europe								
Belarus	EVS 2000	8	6	2	22	14	8	88
Bulgaria	FFS 1997	4	2	2	3	2	1	843
	EVS 1999	3	–	3	–	–	–	60
Romania	EVS 1999	20	20	–	10	7	3	85
Russian Federation	EVS 1999	2	1	1	16	5	11	171
Ukraine	EVS 1999	–	–	–	10	5	5	99

Source: FFS data: UNECE and UNFPA, *Fertility and Family Surveys in Countries of the ECE Region. Standard Country Report*, Economic Studies, No. 10 (various issues) and table 6.7.1; EVS data: European Values Surveys Consortium, national data sets.

Note: Results from Fertility and Family Surveys (FFS) and European Values Surveys (EVS). FFS data for Belarus, Croatia, Romania, the Russian Federation, Slovakia and Ukraine are not available.

Despite the caveats, one can safely conclude that premarital cohabitation is spreading in central Europe. Procreation within this new household type may not be far off or has already started. As might be expected, eastern Europe displays a lag in both respects.

6.3 Which values matter?

The initial article on "the second demographic transition"[523] posited that the new living arrangements, and cohabitation in particular, were the expression of secular and anti-authoritarian sentiments of better educated young cohorts with an egalitarian world view and greater emphasis on "higher order needs" (i.e. self-actualization, expressive values, recognition). This reflects the picture of cohabitants in the Low Countries during the late 1960s and early 1970s. In addition, Belgium and the Netherlands had a plethora of political parties that represented the entire spectrum from "old values" to "new values",[524] and voting behaviour according to living arrangements provided the initial empirical check. At the same time the correlates of Inglehart's "post-materialist" orientation were high on the research agenda of political scientists, and both the European Union Eurobarometer Surveys and the first EVS of 1981 provided data for more detailed empirical verification in several west European countries. Also in the United States statistical associations between value orientations and living arrangements were drawing attention. Moreover, the United States demographers and

---

[523] R. Lesthaeghe and D. van de Kaa, "Twee Demografische Transities?", in R. Lesthaeghe and D. van de Kaa (eds.), *Groei of Krimp?*, book volume of "Mens en Maatschappij" (Deventer, Van Loghum-Slaterus, 1986). The first broader empirical check using the 1981 EVS data can be found in R. Lesthaeghe and D. Meekers, "Value changes and the dimensions of familism in the European Community", *European Journal of Population*, No. 2, 1986, pp. 225-268. The 1990 EVS data served again in R. Lesthaeghe and G. Moors, "Living arrangements, socio-economic position and values among young adults – a pattern description for France, Belgium, Germany and the Netherlands", in D. Coleman (ed.), *Europe's Population in the 1990s* (Oxford, Oxford University Press, 1996), pp. 163-221.

[524] A. Felling, J. Peters and O. Schreuder, *Burgerlijk en Onburgerlijk Nederland* (Deventer, Van Loghum-Slaterus, 1983) contains a thorough exploration of the connections between voting behaviour and value orientations for the late 1970s in the Netherlands. A similar analysis for Belgium including household positions as well is found in R. Lesthaeghe and G. Moors, "De gezinsrelaties: de ontwikkeling en stabilisatie van patronen", in J. Kerkhofs, K. Dobbelaere and L. Voyé (eds.), *De Versnelde Ommekeer* (Tielt, Uitgeverij Lannoo and King Baudouin Foundation, 1992), pp. 19-68.

sociologists had moved on to panel studies in which specific value orientations were recorded at each wave in tandem with the recording of vital events occurring in the intervals between successive waves.[525] As a result, American scholars were able to verify whether or not specific value orientations had predictive power with respect to later household choices, and furthermore, they were able to assess to what extent earlier transitions in household position had led to the accentuation or the adjustment of previously held values and attitudes. In other words, a recursive model emerged with (i) *values-based selection* into alternative living arrangements; and (ii) *event-based values adaptation*. This feedback model of selection and adaptation provides the dynamics of the process, whereas the cross-sectional correlations between values and household positions are merely *footprints* of this recursive mechanism.[526]

As indicated above, the initial set of values that were thought to determine the selection among alternative pathways of household formation mainly dealt with the following dimensions in the west:

- *Secularization*, or the reduction in religious practice, the abandonment of traditional religious beliefs (heaven, sin, …) and a decline in individual sentiments of religiosity (prayer, meditation, …);

- *The "new political left"*, with indicators pertaining to Inglehart's "postmaterialism", voting for Green parties or left-wing liberals, the propensity to protest, distrust in institutions, and anti-authoritarianism more generally;

- *Egalitarianism*, with an emphasis on gender equality, tolerance for minorities, rejection of social class distinctions, and a preoccupation with North-South equity associated with "world citizenship";

- *Unconventional civil morality and ethics*, with greater tolerance for forms of uncivil conduct (e.g. joyriding, drugs, tax evasion, …) as well as for interference in matters of life and death (euthanasia, abortion, suicide);

- *Accentuation of expressive values*, showing an enhanced preoccupation with individuality and self-fulfilment. Typical indicators are the ranking of the traits of "imagination" and "independence" above all other qualities in the education of children, or the preference for a job's intrinsic qualities (challenging, interesting, permitting social contact and initiative) rather than its material advantages (pay, vacations, promotion);

- *Companionship and unconventional marital ethics*, stressing the quality of a relationship (communication, tolerance and understanding, happy sexual relationship) over the conventional and institutional foundations of marriage and parenthood, and the toleration of deviations from strict marital morality (adultery, casual sex, …).

During the 1990s, aspects related to social cohesion and social capital were added to the list. There was a suspicion that traditional families had maintained stronger community ties and a higher degree of involvement in various types of local association, whereas others had relinquished such links in favour of social networks based on personal friendships. These connections have not been adequately researched so far,[527] but in this chapter membership of associations and voluntary work are included as extra items.

At this point it should be stressed that value orientations are not the only influences that are important. Other factors matter and empirical research has found a role for:

Family antecedents: the experience of parental divorce, and/or of family reconstruction after a parental divorce, frequently lead to earlier home leaving, single living, premarital cohabitation and even lone parenthood;[528]

Regional historical contexts: in several European countries, cohabitation and procreation within cohabitation have increased much faster in regions (often rural ones) with a much longer history of tolerance for such forms of family formation (e.g. northern Scandinavia, Austrian alpine regions).[529] In other

---

[525] The most important United States panel studies with adequate measurements of values and attitudes are: the Detroit Intergenerational Panel Study of Parents and Children, the United States National Longitudinal Study of the High School Class of 1972, the United States National Education Longitudinal Study, and the United States National Survey of Families and Households. Panel studies of similar questions came much later in Europe, and only two have adequate data for the present purposes: the Bielefeld Panel Study "Familienentwicklung in Nordrhein-Westfalen", and the Panel Study on Social Integration in the Netherlands.

[526] R. Lesthaeghe and G. Moors, "Life course transitions and value orientations: selection and adaptation", in R. Lesthaeghe (ed.), *Meaning and Choice – Value Orientations and Life Course Decisions*, NIDI-CBGS Monograph No. 37, Netherlands Interdisciplinary Demographic Institute (The Hague), 2002 (forthcoming), chap. 1.

[527] The issues of social capital, membership of associations and social cohesion are mainly studied from a political science perspective, i.e. focusing on the role of such network memberships in fostering democratic values and in creating barriers to the extreme right. Association memberships and social networks are rarely related to household formation and life course transitions.

[528] There is a very extensive literature in both psychology and sociology on the effect of parental household dissolution, particularly in Anglo-Saxon countries where these effects are enhanced, partly as a result of less adequate family support policies than in the rest of western Europe.

[529] J. Kytir, "Unehelich, Vorehelich, Ehelich: Familiengründung im Wandel", *Demografische Informationen 1992-93*, Institut für Demografie, Oesterreichische Akademie der Wissenschaften (Vienna), 1993, pp. 29-40. For the levels of illegitimate fertility for all European provinces at the end of the nineteenth century, see A. Coale and R. Treadway, "A summary of the changing distribution of overall fertility, marital fertility and of proportions married in the provinces of Europe", in A. Coale and S. Cotts Watkins (eds.), *The Decline of Fertility in Europe* (Princeton, N.J., Princeton University Press, 1986), pp. 31-79. These figures illustrate that procreation within consensual unions was already widespread by 1900 in several areas of Austria, Germany, Hungary, Portugal, Spain and Sweden. Most of these areas were rural.

countries, the current emergence of new forms of household formation displays a strong correlation with the regional pattern of the "first demographic transition", i.e. with the onset of fertility control and the weakening of the late Malthusian marriage pattern during the nineteenth century (e.g. Belgium, France, Switzerland);[530]

Diffusion mechanisms: with the passing of time new forms of behaviour gain acceptability and legitimation, even to the point where they are accommodated by the legal system. Increased legitimation is both the motor and the outcome of social diffusion from an "innovative core" to other population segments;

Economic differentiation: new living arrangements may accommodate different economic aspirations and situations. For example, cohabitation may suit the motivation of women to maintain their economic independence, as postulated in neo-classical economic theory. Alternatively, it may be the expression of economic uncertainty, as proposed by Easterlin's relative deprivation theory.[531] In the former case, cohabitation is likely to be found among better-educated women with careers, whereas the latter case would be a dominant trait for lower social strata with less income security. Moreover, cohabitation may be an interim phase that is a correlate of the overall destandardization of the life course, including the destandardization of job and career paths. Obviously, the "crisis-theory" invoked in central and eastern Europe to explain the rise of new household types refers to the latter mechanism;

Policy effects, labour market characteristics and housing conditions: earlier home leaving, single living and premarital cohabitation in the west are more typical of countries with income support policies for young adults in the form of scholarships, cheap student accommodation and transport subsidies.[532] Also the existence of flexible labour markets with an ample supply of part-time jobs contributes to earlier economic independence for younger adults. At the other end of the spectrum, prolonged residence in the parental home is more typical of countries without such policies and/or with expensive housing;[533]

To sum up, the shift towards "unconventional" values, often occurring via a succession of generations, is by no means the only factor that has shaped the "second demographic transition" in the west, but it has been a *non-redundant* factor in sustaining a long-term demographic trend through periods of slower and faster economic growth alike.

6.4 The footprints of selection and adaptation: what to expect?

In this section there is an analysis of the expected effects of values as they influence the choice of path in family formation, and of the ways in which values are reinforced or adapted following such life course events. The overall picture of expectations is summarized in chart 6.4.1. First, on the vertical axis there is a variation between two poles. One pole brings together the values that are non-conformist and more libertarian. These are characterized by expressive values accentuating personality and self-actualization in non-material domains, by the stress on individual autonomy with respect to all choices (morality and ethics included) and, correspondingly, by a rejection of institutional authority. This pole is also a secular one, with tolerance of all types of minorities, but also with a low identification or involvement in local community affairs. The opposite pole is obviously characterized by high conformity and respect for tradition, higher religiosity, respect for ethical and moral values that uphold social cohesion and respect for authority coupled with a greater trust in institutions.

The starting position in chart 6.4.1 is the respondent's residence in the parental household (Respar). At that point the "formative years", or the late adolescent period of values formation, are nearing their completion, and individuals have been subject to the influence of parents, schools and peers. The influence of the latter is often in the opposite direction from that of the other two, and may rise over time.[534] Also, as already indicated, problems in the parental household (discord, separation, divorce) have a major influence on both children's values and options chosen in their life course. It may therefore be expected that the position of young adults is already shifting toward the non-conformist pole prior to leaving home.

During the next steps in the unfolding of the life course, it is expected that leaving home in favour of living alone is predicated on the dominance of the non-conformist set of values, whereas leaving home to get directly married reflects a choice based on conventional

---

[530] R. Lesthaeghe and K. Neels, "From the first to the second demographic transition – an interpretation of the spatial continuity of demographic innovation in France, Belgium and Switzerland", *European Journal of Population*, 2002 (forthcoming).

[531] R. Easterlin, *Birth and Fortune* (Chicago, University of Chicago Press, 1987). See also R. Easterlin, "The conflict between aspirations and resources", *Population and Development Review*, Vol. 2, No. 3, 1976, pp. 417-425 and idem, "Relative economic status and the American fertility swing", in E. Sheldon (ed.), *Family Economic Behavior* (Philadelphia, Lippincot, 1973), pp. 170-223.

[532] OECD, *Preparing Youth for the 21st Century* (Paris, OECD Publications, 1999); R. Lesthaeghe, *Europe's Demographic Issues: Fertility, Household Formation and Replacement Migration*, United Nations Expert Group Meeting on Policy Responses to Population Decline and Ageing (New York), 16-18 October 2000.

[533] T. Castro-Martin, "Delayed childbearing in contemporary Spain – trends and differentials", *European Journal of Population*, Vol. 8, No. 3, 1992, pp. 217-246; P. Miret-Gamundi, "Nuptiality patterns in Spain in the eighties", *Genus*, Vol. 53, No. 3-4, 1997, pp. 185-200; G. Dalla Zuana, M. Atoh et al., "Late marriage among young people: the case of Italy and Japan", *Genus*, Vol. 53, No. 3-4, 1997, pp. 187-232.

[534] D. Alwin, "Historical changes in parental orientations to children", *Sociological Studies of Child Development*, No. 3, 1990, pp. 65-86.

CHART 6.4.1

Flow chart of life-course development and hypothesized changes in value orientations stemming from selection-adaptation mechanism

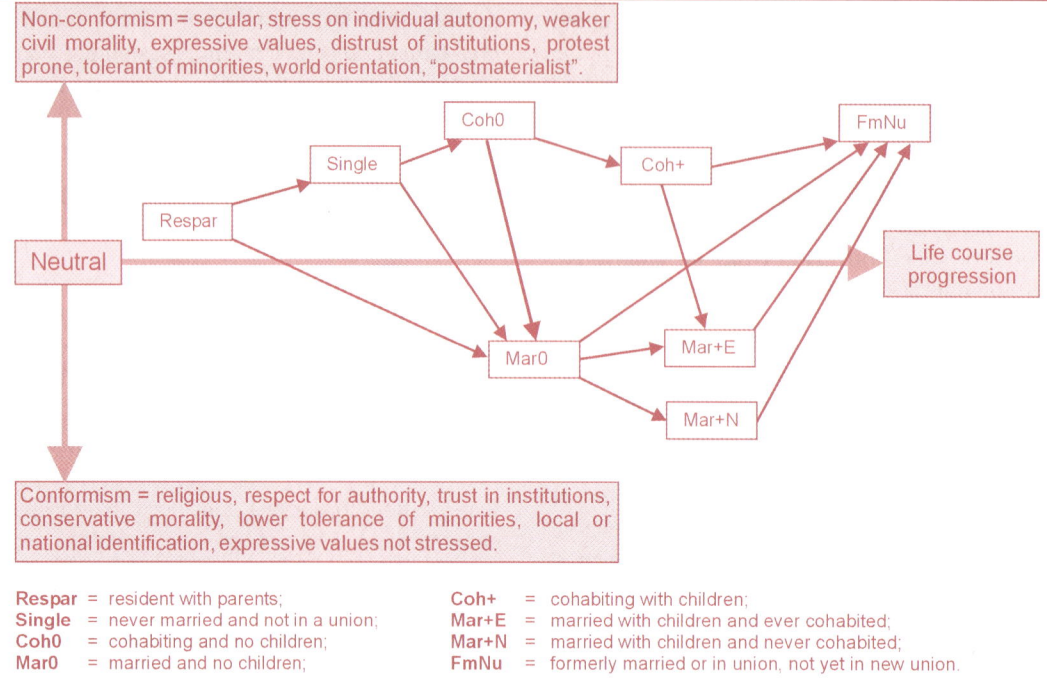

Respar = resident with parents;
Single = never married and not in a union;
Coh0 = cohabiting and no children;
Mar0 = married and no children;
Coh+ = cohabiting with children;
Mar+E = married with children and ever cohabited;
Mar+N = married with children and never cohabited;
FmNu = formerly married or in union, not yet in new union.

values.[535] At the same time, these two options reinforce the values that were responsible for the choice in the first place.[536] Hence, the position of "single" tends toward the non-conformist pole in chart 6.4.1, whereas "married without children" (Mar0) is toward the conformist end.

"Singles" face the option of moving into cohabitation (Coh0) or of marrying (Mar0). The former reinforces non-conformist values.[537] Partners are likely to be chosen for their preference for unconventional values that underpin the choice in favour of cohabitation. The mutually reinforcing orientations of such partners may then enhance the consistency of various values sets more generally, so that childless cohabitants (Coh0) can be expected to score *highest* and most *consistently* on the value orientations associated with pole 1. By contrast, singles who move into marriage may do so because of a higher respect for traditional institutions, out of respect for parental preferences, or because they choose a partner with more conventional attitudes. Once the institution of marriage is accepted, the consistency of values is again reinforced, and a move in the opposite direction, i.e. towards pole 2, can be expected. A similar process would apply to cohabitants who marry prior to parenthood. For them, the reorientation of values associated with a transition to marriage could be quite substantial given that they come from a strongly non-conventional position. However, it is possible that the earlier convictions are not obliterated altogether, and that the *experience of cohabitation leaves a durable imprint*.

The adjustment effects of parenthood are expected to be even stronger than those of marriage. In fact, value shifts in the conformist direction already occur in anticipation of parenthood,[538] the transition from cohabitation to marriage often being made in anticipation

---

[535] F. Kobrin-Goldscheider and C. Goldscheider, *Leaving Home before Marriage – Ethnicity, Familism and Generational Relationships* (Madison, WI, University of Wisconsin Press, 1993); F. Kobrin-Goldscheider and L. Waite, "Nest-leaving patterns and the transition to marriage for young men and women", *Journal of Marriage and the Family*, Vol. 49, 1987, pp. 507-516; F. Kobrin-Goldscheider and J. Davanzo, "Semi-autonomy and leaving home in early adulthood", *Social Forces*, Vol. 65, No. 1, 1986, pp. 187-201; E. Marchena and L. Waite, "Reassessing family goals and attitudes in late adolescence: the effects of natal family experiences and early family formation", in R. Lesthaeghe (ed.), *Meaning and Choice ...*, op. cit., chap. 3.

[536] A. Thornton, W. Axinn et al., "Reciprocal effects of religiosity, cohabitation and marriage", *American Journal of Sociology*, Vol. 98, No. 3, 1992, pp. 628-651; L. Waite and F. Kobrin-Goldscheider, "Non-family living and the erosion of traditional family orientations among young adults", *American Sociological Review*, Vol. 51, 1986, pp. 541-554; G. Moors, "Values and living arrangements: a recursive relationship", in L. Waite et al. (eds.), *Ties that Bind: Perspectives on Marriage and Cohabitation* (Hawthorne, Aldine de Gruyter Publishers, 2001), chap. 11.

[537] J. Barber, W. Axinn and A. Thornton, "The influence of attitudes on family formation processes", in R. Lesthaeghe (ed.), *Meaning and Choice ...*, op. cit., chap. 2; M. Jansen and M. Kalmijn, "Investment in family life – the impact of value orientations on patterns of consumption, production and reproduction in married and cohabiting couples", in R. Lesthaeghe (ed.), *Meaning and Choice ...*, op. cit., chap. 4.

[538] G. Moors, "Reciprocal relations between gender role values and family formation", in R. Lesthaeghe (ed.), *Meaning and Choice ...*, op. cit., chap. 7; M. Jansen and M. Kalmijn, "Emancipatiewaarden en de Levensloop van Jong-volwassen Vrouwen", *Sociologische Gids*, Vol. 47, No. 4, 2000, pp. 293-314.

of the arrival of the first child. Parenthood corresponds with a firm commitment to both partner and child, closes "open futures", and redirects attention to the well-being of the next generation. Moral, civil and ethical values are reaffirmed, and social networks associated with children are activated. Tolerance for deviance diminishes, authority regains prominence, and self-actualization takes second place. Priorities are centred on the "priceless child", and preoccupations shift in favour of those upholding greater social cohesion. In chart 6.4.1, all positions with children are therefore located further toward the conformist pole. Nevertheless, it is hypothesized that the earlier experience of cohabitation acts as a brake on this readjustment. The position of Mar+E (= ever cohabited) therefore remains above that of Mar+N (= never cohabited) on chart 6.4.1.

Finally, a separation or divorce which has not yet been followed by a new partnership (FmNu) causes a complete overhaul of the values structure. New doubts emerge with respect to religion, traditional family values and trust in institutions. The individual is also more likely to become more self-focused, and hence with the expressive values and with individual autonomy. It is therefore hypothesized that the FmNu position shifts toward the non-conformist pole.

The household positions in chart 6.4.1 are incomplete, and so are the types of transition. However, they capture the dominant streams through the life course. Moreover, the EVS only captures sections of the life course, and the sample sizes are too small to separate certain categories into more meaningful ones. For instance, there is no question referring to an earlier divorce or separation, so that the "currently married" cannot be split into the "ever" and "never divorced", and the category Mar0, i.e. married without children, is too small to disaggregate into those who "ever" and "never cohabited". This highlights once more the need for larger samples, and it shows the usefulness of "ever" questions probing for the occurrence of earlier events or life markers.

The overall outcome of this section is that there should be an ordering of individual household positions along the vertical axis of chart 6.4.1, i.e. roughly from "traditional" to "non-conformist". In this ordering, cohabitants without children should score highest on non-conformism, followed by singles and formerly married. Residents in parental households should come next. More towards the opposite pole are married persons without children, cohabiting parents and married parents who had previously cohabited. The most conservative values should be found among married parents who never cohabited. It should also be noted that these expectations about the "footprints" of the recursive life cycle model were formulated *in tempore non suspecto*, i.e. well before the present EVS survey results were available.[539]

6.5 Measurement and profiles: do we find the footprints of selection and adaptation?

In this section the use of 80 specified values is proposed, and these are analysed for respondents aged 18 to 45. The selected values were common to all the country-specific questionnaires of the 1999 EVS. The item profiles according to the household positions of respondents are checked to see whether the expectations just formulated are indeed emerging in all three pooled country data sets. In doing so, the question as to whether the profiles of central and east European countries are similar to those in the west is addressed. Similarity would indicate that the selection and adjustment mechanisms that connect value orientations and life course choices are more universal and not just an idiosyncrasy of western countries.

Firstly, the selection of 80 items was made on the basis of the individual country data sets. In this exploratory analysis use was made of Multiple Classification Analysis (MCA) of over 150 items. For each item the covariates were household position (8 categories), age and age squared (continuous), education level (4 categories), profession (8 categories, including unemployed, housewives and students), gender, and urbanization (2 categories). The selection of the final 80 items was based on: (i) *the topic*, i.e. making sure that the items covered all major domains or subjects, and (ii) the *strength of their association* with household positions, i.e. the least discriminating items were left out.[540] A set of 80 items is still very large, but maintaining multiple items per subject increases measurement validity. The 80 items are listed in table 6.5.1. *All* items are coded as dummy variables, with the value of unity always assigned to the non-conformist or unconventional opinion. Such a *uniform coding direction* facilitates the subsequent inspection of value profiles across covariates and countries.

The list in table 6.5.1 contains nine major subjects. The largest number of items (15) pertains to attitudes related to marriage as an institution, to the qualities needed for the success of a marriage, to the meaning of parenthood and parent-child duties, and to the degree of permissiveness with respect to sexual freedom, divorce and abortion. Secularism is represented by 9 items

---

[539] The "selection-adaptation" hypothesis was also the starting point of a symposium held in October 2000 at the Belgian Academy. The participants were all authors who had documented these recursive effects in their work with panel data. A translation of such effects into cross-sectional profiles is given in R. Lesthaeghe, J. Surkyn and J. Anson, "Household positions and value orientations – an exploration with Belgian and German EVS data", paper presented at the Euresco Conference on The Second Demographic Transition (Bad Herrenalb), 23-28 June 2001, session 4B.

[540] The excluded items were related to the "left-right" dimension in economic and social policies (state and labour union interference versus free enterprise) and economic equity, perceived causes of poverty, overall job satisfaction, political items covering the functioning of democracy, and more detailed attitudes towards elderly people and immigrants. Several items pertaining to female autonomy and gender inequality were also excluded since they were not incorporated in all the national questionnaires.

TABLE 6.5.1

European Values Surveys, 1999: overview of 80 values used in the current analysis

Topics and corresponding items	Item description
Marriage and family: A1-A15	Marriage, an outdated institution (A1); children not necessary for life fulfilment (A2); parents should not sacrifice themselves for children (A3); acceptable: casual sex (A4), adultery (A5), divorce (A6), abortion (A7); important for marriage: tolerance and understanding (A8), sharing chores (A9), talking (A10), time together (A11), happy sexual relations (A12); not very important for the success of marriage: faithfulness (A13), children (A14); single motherhood acceptable (A15).
Religion: A16-A24	Not believing in: god (A16), sin (A17), hell (A18), heaven (A19); no comfort from religion (A20); no moments of prayer or meditation (A21); god not at all important in life (A22); distrust church (A23); religious faith not mentioned as socialization trait (A24).
Civil morality: A25-A36	Acceptable: soft drugs (A25), homosexuality (A26), joyriding (A27), suicide (A28), euthanasia (A29), speeding (A30), drunk driving (A31), accepting bribes (A32), tax cheating (A33), lying (A34), tax evasion by paying cash (A35), claiming unentitled state benefits (A36).
Politics: B1-B11	Distrust in institutions: education system (B1), army (B2), police (B3), justice system (B4), civil service (B5); participated or willing to: participate in unofficial strikes (B6), attend unlawful demonstrations (B7), join boycotts (B8), occupy buildings (B9); no more respect for authority (B10); post-materialist (B11).
Identification: B12-B17	Identification with "Europe and world" (B12), not with "own village or town" (B13), not very or quite proud of own nationality (B14); no priority for national workers (B15); no trust in European Union (B16) or United Nations (B17).
Retreat: B18-B21	Not a member of any voluntary organization (B18); no voluntary work (B19); people cannot be trusted (B20); never discuss politics (B21).
Socialization: C1-C7	Not mentioned as desirable traits in educating children: hard work (C1), obedience (C2), good manners (C3), unselfishness (C4), tolerance and respect (C5); stressed as desirable: independence (C6), imagination (C7).
Work qualities: C8-C15	Not mentioned as desirable job aspects: good hours (C8), promotion (C9); stressed as desirable: respected job (C10), responsible job (C11), meeting people (C12), useful for society (C13), interesting work (C14), enabling initiative (C15).
Social distance: C16-C23	Not wanted as neighbours: large families (C16), right-wing people (C17); no objection to having as neighbours: aids patients (C18), unstable people (C19), those with criminal record (C20), drug addicts (C21), homosexuals (C22), immigrants (western countries) or gypsies (central European countries) (C23).

Note: All items are presented from a "non-conformist" perspective.

indicating a loss of traditional religious beliefs, a low level of individual religious sentiment, and distrust in the churches as institutions. The civil morality set with 12 items captures permissiveness with respect to different forms of deviant behaviour, but also the ethical acceptability of forms of interference in matters of life and death. The political set contains 11 items dealing with distrust of institutions, the propensity to protest, Inglehart's postmaterialism index and the rejection of authority more generally. The social distance or tolerance set is made up of 8 items that indicate the type of persons that are either tolerated as neighbours or considered as undesirable. The expressive values are spread over the socialization and work qualities sets. The former (7 items) show the preference for developing imagination and independence in education rather than conformity and respect for others. The latter (8 items) indicate a similar preference for intrinsic work qualities over material rewards or status. The identification set (6 items) distinguishes a global or larger orientation rather than a local identification or national pride, but with distrust in established international organizations. The last set of 4 indicate a retreat from social and political life, and reveals the absence of any memberships or voluntary work, a distrust of people in general and a lack of any interest in politics. In all further analyses these 80 values will be used without any prior data reduction, such as factor analysis. Hence, no particular structure will be imposed prior to further statistical work.

At this point the value profiles according to household position can be established. It will be recalled that (i) all items are coded in the unconventional or non-conformist direction; and (ii) that controls are present for other covariates (i.e. gender, age, education, profession and urbanity). The data set now takes the form of *net* deviations from the item mean associated with each of the eight household positions. Such net deviations are available for each of the 80 items and for each of the three groups of countries. A positive value of a net deviation from the item mean indicates that a particular household position has a more non-conformist attitude than average for the item concerned. Hence, a single tally of the number of positive deviations for each household position is already highly revealing of the overall profile.

The results of such a tally for each group of countries are displayed in chart 6.5.1. Respondents who are still residing with parents (Respar) have around 40 positive net deviations in the set of 80. This puts them close to the "neutral" line in chart 6.4.1. As expected, a move to single living increases the number of positive net deviations. This holds for all three regions, but the effect is more pronounced in the western countries (60 against

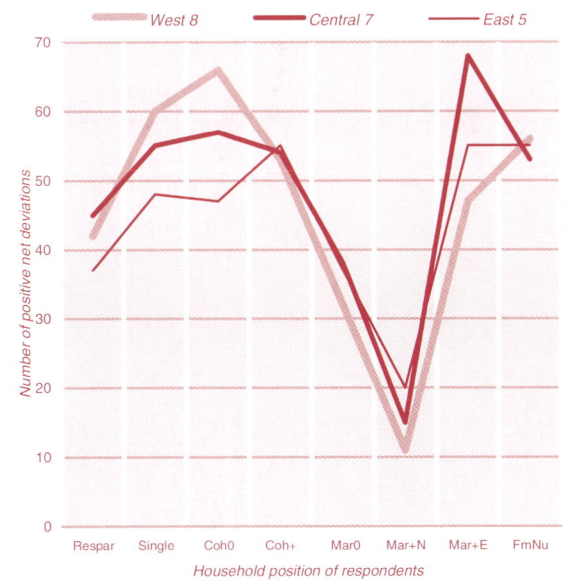

CHART 6.5.1

Number of positive (non-conformist) net deviations for 80 items, European Values Surveys, 1999: pooled results for eight western, seven central and five east European countries

adjusted them to the same degree after parenthood. In other words, the earlier experience of cohabitation leaves a lasting mark, and largely inhibits a return to conformity. The difference between the Mar+N and Mar+E groups is just as large in central and eastern Europe as in the west. In fact, the Mar+E category in central Europe has the highest non-conformity score of all (68 items). Finally, a divorce or separation (FmNu) leads to the expected increase in non-conformity when compared with Mar0 or Mar+N, and with all married groups in western Europe.

So far, the results of the comparisons in chart 6.5.1 indicate unambiguously that there is a systematic association between current household position and earlier life course history on the one hand, and value orientations on the other. The magnitude of selection and adaptation effects vary among the three broader European regions – as they may in individual countries – but the resulting profiles are essentially similar and in line with the hypothesis of a "second demographic transition".

The data of chart 6.5.1 can be disaggregated according to topic. In chart 6.5.2, the set of 80 non-conformity items is divided into three subsets: A, consisting of secularization, marriage and parenthood, and the civil morality items (total = 36); B, with items pertaining to politics, identification and retreat (total = 21); and C, containing the remaining items on socialization traits, work qualities and social distance (total = 23). The tally of the number of positive net deviations is plotted for each of these subsets. The overall picture in chart 6.5.2 is that the pattern of values according to household position for the entire set is essentially replicated in each of the three subsets. This also holds for the three regions. In other words, the outcome displayed in chart 6.5.1 is not produced by a concentration of positive deviations in a particular cluster of items, but is a reflection of non-conformity across most of the 80 values.

In fact, there are only a few anomalies in chart 6.5.2. The most striking is that childless cohabitants (Coh0) in eastern Europe score lower than expected on the items of subset B. A closer inspection of net deviations for individual items reveals that these respondents have high scores on nationalism and trust in institutions (except international ones) and that they also want more respect for authority. On the other hand, they fit the standard picture for childless cohabitants with high scores for the propensity to protest and postmaterialism. It seems that east European childless cohabitants are vocal (as expected), but wish to express their loyalty to national institutions. This latter trait sets them apart from childless cohabitants in western and central Europe.

The overall conclusion from these findings is that the footprints of the proposed selection and adaptation mechanisms of chart 6.4.1 are indeed clearly visible in all three regions, not solely in western Europe, and furthermore, that the footprints are detectable in all domains for which indicators of non-conformity could be

55 and 48). A further move to cohabitation without children (Coh0) increases the overall score for non-conformism even more in western countries (66 items), only slightly in the central European group (57) but not at all in the east European populations (47). The progression to parenthood prior to marriage but within a consensual union (Coh+) has the expected readjustment effect in western and central Europe, but not in eastern Europe. In the latter group, cohabitants with children have the higher score for overall non-conformity (55). The exact reason for this cannot be established, but it is likely that parenthood within cohabitation, i.e. a rare transition in eastern Europe, is a result of a strong preference for non-conformity, which outweighs the adjustment effect associated with parenthood. On the whole, leaving home and cohabitation, with or without children, are clearly associated with higher non-conformity scores, and this pattern holds in all three groups of countries. There is no particular western idiosyncrasy in this respect.

The overall pattern holds even more for the remaining household positions. Married couples without children (Mar0) have positive net deviations for less than half the number of items, and married couples with children who never cohabited (Mar+N) have by far the most conservative attitudes. They both selected this household position because of their initial conformity, and have further reaffirmed their opinions in this direction as a consequence of parenthood. In contrast, married couples with children who previously cohabited before marriage (Mar+E), did not share the values associated with direct marriage, and apparently have not

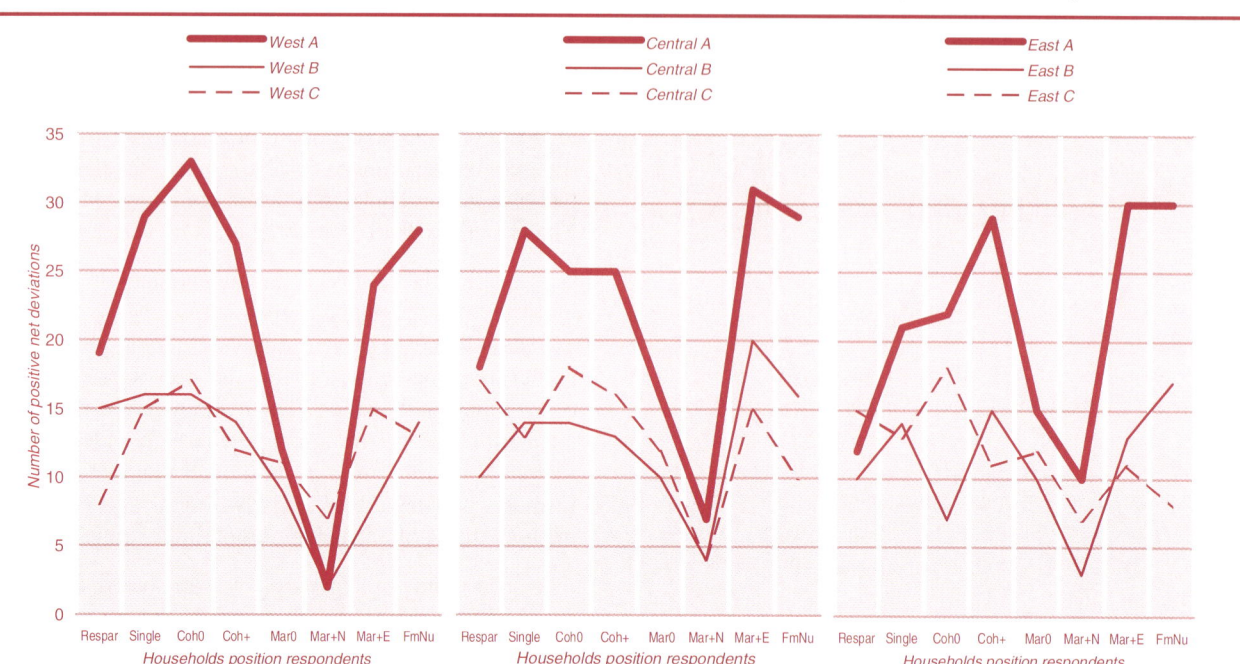

CHART 6.5.2

Number of positive (non-conformist) net deviations for groups of items and countries, European Values Surveys, 1999

Note: A = 36 items including marriage and family, religion and civil morality; B = 21 items including politics, identification and retreat; C = 23 items including socialization, work qualities and social distance.

established. This means that not only unconventional views with respect to marriage, family and parenthood are behind the choices of single living and cohabitation, but that a much broader array of non-conformist attitudes are involved as well.

6.6 Finer distinctions

So far the analysis has relied on simple tallies of net positive deviations generated by the MCA. In what follows, the item-by-item analysis is extended by using the net positive deviations as inputs into a correspondence analysis.[541] The aim is to bring out the *proximities* of value items and household positions by trying to project them on a plane. Since proximities rely on distance, which obviously cannot be negative, the net deviations generated by the MCA are converted into rankings.[542] Hence, the input is now the ranking of a household position (from 1 to 8) on each of the 80 items, a rank of 1 indicating that it has the highest positive net deviation for a particular item. It is recalled that the net deviations are measured *after* controls for gender, age, education, profession and urbanization. The correspondence analysis furthermore shows that two dimensions (hence a plane) suffice to summarize 50 per cent of the information, and that a third dimension would add only about 12 per cent in all three regions.

With 80 items and 8 household positions, the projection of proximities yields a plot with 88 dots. Since all of these need to be identified with labels, such "busy" plots are not easy to read. To overcome this drawback, new figures were prepared using the following procedure:

- The 8 household positions are plotted on their exact location in the plane, but the items are grouped according to their own proximities. The group is then represented by an arrow in the plane starting at the origin. Hence, items in a group are all located near their arrow;

- It turned out in all three regions that 7 groups of items, and hence 7 arrows, could give an adequate description of the item plots. The content of these groups, however, varies across the three regions, and so does the direction of the arrows. It is at this point that finer distinctions between the three groups of countries emerge;

- It is helpful to add the information from the previous section, and to indicate to what extent each household position contributes to the overall non-conformity score from 0 to 80. We have therefore tilted the projection plane, so that a third dimension can be used to indicate the overall non-conformity score of each household position;

---

[541] For the philosophy and technical details, see J.-P. Benzecri, *L'analyse des données – L'analyse des correspondences* (Paris, Dunod, 1973) and M. Greenacre, *Theory and Applications of Correspondence Analysis* (London, Academic Press, 1984). In the current application, the SAS software was used. See SAS Institute Inc., *Statistics and Graphics Guide*, Version 3.1 JMP (Cary, NC), 1995, pp. 105-111.

[542] We owe this useful methodological suggestion to J. Anson, who also put us on the path of correspondence analysis as a powerful tool for visualizing the proximities between household positions and value items.

- The tilted projection plane is located at a non-conformity score of 40. The vertical bars for each household position then indicate the number of items in the non-conventional direction above or below 40 for that household position.

The resulting three-dimensional figures now contain a large amount of information. If a household type has an overall non-conformity score well in excess of 40 and is located near the edges of the plane, then it draws disproportionately on those non-conformity items that are identified by the nearest arrows. In other words, these are the items for which the household position has produced the higher rankings with respect to the net deviations. Conversely, if the household type has a low overall non-conformity score well below 40, it would still have higher rankings on the items identified by the nearest arrows. Household positions that are located close to the origin have higher rankings for all items, and not mainly from a particular group identified by an arrow. When this is coupled with a high overall non-conformity score, this indicates that the household position produced high rankings for a great variety of items, and if such a position near the origin is coupled with a low overall score, then it draws its small set of the higher rankings from all sorts of items as well. Finally, household types that are located at the opposite end of certain arrows draw nothing or almost nothing from the items associated with these arrows. For instance, in chart 6.6.1, the two arrows that point to the left mainly refer to items that refer to relaxed civil, ethical and marital morality, to greater distrust of institutions, to a greater propensity to protest, and more "postmaterialism". The household group of "singles" is located at a small distance from both arrows, which means that these items are highly characteristic of them. Moreover, the singles group scores +20 on overall non-conformity, and the items mentioned above contribute strongly to this. The group Mar+N in chart 6.6.1 is located diametrically opposite, which means that they are far from subscribing to the items concerning relaxed morality, "postmaterialism", etc. Also, they have a large negative score of -29 for overall non-conformity. Hence, the few items of non-conformity that characterize them are disproportionately concentrated among those identified by the nearest arrow, i.e. companionship (talking, time together, ...) and lack of trust in people in general.

The results of the correspondence analysis are shown in charts 6.6.1, 6.6.2 and 6.6.3, respectively, for the three groups of countries. The outcomes for the eight west European populations display the most "classic" profiles and are therefore used as a reference.

Respondents still residing in the parental household (Respar) display the more "youthful" form of non-conformism: family considerations and parenthood are still in a more distant future, and unconventional household forms are fully acceptable to them in tandem with a more relaxed sexual morality. Other expressions of uncivil behaviour that are more typical for younger ages are also tolerated, such as using soft drugs, joyriding or speeding. This is linked to distrust in institutions and to the rejection of the forms of authority with which they are confronted more directly: the educational system, the police, the justice system and the army. But a number of new traits also emerge: home dwellers score low on memberships of voluntary associations and on voluntary work. On the whole, however, this category of respondents has only a modest surplus of unconventional scores (+2), mainly because they do not score as highly on other dimensions such as expressive values in socialization and work, tolerance for ethnic and sexual minorities, world citizenship, the propensity to protest or secularism. These issues still appear to be too remote for them.

Single respondents living on their own carry a number of these non-conformist traits with them, and even reinforce this pattern by adding extra items in the spheres of weaker civil and sexual morality. Their larger score (+20) of non-conformist items is produced by higher scores for secularism, the propensity to protest, postmaterialism and world orientation. Also, "imagination" as a socialization trait comes to the fore. When a move into cohabitation is made (Coh0), these features are again reinforced, probably as a result of both selection and further articulation. From that point onward, the high score for non-conformism (+26) is based on many features of secularism, a tolerance for interference in matters of life and death (abortion, suicide, euthanasia), a further strengthening of expressive values (independence in socialization to the detriment of good manners, jobs permitting initiative) and greater tolerance for the more adult forms of uncivil behaviour (tax evasion, tax cheating). This latter set of values equally typifies cohabitants with children (Coh+), but non-conformist values associated with earlier forms of living arrangements are fading away in these households. This accounts for their lower overall non-conformism score (+13), and for their location at the edge of the projection plane.

The experience of earlier cohabitation apparently has a lasting impact after marriage and parenthood (see Mar+E). This group continues along the previous lines of a high propensity to protest (demonstrations, boycotts) and emphasis on expressive values (responsible and interesting job, obedience not stressed in education). But the refusal of nationalist reflexes is added (lower national pride, no priority for own workers, no exclusion of immigrants, no right wing neighbours). However, the values that are diametrically opposite on the projection plane are refuted. This holds particularly for lax civil and sexual morality, which are incompatible with parenthood. The outcome is a further reduction in the non-conformist score (+7).

Married respondents without children (Mar0) have a deficit on the non-conformist scale (-8), as expected. As a result of small sample sizes this group is also

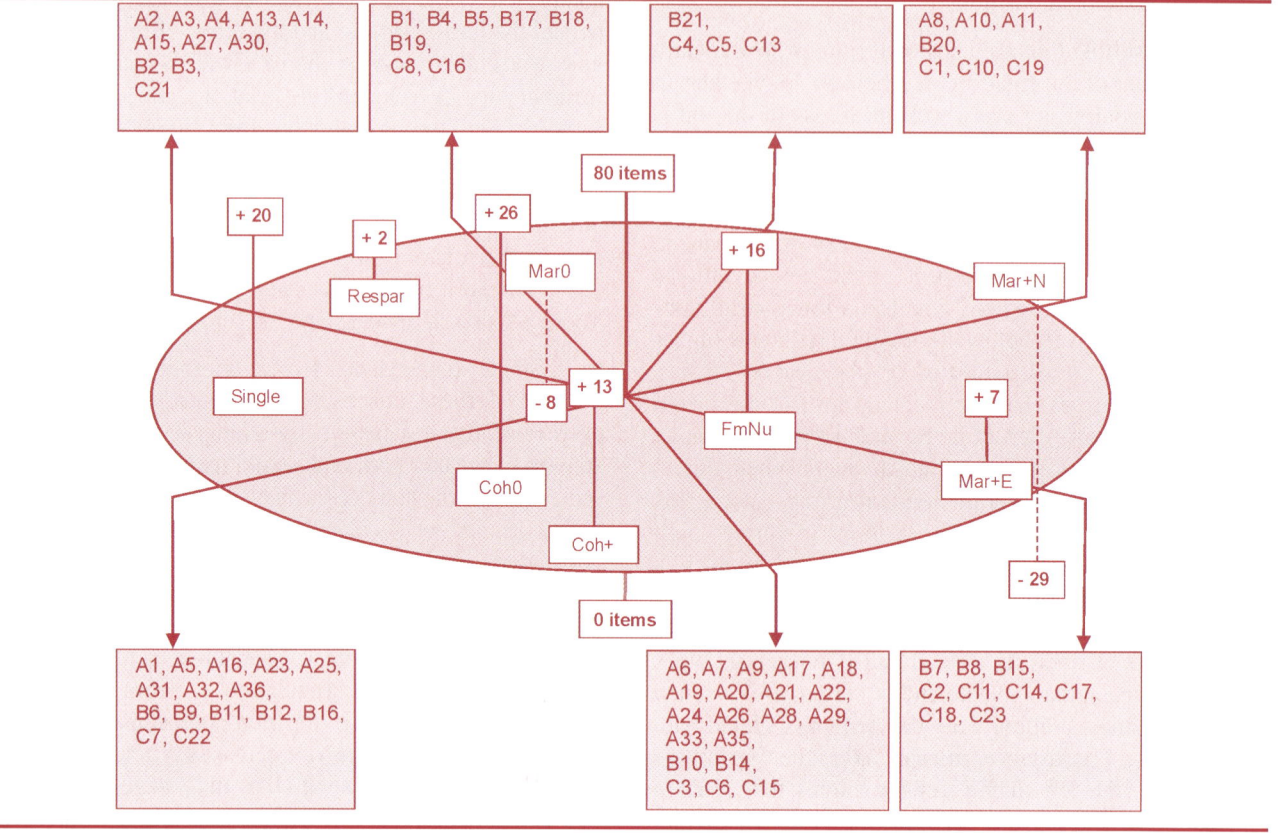

CHART 6.6.1

Correspondence between household positions and 80 non-conformist values, European Values Surveys, 1999: results for eight west European countries
(Pooled samples)

Note: The codes refer to the values (items) listed in table 6.5.1.

undifferentiated with respect to the presence or absence of earlier transitions (living alone, cohabitation) and it remains therefore quite heterogeneous. In chart 6.6.1, however, childless married persons have retained certain characteristics of those staying at home (Respar), such as low membership rates, the absence of voluntary work and a distrust of certain institutions. In addition, they are also characterized by other aspects of low community orientation, such as not stressing tolerance and respect or unselfishness in socialization and by a lack of political interests. They offset these traits with a preference for jobs that are useful for society.

Married respondents with children who never cohabited before marriage (Mar+N) have a very low non-conformist score of -29. They are located at the edge of the projection plane at the opposite edge to singles and childless cohabitants. Hence, they score very low on all the items associated with these two positions. Conversely, if there is a contribution to non-conformism, it stems from stressing companionship in marriage rather than social homogamy, and from a few isolated items such as a reduced trust in people, not stressing hard work in socialization, seeking a respected job but accepting unstable people as neighbours.

Divorced or separated respondents who are not yet in a new union (FmNu), exhibit the expected reversion to much higher scores for non-conformism scores (+16), but they are located close to the origin in the projection plane. This means that they draw on a wider variety of items than all the other household positions. The group they resemble most is that of married parents who had previously cohabited (Mar+E).

The detailed central European profile in chart 6.6.2 displays both similarities and differences when compared to the western picture. Respondents residing in the parental home (Respar) are also moderately non-conformist, with a score of +5, and similar items are equally overrepresented. These pertain to the remoteness of parenthood, a high tolerance of unconventional living arrangements, the acceptability of neighbours with deviant characteristics, lower standards of marital morality, and the classic distrust of the education system and the civil service. A move to living alone, which is a rarer transition in central Europe, adds a number of traits (taking the score to +15) that accentuate the acceptability of uncivil morality. Also conformity in socialization is rejected (no stress on good manners, obedience or tolerance and respect). To this a new feature is added: low national pride and a weaker identification with the

CHART 6.6.2

Correspondence between household positions and 80 non-conformist values, European Values Surveys, 1999: results for seven central European countries
(Pooled samples)

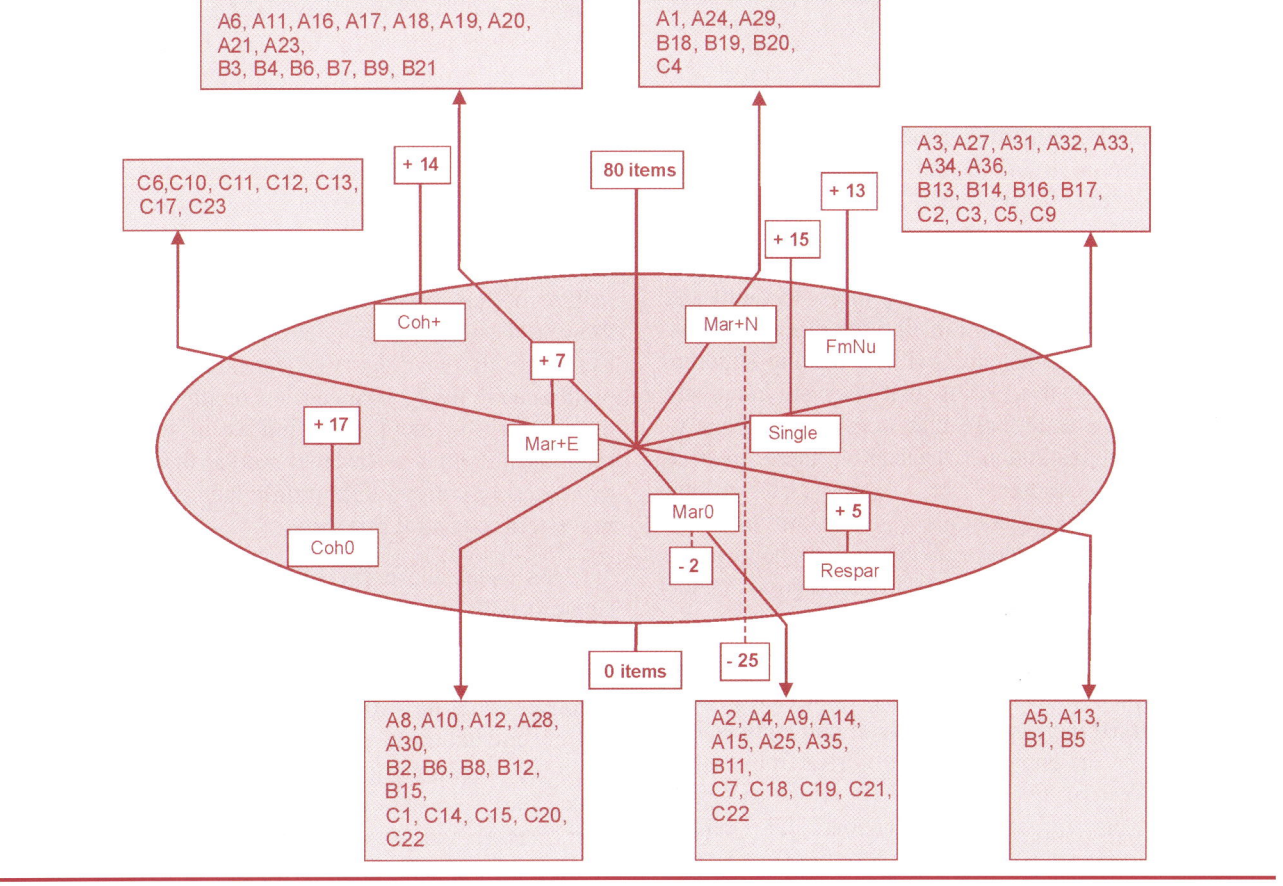

Note: The codes refer to the values (items) listed in table 6.5.1.

local village or town. But this broader outlook is matched by distrust of supranational or international institutions (European Union, United Nations). What are missing among central European singles is a selection for and an articulation of secularism.

Childless cohabitants (Coh0) in central Europe strongly resemble their western counterparts with respect to their propensity to protest (permitting strikes, occupation of buildings, boycotts, etc.), "world citizenship" and tolerance of homosexuals. There is also a weaker parallel with respect to expressive values (independence and imagination in education, expressive job qualities). But again, central European childless cohabitants do not share the pronounced secularism of their western counterparts. Instead, they are already accentuating the value of companionship. The overall score for non-conformism among central European childless cohabitants is also lower (+17) than in the west (+27), largely reflecting their less secular values and less tolerance for expressions of uncivil morality.

Secularism in central Europe shows up rather strikingly for cohabitants with children (Coh+). Other traits typical of western cohabitants are now emerging among central European cohabiting parents: a stronger propensity to protest and a distrust of institutions. Similarly, tolerance of uncivil behaviour has faded away with parenthood.

In central Europe the experience of earlier cohabitation also leaves its mark after marriage and parenthood (see Mar+E). Despite the fact that the score for non-conformist choices (+7) has fallen and that the position of Mar+E has shifted closer to the origin, the earlier experience of cohabitation still tends to be associated with traits typical of cohabiting couples with or without children. They also resemble their western counterparts with greater stress on expressive job characteristics, a greater acceptability of ethnic minorities (in central Europe: gypsies), and an aversion to right wing neighbours.

Also relatively close to the origin of the projection plane are the central European childless married respondents (Mar0). Their degree of conformism is slight (-2). Both features taken together imply that this group as a whole has the most undifferentiated profile of all. As in the west, central European childless married persons do not accentuate the values of parenthood, and are relatively tolerant of deviations from strict marital morality and civil codes of conduct. Similarly, they tend to have a high score on postmaterialism. The main difference is that childless married respondents in central Europe tend to be less choosy about their neighbours and are more willing to accept persons with a deviant profile.

Western and central European married parents who had never cohabited before marriage (Mar+N) obviously share the overall conservative profile, with scores on the non-conformism scale of -29 and -25, respectively. But the two small sets of contributory items have little more in common than a distrust of people in general. Central European married parents who never cohabited extend this pattern to a lack of membership in associations and the absence of voluntary work in all types of associations. Also, unselfishness is not stressed in educating children. More surprising is the finding that they are relatively overrepresented among those who consider marriage to be an outdated institution. Evidently, these married couples with children are older and belong disproportionately to generations which entered marriage and parenthood at an earlier age. They may therefore be expressing some regret about missing the options of the late 1990s.

The divorced and separated respondents who are not yet in a new union (FmNu) in central Europe also display a marked increase in the overall non-conformism score (+13), but its composition is different from that in the west. Western divorcees show no return to higher tolerance levels for uncivil behaviour, but central European divorcees do. They also score high on distrust of the United Nations and the European Union despite their lower national pride and lack of local identification. Except for a lack of focus on promotion, they do not share the stress on intrinsic work values of their western counterparts. The only feature that the western and central European groups of FmNu have in common is a lack of stress on obedience in rearing children.

To sum up, the value profiles of persons according to living arrangements in western and central Europe are very similar both in terms of the overall score for non-conformism and in the more precise composition of these scores. The largest differences, however, are for single people, childless cohabitants and those who were formerly in a union. Single persons and childless cohabitants in central Europe are not selected on the basis of secularism as in the west, but childless cohabitants in central Europe are less tolerant toward expressions of uncivil morality. By contrast, the latter feature emerges strongly among central European divorcees, whereas it has disappeared as a characteristic trait of western divorcees.

Before turning to the details of the east European profiles, it should be recalled that single living, cohabitation and parenthood within consensual unions are all much rarer forms of living arrangement in this area. The allocation of people into these categories could therefore be the outcome of a different process or trajectory than in the other two regions.

East European respondents still residing in the parental home (Respar) have a fairly undifferentiated profile. Their position in chart 6.6.3 is fairly close to the origin of the projection plane, and they score slightly (-3) towards the conformist end of the scale. There is, however, already a slight above-average representation in the direction of secularism, a lack of community involvement (no memberships, no voluntary work), a refusal of authority and a broader world outlook, but matched by a distrust of international organizations. The feature of low community involvement is shared with those in the west who stay at home, but for the rest, east Europeans in this category do not exhibit a high tolerance of alternative living arrangements or childlessness, nor do they share the more lax attitudes in matters of civil morality and ethics. In other words, they start from an overall more conservative profile than their counterparts in western or central Europe.

As usual, the non-conformity score increases with single people living alone (+8). Besides secularism and continued low community involvement (and no discussion of politics), the features of acceptability of unconventional household types, tolerance for deviance, and an orientation to expressive values (interesting job, no stress on hard work and obedience) are now emerging more clearly. Most of these features are shared with both their central and west European counterparts. Only distrust of national institutions is not yet accentuated among east European single persons.

The marked rise in non-conformism associated with cohabitation (Coh0) is also absent in eastern Europe, as already noted with respect to charts 6.4.1 and 6.5.1. The position of childless cohabitants is located at the opposite edge of the projection plane compared with that of earlier household positions. Eastern childless cohabitants share with both western and central cohabitants a higher propensity to protest, postmaterialism and expressive work and socialization values, and, with the central European ones, an orientation towards companionship. What they lack is secularism, tolerance for expressions of uncivil conduct, and distrust of national institutions. These three features surface fully at the next stage, i.e. when the move into unmarried parenthood is made (Coh+). In this respect there is a parallel with central European cohabiting parents who were also differentiated by their high levels of secularization. The overall picture is that procreation within consensual unions in all three parts of Europe is associated with secularism, distrust of institutions, a propensity to protest, an accentuation of expressive values, and a lowering of standards in matters of civil morality. East European cohabiting parents, furthermore, share with their western counterparts a low level of national pride and weaker local identification.

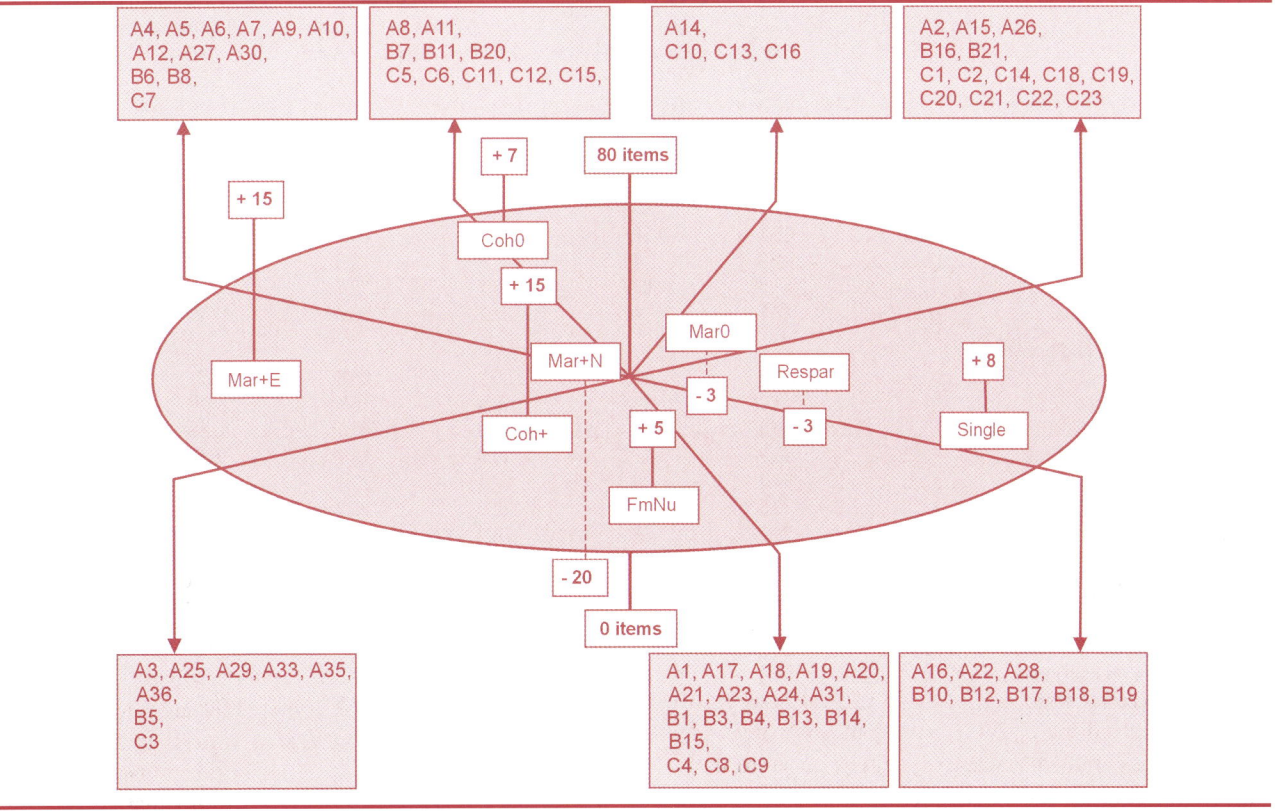

CHART 6.6.3

Correspondence between household positions and 80 non-conformist values, European Values Surveys, 1999: results for five east European countries
(Pooled samples)

Note: The codes refer to the values (items) listed in table 6.5.1.

As in the two other regions, east European married parents who experienced cohabitation before marriage (Mar+E) are clearly distinct from childless married persons (Mar0) and from married parents who did not experience previous cohabitation (Mar+N). This not only pertains to the marked difference in the total non-conformism score, but equally to the underlying value profiles. The positions of Mar0 and Mar+N are close to the origin of the projection plane, indicating that value profiles are not very different. The position of Mar+E, by contrast, reveals a strong accentuation of particular traits: low civil morality (bribery, tax cheating and tax evasion, collecting state benefits to which they are not entitled, soft drugs), distrust of the civil service, lower marital fidelity despite valuing companionship (talking, chores, sex), a propensity to protest (boycotts, demonstrations), and a preference for "imagination" over "good manners" in education.

Finally, east European divorcees who are not yet in another union (FmNu) retain or regain non-conformist traits indicative of secularism, distrust of institutions (education, police, justice) and low national pride. As their position on the projection plane shows, they resemble cohabiting parents in most respects.

On the whole, the most striking differences from the other two regions are that parenthood among currently or formerly cohabiting couples is not associated in eastern Europe with a reduction in their overall degree of non-conformism, nor with a correction in their attitudes towards civil morality in particular. In fact, rather the opposite is true, which suggests that the smaller Coh+ and Mar+E groups in eastern Europe are composed more of respondents with complex and perturbed partnerships and marital histories than in western and central Europe. Unfortunately this hypothesis cannot be checked against the EVS data because of the lack of more detailed retrospective questions on these issues.

6.7 Changes in value orientations during the 1990s

It would of course be totally erroneous to assume that there was no ideational change in central and eastern Europe during the communist period and that everything started to move in 1989. Rather, the events of that year were the culmination of political groundswells that were also grounded in shifting aspirations and value orientations, and were not exclusively a response to deteriorating state efficiency in economic and material spheres. In fact, one of the crucial political elements

TABLE 6.7.1

Trends in selected comparable items among respondents aged 18-49, three groups of countries with transition economies, 1990 and 1999

(Per cent of GDP)

| | Baltic states | | Central Europe | | Eastern Europe | |
	Estonia, Latvia, Lithuania		Czech Republic / Slovakia, Hungary, Poland, Slovenia		Belarus, Bulgaria, Romania, Russian Federation	
	1990	1999	1990	1999	1990	1999
Family						
Women do not need children for life fulfilment	8.3	26.5	23.5	44.6	10.8	25.1
Marriage an outdated institution	11.1	23.8	12.7	19.5	14.3	23.2
Single motherhood acceptable	57.8	83.9	69.4	78.4	78.5	80.7
Parents must not make sacrifices for children	36.1	32.0	22.9	20.7	28.6	30.3
Homosexuality acceptable	2.3	4.5	11.8	16.0	2.2	4.3
Adultery acceptable	62.6	61.0	56.7	52.6	59.4	58.2
Divorce acceptable	17.1	14.0	21.1	22.6	17.2	18.4
Distrust institutions and politics						
No trust in church	8.0	11.7	20.2	26.6	17.5	15.4
No trust in civil service	7.3	13.6	10.4	16.4	24.2	27.6
No trust in police	80.8	70.4	61.3	56.1	67.6	64.1
No trust in justice system	13.2	17.5	10.9	15.7	18.6	21.5
No trust in education system	54.4	33.5	37.1	29.9	48.4	30.4
Never discuss politics	3.5	19.6	12.9	25.3	13.8	25.0
One cannot trust people	75.9	77.9	74.4	78.6	72.9	75.0
Expressiveness						
Independence stressed	67.4	54.4	34.5	61.2	41.0	38.9
Imagination stressed	12.0	10.2	8.3	13.1	15.9	15.7
Obedience not mentioned	82.2	79.8	71.4	74.7	79.6	79.0
Good manners not mentioned	44.1	39.1	34.5	29.3	25.6	30.0
Unselfishness not mentioned	74.4	82.4	73.5	72.5	75.7	84.2
Civil morality						
Fraudulent claiming of benefits justified	40.2	54.5	59.1	47.8	38.8	46.7
Taking soft drugs justified	12.2	16.8	21.1	29.7	14.6	18.2
Accepting a bribe justified	38.0	37.4	33.2	43.5	28.0	39.3
Tax cheating justified	51.0	64.4	58.2	59.2	57.5	61.1
Identification						
Identification: Europe + world	6.2	8.3	9.9	7.6	11.9	11.7
Identification: not own locality or town	71.0	52.5	61.0	45.3	54.4	52.7
National pride: not proud	13.8	37.1	15.7	12.7	30.6	30.0

Source: Original data sets, European Values Surveys Consortium and World Values Studies.

Note: Each individual country has a weight of unity; in 1999 the data for the Czech Republic and Slovakia were merged to make them comparable to the 1990 data, which covered the whole of Czechoslovakia.

---

leading to the events of 1989 was the long-standing quest for the rebirth of a "civil society".[543] This envisaged the contraction of the party-state and the creation of political space for voluntary civic organizations such as independent labour unions, professional organizations, student associations, church groups, a free press, etc. Hence, the quest for political autonomy and grass-roots democracy was rising steadily prior to 1989. During the 1990s, however, not much was left of the "civil society" discourse: economic restructuring led to social disruption, uncertainty and rising inequality. Individual autonomy and freedom of choice were restored, including that of opting for other forms of living arrangements, but the ideal of an "energized population" supportive of social cohesion proved to be illusory.[544] Instead, central and east European societies became much more atomistic and individualistic, and they faced problems of inclusion and exclusion similar to or more serious than those in western countries.

The features of this transformation can also be traced in the opinions and attitudes recorded in the 1990 and 1999 EVS rounds. To document this, an attempt was made to search the 1990 data sets for comparable items.[545] The 1990 questionnaires in the various countries were not yet as standardized as those of 1999, with the result that the set of comparable items across countries and for the two points in time is much more limited than the set of 80 used so far. Nevertheless, table 6.7.1 reports on 26 items that passed the test; they pertain

---

[543] For a succinct overview see J. Ehrenberg, *Civil Society – The Critical History of an Idea* (New York, New York University Press, 1999), especially chap. 7.

[544] Ibid, p. 199.

[545] Not only the wording has to be identical in both rounds, but also the composition and scoring options need to match perfectly.

to family values, trust in institutions, civil morality, socialization values and identification. Given the small national sample sizes, the results are again presented for groups of countries.

The most striking changes emerging from table 6.7.1 are those for several family items, and especially for those that are directly related to the tolerance of new living arrangements and to procreation. In each of the three groups of countries (the Baltics, central and eastern Europe), there is a substantial rise in the number of women who do not need children for life fulfilment, who regard marriage as an outdated institution, and who consider that motherhood for women without a partner or husband is acceptable. The increases in the proportions with these opinions are of the order of 10 to 25 percentage points for the period 1990-1999. Also the tolerance of homosexuality has increased, whereas the acceptability of adultery and divorce has remained stable. Evidently, the presumed rise in premarital cohabitation displayed in table 6.2.2 is matched by a similar rise in the acceptance of non-conformist household types.

The other items that display a rise in *each of the three groups* of countries are related to:

- Greater distrust of several institutions, especially the church, the civil service and the justice system (but not the police or the education system). This is coupled with much higher proportions stating that they never discuss politics, and to a more modest increase in the percentages displaying distrust of people in general;
- Greater tolerance for several forms of uncivil behaviour, and especially for fraudulent claims to state benefits, tax cheating, using soft drugs and accepting bribes. Among the socialization items, "unselfishness" is more frequently absent among the traits that were given priority. And among the identification items, there is a decline in identification with the national or supranational levels in favour of stronger local links.

Aside from these general trends, there are also some remarkable shifts in specific groups of countries. Thus, independence and imagination as socialization traits have gained ground in central Europe, and there is a clear fall in national pride in the Baltic states.

To sum up, the general pattern displayed in table 6.7.1 indicates that the acceptability of non-conventional household forms and life courses is clearly on the rise, and that this tolerance is imbedded in a more general "atomization" of society. Individuals are free to choose, but have to do so at their own risk and with their own coping strategies. As a colleague noted: "the second demographic transition is not kind to all".[546]

---

[546] Kathleen Kiernan's comment at the ESF Conference on The Second Demographic Transition (Bad Herrenalb), 23-28 June 2001.

6.8 Conclusions

The EVS for central Europe indicate that new forms of household formation have gained ground during the 1990s, and that their acceptability and legitimacy have increased as well. However, the precise orders of magnitude need validation via other and especially larger surveys. The trend toward unconventional living arrangements is less pronounced in eastern Europe, as expected, but here too the tolerance for such forms is increasing.

The cross-sectional "footprints" of the selection-adaptation model are found in all cases, including eastern Europe, and the overall profiles of non-conformity according to living arrangements are following the western pattern to a remarkable degree. In all three groups of countries, those who never cohabited and moved directly into marriage and parenthood have by far the most conservative profiles, whereas cohabitants and divorced persons are the most non-conformist. Similarly, an earlier experience of cohabitation leaves a durable imprint in the direction of non-conformity as well. This effect is stronger than anticipated in all three regions.

A more detailed analysis at the item level reveals several differences between the groups of countries. For instance, the selection of cohabitants on the basis of secularism, which is still highly typical in the west, is not as pronounced in central and eastern Europe. This can be attributed to historical factors, and more particularly to the secular tradition stemming from the communist era. Also, childless cohabitants in central and eastern Europe are less tolerant of expressions of uncivil morality. Instead, this has become a more pronounced trait among divorcees and those selected cohabitants who progress to parenthood without prior marriage.

Other items that were comparable over time indicate that the rise in new living arrangements and their value profiles are associated with a weakening of social cohesion and the "atomization" of society. The latter factor is obviously linked to the economic crisis of the 1990s, but it is also a trait associated with capitalism more generally, and therefore a lasting characteristic. Hence, an economic recovery is unlikely to alter the demographic trend in a fundamental way, since the "second horse" in our metaphor is very likely to take over the running. In short, a capitalist restructuring leads to greater individual autonomy in the ideational sphere, and this in its turn means more convergence of family formation patterns to the western types. Rather than the economic crisis per se, it is the entire restructuring of society that is the accelerator of the ideational and demographic changes.

This is not to say that all central European countries will end up with household patterns that are perfect copies of western ones. There is substantial variation within the west, and such heterogeneity can also be expected to emerge among the central European countries. But, the restoration of more stable and conventional patterns of household formation with early procreation no longer seems to be an option.

The main conclusion of this exploration of the 1999 EVS data is that many features of the "second demographic transition", including the change in the structure of values along the lines of the "selection-adaptation" model of life-course progression, are now clearly visible in central Europe. In terms of actual behaviour, eastern Europe has not yet reached the "take-off" phase, but in terms of the changes in values since 1990, new patterns of household formation have gained acceptability. Hence, a further diffusion of the features of the "second demographic transition" to eastern Europe should no longer come as a surprise.

In terms of fertility trends, this implies that the spread of new patterns of household formation will continue to be one of the causes of the postponement of parenthood. In the short run this implies the prolongation of a period of very low fertility. However, when the younger generations reach, say, the age of 30, some catching up of fertility may occur and this is likely to cause a modest rise in period fertility measures. The exact degree of fertility "recuperation" after the age of 30 will be a crucial element, and it is highly likely that it will vary substantially between the various central and east European populations. It is also possible that those countries with the faster rate of transition in household structures will be the first to move to the fertility recuperation stage at older ages, and hence to be the first to recover to more acceptable levels of subreplacement fertility. At this point, an economic recovery will also help, but in the meantime the "second demographic transition" will have become a fact of life for much of Europe.

PART FOUR

STATISTICAL APPENDIX

For the user's convenience, as well as to lighten the text, the *Economic Survey of Europe* includes a set of appendix tables showing time series for the main economic indicators over a longer period. The data are presented in two sections, following the structure of the text: *Appendix A* provides macroeconomic indicators for the market economies in western Europe, North America and Japan for 1987-2001, *Appendix B* does the same for the east European countries, the Baltic states and the Commonwealth of Independent States for 1980-2001.

Re-estimated historical series are not yet available for all the transition economies, and longer time series could in some instances be obtained only by splicing older data with the new statistics (as explained in the notes to the tables). Historical series for Czechoslovakia, the former SFR of Yugoslavia and the Soviet Union can be found in previous issues of this *Survey*. For the economies of western Europe and North America data for the more recent years may also be subject to revision as more comprehensive benchmark figures become available.

Data were compiled from international and national statistical sources. Details on recent changes in national accounts methodology were provided in chapter 7 of the *Economic Survey of Europe, 2000 No. 1*. Aggregates are UNECE secretariat calculations, using PPPs obtained from the 1996 European Comparison Programme. Greece has become a member of the euro area at the beginning of 2001. In order to ensure continuity of time series and comparability with the text tables, Greece has been included in the euro area aggregates for all years shown in the appendix tables.

The figures for 2001 are based on data available at mid-March 2002.

APPENDIX TABLE A.1

Real GDP in western Europe, North America and Japan, 1987-2001

(Percentage change over preceding year)

	1987	1988	1989	1990	1991	1992	1993	1994	1995	1996	1997	1998	1999	2000	2001
France [a]	2.5	4.6	4.2	2.6	1.0	1.5	-0.9	2.1	1.7	1.1	1.9	3.4	3.0	3.6	2.0
Germany [b]	1.7	3.7	3.5	3.2	2.8	2.2	-1.1	2.3	1.7	0.8	1.4	2.0	1.8	3.0	0.6
Italy	3.0	3.9	2.9	2.0	1.4	0.8	-0.9	2.2	2.9	1.1	2.0	1.8	1.6	2.9	1.8
Austria	1.6	3.4	4.2	4.7	3.3	2.3	0.4	2.6	1.6	2.0	1.6	3.5	2.8	3.0	1.1
Belgium	2.8	4.6	3.9	2.9	1.8	1.6	-1.5	2.8	2.6	1.2	3.6	2.2	3.0	4.0	1.3
Finland	4.2	4.7	5.1	–	-6.3	-3.3	-1.1	4.0	3.8	4.0	6.3	5.3	4.1	5.6	0.7
Greece	-2.3	4.3	3.8	–	3.1	0.7	-1.6	2.0	2.1	2.4	3.6	3.4	3.4	3.8	4.1
Ireland	4.7	5.2	5.8	8.5	1.9	3.3	2.7	5.8	10.0	7.8	10.8	8.6	10.8	11.5	6.5
Luxembourg	2.3	10.4	9.8	2.2	6.1	4.5	8.7	4.2	2.8	3.6	9.0	5.8	6.0	7.5	4.0
Netherlands	1.4	3.1	5.0	4.1	2.5	1.7	0.9	2.6	2.9	3.0	3.8	4.3	3.7	3.5	1.5
Portugal	6.4	7.5	6.4	4.0	4.4	1.1	-2.0	1.0	4.3	3.8	3.9	4.5	3.4	3.4	1.7
Spain	5.5	5.1	4.8	3.8	2.5	0.9	-1.0	2.4	2.8	2.4	4.0	4.3	4.1	4.1	2.8
Euro area [c]	2.6	4.2	3.9	2.9	2.0	1.5	-0.8	2.3	2.4	1.5	2.4	2.9	2.7	3.4	1.6
United Kingdom	4.5	5.2	2.2	0.8	-1.4	0.2	2.5	4.7	2.9	2.6	3.4	3.0	2.1	3.0	2.3
Denmark	–	1.2	0.2	1.0	1.1	0.6	–	5.5	2.8	2.5	3.0	2.5	2.3	3.0	1.3
Sweden [a]	3.3	2.6	2.7	1.1	-1.1	-1.7	-1.8	4.1	3.7	1.1	2.1	3.6	4.5	3.6	1.4
European Union [d]	2.9	4.3	3.5	2.5	1.4	1.2	-0.3	2.8	2.5	1.7	2.6	2.9	2.6	3.4	1.7
Cyprus	7.0	8.5	7.9	7.4	0.6	9.8	0.7	5.9	6.1	1.9	2.4	5.0	4.6	5.1	3.7
Iceland	8.5	-0.1	0.3	1.2	0.7	-3.3	0.6	4.5	0.1	5.2	4.8	4.5	4.3	3.6	1.1
Israel [a]	6.2	3.4	1.4	6.3	5.7	6.8	3.4	6.9	8.3	4.5	3.3	2.7	2.6	6.4	-0.5
Malta	4.1	8.4	8.2	6.3	6.3	4.7	4.5	5.7	6.2	4.0	4.8	3.4	4.1	5.4	-0.3
Norway [a]	2.0	-0.1	0.9	2.0	3.1	3.3	2.7	5.5	3.8	4.9	4.7	2.4	1.1	2.3	1.4
Switzerland	0.7	3.1	4.3	3.7	-0.8	-0.1	-0.5	0.5	0.5	0.3	1.7	2.4	1.6	3.0	1.3
Turkey	9.5	2.1	0.3	9.3	0.9	6.0	8.0	-5.5	7.2	7.0	7.5	3.1	-4.7	7.2	-7.3
Western Europe	3.2	4.1	3.4	2.9	1.3	1.4	0.1	2.4	2.7	1.9	2.8	2.9	2.2	3.5	1.3
Canada [a]	4.2	4.9	2.6	0.2	-2.1	0.9	2.4	4.7	2.8	1.6	4.3	3.9	5.1	4.4	1.5
United States [a]	3.4	4.2	3.5	1.8	-0.5	3.0	2.7	4.0	2.7	3.6	4.4	4.3	4.1	4.1	1.2
North America	3.5	4.2	3.4	1.6	-0.6	2.9	2.6	4.1	2.7	3.4	4.4	4.3	4.2	4.2	1.2
Japan	4.5	6.5	5.3	5.3	3.1	0.9	0.4	1.0	1.6	3.5	1.8	-1.1	0.7	2.4	-0.5
Total above	3.5	4.5	3.7	2.7	0.8	2.0	1.2	2.9	2.5	2.8	3.3	2.8	2.8	3.6	1.0
Memorandum items:															
4 major west European economies [e]	2.8	4.3	3.2	2.3	1.2	1.3	-0.2	2.8	2.2	1.3	2.1	2.5	2.1	3.1	1.6
Western Europe and North America	3.3	4.2	3.4	2.2	0.4	2.1	1.4	3.2	2.7	2.7	3.6	3.6	3.2	3.9	1.2

Source: Eurostat, New Cronos Database; OECD, *National Accounts* (Paris), various issues; national statistics.

Note: All aggregates exclude Israel. Growth rates of regional aggregates have been calculated as weighted averages of growth rates in individual countries. Weights were derived from 1996 GDP data converted from national currency units into dollars using 1996 purchasing power parities. 1993 SNA/ESA95 definitions except for Switzerland and Turkey.

[a] Annual changes are calculated from chained national currency series.

[b] Data before 1991 refer to West Germany.

[c] Twelve countries above.

[d] Fifteen countries above.

[e] France, Germany, Italy and the United Kingdom.

APPENDIX TABLE A.2

Real private consumption expenditure in western Europe, North America and Japan, 1987-2001

(Percentage change over preceding year)

	1987	1988	1989	1990	1991	1992	1993	1994	1995	1996	1997	1998	1999	2000	2001
France [a]	3.0	2.7	3.0	2.7	0.7	0.9	-0.4	1.2	1.2	1.3	0.2	3.4	3.1	2.9	2.9
Germany [b]	3.4	2.7	2.8	5.4	5.6	2.7	0.1	1.1	2.1	1.0	0.6	1.8	3.1	1.4	1.1
Italy	3.8	4.0	3.7	2.1	2.9	1.9	-3.7	1.5	1.7	1.2	3.2	3.2	2.4	2.7	1.1
Austria	2.6	3.1	4.3	4.5	2.5	3.0	0.8	2.4	2.6	3.2	1.7	2.8	2.7	2.5	1.4
Belgium	1.8	3.7	3.9	3.2	3.1	2.2	-1.0	2.0	0.7	1.2	2.0	2.9	2.1	3.8	2.0
Finland	5.1	5.3	4.6	-0.6	-3.8	-4.4	-3.1	2.6	4.4	4.2	3.5	5.1	4.0	2.2	1.4
Greece	2.7	6.1	6.3	2.6	2.9	2.3	-0.8	1.9	2.5	2.4	2.7	3.5	2.9	3.2	3.1
Ireland	3.3	4.5	6.5	1.4	1.8	2.9	2.9	4.4	4.1	6.4	7.4	7.3	8.3	10.0	6.2
Luxembourg	4.6	4.6	5.1	5.7	6.3	-0.9	1.7	2.4	0.6	3.7	3.6	4.0	2.1	3.1	3.7
Netherlands	2.7	0.6	3.3	3.9	2.7	0.8	0.5	0.9	3.0	4.0	3.0	4.8	4.5	3.7	1.3
Portugal	5.3	6.8	2.9	6.4	4.2	4.7	1.1	1.0	0.5	3.2	3.4	5.1	4.8	2.6	1.1
Spain	6.0	4.9	5.4	3.5	2.9	2.2	-1.9	1.1	1.7	2.2	3.2	4.5	4.7	4.0	2.7
Euro area [c]	3.6	3.4	3.6	3.6	3.2	1.9	-1.0	1.3	1.8	1.7	1.8	3.1	3.3	2.7	1.8
United Kingdom	5.3	7.5	3.3	1.0	-1.5	0.6	3.2	3.3	1.9	3.8	3.8	3.8	4.2	4.1	3.8
Denmark	-2.2	-2.1	-0.1	0.1	1.6	1.9	0.5	6.5	1.2	2.5	2.9	2.3	0.2	-0.3	1.2
Sweden [a]	5.3	2.6	1.2	-0.4	1.0	-1.3	-3.0	1.8	0.6	1.4	2.0	2.7	3.9	4.5	0.9
European Union [d]	3.8	3.9	3.4	3.0	2.3	1.7	-0.4	1.7	1.8	2.0	2.1	3.2	3.4	2.9	2.1
Cyprus	5.5	10.5	6.9	9.0	9.9	3.2	-4.8	5.0	10.3	3.5	4.0	8.4	3.1	8.0	4.0
Iceland	16.2	-3.8	-4.2	0.5	2.9	-3.1	-4.7	2.9	2.2	5.4	5.5	10.0	6.9	4.0	2.5
Israel [a]	8.9	4.5	0.4	5.6	7.2	8.0	7.3	9.5	5.8	5.2	3.9	4.3	3.2	6.6	3.1
Malta	0.5	9.0	9.2	3.8	3.8	4.3	0.8	2.3	10.5	7.1	1.6	2.5	6.1	6.7	2.5
Norway [a]	-0.8	-2.0	-0.6	0.7	1.5	2.2	2.2	4.0	3.4	5.3	3.6	3.4	2.2	2.4	2.2
Switzerland	1.8	1.4	2.7	1.0	1.9	–	-1.7	0.5	-0.4	0.3	1.8	2.6	2.3	2.3	2.1
Turkey	-0.3	1.2	-1.0	13.1	2.7	3.2	8.6	-5.4	4.8	8.5	8.4	0.6	-2.6	6.4	-6.9
Western Europe	3.5	3.6	3.1	3.4	2.3	1.7	0.1	1.4	1.9	2.3	2.4	3.1	3.1	3.0	1.7
Canada [a]	4.1	4.3	3.4	1.2	-1.6	1.6	1.8	3.0	2.1	2.6	4.6	3.0	3.4	3.6	2.5
United States [a]	3.3	4.0	2.7	1.8	-0.2	2.9	3.4	3.8	3.0	3.2	3.6	4.8	5.0	4.8	3.1
North America	3.4	4.1	2.7	1.8	-0.3	2.8	3.2	3.7	2.9	3.1	3.6	4.6	4.9	4.7	3.0
Japan	4.1	5.1	4.7	4.4	2.7	2.6	1.8	2.6	1.4	2.4	0.8	0.1	1.2	0.6	0.3
Total above	3.6	4.1	3.2	2.9	1.3	2.3	1.7	2.6	2.3	2.7	2.7	3.3	3.5	3.4	2.0
Memorandum items:															
4 major west European economies [e]	3.8	4.1	3.2	3.1	2.3	1.7	-0.2	1.7	1.8	1.7	1.8	2.9	3.2	2.6	2.1
Western Europe and North America	3.4	3.9	2.9	2.6	1.0	2.2	1.7	2.6	2.4	2.7	3.0	3.9	4.0	3.9	2.3

Source: Eurostat, New Cronos Database; OECD, *National Accounts* (Paris), various issues; national statistics.

Note: See appendix table A.1.

[a] Annual changes are calculated from chained national currency series.

[b] Data before 1991 refer to West Germany.

[c] Twelve countries above.

[d] Fifteen countries above.

[e] France, Germany, Italy and the United Kingdom.

APPENDIX TABLE A.3

Real general government consumption expenditure in western Europe, North America and Japan, 1987-2001
(Percentage change over preceding year)

	1987	1988	1989	1990	1991	1992	1993	1994	1995	1996	1997	1998	1999	2000	2001
France [a]	2.2	3.2	1.6	2.5	2.7	3.8	4.6	0.7	-0.1	2.3	2.1	-0.1	2.0	2.3	2.1
Germany [b]	2.1	2.2	-1.3	2.3	0.8	5.0	0.1	2.4	1.5	1.8	0.3	1.2	1.6	1.2	1.7
Italy	4.8	4.0	0.2	2.5	1.7	0.6	-0.2	-0.9	-2.2	1.1	0.3	0.3	1.4	1.7	2.3
Austria	0.1	1.1	1.7	2.3	3.2	3.5	3.7	3.0	1.3	1.2	-1.5	2.8	2.2	0.9	-0.2
Belgium	2.6	-0.7	1.1	-0.4	3.6	1.4	-0.2	1.4	1.3	2.4	0.3	1.5	3.2	2.5	2.4
Finland	4.4	1.9	2.2	4.0	2.1	-2.4	-4.2	0.3	2.0	2.5	4.1	1.7	1.9	-0.2	1.7
Greece	0.2	-5.5	5.4	0.6	-1.5	-3.0	2.6	-1.1	5.6	0.9	3.0	1.7	-0.1	2.3	1.8
Ireland	-4.8	-5.0	-1.3	5.4	2.7	3.0	0.1	4.1	3.9	3.3	5.3	5.7	6.3	5.4	6.0
Luxembourg	4.7	4.9	3.9	3.1	3.9	1.5	3.7	2.0	5.7	5.5	3.0	1.4	7.7	4.7	3.4
Netherlands	2.8	1.9	2.0	2.3	3.0	2.8	1.6	1.5	1.3	-0.4	3.2	3.6	2.8	1.9	3.4
Portugal	3.8	8.6	6.4	4.2	9.6	-0.9	-0.2	4.3	1.0	3.4	2.2	3.8	4.5	2.5	1.9
Spain	9.2	3.6	8.3	6.3	6.0	3.5	2.7	0.5	2.4	1.3	2.9	3.7	4.2	4.0	3.1
Euro area [c]	3.3	2.7	1.3	2.7	2.4	2.9	1.4	1.1	0.6	1.6	1.3	1.3	2.1	2.0	2.2
United Kingdom	–	0.2	1.0	2.2	3.0	0.7	-0.7	1.0	1.7	1.2	0.1	1.5	2.8	1.9	2.4
Denmark	2.1	-0.2	-0.8	-0.2	0.6	0.8	4.1	3.0	2.1	3.4	0.8	3.1	1.8	0.6	1.7
Sweden [a]	1.2	1.1	3.0	2.5	3.4	0.2	-0.1	-0.9	-0.6	0.9	-1.2	3.2	1.7	-0.9	1.1
European Union [d]	2.7	2.2	1.3	2.6	2.5	2.5	1.1	1.1	0.8	1.6	1.0	1.4	2.2	1.9	2.2
Cyprus	5.3	10.5	1.9	17.4	3.9	13.8	-14.3	4.1	2.9	12.6	4.0	7.3	-8.7	2.7	9.7
Iceland	6.5	4.7	3.0	4.4	3.1	-0.7	2.3	4.0	1.8	1.2	2.5	3.4	4.9	3.5	2.5
Israel [a]	18.3	-2.5	-8.6	7.7	4.1	1.4	4.2	-0.2	1.8	5.3	1.8	2.4	3.1	1.1	3.2
Malta	9.1	6.0	12.7	5.7	10.9	8.9	6.0	6.4	8.5	8.4	-1.1	-4.0	-0.6	5.2	3.3
Norway [a]	4.6	-0.1	1.9	4.9	4.3	5.3	2.2	1.4	0.3	2.8	1.9	3.8	3.3	1.4	1.5
Switzerland	1.7	4.5	5.4	5.4	3.5	0.7	-0.1	2.0	-0.1	2.0	–	1.3	0.5	-0.4	0.1
Turkey	9.4	-1.1	0.8	8.0	3.7	3.6	8.6	-5.5	6.8	8.6	4.1	7.8	6.5	7.1	-5.5
Western Europe	3.1	2.1	1.4	3.0	2.6	2.6	1.4	0.8	1.1	1.9	1.2	1.8	2.4	2.1	1.8
Canada [a]	1.4	4.5	2.7	3.5	2.9	0.9	–	-1.3	-0.6	-1.4	-0.8	1.8	2.6	2.2	2.2
United States [a,e]	3.0	1.2	2.8	3.3	1.2	0.5	-0.8	0.1	0.4	1.1	2.4	1.9	3.3	2.7	3.6
North America	2.8	1.4	2.7	3.3	1.3	0.5	-0.7	–	0.4	0.9	2.1	1.9	3.2	2.6	3.4
Japan	3.5	3.4	2.9	2.5	3.2	2.7	3.2	2.9	4.3	2.8	1.3	1.9	4.5	4.6	3.1
Total above	3.0	2.0	2.2	3.0	2.2	1.7	0.8	0.8	1.3	1.6	1.6	1.9	3.1	2.7	2.7
Memorandum items:															
4 major west European economies [f]	2.3	2.4	0.2	2.4	1.9	2.8	0.9	1.0	0.4	1.6	0.7	0.8	1.9	1.7	2.1
Western Europe and North America	2.9	1.7	2.1	3.1	2.0	1.5	0.3	0.4	0.7	1.4	1.6	1.8	2.8	2.4	2.6

Source: Eurostat, New Cronos Database; OECD, *National Accounts* (Paris), various issues; national statistics.

Note: See appendix table A.1.

[a] Annual changes are calculated from chained national currency series.

[b] Data before 1991 refer to West Germany.

[c] Twelve countries above.

[d] Fifteen countries above.

[e] Includes final consumption expenditure and gross investment.

[f] France, Germany, Italy and the United Kingdom.

APPENDIX TABLE A.4

Real gross domestic fixed capital formation in western Europe, North America and Japan, 1987-2001

(Percentage change over preceding year)

	1987	1988	1989	1990	1991	1992	1993	1994	1995	1996	1997	1998	1999	2000	2001
France [a]	6.0	9.5	7.3	3.3	-1.5	-1.6	-6.4	1.5	2.0	–	-0.1	7.0	6.2	6.2	2.8
Germany [b]	2.5	4.6	5.6	9.0	9.8	4.5	-4.4	4.0	-0.6	-0.8	0.6	3.0	4.2	2.3	-4.8
Italy	4.2	6.7	4.2	4.0	1.0	-1.4	-10.9	0.1	6.0	3.6	2.1	4.0	5.7	6.5	2.4
Austria	3.8	7.4	4.1	6.2	6.6	0.6	-0.9	4.6	1.3	2.2	2.0	3.4	1.5	5.1	-0.2
Belgium	6.2	15.7	12.6	8.5	-4.1	1.7	-3.1	-0.1	5.6	1.3	6.8	4.3	3.3	2.6	-0.6
Finland	4.9	11.0	13.0	-4.6	-18.6	-16.7	-16.6	-2.7	10.6	8.4	11.9	9.3	3.0	4.8	2.1
Greece	-5.6	2.6	6.1	4.5	4.2	-3.5	-4.0	-3.1	4.1	8.4	6.8	10.6	6.2	7.8	9.1
Ireland	-1.1	5.2	10.1	13.4	-7.0	–	-5.1	11.8	13.4	16.6	17.8	15.7	13.5	7.3	3.4
Luxembourg	17.9	15.0	7.0	2.7	31.6	-9.0	28.4	-14.9	2.9	1.7	14.3	2.8	19.6	-3.0	5.8
Netherlands	0.8	5.3	5.1	2.5	0.4	0.7	-3.2	2.1	3.9	6.3	6.6	4.2	7.8	3.8	-1.3
Portugal	18.0	14.8	3.7	7.6	3.3	4.5	-5.5	2.7	6.6	6.2	13.9	11.2	7.1	5.3	-1.0
Spain	12.2	13.6	12.0	6.5	1.7	-4.1	-8.9	1.9	7.7	2.1	5.0	9.7	8.8	5.7	2.5
Euro area [c]	4.8	7.8	6.7	5.8	2.9	0.1	-6.5	2.0	3.3	1.8	2.7	5.4	5.7	4.7	0.1
United Kingdom	9.3	14.9	6.0	-2.6	-8.2	-0.9	0.3	4.7	3.1	4.7	7.1	13.2	0.9	4.9	2.0
Denmark	-0.8	-3.2	-0.6	-2.2	-3.4	-2.1	-3.8	7.7	11.6	3.9	10.9	10.0	1.0	10.7	-2.3
Sweden [a]	8.0	6.4	12.1	0.2	-8.6	-11.6	-15.0	6.1	9.4	5.0	-1.1	8.5	9.6	5.0	3.2
European Union [d]	5.5	8.7	6.6	4.2	0.8	-0.3	-5.6	2.6	3.5	2.4	3.4	6.8	4.9	4.8	0.4
Cyprus	4.5	10.6	20.0	-2.8	-1.6	16.2	-12.8	-2.5	-1.7	7.4	-4.5	8.0	-1.4	7.0	3.4
Iceland	18.8	-0.2	-7.9	3.0	3.3	-11.1	-10.7	0.6	-1.1	25.7	9.6	26.6	-0.8	11.1	1.0
Israel [a]	6.1	1.6	-2.2	25.3	41.9	5.2	5.3	8.4	6.6	9.0	-0.9	-3.6	0.4	1.1	-8.9
Malta	30.7	6.1	1.0	17.9	–	-0.2	11.1	8.5	17.8	-8.4	-4.5	-3.4	4.0	18.1	-13.7
Norway [a]	0.3	-1.8	-6.9	-10.8	-0.4	-3.1	4.3	4.5	3.4	9.9	13.9	10.6	-8.2	-1.1	-5.9
Switzerland	4.0	8.1	5.3	3.8	-2.9	-6.6	-2.7	6.5	1.8	-2.4	1.5	4.5	3.7	5.8	-1.3
Turkey	45.1	-1.0	2.2	15.9	0.4	6.4	26.4	-16.0	9.1	14.1	14.8	-3.9	-15.7	16.5	-23.0
Western Europe	7.3	8.1	6.1	4.6	0.6	-0.2	-3.9	1.8	3.7	3.0	4.0	6.3	3.7	5.3	-0.8
Canada [a]	10.5	9.4	5.6	-4.0	-5.5	-2.7	-2.0	7.5	-2.1	4.4	15.2	2.4	7.3	6.7	1.4
United States [a,e]	..	3.6	2.7	-1.8	-6.9	6.5	8.1	9.1	6.1	9.3	9.6	11.4	7.8	7.6	-1.9
North America	..	4.1	3.0	-2.0	-6.8	5.7	7.3	9.0	5.4	8.9	10.0	10.6	7.8	7.5	-1.7
Japan	9.4	12.0	8.6	8.8	2.2	-2.5	-3.1	-1.4	0.3	6.8	1.0	-4.0	-0.8	3.2	-1.7
Total above	..	7.0	5.2	2.5	-2.3	1.9	0.9	4.3	3.9	6.1	6.1	6.5	4.7	5.9	-1.3
Memorandum items:															
4 major west European economies [f]	5.1	8.4	5.8	4.1	1.3	0.6	-5.3	2.7	2.3	1.6	2.2	6.4	4.3	4.7	–
Western Europe and North America	..	6.1	4.5	1.3	-3.1	2.8	1.7	5.4	4.6	5.9	7.0	8.5	5.8	6.4	-1.2

Source: Eurostat, New Cronos Database; OECD, *National Accounts* (Paris), various issues; national statistics.

Note: See appendix table A.1.

[a] Annual changes are calculated from chained national currency series.

[b] Data before 1991 refer to West Germany.

[c] Twelve countries above.

[d] Fifteen countries above.

[e] Includes only gross domestic fixed investment of the private sector; the government sector gross investment is included in the final consumption expenditure of general government.

[f] France, Germany, Italy and the United Kingdom.

APPENDIX TABLE A.5

Real total domestic expenditures in western Europe, North America and Japan, 1987-2001

(Percentage change over preceding year)

	1987	1988	1989	1990	1991	1992	1993	1994	1995	1996	1997	1998	1999	2000	2001
France [a]	3.2	4.7	3.9	2.8	0.5	0.8	-1.6	2.1	1.6	0.7	0.7	4.0	3.1	3.9	1.7
Germany [b]	2.6	3.6	2.7	4.8	6.2	2.8	-1.1	2.3	1.7	0.3	0.6	2.4	2.6	2.0	-1.0
Italy	4.3	4.1	3.1	2.7	2.1	0.9	-5.1	1.7	2.0	0.9	2.7	3.1	3.0	2.1	1.6
Austria	2.4	3.3	3.7	4.4	3.5	2.3	0.6	3.5	2.6	1.9	1.5	2.7	2.8	2.5	0.6
Belgium	3.6	4.8	4.5	3.0	1.6	1.8	-1.5	1.9	2.0	0.9	2.8	3.3	2.2	3.8	1.6
Finland	5.7	6.6	7.0	-0.5	-7.9	-5.7	-5.5	3.7	4.4	2.9	6.0	5.8	2.0	3.6	0.7
Greece	-2.7	5.9	5.3	2.2	3.5	-0.5	-1.0	1.1	3.5	3.3	3.5	4.6	2.6	5.1	4.2
Ireland	-0.4	1.9	6.9	6.3	0.2	-0.1	1.0	5.1	6.7	7.4	9.4	10.3	6.6	9.2	5.2
Luxembourg	5.3	6.5	8.0	3.2	8.1	-1.2	8.8	-0.2	3.2	4.1	6.7	3.1	7.5	2.4	3.8
Netherlands	1.4	2.2	5.0	3.4	2.2	1.2	-1.6	2.0	3.5	2.8	3.9	4.8	4.2	3.0	1.3
Portugal	9.0	10.1	4.9	5.3	6.1	3.4	-2.1	1.5	4.1	3.3	5.1	6.7	5.5	3.0	0.7
Spain	7.9	6.8	7.3	4.6	3.0	1.0	-3.3	1.5	3.1	1.9	3.5	5.7	5.6	4.2	2.9
Euro area [c]	3.6	4.4	4.0	3.7	3.1	1.4	-2.3	2.0	2.2	1.1	2.0	3.7	3.3	3.0	1.0
United Kingdom	4.9	8.1	2.9	-0.3	-2.5	0.9	2.2	3.8	2.0	3.1	3.9	5.1	3.4	3.6	3.0
Denmark	-2.0	0.2	-0.1	-0.7	-0.1	0.9	-0.3	7.0	4.2	2.2	4.9	4.0	-0.5	2.6	0.5
Sweden [a]	4.3	3.0	4.0	0.7	-1.6	-1.8	-4.6	3.1	1.9	0.7	0.8	4.3	3.5	3.5	1.0
European Union [d]	3.7	4.9	3.7	2.9	2.0	1.3	-1.6	2.4	2.2	1.4	2.3	3.9	3.2	3.1	1.3
Cyprus	5.6	11.3	9.7	6.3	5.1	9.3	-9.5	7.4	7.8	4.8	1.3	8.8	0.4	6.8	3.9
Iceland	15.7	-0.7	-4.4	1.5	4.6	-4.6	-4.2	2.4	2.2	7.3	5.7	12.4	4.6	5.4	2.2
Israel [a]	9.0	2.5	-2.8	9.3	12.1	5.3	6.8	5.3	6.7	5.7	2.0	1.2	2.1	3.2	3.7
Malta	3.5	11.5	8.3	7.6	4.9	0.1	4.8	5.9	9.5	2.8	-0.1	-1.1	5.8	10.6	-4.8
Norway [a]	-0.6	-2.5	-1.6	–	1.1	1.7	3.1	4.0	4.3	4.2	6.4	5.4	-0.6	2.2	-0.6
Switzerland	1.8	2.4	4.3	3.8	-0.5	-2.8	-1.5	2.3	1.2	0.2	1.5	4.4	1.5	2.8	1.7
Turkey	8.9	-1.3	1.5	14.6	-0.6	5.6	14.2	-12.5	11.4	7.6	9.0	0.6	-3.7	9.6	-15.8
Western Europe	3.9	4.5	3.6	3.4	1.8	1.4	-0.8	1.8	2.6	1.7	2.7	3.8	2.8	3.4	0.5
Canada [a]	4.9	5.4	4.0	-0.3	-1.9	0.5	1.6	3.2	1.8	1.2	6.1	2.3	4.0	4.5	0.9
United States [a]	3.7	3.2	2.9	1.4	-1.1	3.1	3.2	4.4	2.4	3.7	4.7	5.5	5.0	4.8	1.1
North America	3.8	3.4	3.0	1.3	-1.1	2.9	3.0	4.3	2.4	3.5	4.8	5.2	4.9	4.8	1.1
Japan	5.3	7.3	5.6	5.3	2.7	0.6	0.3	1.2	2.1	4.0	0.9	-1.5	0.8	1.9	0.2
Total above	4.1	4.5	3.6	2.8	0.7	1.9	1.0	2.8	2.4	2.8	3.3	3.6	3.4	3.7	0.7
Memorandum items:															
4 major west European economies [e]	3.6	4.9	3.1	2.8	2.1	1.5	-1.4	2.4	1.8	1.1	1.8	3.5	3.0	2.8	1.1
Western Europe and North America	3.8	3.9	3.3	2.4	0.4	2.1	1.1	3.0	2.5	2.6	3.7	4.5	3.9	4.1	0.8

Source: Eurostat, New Cronos Database; OECD, *National Accounts* (Paris), various issues; national statistics.

Note: See appendix table A.1.

[a] Annual changes are calculated from chained national currency series.

[b] Data before 1991 refer to West Germany.

[c] Twelve countries above.

[d] Fifteen countries above.

[e] France, Germany, Italy and the United Kingdom.

APPENDIX TABLE A.6

Real exports of goods and services in western Europe, North America and Japan, 1987-2001

(Percentage change over preceding year)

	1987	1988	1989	1990	1991	1992	1993	1994	1995	1996	1997	1998	1999	2000	2001
France [a]	3.4	8.7	10.0	4.8	5.9	5.4	–	7.7	7.7	3.5	11.8	8.3	3.9	13.3	1.1
Germany [b]	0.5	5.1	9.8	0.5	-2.9	-0.8	-5.5	7.6	5.7	5.1	11.2	6.8	5.6	13.2	4.7
Italy	4.5	5.1	7.8	7.5	-1.4	7.3	9.0	9.8	12.6	0.6	6.4	3.4	0.3	11.7	0.8
Austria	2.3	9.8	9.7	7.8	5.2	1.5	-1.4	5.6	3.0	5.2	12.4	7.9	8.7	12.2	5.3
Belgium	5.0	9.6	8.3	4.6	3.1	3.6	-0.4	8.4	5.7	2.9	6.1	5.8	5.0	9.7	0.7
Finland	2.9	3.5	1.6	1.2	-7.3	10.3	16.7	13.1	8.6	5.8	14.1	8.9	6.8	18.2	-0.7
Greece	5.9	-2.1	1.9	-3.5	4.1	10.0	-2.6	7.4	3.0	3.5	20.0	5.3	8.1	5.6	5.7
Ireland	13.7	9.0	10.3	8.7	5.7	13.9	9.7	15.1	20.0	12.2	17.4	21.4	15.7	17.8	9.1
Luxembourg	4.4	11.7	8.1	3.4	6.7	4.8	2.8	4.4	6.1	5.4	13.4	12.9	13.3	16.4	4.7
Netherlands	4.1	8.9	7.9	5.1	5.6	2.4	5.7	9.7	8.8	4.6	8.8	7.4	5.4	9.5	2.3
Portugal	11.2	8.2	12.2	9.5	1.2	3.2	-3.3	8.4	8.8	7.1	7.1	9.2	3.2	8.1	6.2
Spain	5.3	3.8	1.4	4.7	8.2	7.5	7.8	16.7	9.4	10.4	15.3	8.2	7.6	9.6	3.4
Euro area [c]	3.4	6.2	8.2	4.0	1.6	4.1	1.5	9.2	8.2	4.4	10.7	6.8	4.6	12.0	2.8
United Kingdom	6.1	0.6	4.5	5.4	-0.1	4.3	4.4	9.2	9.0	8.2	8.3	3.0	5.4	10.3	-0.1
Denmark	4.3	11.2	4.2	6.2	6.1	-0.9	-1.5	7.0	2.9	4.3	4.1	4.3	10.8	11.5	4.1
Sweden [a]	4.3	2.8	3.2	1.8	-1.9	2.2	8.3	14.1	11.3	3.5	13.7	8.4	6.5	10.3	-0.1
European Union [d]	3.8	5.3	7.4	4.2	1.3	4.0	2.0	9.3	8.3	5.0	10.3	6.2	4.9	11.7	2.3
Cyprus	13.7	13.5	16.8	7.9	-8.4	18.7	-1.3	7.9	4.6	4.4	0.5	-2.4	6.3	9.0	4.9
Iceland	3.3	-3.6	2.9	–	-5.9	-1.9	7.0	9.9	-2.1	9.9	5.7	2.2	5.5	2.6	-1.0
Israel [a]	10.2	-1.5	4.0	2.0	-2.6	14.1	9.9	12.8	8.4	4.9	8.0	6.6	11.6	23.9	-13.1
Malta	12.6	6.1	10.7	13.3	7.5	9.7	5.3	7.1	5.4	-5.9	4.0	8.1	8.2	5.6	-7.3
Norway [a]	1.1	6.4	11.0	8.6	6.1	5.2	3.2	8.7	4.3	9.3	6.1	0.3	2.8	2.7	5.3
Switzerland	2.2	7.0	6.9	2.6	-2.5	3.6	2.6	2.2	2.4	3.0	8.7	5.3	6.3	11.8	-0.5
Turkey	26.4	18.4	-0.3	2.5	3.7	11.0	7.7	15.2	8.0	22.0	19.1	12.0	-7.0	19.3	5.0
Western Europe	4.8	6.0	7.1	4.2	1.4	4.3	2.3	9.4	8.1	5.8	10.6	6.4	4.3	11.9	2.4
Canada [a]	2.9	8.9	1.0	4.7	1.8	7.2	10.8	12.7	8.5	5.6	8.3	8.9	9.9	7.6	-3.7
United States [a]	11.2	16.1	11.8	8.7	6.5	6.2	3.3	8.9	10.3	8.2	12.3	2.1	3.2	9.5	-4.6
North America	10.5	15.5	10.9	8.4	6.1	6.2	4.0	9.3	10.1	7.9	11.9	2.7	3.8	9.3	-4.5
Japan	-0.5	5.9	9.1	7.0	4.1	3.9	-0.1	3.5	4.1	6.5	11.2	-2.3	1.4	12.4	-6.6
Total above	6.4	10.0	9.0	6.4	3.8	5.1	2.6	8.4	8.3	6.8	11.3	3.5	3.6	10.9	-1.9
Memorandum items:															
4 major west European economies [e]	3.3	4.9	8.2	4.1	–	3.6	1.2	8.5	8.4	4.4	9.6	5.6	4.0	12.3	2.0
Western Europe and North America	7.7	10.7	9.0	6.3	3.8	5.3	3.2	9.3	9.1	6.9	11.3	4.5	4.1	10.6	-1.1

Source: Eurostat, New Cronos Database; OECD, *National Accounts* (Paris), various issues; national statistics.

Note: See appendix table A.1. Data on national accounts basis.

[a] Annual changes are calculated from chained national currency series.

[b] Data before 1991 refer to West Germany.

[c] Twelve countries above.

[d] Fifteen countries above.

[e] France, Germany, Italy and the United Kingdom.

APPENDIX TABLE A.7

Real imports of goods and services in western Europe, North America and Japan, 1987-2001

(Percentage change over preceding year)

	1987	1988	1989	1990	1991	1992	1993	1994	1995	1996	1997	1998	1999	2000	2001
France [a]	7.7	8.8	8.0	5.5	3.1	1.8	-3.7	8.2	8.0	1.6	6.9	11.6	4.2	15.4	-0.2
Germany [b]	4.2	4.9	7.8	5.2	9.6	1.5	-5.5	7.4	5.6	3.1	8.3	8.9	8.5	10.0	0.1
Italy	12.2	5.9	8.9	11.5	2.3	7.4	-10.9	8.1	9.7	-0.3	10.1	8.9	5.3	9.4	0.2
Austria	4.8	9.3	8.0	6.9	5.8	1.4	-1.1	8.2	5.6	4.9	12.0	5.9	8.8	11.1	4.4
Belgium	6.7	10.4	9.6	4.8	2.9	4.1	-0.4	7.3	4.9	2.5	5.1	7.5	4.1	9.7	1.0
Finland	9.2	10.9	9.0	-0.8	-13.5	0.6	1.3	12.8	7.8	6.4	11.3	8.5	4.0	16.2	-1.0
Greece	2.1	7.3	10.5	8.4	5.8	1.1	0.6	1.5	8.9	7.0	14.2	9.2	3.6	9.7	5.6
Ireland	6.2	4.9	13.5	5.1	2.4	8.2	7.5	15.5	16.4	12.5	16.8	25.8	11.9	16.6	8.5
Luxembourg	7.5	8.2	6.6	4.5	9.0	-0.8	2.8	-0.1	6.9	6.1	11.8	11.5	15.6	13.8	4.7
Netherlands	4.2	6.9	8.2	3.6	5.1	1.4	0.7	9.4	10.6	4.4	9.5	8.5	6.3	9.4	2.2
Portugal	23.1	18.0	5.9	14.5	7.2	10.7	-3.3	8.8	7.4	5.0	10.0	14.2	8.7	6.0	2.7
Spain	24.8	16.1	17.7	9.6	10.3	6.8	-5.2	11.4	11.1	8.0	13.2	13.3	12.8	9.8	3.7
Euro area [c]	9.2	8.0	9.3	7.1	5.7	3.6	-5.1	8.3	8.0	3.0	9.3	10.0	7.0	11.0	1.0
United Kingdom	7.9	12.8	7.4	0.5	-4.5	6.8	3.3	5.7	5.4	9.6	9.7	9.6	8.9	10.9	-0.2
Denmark	-3.1	8.3	4.1	1.2	3.0	-0.4	-2.7	12.3	7.3	3.5	10.0	8.9	3.3	11.2	2.5
Sweden [a]	7.6	4.5	7.7	0.7	-4.9	1.5	-2.2	12.2	7.2	3.0	12.5	11.2	4.4	11.5	-1.1
European Union [d]	8.8	8.6	8.9	5.8	3.8	4.0	-3.6	8.1	7.5	4.1	9.4	10.0	7.2	11.0	0.8
Cyprus	5.5	13.4	20.4	5.8	2.2	18.2	-18.1	8.2	11.5	7.0	0.4	6.6	-1.9	10.0	3.8
Iceland	23.3	-4.6	-10.3	1.0	5.3	-5.9	-7.7	4.2	4.0	16.7	8.5	23.3	6.1	7.0	2.0
Israel [a]	19.6	-2.8	-5.0	9.5	16.0	8.8	14.1	10.9	4.5	7.7	3.3	1.7	14.8	12.2	-6.4
Malta	12.3	11.1	11.1	15.7	5.4	3.0	5.9	7.5	10.0	-5.9	-1.7	2.5	10.1	11.2	-11.5
Norway [a]	-6.5	-2.4	2.2	2.5	0.2	0.7	4.4	4.9	5.6	8.0	11.3	8.0	-1.6	2.5	0.3
Switzerland	5.8	4.7	6.6	2.8	-1.4	-4.7	-0.4	8.0	4.5	2.7	8.3	10.8	5.9	11.2	0.4
Turkey	23.0	-4.5	6.9	33.0	-5.2	10.9	35.8	-21.9	29.6	20.5	22.4	2.3	-3.7	25.4	-19.0
Western Europe	9.2	7.8	8.6	7.0	3.2	4.1	-1.6	6.6	8.5	4.9	10.0	9.6	6.5	11.6	-0.2
Canada [a]	5.3	13.5	5.9	2.0	2.5	4.7	7.4	8.0	5.7	5.1	14.2	4.9	7.3	8.1	-5.7
United States [a]	6.1	3.8	4.0	3.8	-0.5	6.6	9.1	12.0	8.2	8.6	13.7	11.8	10.5	13.4	-2.7
North America	6.0	4.6	4.1	3.7	-0.3	6.5	8.9	11.6	8.0	8.3	13.7	11.2	10.2	12.9	-3.0
Japan	11.3	19.5	15.7	7.0	-1.1	-0.7	-1.4	7.8	12.8	13.2	1.2	-6.8	3.0	9.6	-0.6
Total above	8.2	8.3	7.8	5.6	1.1	4.4	2.9	8.9	9.0	7.6	10.2	7.7	7.5	11.8	-1.4
Memorandum items:															
4 major west European economies [e]	7.6	7.7	8.0	5.7	3.4	4.0	-4.4	7.4	7.0	3.4	8.7	9.7	6.9	11.3	–
Western Europe and North America	7.6	6.2	6.4	5.3	1.5	5.3	3.7	9.1	8.2	6.6	11.9	10.4	8.4	12.3	-1.6

Source: Eurostat, New Cronos Database; OECD, *National Accounts* (Paris), various issues; national statistics.

Note: See appendix table A.1. Data on national accounts basis.

[a] Annual changes are calculated from chained national currency series.

[b] Data before 1991 refer to West Germany.

[c] Twelve countries above.

[d] Fifteen countries above.

[e] France, Germany, Italy and the United Kingdom.

APPENDIX TABLE A.8

Industrial output in western Europe, North America and Japan, 1987-2001
(Percentage change over preceding year)

	1987	1988	1989	1990	1991	1992	1993	1994	1995	1996	1997	1998	1999	2000	2001
France	1.2	4.6	3.7	3.1	-0.3	-1.0	-3.8	4.4	2.3	0.9	3.8	5.2	2.0	3.5	0.9
Germany [a]	0.4	3.6	4.9	5.2	3.5	-2.4	-7.9	3.2	0.8	0.8	3.7	4.1	1.5	6.2	0.3
Italy	2.6	6.9	3.9	6.3	-0.4	-1.3	-2.1	6.2	5.0	-1.9	3.8	1.1	–	4.8	-1.2
Austria	1.0	4.4	5.8	6.8	1.9	-1.2	-1.5	4.0	4.9	1.0	6.4	8.2	6.0	8.9	0.2
Belgium	2.1	5.8	3.4	1.5	-1.9	-0.4	-5.1	2.1	6.5	0.5	4.7	3.4	0.9	5.3	-0.7
Finland	5.0	3.2	3.6	-0.6	-8.7	1.3	5.6	11.3	7.3	3.6	10.1	7.4	6.1	11.5	-1.1
Greece	-1.2	5.1	1.8	-2.5	-1.0	-1.1	-2.9	1.3	1.8	1.2	1.3	7.1	3.9	0.5	2.1
Ireland	8.9	10.7	11.6	4.7	3.3	9.1	5.6	11.9	20.5	8.1	17.5	19.8	14.8	15.4	9.7
Luxembourg	-0.6	8.7	7.8	2.6	0.4	-0.8	-4.3	5.9	2.0	0.1	5.8	-0.1	11.5	4.3	2.4
Netherlands	1.1	0.1	5.1	2.4	1.8	-0.2	-1.1	4.9	4.1	2.4	0.2	2.2	1.9	3.8	-0.9
Portugal	4.4	3.8	6.7	9.0	–	-2.3	-5.2	-0.2	11.6	5.3	2.6	5.7	3.0	0.5	2.4
Spain	4.6	3.1	5.1	-0.3	-0.7	-3.1	-4.7	7.7	4.8	-1.3	6.9	5.5	2.6	4.4	-1.4
Euro area [b]	1.6	4.1	4.6	2.8	0.7	-1.7	-4.8	4.5	3.4	0.4	4.2	4.3	2.0	5.5	0.2
United Kingdom	4.1	5.2	2.1	–	-3.3	0.4	2.1	5.2	1.8	1.2	1.1	1.0	0.8	1.8	-2.3
Denmark	-3.6	2.5	2.4	2.4	–	3.5	-2.2	10.3	4.2	2.0	5.9	1.9	2.7	6.2	1.7
Sweden	2.8	2.9	2.9	0.3	-5.2	-1.7	-0.4	11.5	9.8	1.0	6.5	4.2	3.1	8.4	-1.7
European Union [c]	1.9	4.2	4.3	2.4	-0.1	-1.4	-3.6	5.1	3.4	0.6	3.9	3.7	1.8	5.0	-0.2
Israel	4.9	-3.1	-1.6	8.0	6.8	8.2	6.9	7.4	8.4	5.4	1.8	2.8	1.4	10.1	-5.7
Norway	6.6	2.9	9.3	2.5	2.5	5.6	3.6	7.0	5.9	5.4	3.4	-0.6	-0.2	2.9	0.1
Switzerland	1.2	7.8	1.5	4.8	0.5	-1.0	-1.8	4.3	2.0	–	4.6	3.6	3.6	8.3	4.0
Turkey	10.5	1.6	3.6	9.5	2.7	5.0	8.0	-6.2	12.7	7.5	10.7	1.2	-3.7	6.1	-9.3
Western Europe	2.3	4.1	4.2	2.7	0.1	-1.1	-3.0	4.5	3.8	0.9	4.3	3.5	1.5	5.1	-0.5
Canada	4.2	6.6	-0.3	-2.7	-3.6	1.3	4.8	6.2	4.6	1.2	5.6	3.4	5.6	5.5	-2.8
United States	4.6	4.5	1.8	-0.2	-2.0	3.1	3.4	5.5	4.8	4.6	6.9	5.1	3.7	4.5	-3.8
North America	4.6	4.7	1.6	-0.4	-2.1	3.0	3.6	5.6	4.8	4.3	6.8	5.0	3.8	4.6	-3.8
Japan	3.4	9.4	5.8	4.2	1.9	-5.7	-3.5	1.3	3.3	2.3	3.5	-6.5	0.8	5.7	-7.3
Total above	3.4	5.3	3.5	1.8	-0.4	-0.6	-0.7	4.3	4.1	2.5	5.2	2.4	2.4	5.0	-3.0
Memorandum items:															
4 major west European economies [d]	1.8	4.8	3.9	4.0	0.5	-1.4	-3.9	4.5	2.3	0.2	3.2	3.0	1.1	4.4	-0.4
Western Europe and North America	3.4	4.4	3.0	1.3	-0.9	0.8	–	5.0	4.3	2.6	5.5	4.3	2.7	4.8	-2.2

Source: National statistics; Eurostat, New Cronos Database; OECD, *Main Economic Indicators* (Paris), various issues; UNECE secretariat estimates.

Note: Except for the EU and the euro area, industrial output indices for regional aggregates have been calculated as weighted averages of the indices of the constituent countries; the EU and euro region aggregates are provided by Eurostat. Weights were derived from 1995 value added originating in industry converted from national currency units into dollars using 1995 GDP purchasing power parities.

[a] West Germany, 1987-1991.

[b] Twelve countries above.

[c] Fifteen countries above.

[d] France, Germany, Italy and the United Kingdom.

APPENDIX TABLE A.9

Total employment in western Europe, North America and Japan, 1987-2001
(Percentage change over preceding year)

	1987	1988	1989	1990	1991	1992	1993	1994	1995	1996	1997	1998	1999	2000	2001
France	-0.1	0.7	0.9	0.9	-0.1	-0.3	-1.3	-0.1	0.9	0.3	0.5	1.3	1.8	2.5	1.6
Germany [a]	0.7	0.8	1.5	3.0	2.5	-1.5	-1.4	-0.2	0.2	-0.3	-0.2	1.1	1.2	1.6	0.1
Italy	0.2	1.1	0.7	1.6	1.9	-0.5	-2.5	-1.5	-0.1	0.6	0.4	1.0	1.1	1.6	1.5
Austria	–	0.6	1.3	1.6	1.4	0.2	-0.6	-0.1	–	-0.6	0.5	0.8	1.4	0.9	0.2
Belgium	0.6	1.7	1.2	0.9	0.1	-0.5	-0.7	-0.4	0.7	0.4	0.8	1.2	1.4	1.6	1.2
Finland	0.5	1.2	0.7	-0.5	-5.7	-7.2	-6.2	-1.1	1.6	1.4	3.3	2.1	2.7	1.8	1.4
Greece	0.3	1.3	0.6	0.6	-1.4	1.0	1.2	1.7	0.5	-0.4	-0.5	4.1	-0.8	-0.3	1.1
Ireland	0.8	0.6	1.0	2.7	0.4	1.1	1.3	4.3	4.3	3.6	5.6	8.6	6.0	4.7	2.3
Luxembourg	2.7	3.0	3.5	4.1	4.1	2.5	1.8	2.5	2.6	2.7	3.2	4.3	5.0	5.6	5.5
Netherlands	1.6	2.0	2.6	2.9	1.8	1.6	–	0.7	1.5	2.3	3.2	2.6	2.5	2.3	2.0
Portugal	2.3	2.5	1.9	1.9	2.8	-0.9	-1.9	-0.2	-0.7	0.5	1.7	2.7	1.7	2.0	1.5
Spain	4.8	3.5	3.6	3.8	1.2	-1.4	-2.8	-0.5	1.9	1.3	2.9	3.6	3.5	3.1	2.3
Euro area [b]	0.9	1.3	1.5	2.1	1.3	-0.9	-1.7	-0.4	0.6	0.5	0.8	1.7	1.7	1.8	1.2
United Kingdom [c]	2.6	4.3	2.4	0.3	-3.0	-2.1	-0.4	1.0	1.4	1.1	2.0	1.1	1.3	1.2	0.7
Denmark	0.4	-0.7	-0.7	-0.7	-0.6	-0.8	-1.5	1.4	0.5	0.7	1.2	1.2	1.1	0.7	0.6
Sweden	0.8	1.4	1.5	1.0	-2.0	-4.3	-5.3	-0.8	1.4	-0.6	-1.1	1.2	2.2	2.1	1.8
European Union [d]	1.2	1.8	1.6	1.7	0.4	-1.2	-1.8	-0.2	0.7	0.6	1.0	1.6	1.6	1.7	1.1
Cyprus [e]	3.1	4.7	3.9	2.8	0.6	4.1	-0.1	2.8	3.4	1.0	-0.2	1.8	0.3	–	–
Iceland [e]	5.8	-3.0	-1.5	-1.1	-0.1	-1.4	-0.8	0.5	0.9	3.1	1.0	3.4	2.7	2.0	0.5
Israel	2.6	3.5	0.5	2.1	6.1	4.2	6.1	6.9	5.2	2.4	1.4	1.6	3.1	4.0	2.3
Malta [f]	5.9	2.5	0.9	0.8	2.5	1.0	0.5	-1.3	3.3	1.0	0.5	0.4	0.6	2.1	-2.6
Norway	2.0	-0.5	-2.8	-0.8	-0.7	-0.3	0.2	1.3	2.1	2.1	2.9	2.3	0.7	0.5	0.3
Switzerland	2.5	2.6	2.7	3.2	1.8	-1.5	-0.8	-0.3	0.3	0.1	0.1	1.0	0.7	1.1	1.2
Turkey	2.3	1.5	2.6	1.7	1.7	0.2	0.2	2.8	3.7	2.0	-2.5	2.8	2.2	2.7	-3.0
Western Europe	1.4	1.7	1.7	1.7	0.6	-1.0	-1.4	0.2	1.1	0.8	0.6	1.7	1.7	1.8	0.6
Canada	2.7	3.2	2.1	–	-1.8	-0.7	0.8	2.0	1.9	0.8	2.3	2.6	2.8	2.6	1.1
United States	2.6	2.3	2.0	1.3	-0.9	0.7	1.5	2.3	1.5	1.4	2.2	1.5	1.5	1.3	-0.1
North America	2.6	2.3	2.1	1.1	-1.0	0.5	1.4	2.3	1.5	1.4	2.3	1.6	1.7	1.4	–
Japan	0.4	1.2	1.5	1.7	2.0	1.1	0.4	0.1	0.2	0.5	1.1	-0.7	-0.8	-0.2	-0.2
Total above	1.6	1.9	1.8	1.5	0.3	-0.1	-0.1	0.9	1.1	0.9	1.3	1.3	1.3	1.3	0.3
Memorandum items:															
4 major west European economies [g]	0.9	1.7	1.4	1.6	0.5	-1.2	-1.4	-0.2	0.6	0.4	0.6	1.1	1.3	1.7	0.9
Western Europe and North America	1.9	2.0	1.8	1.5	0.0	-0.4	-0.2	1.1	1.3	1.0	1.3	1.7	1.7	1.6	0.4

Source: Eurostat, New Cronos Database; OECD, *National Accounts* and *Economic Outlook*, latest issues; national statistics.

Note: Total employment is defined as the number of persons engaged in some productive activity within resident production units (national accounts concept). The labour force survey concept (based on resident household surveys) is used for Canada, Israel, Turkey, the United Kingdom and the United States; Austria (up to 1987); Portugal (up to 1990); Ireland, the Netherlands and Spain (up to 1994). All aggregates exclude Israel.

[a] West Germany, 1987-1991.
[b] Twelve countries above.
[c] Number of jobs.
[d] Fifteen countries above.
[e] Full-time equivalent.
[f] Full-time occupied at the end of the year.
[g] France, Germany, Italy and the United Kingdom.

APPENDIX TABLE A.10

Standardized unemployment rates [a] in western Europe, North America and Japan, 1987-2001

(Per cent of civilian labour force)

	1987	1988	1989	1990	1991	1992	1993	1994	1995	1996	1997	1998	1999	2000	2001
France	10.1	9.6	9.0	8.6	9.1	10.0	11.3	11.8	11.3	11.9	11.8	11.4	10.7	9.3	8.6
Germany [b]	6.3	6.2	5.6	4.8	4.2	6.6	7.9	8.4	8.2	8.9	9.9	9.3	8.6	7.9	7.9
Italy	9.6	9.7	9.7	8.9	8.5	8.7	10.1	11.0	11.5	11.5	11.6	11.7	11.2	10.4	9.5
Austria	3.8	3.6	3.1	3.2	3.5	3.6	3.9	3.8	3.9	4.4	4.4	4.5	3.9	3.7	3.6
Belgium	9.8	8.8	7.4	6.6	6.4	7.1	8.6	9.8	9.7	9.5	9.2	9.3	8.6	6.9	6.6
Finland	4.9	4.2	3.1	3.2	6.6	11.7	16.4	16.6	15.4	14.6	12.7	11.4	10.2	9.8	9.1
Greece	6.7	6.8	6.7	6.4	7.0	7.9	8.6	8.9	9.2	9.6	9.8	10.9	11.6	10.9	10.2
Ireland	16.6	16.2	14.7	13.4	14.7	15.4	15.6	14.3	12.3	11.7	9.9	7.5	5.6	4.2	3.8
Luxembourg	2.5	2.0	1.8	1.7	1.7	2.1	2.6	3.2	2.9	3.0	2.7	2.7	2.4	2.4	2.4
Netherlands	7.6	7.2	6.6	5.8	5.5	5.3	6.2	6.8	6.6	6.0	4.9	3.8	3.2	2.8	2.4
Portugal	7.2	5.8	5.2	4.8	4.2	4.3	5.6	6.9	7.3	7.3	6.8	5.1	4.5	4.1	4.1
Spain	20.4	19.4	17.1	16.1	16.2	18.3	22.5	23.9	22.7	22.0	20.6	18.6	15.7	14.0	13.0
Euro area [c]	10.0	9.6	8.9	8.2	8.2	9.1	10.6	11.3	11.1	11.3	11.3	10.7	9.8	8.8	8.3
United Kingdom	10.3	8.5	7.1	6.9	8.6	9.8	10.2	9.4	8.5	8.0	6.9	6.2	5.9	5.4	5.1
Denmark	5.0	5.7	6.8	7.2	7.9	8.6	9.5	7.7	6.7	6.3	5.2	4.9	4.8	4.4	4.3
Sweden	2.2	1.8	1.6	1.7	3.1	5.6	9.1	9.4	8.8	9.6	9.9	8.3	7.2	5.9	5.1
European Union [d]	9.7	9.1	8.3	7.7	8.2	9.2	10.5	10.9	10.5	10.6	10.4	9.8	9.0	8.1	7.6
Cyprus [e]	3.4	2.8	2.3	1.8	3.0	1.8	2.6	2.7	2.6	3.1	3.4	3.4	3.6	3.5	3.5
Iceland	0.5	0.6	1.6	1.8	2.6	4.3	5.3	5.4	4.9	3.8	3.9	2.7	2.1	2.3	1.4
Israel [f]	6.1	6.4	8.9	9.6	10.6	11.2	10.0	7.8	6.9	6.7	7.7	8.5	8.9	8.8	9.2
Malta [g]	4.4	4.0	3.7	3.8	3.6	4.0	4.5	4.1	3.8	4.4	5.0	5.1	5.3	4.5	4.9
Norway	2.1	3.2	5.0	5.3	5.6	6.0	6.1	5.5	5.0	4.9	4.1	3.3	3.3	3.5	3.6
Switzerland	0.7	0.6	0.5	0.5	2.0	3.1	4.0	3.8	3.5	3.9	4.2	3.5	3.0	2.6	2.6
Turkey [f]	8.3	8.4	8.7	8.2	7.9	7.9	7.6	8.1	6.9	6.1	6.4	6.8	7.6	6.6	7.9
Western Europe	9.3	8.8	8.1	7.6	8.0	8.9	10.0	10.4	9.9	9.9	9.7	9.2	8.6	7.7	7.5
Canada	8.8	7.8	7.6	8.1	10.3	11.2	11.4	10.4	9.4	9.6	9.1	8.3	7.6	6.8	7.2
United States	6.2	5.5	5.3	5.5	6.7	7.4	6.8	6.1	5.6	5.4	4.9	4.5	4.2	4.0	4.8
North America	6.5	5.7	5.5	5.8	7.1	7.8	7.3	6.5	6.0	5.8	5.3	4.9	4.5	4.3	5.0
Japan	2.8	2.3	2.3	2.1	2.1	2.2	2.5	2.9	3.1	3.4	3.4	4.1	4.7	4.7	5.0
Total above	7.2	6.6	6.2	6.0	6.7	7.4	7.8	7.8	7.4	7.4	7.1	6.8	6.5	6.0	6.2
Memorandum items:															
4 major west European economies [h]	9.0	8.4	7.7	7.1	7.2	8.5	9.6	9.9	9.6	9.8	9.9	9.5	8.9	8.1	7.7
Western Europe and North America	8.1	7.5	7.0	6.8	7.6	8.4	8.8	8.7	8.2	8.1	7.8	7.3	6.9	6.2	6.4

Source: OECD, *Main Economic Indicators* and *Quarterly Labour Force Statistics* (Paris), latest issues; Eurostat, New Cronos Database; national statistics.

Note: All aggregates exclude Israel. Comparisons with previous years are limited by changes in methodology in Austria (1993), Iceland (1991), Israel (1995), Norway (1989), Switzerland (1991) and the United States (1994).

[a] Eurostat-OECD definition except for Austria (1987-1992), Cyprus, Iceland (1987-1990), Israel, Malta, Switzerland (1987-1990) and Turkey.

[b] West Germany, 1987-1991.

[c] Twelve countries above.

[d] Fifteen countries above.

[e] Registered unemployment rate, average of monthly data.

[f] Definitions comply with ILO guidelines but do not follow the Eurostat-OECD standards.

[g] Registered unemployment rate at the end of the year.

[h] France, Germany, Italy and the United Kingdom.

APPENDIX TABLE A.11

Consumer prices in western Europe, North America and Japan, 1987-2001

(Percentage change over previous year)

	1987	1988	1989	1990	1991	1992	1993	1994	1995	1996	1997	1998	1999	2000	2001
France	3.3	2.7	3.5	3.5	3.2	2.4	2.1	1.7	1.8	2.0	1.2	0.8	0.5	1.7	1.6
Germany[a]	0.3	1.3	2.8	2.7	4.0	5.1	4.4	2.8	1.7	1.4	1.9	0.9	0.6	1.9	2.5
Italy	4.8	5.1	6.3	6.5	6.3	5.3	4.6	4.1	5.2	4.0	2.0	2.0	1.7	2.5	2.8
Austria	1.5	1.9	2.6	3.3	3.3	4.0	3.6	3.0	2.2	1.5	1.3	0.9	0.6	2.3	2.6
Belgium	1.6	1.2	3.1	3.4	3.2	2.4	2.8	2.4	1.5	2.1	1.6	0.9	1.1	2.5	2.5
Finland	4.1	5.1	6.6	6.1	4.3	2.9	2.2	1.1	0.8	0.6	1.2	1.4	1.2	3.4	2.6
Greece	16.4	13.5	13.7	20.4	19.5	15.9	14.4	10.9	8.9	8.2	5.5	4.8	2.6	3.2	3.4
Ireland	3.1	2.1	4.1	3.3	3.2	3.1	1.4	2.4	2.5	1.7	1.4	2.4	1.6	5.6	4.9
Luxembourg	-0.1	1.4	3.4	3.3	3.1	3.2	3.6	2.2	1.9	1.3	1.4	1.0	1.0	3.2	2.7
Netherlands	-0.7	0.7	1.1	2.4	3.2	3.2	2.6	2.8	1.9	2.0	2.2	2.0	2.2	2.5	4.5
Portugal	9.4	9.7	12.6	13.4	10.5	9.5	6.7	5.4	4.2	3.1	2.3	2.8	2.3	2.9	4.3
Spain	5.2	4.8	6.8	6.7	5.9	5.9	4.6	4.7	4.7	3.6	2.0	1.8	2.3	3.4	3.6
Euro area[b]	2.6	3.0	4.3	4.6	4.8	4.7	4.0	3.2	3.0	2.6	1.9	1.4	1.2	2.4	2.7
United Kingdom	4.2	4.9	7.8	9.5	5.9	3.7	1.6	2.5	3.4	2.5	3.1	3.4	1.6	2.9	1.8
Denmark	4.0	4.5	4.8	2.6	2.4	2.1	1.3	2.0	2.1	2.1	2.2	1.8	2.5	2.9	2.4
Sweden	4.2	6.1	6.6	10.4	9.7	2.6	4.7	2.4	2.9	0.8	0.9	0.4	0.3	1.3	2.6
European Union[c]	2.9	3.4	4.9	5.5	5.0	4.4	3.6	3.1	3.1	2.5	2.1	1.8	1.3	2.5	2.6
Cyprus	2.8	3.4	3.8	4.5	5.0	6.5	4.9	4.7	2.6	2.9	3.6	2.2	1.7	4.3	2.0
Iceland	18.3	25.7	20.8	15.5	6.8	4.0	4.1	1.6	1.6	2.3	1.8	1.7	3.2	5.1	6.4
Israel	19.9	16.3	20.2	17.2	19.0	12.0	11.0	12.3	10.1	11.3	9.0	5.4	5.2	1.1	1.1
Malta	0.5	0.9	0.9	3.0	2.5	1.6	4.1	4.1	4.0	2.5	3.2	2.2	2.1	2.3	2.9
Norway	8.7	6.7	4.5	4.1	3.4	2.3	2.3	1.4	2.4	1.3	2.6	2.3	2.3	3.1	3.0
Switzerland	1.4	1.9	3.1	5.4	5.9	4.0	3.3	0.9	1.8	0.8	0.5	0.0	0.8	1.6	1.0
Turkey	37.8	69.4	63.3	60.3	65.8	70.1	66.2	105.2	89.1	80.4	85.7	84.6	64.9	54.9	54.4
Western Europe	3.0	3.4	4.9	5.5	5.0	4.4	3.6	3.0	3.0	2.5	2.1	1.7	1.3	2.4	2.5
Canada	4.3	4.1	5.0	4.8	5.6	1.5	1.9	0.2	2.2	1.6	1.6	1.0	1.7	2.7	2.5
United States	3.6	4.1	4.8	5.4	4.2	3.0	3.0	2.6	2.8	3.0	2.3	1.6	2.2	3.4	2.8
North America	3.7	4.1	4.8	5.4	4.3	2.9	2.9	2.4	2.8	2.9	2.2	1.5	2.2	3.3	2.8
Japan	–	0.7	2.4	3.1	3.2	1.7	1.3	0.7	-0.1	0.1	1.8	0.6	-0.3	-0.7	-0.5
Total above	2.8	3.3	4.4	5.0	4.4	3.3	2.9	2.4	2.5	2.3	2.1	1.5	1.5	2.4	2.3
Memorandum items:															
4 major west European economies[d]	2.7	3.1	4.7	5.2	4.7	4.2	3.3	2.7	2.9	2.4	2.1	1.7	1.0	2.3	2.2
Western Europe and North America	3.4	3.8	4.8	5.4	4.6	3.6	3.2	2.7	2.9	2.7	2.2	1.6	1.8	2.9	2.7

Source: National statistics.

Note: All aggregates exclude Israel and Turkey. Consumer price indexes for regional aggregates have been calculated as weighted averages of constituent country indices. Weights were derived from 1995 private final consumption expenditure converted from national currency units into a common currency using 1995 purchasing power parities.

[a] West Germany, 1987-1991.

[b] Twelve countries above.

[c] Fifteen countries above.

[d] France, Germany, Italy and the United Kingdom.

APPENDIX TABLE B.1

Real GDP/NMP in eastern Europe, the Baltic states and the CIS, 1980, 1988-2001

(Indices, 1989=100)

	1980	1988	1989	1990	1991	1992	1993	1994	1995	1996	1997	1998	1999	2000	2001
Eastern Europe	88.7	100.8	100.0	93.2	82.9	79.3	79.0	82.1	86.9	90.2	92.1	94.0	95.6	99.1	102.1
Albania	79.4	91.0	100.0	90.0	64.8	60.1	65.9	71.4	80.9	88.2	82.0	88.6	95.0	102.4	109.5
Bosnia and Herzegovina	..	..	..	..	..	..	..	..	..	..	..	..	..	..	..
Bulgaria	76.2	101.9	100.0	90.9	83.3	77.2	76.1	77.5	79.7	71.6	66.6	68.9	70.6	74.7	78.3
Croatia	99.0	101.6	100.0	92.9	73.3	64.7	59.5	63.0	67.3	71.3	76.2	78.1	77.8	80.7	84.2
Czech Republic	..	95.7	100.0	98.8	87.3	86.9	86.9	88.9	94.1	98.2	97.4	96.3	95.9	98.7	102.2
Hungary	86.3	99.3	100.0	96.5	85.0	82.4	81.9	84.4	85.6	86.8	90.7	95.1	99.1	104.3	108.3
Poland	91.1	99.8	100.0	88.4	82.2	84.4	87.6	92.1	98.6	104.5	111.7	117.1	121.8	126.7	128.1
Romania	88.5	106.2	100.0	94.4	82.2	75.0	76.2	79.2	84.8	88.2	82.8	78.8	77.9	79.3	83.5
Slovakia	..	99.0	100.0	97.5	83.3	77.9	75.1	78.7	84.0	89.3	94.8	98.7	100.6	102.8	106.2
Slovenia	98.9	100.5	100.0	91.9	83.7	79.1	81.4	85.7	89.3	92.4	96.6	100.3	105.5	110.4	113.7
The former Yugoslav Republic of Macedonia	93.3	98.1	100.0	89.8	84.3	78.7	72.8	71.6	70.8	71.6	72.6	75.1	78.4	82.0	78.2
Yugoslavia [a]	95.7	98.8	100.0	92.1	81.4	58.7	40.6	41.7	44.2	46.8	50.3	51.5	42.4	45.1	47.9
Baltic states	67.8	96.0	100.0	97.8	89.9	67.9	58.2	55.2	56.5	58.8	63.7	66.7	65.6	69.2	73.4
Estonia	74.5	93.8	100.0	91.9	82.7	71.0	65.0	63.7	66.6	69.2	76.5	80.3	79.8	85.3	89.8
Latvia	68.5	93.6	100.0	102.9	92.2	60.1	51.1	51.5	51.0	52.7	57.3	59.5	60.1	64.3	69.1
Lithuania	64.7	98.4	100.0	96.7	91.2	71.8	60.2	54.3	56.1	58.7	63.0	66.2	63.6	66.1	69.9
CIS [b]	77.5	98.1	100.0	96.8	90.9	78.0	70.4	60.3	56.9	55.0	55.6	54.0	56.4	61.1	64.9
Armenia	73.5	92.2	100.0	94.5	83.4	48.6	44.3	46.7	49.9	52.8	54.6	58.6	60.5	64.1	70.3
Azerbaijan	79.6	109.7	100.0	88.3	87.7	67.9	52.2	41.9	37.0	37.4	39.6	43.6	46.8	52.0	57.1
Belarus	65.7	92.4	100.0	98.1	96.9	87.6	81.0	70.8	63.4	65.2	72.6	78.7	81.4	86.1	89.6
Georgia	79.4	103.6	100.0	84.9	67.0	36.9	26.1	23.4	24.0	26.7	29.5	30.3	31.2	31.9	33.3
Kazakhstan	87.0	100.1	100.0	99.0	88.2	83.5	75.8	66.2	60.8	61.1	62.1	60.9	62.6	68.7	77.8
Kyrgyzstan	69.1	95.6	100.0	104.8	96.5	83.2	70.3	56.2	53.1	56.9	62.5	63.9	66.2	69.8	73.5
Republic of Moldova	72.1	91.9	100.0	97.6	80.5	57.2	56.5	39.0	38.5	36.2	36.8	34.4	33.2	34.0	36.0
Russian Federation	78.1	98.4	100.0	97.0	92.2	78.8	71.9	62.8	60.2	58.2	58.7	55.8	58.8	64.1	67.3
Tajikistan	80.8	06.9	100.0	100.2	91.7	62.1	52.0	40.9	35.8	29.8	30.3	32.0	33.1	35.9	39.5
Turkmenistan	80.7	07.5	100.0	101.8	97.0	82.5	83.7	69.2	64.2	68.5	60.7	65.0	76.0	89.4	107.7
Ukraine	75.0	95.2	100.0	96.4	88.0	79.3	68.0	52.4	46.0	41.4	40.2	39.4	39.3	41.6	45.4
Uzbekistan	76.0	97.0	100.0	99.2	98.7	87.7	85.7	81.2	80.5	81.9	86.1	89.9	93.9	97.6	102.0
Total above	80.3	98.7	100.0	95.9	88.8	78.1	72.4	65.9	64.8	64.3	65.3	64.7	66.9	71.2	74.8
Memorandum items:															
CETE-5	88.6	99.6	100.0	93.3	84.1	83.8	85.0	88.5	93.5	97.9	102.4	105.8	109.0	113.2	115.9
SETE-7	88.8	102.9	100.0	93.1	80.8	71.4	68.5	70.9	75.3	76.7	74.1	73.2	71.9	74.4	78.1
Former GDR	..	..	100.0	84.5	68.3	73.3	80.1	87.9	91.9	94.9	96.7	97.7	99.0	101.1	100.4

Source: UNECE Common Database, derived from national and Interstate Statistical Committee of the CIS statistics; *DIW Wochenbericht,* No. 43 (Vienna), 2001 for the former GDR since 1998.

Note: Data for the east European countries are based on a GDP measure, except where otherwise mentioned. For the countries of the former Soviet Union, NMP data for 1980-1990 were chain-linked to GDP data from 1990. Country indices were aggregated with previous year PPP-based weights obtained from the European Comparison Programme for 1996.

[a] Gross material product (1980-1989 for Croatia). Yugoslavia: since 1999, without Kosovo and Metohia.

[b] Net material product for 1980-1990 (until 1992 in the case of Turkmenistan).

APPENDIX TABLE B.2

Real total consumption expenditure in eastern Europe, the Baltic states and the CIS, 1980, 1988-2001

(Indices, 1989=100 or earliest year available thereafter)

	1980	1988	1989	1990	1991	1992	1993	1994	1995	1996	1997	1998	1999	2000	2001
Bulgaria	..	..	100.0	100.6	92.3	89.4	86.2	82.3	80.7	75.3	64.0	68.8	72.0	75.3	..
Croatia	..	..	..	..	100.0	87.2	85.3	92.0	106.6	106.5	117.1	118.2	116.2	119.3	..
Czech Republic	..	93.1	100.0	104.9	85.5	88.4	90.2	94.5	97.2	103.7	104.3	102.1	103.5	104.6	107.3
Hungary	92.2	102.0	100.0	97.3	92.2	92.8	97.9	95.6	89.3	86.1	88.1	91.7	95.6	99.4	..
Poland	108.0	114.7	100.0	88.3	94.9	98.2	103.0	107.0	110.5	118.4	125.6	130.8	136.6	139.9	142.4
Romania	83.9	90.6	100.0	108.9	96.0	90.7	91.8	95.3	105.5	112.9	108.1	109.3	106.5	107.7	113.8
Slovakia	..	92.1	100.0	103.3	76.9	75.6	74.2	72.2	74.4	83.0	87.2	91.8	89.8	87.4	91.2
Slovenia	..	..	..	100.0	91.6	88.8	99.1	102.6	110.2	112.7	116.3	120.9	127.7	129.5	..
The former Yugoslav Republic of Macedonia	..	..	..	100.0	93.9	84.2	89.7	95.9	94.3	96.5	98.5	101.8	105.6	..	..
Estonia	..	..	..	..	..	..	100.0	102.1	110.9	116.7	124.7	131.1	131.9	139.7	..
Latvia	..	..	..	100.0	76.7	49.2	46.5	47.4	47.0	50.8	52.7	56.0	58.2	60.4	..
Lithuania	..	..	..	..	..	..	..	..	100.0	108.2	116.4	125.4	122.4	126.8	..
Armenia	..	..	..	100.0	97.4	84.9	66.4	68.9	74.5	76.8	81.7	85.4	85.7	91.1	..
Azerbaijan	..	..	..	..	..	100.0	77.4	62.2	60.4	65.3	72.2	80.4	88.1	96.9	..
Belarus	..	..	..	100.0	93.5	84.1	78.8	70.0	63.4	65.7	72.3	80.9	87.7	94.2	..
Georgia	..	..	..	100.0	79.2	77.1	45.4	42.4	46.1	..	..	..	..	..	..
Kazakhstan	..	..	..	100.0	96.8	96.2	84.9	67.7	55.0	51.3	51.8	50.4	51.1	54.7	..
Kyrgyzstan	..	..	..	100.0	83.5	72.8	64.3	51.8	43.4	46.2	42.4	48.8	49.3	47.9	..
Republic of Moldova	..	..	..	..	..	..	100.0	82.6	90.3	99.7	111.5	109.3	92.0	107.8	117.5
Russian Federation	..	..	..	100.0	93.9	89.0	88.1	85.4	83.1	80.5	82.9	81.7	79.7	85.6	90.9
Ukraine	..	..	..	100.0	94.7	88.7	72.1	65.1	62.7	57.5	56.4	56.4	54.3	55.4	60.3

Source: UNECE Common Database, derived from national and Interstate Statistical Committee of the CIS statistics.

APPENDIX TABLE B.3

Real gross fixed capital formation in eastern Europe, the Baltic states and the CIS, 1980, 1988-2001

(Indices, 1989=100 or earliest year available thereafter)

	1980	1988	1989	1990	1991	1992	1993	1994	1995	1996	1997	1998	1999	2000	2001
Bulgaria	..	..	..	100.0	80.0	74.1	61.2	61.9	71.8	56.6	43.0	57.2	71.7	77.6	..
Croatia	..	..	..	..	100.0	88.5	94.5	93.6	108.2	148.8	183.5	189.1	187.0	180.5	..
Czech Republic	..	99.4	100.0	97.9	71.1	82.8	83.0	90.5	108.5	117.3	113.9	114.0	113.3	118.1	126.4
Hungary	114.7	100.6	100.0	92.9	83.1	81.0	82.6	92.9	88.9	94.8	103.6	117.3	124.2	133.8	..
Poland	124.6	126.5	100.0	75.2	71.9	73.6	75.7	82.6	96.2	115.2	140.1	160.0	170.9	175.5	157.5
Romania	163.7	157.6	100.0	64.4	44.0	48.9	52.9	63.9	68.3	72.2	73.4	69.2	65.9	68.9	73.5
Slovakia	..	..	..	100.0	74.8	71.5	67.7	64.3	67.7	89.4	100.1	111.2	90.3	89.7	100.1
Slovenia	..	..	..	100.0	88.5	77.1	85.4	97.4	113.8	123.9	138.2	153.9	183.2	183.6	..
The former Yugoslav Republic of Macedonia	..	..	..	100.0	95.8	79.9	73.6	67.3	74.1	79.0	75.6	73.6	72.5	..	..
Estonia	..	..	..	..	..	..	100.0	106.2	110.5	123.1	144.6	161.0	137.5	140.2	..
Latvia	..	..	..	100.0	36.1	25.7	21.6	21.8	23.7	29.0	35.0	50.4	48.4	58.0	..
Lithuania	..	..	..	..	..	..	..	..	100.0	117.6	143.6	157.8	147.9	142.2	..
Armenia	..	..	..	100.0	67.0	8.6	7.9	11.5	9.5	10.5	10.7	12.0	12.1	13.5	..
Azerbaijan	..	..	..	..	..	100.0	61.0	115.3	94.5	199.8	333.7	483.9	474.2	486.5	..
Belarus	..	..	..	100.0	104.2	84.8	78.3	67.6	47.6	46.1	56.2	61.8	59.3	60.7	..
Georgia	..	..	..	100.0	67.3	49.2	18.5	133.4	219.9	..	..	..	..	..	..
Kazakhstan	..	..	..	100.0	74.2	61.9	44.2	39.2	24.3	18.5	19.1	17.8	17.9	20.4	..
Kyrgyzstan	..	..	..	100.0	89.4	63.2	49.4	35.1	56.4	49.1	34.6	34.0	43.6	55.3	..
Republic of Moldova	..	..	..	..	..	..	100.0	56.5	50.8	63.8	60.4	66.0	50.8	46.4	44.9
Russian Federation	..	..	..	100.0	84.5	49.4	36.7	27.1	25.1	20.8	19.1	17.3	18.2	20.8	22.5
Ukraine	..	..	..	100.0	79.1	67.4	46.8	27.6	19.1	14.8	15.1	15.5	15.5	17.4	18.8

Source: UNECE Common Database, derived from national and Interstate Statistical Committee of the CIS statistics.

APPENDIX TABLE B.4

Real gross industrial output in eastern Europe, the Baltic states and the CIS, 1980, 1988-2001

(Indices, 1989=100)

	1980	1988	1989	1990	1991	1992	1993	1994	1995	1996	1997	1998	1999	2000	2001
Eastern Europe	82.8	100.6	100.0	85.9	70.2	63.0	61.5	65.5	70.3	73.9	77.5	78.6	78.8	85.4	88.2
Albania	77.0	95.2	100.0	86.7	50.4	35.2	31.7	25.8	23.9	18.1	18.6	22.7	26.3	29.4	23.6
Bosnia and Herzegovina	106.0	98.1	100.0	101.8	76.9	25.5	2.0	1.7	2.8	5.2	7.0	8.7	9.6	10.5	11.8
Bulgaria	71.3	101.1	100.0	83.2	66.4	54.2	48.8	54.0	56.4	59.3	53.4	49.1	44.6	47.1	47.5
Croatia	88.7	100.6	100.0	88.7	63.4	54.2	51.0	49.6	49.7	51.3	54.8	56.8	56.0	57.0	60.4
Czech Republic	81.5	98.5	100.0	96.6	75.7	69.8	66.1	67.4	73.3	74.8	78.1	79.4	76.9	81.1	86.6
Hungary	92.9	105.3	100.0	90.7	74.0	66.8	69.5	76.2	79.7	82.4	91.5	103.0	113.7	135.0	140.5
Poland	86.3	100.5	100.0	75.8	69.7	71.7	76.3	85.5	93.8	101.6	113.3	117.3	121.5	129.6	129.6
Romania	76.9	101.9	100.0	81.9	63.3	49.4	50.1	51.7	56.6	60.1	55.8	48.1	47.1	50.9	55.1
Slovakia	76.7	100.8	100.0	94.0	75.9	68.6	66.1	69.3	75.1	76.9	77.9	80.9	78.5	85.8	90.6
Slovenia	90.3	98.9	100.0	89.5	78.4	68.1	66.1	70.4	71.8	72.5	73.2	75.9	75.6	80.2	82.6
The former Yugoslav Republic of Macedonia	72.1	95.6	100.0	89.4	74.0	62.3	53.7	48.0	42.9	44.3	45.0	47.0	45.8	47.4	42.6
Yugoslavia	80.0	98.4	100.0	88.0	72.5	57.0	35.7	36.2	37.6	40.4	44.2	45.8	35.2	39.2	39.2
Baltic states	72.1	96.6	100.0	98.6	95.1	64.7	44.4	36.3	37.1	38.9	42.1	44.6	41.0	43.9	49.4
Estonia	78.5	99.3	100.0	100.0	92.8	59.8	48.6	47.1	48.0	49.4	56.6	59.0	57.0	64.4	69.2
Latvia	72.5	97.1	100.0	100.8	100.2	65.6	44.6	40.1	38.7	40.8	46.4	47.9	45.3	47.4	51.4
Lithuania	70.0	95.6	100.0	97.4	94.0	65.8	43.2	31.7	33.4	35.0	36.2	39.2	34.8	36.6	42.8
CIS	73.4	98.2	100.0	99.9	93.1	78.2	68.4	53.6	50.7	49.0	50.3	48.7	53.2	59.4	63.3
Armenia	76.3	109.1	100.0	92.5	85.4	44.2	39.5	41.6	42.2	42.8	43.2	42.3	44.6	47.4	49.2
Azerbaijan	76.1	99.3	100.0	93.7	85.4	59.4	47.7	35.9	28.2	26.3	26.4	27.0	28.0	29.9	31.4
Belarus	61.1	95.6	100.0	102.1	101.1	91.8	83.2	71.0	62.7	64.9	77.1	86.7	95.6	103.0	108.6
Georgia	70.6	99.3	100.0	94.3	73.0	39.6	25.0	15.2	13.2	14.1	15.2	15.0	16.1	17.1	16.9
Kazakhstan	72.4	97.6	100.0	99.2	98.3	84.7	72.2	51.9	47.7	47.8	49.7	48.5	49.8	57.5	65.3
Kyrgyzstan	66.7	95.1	100.0	99.4	99.1	73.5	56.3	35.5	27.8	27.8	38.8	40.9	39.1	41.4	43.7
Republic of Moldova	68.7	94.6	100.0	103.2	91.7	66.9	67.1	48.5	46.6	43.6	43.6	37.0	32.7	35.3	40.3
Russian Federation	74.4	98.6	100.0	99.9	91.9	75.4	64.7	51.2	49.5	47.5	48.5	46.0	51.0	57.1	59.9
Tajikistan	72.9	98.2	100.0	101.2	97.6	73.9	68.1	50.8	43.9	33.4	32.7	35.4	37.4	41.2	47.4
Turkmenistan	75.4	96.9	100.0	103.2	108.2	92.0	95.7	72.1	67.5	81.0	63.1	64.4	74.1	96.3	106.9
Ukraine	72.6	97.3	100.0	99.9	95.1	89.0	81.9	59.5	52.4	49.7	49.6	49.1	51.0	57.4	65.5
Uzbekistan	68.5	96.5	100.0	101.8	103.3	96.4	99.9	101.5	101.6	104.2	108.5	112.4	119.2	126.9	134.2
Total above	76.5	99.0	100.0	95.2	85.4	72.7	65.5	57.2	57.0	57.1	59.2	58.7	61.5	67.8	71.4
Memorandum items:															
CETE-5	85.4	100.7	100.0	87.0	73.3	69.8	70.6	76.3	82.4	86.5	93.7	97.8	100.2	108.9	112.0
SETE-7	78.6	100.6	100.0	84.1	65.4	52.0	47.0	48.3	51.1	53.9	51.8	48.0	44.9	48.1	50.2
Former GDR	75.2	97.7	100.0	72.7	37.0	34.7	34.9	38.1	39.9	41.7	44.6	48.2	51.7	57.2	59.3

Source: UNECE Common Database, derived from national and Interstate Statistical Committee of the CIS statistics.

Note: For the countries of the former Soviet Union, Soviet data for 1980-1990 were chain-linked to national or CIS data from 1990. Country indices were aggregated with previous year PPP-based weights on the basis of data obtained from the European Comparison Programme for 1996.

APPENDIX TABLE B.5

Total employment in eastern Europe, the Baltic states and the CIS, 1980, 1988-2001

(Indices, 1989=100)

	1980	1988	1989	1990	1991	1992	1993	1994	1995	1996	1997	1998	1999	2000	2001
Eastern Europe	96.7	99.9	100.0	97.1	90.7	85.4	82.7	82.6	82.3	82.9	83.0	83.3	81.2	80.3	..
Albania	77.9	97.6	100.0	99.2	97.5	76.0	72.7	80.7	80.4	77.5	76.9	75.4	74.0	74.2	..
Bosnia and Herzegovina	..	..	100.0	97.1	58.1	22.1	9.9	9.1	10.1	22.5	34.4	36.4	37.6	37.9	..
Bulgaria	100.0	102.4	100.0	93.9	81.6	75.0	73.8	74.3	75.2	75.3	72.3	72.2	70.7	67.4	..
Croatia	87.4	100.4	100.0	97.1	89.2	79.3	76.6	74.8	73.9	74.5	73.9	78.8	78.5	77.6	..
Czech Republic	95.3	99.4	100.0	99.1	93.6	91.2	89.7	90.4	92.8	93.4	91.6	90.4	88.1	86.3	..
Hungary [a]	104.2	100.7	100.0	96.7	86.7	78.1	73.2	71.8	70.4	69.8	69.8	70.7	72.9	73.6	73.8
Poland	102.0	100.1	100.0	95.8	90.1	86.3	84.3	85.1	86.7	88.3	90.8	92.9	90.4	88.3	..
Romania [b]	94.6	98.7	100.0	99.0	98.5	95.5	91.9	91.5	86.7	85.7	82.4	80.5	76.9	78.8	..
Slovakia [c]	90.8	99.8	100.0	98.2	85.9	86.8	87.7	84.0	85.7	88.6	88.1	87.8	85.1	83.9	..
Slovenia	84.0	101.3	100.0	96.1	88.7	83.8	81.3	79.3	79.1	78.7	78.6	78.7	80.2	81.2	..
The former Yugoslav Republic of Macedonia	81.2	99.7	100.0	98.2	90.7	86.4	81.5	76.6	69.0	65.8	61.8	60.1	61.1	60.3	..
Yugoslavia	83.4	99.8	100.0	97.0	94.1	90.9	88.3	86.5	85.3	84.8	89.9	89.7	82.4	80.2	80.3
Baltic states	..	99.6	100.0	98.5	98.9	94.4	89.0	83.2	80.6	80.1	80.9	80.5	79.5	77.9	..
Estonia	..	97.6	100.0	98.6	96.4	91.4	84.5	82.7	78.3	77.0	77.4	76.4	73.3	72.6	73.4
Latvia	97.0	100.5	100.0	100.1	99.3	92.0	85.6	77.0	74.3	72.3	73.7	74.1	73.8	73.8	..
Lithuania	93.4	99.8	100.0	97.3	99.7	97.5	93.4	88.0	86.4	87.2	87.7	87.0	86.6	83.3	..
CIS	93.8	98.8	100.0	100.2	98.9	96.6	94.2	91.4	90.5	89.7	88.6	87.7	87.2	..	..
Armenia	86.6	101.4	100.0	102.4	105.0	99.2	97.0	93.5	92.8	90.2	86.2	84.0	81.6	80.3	80.4
Azerbaijan	62.7	..	100.0	100.9	101.7	101.4	101.2	98.9	98.4	100.5	100.7	100.9	100.9	100.9	101.2
Belarus	95.4	99.5	100.0	99.1	96.6	94.1	92.9	90.4	84.8	84.0	84.1	85.0	85.5	85.4	85.3
Georgia	92.7	101.1	100.0	102.3	93.3	73.5	66.4	64.8	79.0	75.4	82.7	84.6	77.0	..	..
Kazakhstan	86.2	96.0	100.0	101.3	100.1	98.3	89.9	85.4	85.0	84.6	84.0	79.5	79.2	80.5	82.6
Kyrgyzstan	81.9	98.7	100.0	100.5	99.6	105.6	96.6	94.6	94.4	95.0	97.1	98.0	101.5	101.7	102.0
Republic of Moldova [d]	97.3	98.9	100.0	99.1	99.0	98.0	80.7	80.4	80.0	79.4	78.7	78.5	71.5	72.5	71.7
Russian Federation	96.9	99.9	100.0	99.6	97.7	95.3	93.7	90.6	87.8	87.2	85.6	84.4	84.6	85.1	86.0
Tajikistan	76.7	96.9	100.0	103.2	104.9	101.6	98.7	98.7	98.6	92.1	95.3	95.6	92.5	92.9	93.7
Turkmenistan	79.8	101.2	100.0	103.4	107.0	110.5	114.0	118.5	122.5	124.7	127.2	128.8	133.6	..	..
Ukraine	99.7	100.0	100.0	100.0	98.3	96.4	94.2	90.6	93.3	91.4	88.9	87.9	85.9	83.7	82.6
Uzbekistan	75.4	95.9	100.0	104.2	109.2	108.7	108.5	109.9	110.8	112.3	113.8	115.4	116.5	117.8	..
Total above	94.6	99.1	100.0	99.3	96.6	93.4	90.9	88.8	88.0	87.6	86.9	86.3	85.4	..	..
Memorandum items:															
CETE-5	99.7	100.1	100.0	96.7	89.8	85.8	83.6	83.5	84.7	85.8	86.8	87.9	86.3	84.9	..
SETE-7	92.3	99.7	100.0	97.6	91.9	84.9	81.5	81.4	79.0	78.8	77.7	77.0	74.1	74.1	..

Source: UNECE Common Database, derived from national and Interstate Statistical Committee of the CIS statistics.

[a] End of year, up to 1992; since 1992, annual average.

[b] End of year.

[c] End of year, up to 1993; since 1993, annual average.

[d] Excluding Transdniestria since 1993.

APPENDIX TABLE B.6

Employment in industry in eastern Europe, the Baltic states and the CIS, 1989-2001
(Indices, 1989=100)

	1989	1990	1991	1992	1993	1994	1995	1996	1997	1998	1999	2000	2001
Eastern Europe [a]	100.0	95.6	86.7	76.7	71.9	69.5	68.6	68.3	66.2	65.0	60.8	..	..
Albania	..	..	..	..	..	..	..	..	..	..	..	..	..
Bosnia and Herzegovina	100.0	98.3	60.3	23.3	8.3	9.2	18.7	22.0	26.8	28.9	28.9	28.9	..
Bulgaria	100.0	91.0	74.7	64.8	59.5	57.3	56.0	55.4	53.0	50.8	46.1	41.2	..
Croatia	100.0	102.4	84.8	70.3	70.4	67.3	59.5	59.0	56.6	60.0	58.3	56.7	..
Czech Republic	100.0	95.8	92.2	85.1	80.9	76.6	77.0	76.4	76.1	74.9	72.5	70.6	..
Hungary	100.0	97.0	85.4	77.3	69.1	66.0	62.5	61.9	63.0	65.9	66.4	65.6	66.8
Poland	100.0	93.7	86.1	76.7	74.4	73.8	76.2	75.6	75.8	75.0	69.5	65.3	..
Romania [b]	100.0	96.5	91.6	79.5	73.0	69.4	65.4	66.0	59.0	55.8	49.5	48.3	..
Slovakia	100.0	95.7	88.2	78.7	74.0	71.7	71.6	71.6	70.2	67.3	65.4	63.3	..
Slovenia	100.0	95.1	85.2	84.6	78.1	75.1	72.2	71.5	68.5	67.8	66.7	66.3	..
The former Yugoslav Republic of Macedonia	100.0	95.3	87.3	81.6	77.5	72.9	63.1	59.0	54.4	52.5	55.4	52.9	..
Yugoslavia	100.0	100.9	92.1	87.2	85.0	82.7	80.6	78.8	76.3	73.8	66.3	..	..
Baltic states	100.0	96.8	95.5	87.0	74.2	63.0	60.8	58.0	57.7	55.8	54.0	54.2	..
Estonia	100.0	96.9	93.0	85.8	73.6	70.7	76.4	73.0	68.8	67.5	63.4	65.6	65.7
Latvia	100.0	97.0	92.1	81.4	69.0	56.3	53.1	50.1	51.9	47.6	45.7	46.7	..
Lithuania	100.0	96.7	99.0	91.4	78.0	64.5	59.5	57.0	57.1	56.5	55.7	54.6	..
CIS	100.0	98.2	96.7	91.8	88.3	79.7	73.2	69.2	63.9	62.6	..	..	..
Armenia	100.0	102.6	95.0	84.0	75.2	73.7	62.8	52.9	47.5	43.4	40.5	37.3	37.3
Azerbaijan	100.0	97.1	94.9	88.9	81.1	77.4	72.9	58.6	50.1	52.0	53.6	51.6	52.0
Belarus	100.0	98.6	96.9	92.1	88.5	84.4	75.2	74.4	74.5	75.5	76.2	75.9	75.7
Georgia	100.0	104.2	92.6	66.0	56.5	51.6	46.8	33.6	25.8	26.4	..	..	..
Kazakhstan	100.0	98.5	99.9	96.2	83.6	76.9	69.6	66.9	59.0	57.8	57.9	54.7	51.2
Kyrgyzstan	100.0	99.9	92.7	89.5	80.5	72.0	61.2	54.6	51.2	50.1	47.4	44.0	43.2
Republic of Moldova [c]	100.0	102.4	95.1	93.1	55.0	52.1	44.7	43.8	42.9	40.8	35.9	36.1	37.0
Russian Federation	100.0	97.7	96.1	91.3	89.8	80.7	74.0	70.4	64.5	63.1	62.6	64.0	64.9
Tajikistan	100.0	102.5	100.8	98.2	86.1	81.9	71.9	71.1	62.5	60.1	52.3	47.7	47.6
Turkmenistan	100.0	104.2	100.8	101.0	110.7	110.5	115.4	119.9	132.9	150.2	152.9	..	..
Ukraine	100.0	98.1	97.3	92.7	87.9	78.3	72.2	66.8	61.2	59.3	54.4	51.3	51.0
Uzbekistan	100.0	101.5	102.5	101.5	103.2	90.1	92.3	93.5	93.7	94.1	94.9	..	..
Total above [a]	100.0	97.4	93.5	86.9	82.8	76.1	71.5	68.7	64.5	63.3	..	..	..
Memorandum items:													
CETE-5	100.0	94.9	87.5	79.1	75.1	73.0	73.6	73.1	73.1	72.6	69.2	66.4	..
SETE-7 [a]	100.0	96.5	85.9	73.8	68.2	65.3	62.5	62.4	58.1	56.0	50.7	..	..

Source: UNECE Common Database, derived from national and Interstate Statistical Committee of the CIS statistics.

[a] Excluding Albania.

[b] End of year.

[c] Excluding Transdniestria since 1993.

APPENDIX TABLE B.7

Registered unemployment in eastern Europe, the Baltic states and the CIS, 1990-2001
(Per cent of labour force, end of period)

	1990	1991	1992	1993	1994	1995	1996	1997	1998	1999	2000	2001
Eastern Europe	..	9.6	12.4	14.0	13.6	12.5	11.7	11.9	12.6	14.6	15.2	..
Albania	9.5	9.2	27.0	22.0	18.0	12.9	12.3	14.9	17.6	18.2	16.9	15.0*
Bosnia and Herzegovina	..	..	..	..	..	..	..	39.0	38.7	39.0	39.4	40.0*
Bulgaria	1.8	11.1	15.3	16.4	12.8	11.1	12.5	13.7	12.2	16.0	17.9	17.3
Croatia	..	14.1	17.8	16.6	17.3	17.6	15.9	17.6	18.6	20.8	22.6	23.1
Czech Republic	0.7	4.1	2.6	3.5	3.2	2.9	3.5	5.2	7.5	9.4	8.8	8.9
Hungary	1.7	7.4	12.3	12.1	10.9	10.4	10.5	10.4	9.1	9.6	8.9	8.0
Poland	6.5	12.2	14.3	16.4	16.0	14.9	13.2	10.3	10.4	13.1	15.1	17.4
Romania	1.3	3.0	8.2	10.4	10.9	9.5	6.6	8.8	10.3	11.5	10.5	8.6
Slovakia	1.6	11.8	10.4	14.4	14.8	13.1	12.8	12.5	15.6	19.2	17.9	18.6
Slovenia	..	10.1	13.3	15.5	14.2	14.5	14.4	14.8	14.6	13.0	12.0	11.8
The former Yugoslav Republic of Macedonia	..	24.5	26.2	27.7	30.0	36.6	38.8	41.7	41.4	43.8	44.9	..
Yugoslavia [a]	..	21.0	24.6	24.0	23.9	24.7	26.1	25.6	27.2	27.4	26.6	27.9
Baltic states	..	..	2.1	4.5	5.3	6.6	6.4	6.3	7.3	9.1	10.0	10.1
Estonia [b]	..	..	1.6	5.0	5.1	5.0	5.6	4.6	5.1	6.7	7.3	7.2
Latvia	..	..	2.3	5.8	6.5	6.6	7.2	7.0	9.2	9.1	7.8	7.7
Lithuania	..	..	3.5	3.4	4.5	7.3	6.2	6.7	6.9	10.0	12.6	12.9
CIS	..	..	2.7	3.6	4.4	5.8	6.6	7.6	9.0	8.3	7.0	6.2
Armenia	..	..	3.5	6.3	6.0	8.1	9.7	11.0	8.9	11.5	10.9	9.8
Azerbaijan	..	..	0.2	0.7	0.9	1.1	1.1	1.3	1.4	1.2	1.2	1.3
Belarus	..	..	0.5	1.3	2.1	2.7	4.0	2.8	2.3	2.0	2.1	2.3
Georgia	..	..	0.3	2.0	3.8	3.4	3.2	8.0	4.2	5.6	..	..
Kazakhstan	..	..	0.4	0.6	1.0	2.1	4.1	3.9	3.7	3.9	3.7	2.8
Kyrgyzstan	..	..	0.1	0.2	0.8	3.0	4.5	3.1	3.1	3.0	3.1	3.1
Republic of Moldova	..	..	0.7	0.7	1.0	1.4	1.5	1.7	1.9	2.1	1.8	1.7
Russian Federation [c]	..	..	5.2	6.1	7.8	9.0	10.0	11.2	13.3	12.2	9.8	9.0
Tajikistan	..	..	0.4	1.1	1.8	1.8	2.4	2.8	2.9	3.1	3.0	2.6
Turkmenistan	..	..	..	..	..	..	..	..	..	..	..	..
Ukraine	..	..	0.3	0.4	0.3	0.6	1.5	2.8	4.3	4.3	4.2	3.7
Uzbekistan	..	..	0.1	0.2	0.3	0.3	0.3	0.3	0.4	0.5	0.6	0.4
Memorandum items:												
CETE-5	..	9.7	11.3	13.3	12.9	12.0	11.2	9.8	10.2	12.5	13.4	14.6
SETE-7	..	9.3	14.2	15.1	14.6	13.7	12.5	14.3	15.4	17.1	17.8	..
Russian Federation [d]	..	0.1	0.8	1.1	2.1	3.2	3.4	2.8	2.7	1.7	1.4	1.6
Former-GDR	..	..	13.5	15.4	13.5	14.9	15.9	19.4	17.4	17.7	17.2	17.6

Source: UNECE Common Database, derived from national and Interstate Statistical Committee of the CIS statistics.

Note: Aggregates for eastern European countries till 1997 exclude Bosnia and Herzegovina, that for CIS excludes Turkmenistan.

[a] Since 1999, excluding Kosovo and Metohia.

[b] Job seekers till October 2000, thereafter – registered unemployed as percentage of the labour force.

[c] Based on Russian Federation Goskomstat's monthly estimates according to the ILO definition, i.e. including all persons not having employment but actively seeking work.

[d] Registered unemployment.

APPENDIX TABLE B.8

Consumer prices in eastern Europe, the Baltic states and the CIS, 1990-2001

(Annual average, percentage change over preceding year)

	1990	1991	1992	1993	1994	1995	1996	1997	1998	1999	2000	2001
Albania	..	35.5	193.1	85.0	21.5	8.0	12.7	33.1	20.3	-0.1	–	3.1
Bosnia and Herzegovina	594.0	116.2	64 218.3	38 825.1	553.5	-12.1	-21.2	11.8	4.9	-0.6	1.7	1.8
Bulgaria	23.8	338.5	91.3	72.9	96.2	62.0	121.7	1 058.3	18.7	2.6	10.2	7.3
Croatia [a]	597.1	124.2	663.6	1 516.6	97.5	2.0	3.6	3.7	5.9	4.3	6.4	5.0
Czech Republic	9.9	56.7	11.1	20.8	10.0	9.1	8.9	8.4	10.6	2.1	3.9	4.7
Hungary	28.9	35.0	23.0	22.6	19.1	28.5	23.6	18.4	14.2	10.1	9.9	9.2
Poland	585.8	70.3	45.3	36.9	33.2	28.1	19.8	15.1	11.7	7.4	10.2	5.5
Romania	5.1	170.2	210.7	256.2	137.1	32.2	38.8	154.9	59.3	45.9	45.7	34.5
Slovakia	10.4	61.2	10.2	23.1	13.4	10.0	6.1	6.1	6.7	10.5	12.0	7.3
Slovenia	551.6	115.0	207.3	31.7	21.0	13.5	9.9	8.4	8.1	6.3	9.0	8.6
The former Yugoslav Republic of Macedonia [a]	596.6	110.8	1 511.0	352.0	126.6	16.4	2.5	0.9	-1.4	-1.3	6.6	5.2
Yugoslavia	580.0	122.0	8 926.0	2.2E+14	7.9E+10	71.8	90.5	23.2	30.4	44.1	77.5	90.4
Estonia	18.0	202.0	1 078.2	89.6	47.9	28.9	23.1	11.1	10.6	3.5	3.9	5.8
Latvia	10.9	172.2	951.2	109.1	35.7	25.0	17.7	8.5	4.7	2.4	2.8	2.4
Lithuania	9.1	216.4	1 020.5	410.1	72.0	39.5	24.7	8.8	5.1	0.8	1.0	1.5
Armenia	6.9	174.1	728.7	3 731.8	4 964.0	175.5	18.7	13.8	8.7	0.7	-0.8	3.2
Azerbaijan	6.1	106.6	912.6	1 129.7	1 663.9	411.5	19.8	3.6	-0.8	-8.6	1.8	1.5
Belarus	4.7	94.1	971.2	1 190.9	2 219.6	709.3	52.7	63.9	73.2	293.7	168.9	61.4
Georgia	4.2	78.7	1 176.9	4 084.9	22 286.1	261.4	39.4	7.1	3.5	19.3	4.2	4.6
Kazakhstan	5.6	114.5	1 504.3	1 662.7	1 880.1	176.3	39.2	17.5	7.3	8.4	13.4	8.5
Kyrgyzstan	5.5	113.9	854.6	1 208.7	278.1	42.9	31.3	23.4	10.3	35.7	18.7	7.0
Republic of Moldova	5.7	114.4	1 308.0	1 751.0	486.4	29.9	23.5	11.8	7.7	39.3	31.3	9.8
Russian Federation	5.2	160.0	1 528.7	875.0	309.0	197.4	47.8	14.7	27.8	85.7	20.8	21.6
Tajikistan	5.9	112.9	822.0	2 884.8	350.3	682.1	422.4	85.4	43.1	27.5	32.9	38.6
Turkmenistan	5.7	88.5	483.2	3 128.4	2 562.1	1 105.3	714.0	83.7	16.8	..	..	..
Ukraine	5.4	94.0	1 485.8	4 734.9	891.2	376.7	80.2	15.9	10.6	22.7	28.2	12.0
Uzbekistan	5.8	97.3	414.5	1 231.8	1 550.0	76.5	54.0	58.8	17.7	29.0	24.9	..

Source: UNECE Common Database, derived from national statistics.

Note: From 1992 onwards indices derived from monthly data except for Armenia, Georgia, Hungary, Slovenia and Yugoslavia (from 1993); Turkmenistan (from 1995); and Uzbekistan (from 1996); retail prices for Bulgaria and the CIS countries for 1989.

[a] Retail prices.

APPENDIX TABLE B.9

Producer price indices in eastern Europe, the Baltic states and the CIS, 1990-2001

(Annual average, percentage change over preceding year)

	1990	1991	1992	1993	1994	1995	1996	1997	1998	1999	2000	2001
Albania	..	..	..	..	..	..	..	..	..	..	..	..
Bosnia and Herzegovina	..	129.5	70 374.7	10 967.6	1 184.8	68.7	-4.8	3.2	3.6	4.3	0.9	2.3
Bulgaria	14.7	296.4	56.1	28.3	59.1	48.9	129.7	888.1	20.0	3.2	16.9	6.7
Croatia	455.3	146.3	826.0	1 510.4	77.7	0.8	1.3	3.7	-1.5	2.5	9.6	3.4
Czech Republic	2.5	70.4	10.8	9.3	5.1	7.7	4.9	5.1	4.9	1.1	5.1	3.0
Hungary	22.0	32.6	12.3	14.1	12.3	28.5	22.3	20.9	11.4	5.0	11.4	5.7
Poland	622.4	40.9	28.0	32.6	31.0	26.0	13.4	12.2	7.2	5.7	7.8	1.8
Romania	26.9	220.1	184.8	165.0	140.7	35.3	50.0	144.0	33.2	41.2	53.4	41.0
Slovakia	5.2	68.9	5.3	17.0	10.0	9.1	4.0	4.6	3.3	3.7	9.8	6.7
Slovenia	390.4	124.1	215.7	23.3	17.8	12.4	6.7	6.1	6.0	2.2	7.7	9.0
The former Yugoslav Republic of Macedonia	394.0	112.0	2 193.5	258.6	88.7	4.7	-0.1	3.5	4.0	-0.2	9.0	2.6
Yugoslavia	468.0	124.0	8 993.0	1.4E+13	7.9E+10	75.7	88.8	20.6	25.9	43.3	105.4	84.4
Estonia	19.3	208.4	1 208.0	75.2	36.1	25.5	14.7	8.8	4.1	-1.3	4.9	4.5
Latvia	..	192.0	1 310.0	117.1	17.0	12.0	13.8	4.3	2.0	-4.0	0.8	1.6
Lithuania	..	148.2	1 510.0	391.7	44.7	28.7	16.5	5.0	-4.6	3.3	17.8	-1.5
Armenia	2.0	120.0	947.0	892.0	4 394.4	187.8	36.3	21.7	6.0	4.1	-0.4	1.1
Azerbaijan	..	179.5	7 453.6	1 974.0	3 971.6	1 340.1	70.6	11.4	-5.5	-1.3	9.4	-2.5
Belarus	2.1	151.1	1 939.2	1 536.3	3 362.1	538.6	35.7	89.4	72.8	355.7	185.6	71.8
Georgia	..	..	..	..	..	..	..	..	2.2	15.5	5.8	3.6
Kazakhstan	..	193.0	2 465.1	1 042.8	2 918.5	139.7	23.8	15.6	0.8	19.0	38.0	0.4
Kyrgyzstan	..	160.0	1 664.0	831.0	72.9	34.0	29.1	34.8	12.7	51.7	31.7	12.1
Republic of Moldova	..	130.0	1 210.9	1 078.5	711.7	52.2	30.2	14.9	9.7	47.1	33.6	12.6
Russian Federation	3.9	240.0	3 280.0	900.0	340.0	237.6	50.8	15.0	7.0	59.1	46.5	19.1
Tajikistan	..	163.0	1 316.5	1 080.0	665.5	351.7	340.7	103.7	28.4	45.6	39.0	25.1
Turkmenistan	..	211.0	994.0	1 610.0	911.0	296.5	2 974.9	260.6	-30.5	..	..	..
Ukraine	..	163.4	4 128.5	9 667.5	1 134.5	488.9	52.1	7.8	13.2	31.1	20.8	8.6
Uzbekistan	..	147.0	1 296.0	1 119.0	2 162.6	792.5	128.5	53.9	40.0	38.0	61.1	..

Source: UNECE Common Database, derived from national statistics.

Note: From 1994 onwards indices derived from monthly data except: Bosnia and Herzegovina, Croatia, Czech Republic, Poland, The former Yugoslav Republic of Macedonia (from 1992); Hungary, Romania, Slovakia, Slovenia (from 1993); Turkmenistan, Yugoslavia (from 1995).

APPENDIX TABLE B.10

Nominal gross wages in industry in eastern Europe, the Baltic states and the CIS, 1990-2001

(Annual average, percentage change over preceding year) [a]

	1990	1991	1992	1993	1994	1995	1996	1997	1998	1999	2000	2001
Albania [b]	..	..	..	69.5	34.5	29.3	..	..	20.4	10.4	17.7	..
Bosnia and Herzegovina [c]	..	..	..	..	..	..	253.6	50.8	31.1	15.4	12.1	..
Bulgaria	20.8	175.5	132.8	55.1	53.9	57.7	90.4	882.9	34.5	..	..	..
Croatia [c]	453.7	40.7	466.7	1 444.1	130.5	44.0	11.8	16.3	10.6	10.1	7.9	..
Czech Republic	3.0	16.5	21.0	22.6	16.9	18.3	17.7	11.9	10.7	6.7	7.1	..
Hungary [b]	27.2	33.4	24.3	24.9	23.3	19.1	21.7	21.8	16.6	13.5	15.0	14.4
Poland [c]	365.5	64.0	41.2	37.8	39.8	30.1	25.8	19.9	14.1	35.7	9.6	..
Romania [c]	9.7	125.0	173.5	210.7	139.4	55.3	56.7	99.5	49.3	40.7	48.9	..
Slovakia [b]	2.4	17.2	16.9	23.1	17.7	15.2	14.6	9.3	9.7	7.9	9.3	..
Slovenia	361.5	68.4	196.7	45.1	27.1	17.1	14.0	12.1	10.7	9.3	11.6	..
The former Yugoslav Republic of Macedonia [c]	433.6	79.2	1 083.7	454.0	105.8	11.1	3.6	2.3	3.1	1.5	6.2	..
Yugoslavia [c]	400.0	100.6	4 886.4	-62.5	229.6	74.1	74.5	41.7	39.6	24.6	113.5	..
Estonia [d]	..	122.2	570.3	93.3	71.3	34.6	23.8	19.7	13.2	0.9	..	..
Latvia	..	104.5	609.8	112.0	60.0	24.1	14.9	21.6	6.5	4.0	3.2	..
Lithuania	15.3	183.9	632.5	246.1	70.0	43.2	34.0	21.7	12.0	-0.4	-0.7	..
Armenia	5.9	31.8	352.3	739.1	3 640.2	210.9	62.3	41.7	20.5	15.2	19.5	..
Azerbaijan	4.4	82.9	870.4	700.8	575.7	354.8	53.0	58.2	19.8	18.1	33.3	..
Belarus	14.1	112.1	910.9	1 073.8	69.6	618.7	58.5	96.8	109.4	323.9	197.1	109.3
Georgia	6.0	33.9	457.3	2 114.7	23 327.5	87.5	145.4	12.7	26.8	23.2	..	..
Kazakhstan	11.0	80.5	1 053.1	993.1	1 542.5	178.2	30.9	22.5	7.8	21.6	26.1	..
Kyrgyzstan	0.5	62.1	638.3	745.6	176.3	54.8	29.1	60.3	20.0	43.2	-10.0	..
Republic of Moldova	14.2	90.1	815.0	773.0	284.4	43.8	32.2	24.9	16.6	22.5	31.8	..
Russian Federation	13.0	94.9	1 065.7	798.2	260.2	131.4	64.3	21.6	14.3	52.2	48.8	51.9
Tajikistan	8.0	76.2	550.2	917.5	142.2	145.2	425.3	71.9	96.4	26.4	31.5	..
Turkmenistan	..	83.8	1 010.1	1 467.7	622.2	686.2	915.3	121.5	27.1	..	..	..
Ukraine	39.0	107.9	1 419.0	2 286.1	737.8	408.5	71.5	13.5	5.9	18.1	40.2	..
Uzbekistan	..	82.3	706.1	1 159.3	806.8	278.4	99.8	76.3	60.3	50.0	..	..

Source: UNECE Common Database, derived from national statistics.

[a] Calculated from reported annual average wages.

[b] Gross wages in total economy. For Hungary for 1990-1992; for Slovakia for 1990-1991.

[c] Net wages in industry. For Poland and Romania for 1990-1992.

[d] Manufacturing for 1991-1993.

APPENDIX TABLE B.11

Merchandise exports of eastern Europe, the Baltic states and the CIS, 1980, 1989-2001
(Billion dollars)

	1980	1989	1990	1991	1992	1993	1994	1995	1996	1997	1998	1999	2000	2001
Eastern Europe	56.367	63.850	61.733	57.241	59.333	62.675	72.937	94.777	100.206	107.428	119.174	117.842	132.990	147.197
Albania	0.320	0.302	0.231	0.101	0.072	0.123	0.139	0.202	0.213	0.137	0.207	0.352	0.261	0.294*
Bulgaria	7.160	6.651	5.232	3.433	3.992	3.769	3.935	5.345	4.890	4.940	4.194	4.006	4.825	5.099
Czechoslovakia	10.475	11.988	10.728	11.319										
Czech Republic	..	..	..	..	8.767	14.463	15.882	21.273	22.180	22.779	26.351	26.242	28.996	33.369
Slovakia	..	..	..	..	3.500	5.458	6.714	8.585	8.822	9.640	10.775	10.277	11.908	12.691
Hungary	8.609	9.673	9.731	10.226	10.681	8.921	10.701	12.867	15.704	19.100	23.005	25.012	28.092	30.498
Poland	13.071	14.665	18.291	14.912	13.187	14.202	17.240	22.887	24.440	25.756	28.229	27.404	31.651	36.092
Romania	9.217	8.076	4.570	4.266	4.363	4.892	6.151	7.910	8.085	8.431	8.302	8.503	10.369	11.385
Yugoslavia (SFR)	7.514	12.496	12.950	12.984	14.772									
Bosnia and Herzegovina	..	2.100	1.850	..				0.024	0.058	0.193	0.352	0.518	0.675	0.799
Croatia	..	2.600	4.020	3.310	4.353	3.709	4.260	4.633	4.512	4.171	4.541	4.303	4.432	4.659
Slovenia	1.836	3.408	4.118	3.874	6.681	6.083	6.828	8.316	8.310	8.369	9.050	8.546	8.732	9.252
The former Yugoslav Republic of Macedonia	..	0.654	1.113	1.095	1.199	1.055	1.086	1.204	1.147	1.237	1.311	1.186	1.319	1.155
Yugoslavia	..	4.461	4.651	4.704	2.539	..	..	1.531	1.846	2.677	2.858	1.493	1.730	1.903
Baltic states					2.139	4.197	4.324	5.844	6.877	8.467	8.759	7.665	8.850	9.890
Estonia	..	..	..	..	0.444	0.802	1.305	1.838	2.079	2.934	3.236	2.938	3.176	3.305
Latvia	..	..	..	..	0.843	1.401	0.988	1.304	1.443	1.673	1.812	1.723	1.865	2.002
Lithuania	..	..	..	..	0.852	1.994	2.031	2.705	3.355	3.860	3.711	3.004	3.809	4.583
CIS: total	..	..	..	..	..	..	89.991	110.622	121.936	122.913	104.716	104.179	143.691	143.474*
CIS: non-CIS	..	..	..	..	51.242	52.547	62.652	80.007	87.897	88.698	76.757	82.346	114.949	113.613*
Armenia	..	..	..	..	..	0.156	0.216	0.271	0.290	0.233	0.221	0.232	0.301	0.339
Non-CIS	..	..	..	..	0.026	0.029	0.058	0.101	0.162	0.138	0.140	0.175	0.227	0.252
Azerbaijan	..	..	..	..	1.484	0.725	0.653	0.637	0.631	0.781	0.606	0.929	1.745	2.314
Non-CIS	..	..	..	..	0.754	0.351	0.378	0.352	0.341	0.403	0.374	0.718	1.510	2.091
Belarus	..	..	..	..	..	..	2.510	4.707	5.652	7.301	7.070	5.909	7.331	7.428
Non-CIS	..	..	..	..	1.194	0.789	1.031	1.777	1.888	1.922	1.910	2.287	2.927	2.956
Georgia	..	..	..	..	..	..	0.156	0.154	0.199	0.240	0.193	0.238	0.330	0.320
Non-CIS	..	..	..	..	0.068	0.069	0.039	0.057	0.070	0.102	0.085	0.131	0.198	0.176
Kazakhstan	..	..	..	..	..	..	3.231	5.250	5.911	6.497	5.436	5.592	9.126	8.647
Non-CIS	..	..	..	..	1.398	1.501	1.357	2.366	2.732	3.515	3.266	4.100	6.750	6.015
Kyrgyzstan	..	..	..	..	0.317	0.396	0.340	0.409	0.505	0.604	0.514	0.454	0.505	0.476
Non-CIS	..	..	..	..	0.077	0.112	0.117	0.140	0.112	0.285	0.283	0.271	0.297	0.308
Republic of Moldova	..	..	..	..	0.470	0.483	0.565	0.746	0.795	0.875	0.632	0.464	0.472	0.570
Non-CIS	..	..	..	..	0.166	0.178	0.159	0.279	0.252	0.267	0.203	0.211	0.196	0.224
Russian Federation [a]	..	..	..	..	..	..	66.862	79.869	86.889	86.627	73.000	73.700	102.796	100.653*
Non-CIS	..	..	..	..	42.376	44.297	53.001	65.607	70.975	69.959	58.800	62.800	89.068	86.213*
Tajikistan	..	..	..	..	0.193	0.350	0.492	0.749	0.770	0.746	0.597	0.689	0.784	0.652
Non-CIS	..	..	..	..	0.109	0.227	0.399	0.497	0.439	0.473	0.394	0.374	0.411	0.440
Turkmenistan	..	..	..	..	..	..	2.145	1.881	1.682	0.751	0.594	1.190	2.500	2.700
Non-CIS	..	..	..	..	0.908	1.049	0.494	0.951	0.610	0.300	0.442	0.700	1.200	1.300
Ukraine	..	..	..	..	7.415	7.817	10.272	13.128	14.401	14.232	12.637	11.582	14.573	16.265
Non-CIS	..	..	..	..	3.297	3.223	4.653	6.168	6.996	8.646	8.435	8.329	10.075	11.589
Uzbekistan	..	..	..	..	..	..	2.549	2.821	4.211	4.026	3.218	3.200	3.230	3.110
Non-CIS	..	..	..	..	0.869	0.721	0.966	1.712	3.321	2.689	2.425	2.250	2.090	2.050
Former Soviet Union	57.942	62.286	59.056	46.660										
Total	114.310	126.136	120.788	103.901	112.714	119.419	167.252	211.243	229.019	238.808	232.649	229.686	285.531	300.561*

Source: UNECE secretariat, based on national statistical publications and direct communications from national statistical offices.

Note: Trade flows reported include the "new trade" among members of the dissolved federal states: former Czechoslovakia (from 1993), the former SFR of Yugoslavia (from 1992) and the former USSR: for the Baltic states (from 1992) and for the CIS. Data excluding the "new trade" were shown in earlier issues of this publication. Changes in the method of recording trade are reflected from 1993 in data for the Czech Republic (inclusion of OPT transactions, etc.), from 1995 in Latvia (imports registered c.i.f.) and Lithuania (change from special to general system), from 1996 in Hungary (inclusion of trade flows of free trade zones), from 1997 in Slovakia (inclusion of OPT transactions, etc.) and from 2000 in Estonia (change from general to special trade system).

As from 1991, all trade values are expressed in dollars at prevailing market exchange rates. For earlier years, values reported in national currencies were adjusted by the ECE secretariat to remove distortions stemming from mutually inconsistent national rouble/dollar cross-rates in the valuation of the then important intra-CMEA trade flows. For details on the revaluation, see the note to table 2.1.3 and the discussion in box 2.1.1 in UNECE, *Economic Bulletin for Europe*, Vol. 43 (1991).

[a] Russian Goskomstat data excluding trade by physical persons (shuttle trade), but including trade flows not crossing the Russian borders such as off-board fish sales and natural gas deliveries under debt repayment agreements with former CMEA countries.

APPENDIX TABLE B.12

Merchandise imports of eastern Europe, the Baltic states and the CIS, 1980, 1989-2001

(Billion dollars)

	1980	1989	1990	1991	1992	1993	1994	1995	1996	1997	1998	1999	2000	2001
Eastern Europe	65.443	61.185	63.408	61.610	68.388	76.285	86.128	117.026	135.887	146.195	159.491	155.447	172.523	187.288*
Albania	0.320	0.385	0.381	0.409	0.524	0.421	0.549	0.650	0.913	0.620	0.795	0.903	1.070	1.308*
Bulgaria	6.321	7.325	5.584	2.700	4.530	5.120	4.272	5.638	5.074	4.932	4.957	5.515	6.507	7.230
Czechoslovakia	10.619	11.772	11.808	10.962										
Czech Republic	..	..	..	..	10.368	14.617	17.427	25.265	27.919	27.563	28.789	28.073	32.110	36.504
Slovakia	..	..	..	..	3.889	6.332	6.634	8.777	11.112	11.622	13.006	11.265	12.660	14.686
Hungary	9.188	8.863	8.797	11.449	11.123	12.648	14.554	15.466	18.144	21.234	25.706	28.008	32.080	33.682
Poland	14.705	12.941	12.619	15.531	16.141	18.758	21.566	29.043	37.137	42.314	47.054	45.911	48.940	50.257
Romania	11.061	5.834	6.889	5.793	6.260	6.522	7.109	10.278	11.435	11.280	11.838	10.395	13.058	15.552
Yugoslavia (SFR)	13.229	14.064	17.330	14.765										
Bosnia and Herzegovina	..	1.850	1.750	..	..	..	..	0.524	1.204	1.555	2.120	2.431	2.290	2.340
Croatia	..	3.750	5.133	3.811	4.346	4.166	5.229	7.510	7.788	9.104	8.383	7.799	7.887	9.044
Slovenia	2.463	3.216	4.727	4.131	6.141	6.501	7.304	9.492	9.421	9.367	10.098	10.083	10.116	10.145
The former Yugoslav Republic of Macedonia	..	0.934	1.531	1.274	1.206	1.199	1.484	1.719	1.627	1.779	1.915	1.773	2.085	1.688
Yugoslavia	..	5.383	6.701	5.548	3.859	..	..	2.665	4.113	4.826	4.830	3.291	3.721	4.837
Baltic states					1.802	4.101	5.251	8.006	10.110	12.809	13.768	11.888	12.901	14.079
Estonia	..	..	..	..	0.406	0.896	1.659	2.540	3.231	4.441	4.786	4.108	4.256	4.291
Latvia	..	..	..	..	0.794	0.961	1.240	1.818	2.320	2.724	3.189	2.946	3.189	3.507
Lithuania	..	..	..	..	0.602	2.244	2.352	3.649	4.559	5.644	5.794	4.834	5.457	6.281
CIS: total					..	..	63.137	79.576	86.960	94.316	81.747	62.005	70.693	81.729*
CIS: non-CIS					42.297	33.696	36.743	45.678	49.205	57.780	50.560	37.571	38.652	48.875*
Armenia	..	..	..	..	..	0.255	0.394	0.674	0.856	0.892	0.902	0.811	0.885	0.869
Non-CIS	..	..	..	..	0.050	0.087	0.188	0.340	0.578	0.593	0.672	0.624	0.711	0.655
Azerbaijan	..	..	..	..	..	0.940	0.629	0.778	0.668	0.961	0.794	1.077	1.036	1.172	1.431
Non-CIS	..	..	..	..	0.333	0.241	0.292	0.440	0.621	0.443	0.673	0.711	0.797	0.986
Belarus	..	..	..	..	..	..	3.066	5.564	6.939	8.689	8.549	6.674	8.574	8.049
Non-CIS	..	..	..	..	0.843	1.119	0.974	1.887	2.369	2.872	2.995	2.385	2.551	2.443
Georgia	..	..	..	..	..	..	0.338	0.385	0.687	0.944	0.884	0.602	0.651	0.684
Non-CIS	..	..	..	..	0.228	0.167	0.066	0.231	0.417	0.603	0.617	0.377	0.423	0.433
Kazakhstan	..	..	..	..	..	..	3.561	3.807	4.241	4.301	4.350	3.687	5.051	6.363
Non-CIS	..	..	..	..	0.469	0.494	1.384	1.154	1.295	1.969	2.290	2.089	2.295	3.057
Kyrgyzstan	..	..	..	..	0.421	0.448	0.317	0.522	0.838	0.709	0.842	0.600	0.554	0.467
Non-CIS	..	..	..	..	0.071	0.112	0.107	0.168	0.351	0.273	0.401	0.341	0.256	0.210
Republic of Moldova	..	..	..	..	0.640	0.628	0.659	0.841	1.072	1.172	1.024	0.587	0.776	0.897
Non-CIS	..	..	..	..	0.179	0.184	0.183	0.272	0.420	0.567	0.584	0.345	0.517	0.557
Russian Federation [a]	..	..	..	..	..	..	38.661	46.709	47.373	53.568	44.600	31.000	33.769	41.237*
Non-CIS	..	..	..	..	36.984	26.807	28.344	33.117	32.798	39.365	32.500	22.200	22.171	30.121*
Tajikistan	..	..	..	..	0.254	0.630	0.547	0.810	0.668	0.750	0.711	0.663	0.675	0.688
Non-CIS	..	..	..	..	0.132	0.374	0.314	0.332	0.285	0.268	0.265	0.148	0.115	0.150
Turkmenistan	..	..	..	..	..	..	1.468	1.364	1.011	1.183	1.008	1.500	1.780	2.250
Non-CIS	..	..	..	..	0.030	0.501	0.782	0.619	0.450	0.531	0.530	1.000	1.100	1.400
Ukraine	..	..	..	..	6.892	9.533	10.745	15.484	17.603	17.128	14.676	11.846	13.956	15.775
Non-CIS	..	..	..	..	2.049	2.652	2.907	5.488	6.427	7.249	6.779	5.103	5.916	6.943
Uzbekistan	..	..	..	..	..	..	2.603	2.748	4.712	4.186	3.125	3.000	2.850	3.020
Non-CIS	..	..	..	..	0.929	0.958	1.202	1.630	3.195	3.047	2.256	2.250	1.800	1.920
Former Soviet Union	52.218	64.983	64.963	45.405										
Total	117.661	126.168	128.371	106.901	112.487	114.082	154.516	204.609	232.957	253.320	255.006	229.340	256.117	283.096*

Source: UNECE secretariat, based on national statistical publications and direct communications from national statistical offices.

Note: See appendix table B.11.

[a] Russian Goskomstat data excluding trade by physical persons (shuttle trade), but including trade flows not crossing the Russian borders such as off-board fish sales and natural gas deliveries under debt repayment agreements with former CMEA countries.

APPENDIX TABLE B.13

Balance of merchandise trade of eastern Europe, the Baltic states and the CIS, 1980, 1989-2001

(Billion dollars)

	1980	1989	1990	1991	1992	1993	1994	1995	1996	1997	1998	1999	2000	2001
Eastern Europe	-9.076	2.665	-1.675	-4.369	-9.055	-13.610	-13.190	-22.250	-35.680	-38.767	-40.317	-37.605	-39.533	-40.091*
Albania	0.000	-0.083	-0.150	-0.308	-0.452	-0.298	-0.410	-0.448	-0.701	-0.483	-0.589	-0.551	-0.809	-1.013*
Bulgaria	0.839	-0.674	-0.352	0.732	-0.538	-1.352	-0.336	-0.293	-0.184	0.008	-0.763	-1.509	-1.683	-2.131
Czechoslovakia	-0.144	0.216	-1.080	0.356	..	..	..	..	..	..	..	..	..	..
Czech Republic	..	..	..	..	-1.601	-0.154	-1.545	-3.992	-5.739	-4.784	-2.438	-1.831	-3.114	-3.135
Slovakia	..	..	..	..	-0.389	-0.874	0.080	-0.192	-2.290	-1.983	-2.231	-0.988	-0.752	-1.994
Hungary	-0.579	0.810	0.934	-1.223	-0.442	-3.727	-3.853	-2.599	-2.440	-2.134	-2.701	-2.996	-3.988	-3.184*
Poland	-1.634	1.724	5.672	-0.619	-2.955	-4.555	-4.326	-6.156	-12.697	-16.558	-18.825	-18.507	-17.289	-14.183
Romania	-1.844	2.242	-2.320	-1.528	-1.897	-1.630	-0.958	-2.368	-3.351	-2.849	-3.536	-1.892	-2.689	-4.167
Yugoslavia (SFR)	-5.715	-1.568	-4.380	-1.780	14.772	..	..	..	..	..	..	..	..	..
Bosnia and Herzegovina	..	0.250	0.100	..	..	..	..	-0.500	-1.146	-1.362	-1.768	-1.913	-1.615	-1.541
Croatia	..	-1.150	-1.113	-0.501	0.007	-0.457	-0.969	-2.877	-3.276	-4.933	-3.842	-3.496	-3.455	-4.384
Slovenia	-0.626	0.192	-0.609	-0.257	0.540	-0.418	-0.476	-1.176	-1.111	-0.998	-1.048	-1.537	-1.384	-0.893
The former Yugoslav Republic of Macedonia	..	-0.280	-0.418	-0.179	-0.007	-0.144	-0.398	-0.515	-0.480	-0.542	-0.604	-0.587	-0.766	-0.533
Yugoslavia	..	-0.922	-2.050	-0.844	-1.320	..	..	-1.134	-2.267	-2.149	-1.972	-1.798	-1.991	-2.934
Baltic states	..	..	..	..	0.337	0.096	-0.927	-2.162	-3.232	-4.343	-5.009	-4.224	-4.051	-4.189
Estonia	..	..	..	..	0.038	-0.094	-0.353	-0.702	-1.152	-1.507	-1.550	-1.170	-1.080	-0.986
Latvia	..	..	..	..	0.049	0.440	-0.252	-0.514	-0.877	-1.051	-1.377	-1.223	-1.323	-1.505
Lithuania	..	..	..	..	0.250	-0.250	-0.322	-0.944	-1.204	-1.784	-2.083	-1.831	-1.648	-1.698
CIS: total	..	..	..	..	..	..	26.854	31.046	34.975	28.597	22.970	42.174	72.998	61.744*
CIS: non-CIS	..	..	..	..	8.945	18.851	25.909	34.329	38.692	30.919	26.197	44.775	76.297	64.738*
Armenia	..	..	..	..	..	-0.099	-0.178	-0.403	-0.566	-0.660	-0.682	-0.580	-0.584	-0.530
Non-CIS	..	..	..	..	-0.024	-0.058	-0.130	-0.239	-0.416	-0.455	-0.532	-0.449	-0.484	-0.403
Azerbaijan	..	..	..	..	0.544	0.096	-0.125	-0.031	-0.330	-0.013	-0.471	-0.107	0.573	0.883
Non-CIS	..	..	..	..	0.421	0.110	0.086	-0.088	-0.280	-0.040	-0.299	0.007	0.713	1.106
Belarus	..	..	..	..	..	..	-0.556	-0.857	-1.287	-1.388	-1.480	-0.765	-1.244	-0.621
Non-CIS	..	..	..	..	0.351	-0.330	0.057	-0.110	-0.481	-0.950	-1.085	-0.098	0.376	0.513
Georgia	..	..	..	..	..	..	-0.182	-0.231	-0.488	-0.704	-0.692	-0.364	-0.321	-0.364
Non-CIS	..	..	..	..	-0.160	-0.098	-0.027	-0.174	-0.347	-0.501	-0.532	-0.246	-0.226	-0.258
Kazakhstan	..	..	..	..	..	..	-0.330	1.443	1.670	2.196	1.086	1.906	4.075	2.284
Non-CIS	..	..	..	..	0.929	1.007	-0.027	1.212	1.437	1.547	0.976	2.011	4.456	2.958
Kyrgyzstan	..	..	..	..	-0.104	-0.052	0.023	-0.113	-0.333	-0.105	-0.328	-0.146	-0.050	0.009
Non-CIS	..	..	..	..	0.005	0.000	0.010	-0.028	-0.239	0.012	-0.118	-0.070	0.041	0.097
Republic of Moldova	..	..	..	..	-0.170	-0.145	-0.094	-0.095	-0.277	-0.297	-0.392	-0.123	-0.305	-0.327
Non-CIS	..	..	..	..	-0.013	-0.006	-0.024	0.007	-0.168	-0.300	-0.381	-0.134	-0.321	-0.333
Russian Federation [a]	..	..	..	..	..	..	28.201	33.160	39.516	33.059	28.400	42.700	69.027	59.416*
Non-CIS	..	..	..	..	5.392	17.490	24.657	32.490	38.177	30.594	26.300	40.600	66.897	56.091*
Tajikistan	..	..	..	..	-0.061	-0.280	-0.055	-0.061	0.102	-0.004	-0.114	0.026	0.109	-0.036
Non-CIS	..	..	..	..	-0.023	-0.147	0.085	0.165	0.154	0.205	0.129	0.225	0.295	0.290
Turkmenistan	..	..	..	..	..	..	0.677	0.517	0.670	-0.432	-0.414	-0.310	0.720	0.450
Non-CIS	..	..	..	..	0.878	0.548	-0.288	0.332	0.160	-0.231	-0.088	-0.300	0.100	-0.100
Ukraine	..	..	..	..	0.523	-1.716	-0.473	-2.356	-3.202	-2.896	-2.038	-0.265	0.617	0.490
Non-CIS	..	..	..	..	1.248	0.571	1.746	0.680	0.569	1.397	1.657	3.227	4.159	4.646
Uzbekistan	..	..	..	..	..	..	-0.054	0.073	-0.501	-0.159	0.093	0.200	0.380	0.090
Non-CIS	..	..	..	..	-0.060	-0.237	-0.236	0.082	0.126	-0.358	0.169	0.000	0.290	0.130
Former Soviet Union	5.724	-2.697	-5.907	1.255										
Total	-3.351	-0.032	-7.583	-3.000	0.227	5.337	12.736	6.634	-3.937	-14.512	-22.357	0.345	29.414	17.465*

Source: UNECE secretariat, based on national statistical publications and direct communications from national statistical offices.

Note: See appendix table B.11.

[a] Russian Goskomstat data excluding trade by physical persons (shuttle trade), but including trade flows not crossing the Russian borders such as off-board fish sales and natural gas deliveries under debt repayment agreements with former CMEA countries.

APPENDIX TABLE B.14

Merchandise trade of eastern Europe and the Russian Federation, by direction, 1980, 1989-2001

(Shares in total trade, per cent)

	1980	1989	1990	1991	1992	1993	1994	1995	1996	1997	1998	1999	2000	2001[a]
Eastern Europe, *to and from:*														
Exports														
World	100.0	100.0	100.0	100.0	100.0	100.0	100.0	100.0	100.0	100.0	100.0	100.0	100.0	100.0
ECE transition economies	48.3	44.4	38.1	28.5	23.0	28.2	26.3	25.8	26.1	26.1	22.4	18.9	19.0	18.8
Former Soviet Union	28.2	25.5	22.3	17.9	12.4	9.8	9.0	8.9	9.4	10.3	7.4	4.9	4.9	5.0
Eastern Europe	20.2	18.9	15.8	10.7	10.7	18.5	17.4	16.9	16.7	15.8	15.0	14.0	14.1	13.9
Developed market economies	33.0	42.6	49.5	59.8	63.0	58.0	62.5	64.5	65.0	66.4	71.4	75.2	74.8	75.3
Developing economies	18.7	13.0	12.4	11.7	14.0	13.8	11.2	9.7	8.9	7.5	6.2	5.9	6.2	5.9
Imports														
World	100.0	100.0	100.0	100.0	100.0	100.0	100.0	100.0	100.0	100.0	100.0	100.0	100.0	100.0
ECE transition economies	49.3	36.4	26.6	25.5	24.7	29.3	26.1	25.3	23.8	22.1	18.3	18.6	22.0	21.6
Former Soviet Union	30.6	23.5	18.3	20.2	17.9	16.5	14.1	13.2	12.7	11.2	8.6	8.5	11.8	11.1
Eastern Europe	18.8	17.7	14.3	8.1	6.8	12.8	12.0	12.1	11.1	10.9	9.6	10.1	10.2	10.5
Developed market economies	37.0	44.0	53.3	58.3	64.4	61.5	65.0	65.8	66.6	68.1	72.4	70.8	67.7	67.8
Developing economies	13.6	19.5	20.1	16.1	10.9	9.2	9.0	8.9	9.6	9.8	9.3	10.6	10.3	10.6
Former Soviet Union/Russian Federation, *to and from:*														
Exports														
World	100.0	100.0	100.0	100.0	100.0	100.0	100.0	100.0	100.0	100.0	100.0	100.0	100.0	100.0
ECE transition economies	35.0	26.6	21.8	25.9	22.3	18.1	15.1	16.8	18.2	19.5	18.1	17.8	20.0	19.4
Eastern Europe	35.0	26.6	21.8	25.9	20.7	16.8	11.7	13.2	14.3	14.9	14.2	13.3	14.5	14.9
Developed market economies	39.7	41.8	49.5	56.5	57.9	59.7	66.6	60.6	58.1	58.6	60.0	58.0	55.6	55.9
Developing economies	25.3	31.6	28.7	17.6	19.9	22.2	18.3	22.6	23.7	21.9	21.9	24.2	24.4	24.7
Imports														
World	100.0	100.0	100.0	100.0	100.0	100.0	100.0	100.0	100.0	100.0	100.0	100.0	100.0	100.0
ECE transition economies	31.5	27.6	24.7	26.0	15.9	10.6	14.1	15.5	12.6	13.7	12.0	9.6	10.9	9.8
Eastern Europe	31.5	27.6	24.7	26.0	15.0	10.0	11.7	12.4	10.6	11.1	9.8	8.2	9.4	8.4
Developed market economies	44.2	50.1	52.9	58.1	62.4	60.6	70.3	69.5	67.8	68.3	68.2	68.3	69.3	67.2
Developing economies	24.3	22.3	22.4	15.9	21.7	28.8	15.6	15.0	19.6	18.0	19.8	22.1	19.8	22.9

Source: UNECE Common Database, derived from national statistics.

Note: Data for 1980-1990 refer to the east European CMEA countries (Bulgaria, Czechoslovakia, German Democratic Republic, Hungary, Poland and Romania) and to the former Soviet Union. Trade data in national currencies were revalued at consistent rouble/dollar cross-rates (see the note to appendix table B.11). As from 1991, eastern Europe covers Bulgaria, former Czechoslovakia (from 1993, Czech Republic and Slovakia including their mutual trade), Hungary, Poland and Romania, and the second panel reflects non-CIS trade of the Russian Federation only.

Partner country grouping has been recently revised with subsequent revisions back to 1980. Thus, the earlier reported "Transition economies" group is now replaced by "ECE transition economies", which covers the Baltic states, CIS and the east European countries including the successor states of the former SFR of Yugoslavia. The "Eastern Europe" partner group now covers Albania, Bulgaria, the Czech Republic, Hungary, Poland, Romania, Slovakia and the successor states of the former SFR of Yugoslavia, which earlier were in the "Other socialist countries" subgroup. The rest of subgroup "Other socialist countries", which in previous series covered China, Cuba, Democratic People's Republic of Korea, Mongolia and Viet Nam, is now included in the "Developing countries" group.

[a] January-September 2001.

APPENDIX TABLE B.15

Exchange rates of eastern Europe, the Baltic states and the CIS, 1980, 1990-2001

(Annual averages, national currency units per dollar)

	Unit[a]	1980	1990	1991	1992	1993	1994	1995	1996	1997	1998	1999	2000	2001
Albania	lek	..	8.90	24.20	75.03	102.06	94.62	93.14	104.33	148.93	150.63	137.69	143.71	143.484
Bulgaria	lev[b]	0.86	0.79	17.45	23.42	27.85	54.13	67.08	177.88	1 681.87	1 760.37	1.8364	2.1233	2.1868
Czechoslovakia	koruna	5.37	18.56	29.56	28.30	..	..	..	..	..	..	..	..	..
Czech Republic	koruna	..	..	..	..	29.15	28.79	26.54	27.14	31.70	32.29	34.57	38.60	38.04
Slovakia	koruna	..	..	..	..	30.80	31.93	29.71	30.68	33.62	35.23	41.36	46.20	48.35
Hungary	forint	32.64	63.21	74.73	78.98	91.91	105.11	125.69	152.65	186.79	214.40	237.15	282.18	286.49
Poland	zloty[c]	3.05	9 500	10 576	13 627	18 136	22 723	2.42	2.70	3.28	3.49	3.96	4.35	4.09
Romania	leu	4.47	22.43	71.84	307.98	760.12	1 654	2 033	3 085	7 183	8 876	15 333	21 709	29 061
Yugoslavia (SFR)	dinar[d]	24.64	11.32	19.64	..	..	..	..	..	..	..	..	..	..
Bosnia and Herzegovina	dinar	..	..	..	..	..	..	..	..	..	..	..	..	2.19
Croatia	kuna[e]	..	..	18.80	264.30	3 577.63	6.00	5.23	5.43	6.10	6.36	7.11	8.28	8.34
Slovenia	tolar	..	..	27.57	81.29	113.24	128.81	118.52	135.37	159.69	166.13	181.77	222.68	242.75
The former Yugoslav Republic of Macedonia	denar[f]	..	..	19.69	508.07	23.26	43.25	38.05	39.92	50.40	54.48	56.90	65.90	68.06
Yugoslavia	dinar[g]	..	10.65	19.73	750.00	..	1.55	4.74	4.96	5.72	9.23	10.94	16.06	..
Estonia	kroon[h]	..	..	..	12.11	13.22	12.98	11.46	12.03	13.88	14.07	14.68	16.97	17.48
Latvia	lats[i]	..	..	..	..	0.67	0.56	0.53	0.55	0.58	0.59	0.59	0.61	0.63
Lithuania	litas[j]	..	..	..	..	4.37	3.98	4.00	4.00	4.00	4.00	4.00	4.00	4.00
Armenia	dram	..	..	..	..	8.66	288.35	405.93	413.47	490.70	504.92	535.06	539.53	555.08
Azerbaijan	manat	..	..	..	..	..	1 169	4 417	4 295	3 987	3 869	4 119	4 474	4 656.5
Belarus	rouble[k]	..	..	..	..	2 177	4 017	11 538	13 472	26 729	58 971	274 512	881 750	1390000
Georgia	lari[l]	..	..	..	..	..	1.10	1.29	1.26	1.30	1.39	2.02	1.98	2.072
Kazakhstan	tenge	..	..	..	..	..	35.54	60.95	67.30	75.43	78.35	119.47	142.13	146.74
Kyrgyzstan	som	..	..	..	..	..	10.86	10.83	12.81	17.37	21.37	39.73	47.79	48.405
Republic of Moldova	leu	..	..	..	..	..	4.07	4.50	4.60	4.62	5.37	10.52	12.43	12.867
Russian Federation	rouble[m]	0.65	0.59	1.74	192.75	927.46	2 204	4 559	5 121	5 785	9.71	24.62	28.13	29.169
Tajikistan	samoni[n]	..	..	..	..	..	..	107.59	292.89	560.64	..	1 235.57	1.83	2.39
Turkmenistan	manat	..	..	..	..	..	19.50	110.42	3 509	4 143	..	5 200	5 200	5 200
Ukraine	hryvnia[o]	..	..	..	..	4 796	31 700	147 314	1.83	1.86	2.51	4.13	5.44	5.372
Uzbekistan	sum[p]	..	..	..	..	932.15	9.96	29.81	40.15	66.43	..	124.64	236.58	423.08
Memorandum item:														
Former GDR	mark[q]	3.30	8.14	1.66	1.56	1.65	1.62	1.43	1.50	1.73	1.76	..	..	..

Source: UNECE Common Database, derived from national, IMF and CIS statistics. Annual averages are unweighted arithmetic averages of monthly values. Change or redenomination of currency is indicated by a vertical bar.

Note: Under the central planning system with its state foreign trade monopoly, exchange rates served primarily statistical and accounting purposes (notably the conversion of foreign trade values for statistics expressed in domestic currency), without direct impact on domestic price formation. Market-based exchange rates and a meaningful link to domestic currency values emerged only with the transformations from 1989 onward. The official exchange rates of the earlier period are therefore not suitable for the conversion to dollars of macroeconomic and other data of these countries expressed in domestic currency. These strictures should be kept in mind in the interpretation and use of the data for the 1980s shown above.

[a] Currency unit of the last period shown. For prior periods, see footnotes.

[b] The leva was redenominated at 1:1,000 from 5 July 1999.

[c] The zloty was redenominated at 1:10,000 from 1 January 1995.

[d] The dinar was redenominated at 1:10,000 from 1 January 1990.

[e] The kuna replaced the Croat dinar on 3 May 1994 at 1:1,000; the 1994 average is shown in kuna terms.

[f] The denar (which had replaced the Yugoslav dinar 1:1 on 26 April 1992) was redenominated 1:100 on 1 May 1993; the 1993 average is shown in terms of that unit.

[g] The dinar was redenominated on 1 July 1992 (1:10), 1 October 1993 (1:1 million), 1 January 1994 (1:1 trillion) and 24 January 1994 (1:13 million). Average annual exchange rates not available for 1993-1994.

[h] The kroon replaced the Soviet rouble in June 1992 with a peg to the deutsche mark (8:1); the average shown for 1992 refers to June-December.

[i] The lats replaced an earlier Latvian rouble at 1:200 on 18 October 1993; the 1993 average is shown in lat terms.

[j] The litas replaced the earlier talonas at 1:100 on 1 June 1993; the 1993 average is shown in litas terms.

[k] The Belarus rouble was redenominated 1:10 on 10 August 1994; the 1994 average here assumes this applied to the entire year. The Belarus rouble was further redenominated at 1:1,000 since January 2000. Annual averages were calculated from end-of-period monthly rates.

[l] The lari replaced the lari-kupon on 25 September 1995; the annual average for 1994 is shown in million lari-kupon, and that for 1995 in lari.

[m] 1980-1991: Soviet rouble/dollar rate used in the conversion of foreign trade data for statistical purposes. The rouble was redenominated at 1:1,000 from 1 January 1998.

[n] A new currency, the samoni, was put into circulation on 30 October 2000. Made up of 100 dirams, the samoni was to be used in parallel to the Tajik rouble for 5 months, with the exchange rate of 1,000 roubles to the samoni.

[o] The hryvnia replaced the former karbovanets on 2 September 1996 at 1:100,000; the average for 1996 is shown in hryvnia terms.

[p] Sum-kupon in 1993.

[q] German Democratic Republic mark through 1990, deutsche mark thereafter.

APPENDIX TABLE B.16

Current account balances of eastern Europe, the Baltic states and the CIS, 1990-2001

(Million dollars)

	1990	1991	1992	1993	1994	1995	1996	1997	1998	1999	2000	2001
Eastern Europe[a]	-5 726	-2 146	-967	-7 826	-2 496	-1 778	-12 948	-14 616	-17 588	-20 726	-17 637	-17200*
Albania	-118	-168	-51	14	-43	-15	-62	-254	-45	-133	-156	-400*
Bosnia and Herzegovina	..	..	..	..	-177	-193	-748	-1 060	-789	-971	-909	-900*
Bulgaria	-1 710	-77	-361	-1 098	-32	-198	164	1 046	-61	-652	-702	-878
Croatia[b]	-621	-589	329	623	854	-1 442	-1 091	-2 325	-1 531	-1 390	-433	-600*
Czech Republic	-122	1 708	-456	456	-787	-1 369	-4 121	-3 564	-1 386	-1 567	-2 273	-2 500*
Hungary[c]	123	267	325	-3 455	-3 911	-2 480	-1 678	-981	-2 298	-2 081	-1 328	-1 105
Poland[c]	716	-1 359	-269	-2 868	677	5 310	-1 371	-4 309	-6 862	-11 558	-9 946	-7 081
Romania	-3 337	-1 012	-1 564	-1 174	-428	-1 774	-2 571	-2 137	-2 968	-1 469	-1 363	-2 349
Slovakia	-767	-786	173	-532	759	511	-1 960	-1 827	-1 982	-980	-713	-1 820*
Slovenia[b]	518	129	926	192	573	-99	31	11	-147	-783	-612	-67
The former Yugoslav Republic of Macedonia[b]	-409	-259	-19	15	-158	-222	-289	-276	-308	-113	-113	-400*
Yugoslavia	..	..	..	..	-400	-1 037	-600	-1 279	-580	-764	-339	-700*
Baltic states	..	..	548	353	-59	-788	-1 400	-1 890	-2 426	-2 095	-1 483	-1 715
Estonia	..	..	36	22	-167	-158	-398	-563	-478	-247	-315	-300*
Latvia	..	..	191	417	201	-16	-279	-345	-650	-654	-493	-765
Lithuania	..	..	321	-86	-94	-614	-723	-981	-1298	-1194	-675	-650*
CIS	..	..	-538	10 267	5 286	4 409	6 139	-4 092	-6 737	23 675	47 530	32 267*
Armenia	..	..	-50	-67	-104	-218	-291	-307	-403	-307	-278	-220*
Azerbaijan	..	153	488	-160	-121	-401	-931	-916	-1 365	-600	-168	150*
Belarus	..	..	131	-435	-444	-458	-516	-788	-866	-194	-296	-400*
Georgia	..	..	-248	-354	-277	-216	-275	-375	-416	-195	-262	-180*
Kazakhstan	..	-1 300	-1 900	-641	-905	-213	-751	-799	-1 236	-236	743	-1 800*
Kyrgyzstan	..	..	-61	-88	-84	-235	-425	-138	-364	-180	-77	-10*
Republic of Moldova	..	..	-152	-155	-82	-95	-192	-275	-334	-51	-126	-130*
Russian Federation[d]	-6 300	2 500	1 142	12 792	8 434	7 484	11 753	2 060	687	24 731	46 291	34 157
Tajikistan	..	..	-53	-208	-170	-89	-70	-56	-108	-36	-62*	-200*
Turkmenistan	-308	447	926	776	84	23	–	-580	-934	-751*	..	..
Ukraine	..	..	-526	-765	-1 163	-1 152	-1 184	-1 335	-1 296	1 658	1 481	1 300*
Uzbekistan	..	..	-236	-429	118	-21	-980	-584	-102	-164	184	-100*
Total above[a e]	..	..	-957	2 793	2 731	1 843	-8 209	-20 598	-26 751	855	28 410	13 352
Memorandum items:												
CETE-5	469	-41	698	-6 207	-2 689	1 872	-9 098	-10 669	-12 674	-16 969	-14 871	-12 573*
SETE-7[a]	-6 195	-2 105	-1 665	-1 620	193	-3 650	-3 850	-3 946	-4 913	-3 757	-2 766	-4 627*
Asian CIS	..	..	-1 133	-1 170	-1 459	-1 370	-3 723	-3 755	-4 928	-2 468	181	-2 660*
Three European CIS[f]	..	..	-547	-1 355	-1 689	-1 705	-1 892	-2 397	-2 496	1 413	1 059	770*

Source: National balance of payments statistics; IMF, *Balance of Payments Statistics* (Washington, D.C.), various issues and *Staff Country Reports* (www.imf.org); UNECE secretariat estimates.

[a] Totals exclude Bosnia and Herzegovina and Yugoslavia.

[b] Excludes transactions with the republics of the former SFR of Yugoslavia: Croatia (1990-1992), Slovenia (1990-1991) and The former Yugoslav Republic of Macedonia (1990-1992).

[c] Convertible currencies. Hungary until 1995; Poland until 1992.

[d] 1990-1992 exclude transactions with the Baltic and CIS countries.

[e] 2000 and 2001 totals include estimates for Turkmenistan.

[f] Belarus, Republic of Moldova and Ukraine.

APPENDIX TABLE B.17

Inflows of foreign direct investment [a] in eastern Europe, the Baltic states and the CIS, 1990-2001

(Million dollars)

	1990	1991	1992	1993	1994	1995	1996	1997	1998	1999	2000	2001
Eastern Europe [b]	480	2 332	3 125	4 193	3 594	9 336	8 003	9 462	15 478	18 748	20 258	19 602
Albania [c]	–	–	20	58	53	70	90	48	45	41	143	180*
Bosnia and Herzegovina	..	..	..	..	–	–	–	–	100	90	150	164
Bulgaria [c]	4	56	42	40	105	90	109	505	537	819	1 002	651
Croatia	–	–	16	120	117	114	511	533	932	1 479	1 115	900*
Czech Republic	132	513	1 004	654	869	2 562	1 428	1 300	3 718	6 324	4 595	4 500*
Hungary [d]	311	1 459	1 471	2 339	1 146	4 454	2 275	2 173	2 036	1 970	1 649	2 443
Poland (cash basis) [c]	10	117	284	580	542	1 132	2 768	3 077	5 129	6 471	8 294	6 929
Romania	–	40	77	94	341	419	263	1 215	2 031	1 041	1 040	1 137
Slovakia	18	82	100	195	269	308	353	220	684	390	2 075	2 000*
Slovenia	4	65	111	113	128	177	194	375	248	181	176	442
The former Yugoslav Republic of Macedonia [c]	–	–	–	–	24	9	11	16	118	32	170	420*
Yugoslavia	..	..	..	..	..	..	–	740	113	112	25	90
Baltic states	..	..	119	238	460	454	685	1 142	1 863	1 139	1 173	1 457*
Estonia	..	..	82	162	215	202	151	267	581	305	387	600*
Latvia	..	..	29	45	214	180	382	521	357	347	408	257
Lithuania	..	..	8	30	31	73	152	355	926	486	379	600*
CIS	..	..	1 777	1 875	1 770	4 065	5 288	8 856	6 726	6 735	5 367	7 021*
Armenia [c]	..	..	–	1	8	25	18	52	221	122	104	70*
Azerbaijan [c]	..	..	–	60	22	330	627	1 125	1 023	510	129	20*
Belarus [c]	..	..	7	18	11	15	73	200	149	444	116	100*
Georgia [c]	..	..	–	–	8	6	40	203	265	82	131	120*
Kazakhstan [d]	..	..	100	228	635	964	1 137	1 321	1 151	1 468	1 245	2 600*
Kyrgyzstan	..	..	–	10	38	96	47	83	109	44	-2	20*
Republic of Moldova [c]	..	25	17	14	12	67	24	79	74	37	138	150*
Russian Federation	–	100	1 454	1 211	690	2 066	2 579	4 865	2 762	3 309	2 714	2 921
Tajikistan [c]	..	..	9	9	12	20	25	30	24	21	22	20*
Turkmenistan [c]	–	–	11	79	103	233	108	108	64	80*	100*	100*
Ukraine	..	..	170	198	159	267	521	623	743	496	595	800*
Uzbekistan [c]	..	..	9	48	73	-24	90	167	140	121	75	100*
Total above [b]	..	..	5 021	6 305	5 825	13 855	13 975	19 460	24 067	26 622	26 798	28 080*
Memorandum items:												
CETE-5	476	2 236	2 970	3 880	2 954	8 633	7 019	7 146	11 815	15 336	16 789	16 314*
SETE-7 [b]	4	96	155	312	640	703	984	2 316	3 663	3 412	3 469	3 288*
Asian CIS	..	..	129	435	899	1 651	2 092	3 089	2 998	2 449	1 804	3 050*
Three European CIS [e]	..	..	194	229	181	349	617	902	966	977	848	1 050*
Poland (accrual basis) [f]	89	291	678	1 715	1 875	3 659	4 498	4 908	6 365	7 270	9 342	..

Source: National balance of payments statistics; IMF, *Balance of Payments Statistics* (Washington, D.C.), various issues and *Staff Country Reports* (www.imf.org); UNECE secretariat estimates.

Note: Changes in coverage are available in UNECE, *Economic Survey of Europe, 2001 No. 1*, chap. 5, box 5.3.1.

[a] Inflows into the reporting country.

[b] Excluding Bosnia and Herzegovina and Yugoslavia.

[c] Net of residents' investments abroad. Bulgaria, 1990-1994; Poland, 1990-1992.

[d] Exclude reinvested profits. Also see the note.

[e] Belarus, Republic of Moldova and Ukraine. Drawings less repayments.

[f] Includes reinvested profits (which are negative in 1998-2000).

OTHER RECENT PUBLICATIONS OF ECONOMIC ANALYSIS FROM THE UNITED NATIONS ECONOMIC COMMISSION FOR EUROPE

- *Economic Survey of Europe, 2001 No. 2*, Sales No. E.01.II.E.26 (December)

    In addition to an assessment of the macroeconomic situation in the UNECE region in the autumn of 2001, this issue contains the papers presented at the Spring Seminar of May 2001 on *Creating a Supportive Environment for Business Enterprise and Economic Growth: Institutional Reform and Governance*. Included are papers by William Lazonick, Paul Hare, Shang-Jin Wei, Antoni Kamiński and Bartlomiej Kamiński, together with the comments of the discussants and a summary of the general discussion.

- *Economic Survey of Europe, 2001 No. 1*, Sales No. E.00.II.E.14 (April)

    This issue contains the secretariat's annual review of developments in 2000 and the outlook for 2001. Special chapters focus on domestic savings in the transition economies, the relationship between foreign direct investment and economic growth and the impact of economic transformation in the transition economies on their exchange rate.

- *Economic Survey of Europe, 2000 No. 2/3*, Sales No. E.00.II.E.28 (December)

    The papers presented at the Spring Seminar of May 2000 on *The Transition Process After Ten Years*, are gathered in this issue, together with the comments of discussants and a summary of the general discussion. The papers cover long-run developments (Ivan Berend), macroeconomic policies and achievements (Stanislaw Gomulka), changes in production structures (Michael Landesmann) and the social costs and consequences (Michael Ellman).

- *Economic Survey of Europe, 2000 No. 1*, Sales No. E.00.II.E.12 (April)

    This issue reviews developments in 1999 and discusses the outlook for 2000. The "new economy", south-east Europe and capital inflows into the transition economies since 1989 are among the topics included in the current analysis, while there are special chapters on "Economic convergence in Europe" and the "Fertility decline in the transition economies, 1989-1998".

- *Economic Survey of Europe, 1999 No. 3*, Sales No. E.99.II.E.4 (November)

    This issue contains the proceedings of the 1999 Spring Seminar on *Demographic Ageing and the Reform of Pension Systems in the ECE Region*. Included are papers by John Eatwell, Lawrence Thompson and Maria Augusztinovics, together with discussants' comments and an introduction and summary of the day's discussion by the secretariat.

- *Economic Survey of Europe, 1999 No. 2*, Sales No. E.99.II.E.3 (July)

    In addition to an updated summary of economic developments in Europe, the CIS and North America to mid-1999, this issue contains the secretariat's analysis of "Postwar reconstruction and development in south-east Europe", which has received considerable attention in the course of the year.

* * * * *

More details about other publications and activities of the United Nations Economic Commission for Europe, which pay special attention to issues concerning the transition economies, can be found at the secretariat's website: http://www.unece.org

* * * * *

To obtain copies of publications contact:

Publications des Nations Unies
Section de Vente et Marketing
Organisation des Nations Unies
CH-1211 Genève 10
Suisse

Tel: (4122) 917 2612 / 917 2606 / 917 2613
Fax: (4122) 917 0027
E-mail: unpubli@unog.ch

United Nations Publications
2 United Nations Plaza
Room DC2-853
New York, NY 10017
USA

Tel: (1212) 963 8302 / (1800) 253 9646
Fax: (1212) 963 3489
E-mail: publications@un.org